CW01619381

Henry Sinclair Hall
Samuel Ratcliff Knight

SOLUTIONS OF THE EXAMPLES IN ELEMENTARY ALGEBRA *FOR SCHOOLS*

Elibron Classics
www.elibron.com

Elibron Classics series.

ISBN 1-4021-5912-9 (paperback)
ISBN 1-4021-2610-7 (hardcover)

This Elibron Classics Replica Edition is an unabridged facsimile of the edition published in 1906 by Macmillan and Co., Ltd., London.

SOLUTIONS OF THE EXAMPLES

IN

ELEMENTARY ALGEBRA

FOR SCHOOLS

SOLUTIONS OF THE EXAMPLES

IN

ELEMENTARY ALGEBRA

FOR SCHOOLS

BY

H. S. HALL, M.A.

FORMERLY SCHOLAR OF CHRIST'S COLLEGE, CAMBRIDGE

AND

S. R. KNIGHT, B.A., M.B., CH.B.

FORMERLY SCHOLAR OF TRINITY COLLEGE, CAMBRIDGE

NEW EDITION

London

MACMILLAN AND CO., LIMITED

NEW YORK: THE MACMILLAN COMPANY

1906

First Edition 1891.
Reprinted 1893, 1894, 1896,
March, May, November, and December, 1897,
1899, 1901, 1902, 1903.
New Edition 1904, 1906.

GLASGOW: PRINTED AT THE UNIVERSITY PRESS
BY ROBERT MACLEHOSE AND CO. LTD.

PREFACE.

THIS Key has been undertaken in response to numerous requests from teachers who have introduced the Elementary Algebra into their classes, and for such readers it is mainly intended; but it is hoped that, if judiciously used, the solutions may also be found serviceable by that large and increasing class of students who read Mathematics without the assistance of a teacher.

In solving the examples we have received very material assistance from Mr E. J. Brooksmith, of the Royal Military Academy, Woolwich, and we gratefully acknowledge our obligations to him.

H. S. HALL.
S. R. KNIGHT.

Dec. 1890.

In the Seventh and Eighth Editions of the Elementary Algebra many new Examples were introduced. The solutions to these will be found in the Appendix. In particular, the Examples on Graphs in Chapter XLVII. have been dealt with very fully. For many of these solutions I am indebted to Mr. H. C. Beaven of Clifton College, whose valuable help I gratefully acknowledge.

H. S. HALL.

July 1904.

ELEMENTARY ALGEBRA.

EXAMPLES. I. d. PAGE 8.

1. $6a+5b-8c+9d=6.2+5.3-8.1+0=12+15-8=27-8=19.$

2. $3a-4b+6c+5d=3.2-4.3+6.1+0=6-12+6=0.$

3. $5a+3c-2b+6d=5.2+3.1-2.3+0=10+3-6=13-6=7.$

4. $ab+bc+ca-da=2.3+3.1+1.2-0=6+3+2=11.$

5. $6ab-3cd+2da-5cb+2db=6.2.3-0+0-5.1.3+0=36-15=21.$

6. $abc+bcd+cda+dab=2.3.1+0+0+0=6.$

7. $3abc-2bcd+2cda-4dab=3.2.3.1-0+0-0=18.$

8. $2bc+3cd-4da+5ab=2.3.1+0-0+5.2.3=6+30=36.$

9. $3bcd+5cda-7dab+abc=0+0-0+2.3.1=6.$

10. $a^2+b^2+c^2+d^2=2^2+3^2+1^2+0=4+9+1=14.$

11. $2a^2+3b^3-4c^4=2.2^2+3.3^3-4.1^4=2.4+3.27-4.1=8+81-4=85.$

12. $a^4+b^4-c^4=2^4+3^4-1^4=16+81-1=96.$

13. $a^3+b^3+c^3+d^3=1^3+2^3+3^3+0=1+8+27=36.$

14. $\frac{1}{2}bc^3-a^3-b^3-\frac{3}{4}ab^3c=\frac{1}{2}.2.3^3-1^3-2^3-\frac{3}{4}.1.2^3.3=27-1-8-\frac{3}{4}.1.8.3$
$=27-1-8-18=27-27=0.$

15. $3abc-b^2c-6a^3=3.1.2.3-2^2.3-6.1^3=18-4.3-6=18-12-6=0.$

16. $2a^2+2b^2+2c^2+2d^2-2bc-2cd-2da-2ab$
$=2.1^2+2.2^2+2.3^2+0-2.2.3-0-0-2.1.2$
$=2.1+2.4+2.9-12-4=2+8+18-12-4=12.$

17. $c^3+\frac{4}{5}ad^4-3a^3+b^2d=3^3+0-3.1^3+0=27-3=24.$

18. $a^2+2b^2+2c^2+d^2+2ab+2bc+\frac{3}{7}cd$
$=1^2+2.2^2+2.3^2+0+2.1.2+2.2.3+0$
$=1+2.4+2.9+4+12=1+8+18+4+12=43.$

19. $2c^2+2a^2+2b^2-4cb+6acbd=2.3^2+2.1^2+2.2^2-4.3.2+0$
$=2.9+2.1+2.4-24=18+2+8-24=4.$

20. $13a^2+\frac{11}{9}c^4+20ab-16ac-16bc$
$=13.1^2+\frac{11}{9}.3^4+20.1.2-16.1.3-16.2.3$
$=13.1+\frac{11}{9}.81+40-48-96=13+99+40-48-96=8.$

21. $6ab-\frac{4}{3}ac^2-2a+\frac{1}{8}b^4-3d+\frac{4}{9}c^3=6.1.2-\frac{4}{3}.1.3^2-2.1+\frac{1}{8}.2^4-0+\frac{4}{9}.3^3$
$=12-\frac{4}{3}.1.9-2+\frac{1}{8}.16+\frac{4}{9}.27$
$=12-12-2+2+12=12.$

22. $a^2-c^2+b^2-d^2+2ab-2cd=1^2-3^2+2^2-0+2.1.2-0=1-9+4+4=0.$

23. $2ab-\frac{3}{4}b^3+3ac-2c-d+\frac{4}{15}ad=2.1.2-\frac{3}{4}.2^3+3.1.3-2.3-0+0$
$=4-\frac{3}{4}.8+9-6=4-6+9-6=1.$

24. $125b^4c-9d^5+3abc^2d=125.2^4.3-0+0=125.16.3=6000.$

25. $c^2(y-x)-b^2(c-a)=9.2-1.1=18-1=17.$

26. $(2a-c)(x+2y-z)=(4-3)(4+2.6-0)=1.16=16.$

27. $\frac{2}{3}(c^2-z^2)+\frac{3}{5}(y^2-x^2)=\frac{2}{3}(9-0)+\frac{3}{5}(36-16)=\frac{2}{3}.9+\frac{3}{5}.20=18.$

28. $\frac{4}{9}(cy-2c^2)+\frac{3}{7}(xy-bc)=\frac{4}{9}(3.6-2.9)+\frac{3}{7}(4.6-1.3)$
$=\frac{4}{9}(18-18)+\frac{3}{7}(24-3)$
$=0+3.3=9.$

29. $\frac{a^2}{b^2}+\frac{b^2}{a^2}-\frac{2y}{x^2}=\frac{4}{1}+\frac{1}{4}-\frac{2.6}{4^2}=4+\frac{1}{4}-\frac{3}{4}=3\frac{1}{2}.$

30. $\frac{a^2}{b^2}.c^2+\frac{a^2}{b^2}+c^2=4.9+4+9=49.$

31. $\frac{(a+y)^2}{(x-z)^2}-\frac{6(c^2-a)}{7(a^2+x)}=\frac{(2+6)^2}{(4-0)^3}-\frac{6(9-2)}{7(4+4)}$
$=\frac{8^2}{4^3}-\frac{6.7}{7.8}=\frac{64}{64}-\frac{3}{4}=1-\frac{3}{4}=\frac{1}{4}.$

32. $\frac{a^2-b^2}{a^2b^2}-\frac{(a+b+z)^2}{(b+c-z)^2}=\frac{4-1}{4.1}-\frac{(2+1+0)^2}{(1+3-0)^2}$
$=\frac{3}{4}-\frac{3^2}{4^2}=\frac{3}{4}-\frac{9}{16}=\frac{3}{16}.$

33. $\frac{(a+b)^2}{(y-c)^2}-\frac{a(y-z)}{c(x+z)}=\frac{(2+1)^2}{(6-3)^2}-\frac{2(6-0)}{3(4+0)}=\frac{3^2}{3^2}-\frac{2.6}{3.4}=1-1=0.$

34. $\frac{(a+b+c)^2}{c(y-z)}-\frac{4(c-a)^3}{3(a+y)}=\frac{(2+1+3)^2}{3(6-0)}-\frac{4(3-2)^3}{3(2+6)}$
$=\frac{6^2}{3.6}-\frac{4.1^3}{3.8}=2-\frac{1}{6}=1\frac{5}{6}.$

[*For Examples I. e. and Misc. Examples I., see Appendix, p. 333.*]

EXAMPLES. V. b. PAGE 30.

21. $3a^2+bx-4cy=3.(-4)^2+(-3).4-4.(-1).1=3.16-3.4+4.1.1$
$=48-12+4=40.$

22. $2ab^2-3bc^2+2fx=2.(-4).(-3)^2-3.(-3).(-1)^2+0$
$=-2.4.9+3.3.1=-72+9=-63.$

23. $fa^2-2b^3-cx^3=0-2.(-3)^3-(-1).4^3=-2.(-27)+1.64$
$=54+64=118.$

24. $3a^2y^3-5b^2x-2c^3=3.(-4)^2.1^3-5.(-3)^2.4-2.(-1)^3$
$=3.16.1-5.9.4-2.(-1)=48-180+2=-130.$

25. $2a^3-3b^3+7cy^4=2.(-4)^3-3.(-3)^3+7.(-1).1^4$
$=2.(-64)-3.(-27)+7.(-1).1=-128+81-7=-54.$

26. $3b^2y^4-4b^2f-6c^4x=3.(-3)^2.1^4-0-6.(-1)^4.4$
$=3.9.1-6.1.4=27-24=3.$

27. $2\sqrt{ac}-3\sqrt{xy}+\sqrt{b^2c^4}=2\sqrt{(-4).(-1)}-3\sqrt{4.1}+\sqrt{(-3)^2.(-1)^4}$
$=2\sqrt{4}-3\sqrt{4}+\sqrt{9.1}=2.2-3.2+3=1.$

28. $\sqrt{acx}-2\sqrt{b^2y}-6\sqrt{c^2y}=3\sqrt{(-4).(-1).4}-2\sqrt{(-3)^2.1}-6\sqrt{(-1)^2.1}$
$=3\sqrt{16}-2\sqrt{9}-6\sqrt{1}=3.4-2.3-6.1=0.$

29. $7\sqrt{a^2x}-3\sqrt{b^4c^2}+5\sqrt{f^2x}=7\sqrt{(-4)^2.4}-3\sqrt{(-3)^4.(-1)^2}+0$
$=7\sqrt{16.4}-3\sqrt{81.1}=7\sqrt{64}-3\sqrt{81}$
$=7.8-3.9=56-27=29.$

30. $3c\sqrt{(3bc)}-5\sqrt{(4c^2y^3)}-2cy\sqrt{(3bc^5)}$
$=3.(-1)\sqrt{3.(-3).(-1)}-5\sqrt{4.(-1)^2 1^3}-2.(-1).1.\sqrt{3.(-3).(-1)^5}$
$=-3\sqrt{9}-5\sqrt{4}+2\sqrt{9}=-3.3-5.2+2.3=-9-10+6=-13.$

EXAMPLES. V. e. PAGE 34.

1.
$$\begin{array}{l} a+b+c \\ a+b-c \\ \hline a^2+ab+ac \\ \quad +ab+b^2+bc \\ \qquad\quad -ac-bc-c^2 \\ \hline a^2+2ab+b^2-c^2 \end{array}$$

2.
$$\begin{array}{l} a-2b+c \\ a+2b-c \\ \hline a^2-2ab+ac \\ \quad +2ab-4b^2+2bc \\ \qquad\quad -ac+2bc-c^2 \\ \hline a^2-4b^2+4bc-c^2 \end{array}$$

3.
$$\begin{array}{l} a^2-ab+b^2 \\ a^2+ab+b^2 \\ \hline a^4-a^3b+a^2b^2 \\ \quad +a^3b-a^2b^2+ab^3 \\ \qquad\quad +a^2b^2-ab^3+b^4 \\ \hline a^4+a^2b^2+b^4 \end{array}$$

4.
$$\begin{array}{l} x^2+3y^2 \\ x+4y \\ \hline x^3+3xy^2 \\ \quad +4x^2y+12y^3 \\ \hline x^3+4x^2y+3xy^2+12y^3 \end{array}$$

5.
$$\begin{array}{l} x^3 - 2x^2 + 8 \\ x + 2 \\ \hline x^4 - 2x^3 + 8x \\ \quad + 2x^3 - 4x^2 + 16 \\ \hline x^4 - 4x^2 + 8x + 16 \end{array}$$

6.
$$\begin{array}{l} x^4 - x^2y^2 + y^4 \\ x^2 + y^2 \\ \hline x^6 - x^4y^2 + x^2y^4 \\ \quad + x^4y^2 - x^2y^4 + y^6 \\ \hline x^6 + y^6 \end{array}$$

7.
$$\begin{array}{l} x^2 + xy + y^2 \\ x - y \\ \hline x^3 + x^2y + xy^2 \\ \quad - x^2y - xy^2 - y^3 \\ \hline x^3 - y^3 \end{array}$$

8.
$$\begin{array}{l} a^2 - 2ax + 4x^2 \\ a^2 + 2ax + 4x^2 \\ \hline a^4 - 2a^3x + 4a^2x^2 \\ \quad + 2a^3x - 4a^2x^2 + 8ax^3 \\ \qquad + 4a^2x^2 - 8ax^3 + 16x^4 \\ \hline a^4 + 4a^2x^2 + 16x^4 \end{array}$$

9.
$$\begin{array}{l} 16a^2 + 12ab + 9b^2 \\ 4a - 3b \\ \hline 64a^3 + 48a^2b + 36ab^2 \\ \quad - 48a^2b - 36ab^2 - 27b^3 \\ \hline 64a^3 - 27b^3 \end{array}$$

10.
$$\begin{array}{l} x^3 - ax^2 + a^2x - a^3 \\ x + a \\ \hline x^4 - ax^3 + a^2x^2 - a^3x \\ \quad ax^3 - a^2x^2 + a^3x - a^4 \\ \hline x^4 - a^4 \end{array}$$

11.
$$\begin{array}{l} x^2 + x - 2 \\ x^2 + x - 6 \\ \hline x^4 + x^3 - 2x^2 \\ \quad + x^3 + x^2 - 2x \\ \qquad - 6x^2 - 6x + 12 \\ \hline x^4 + 2x^3 - 7x^2 - 8x + 12 \end{array}$$

12.
$$\begin{array}{l} 2x^3 - 3x^2 + 2x \\ 2x^2 + 3x + 2 \\ \hline 4x^5 - 6x^4 + 4x^3 \\ \quad + 6x^4 - 9x^3 + 6x^2 \\ \qquad + 4x^3 - 6x^2 + 4x \\ \hline 4x^5 - x^3 + 4x \end{array}$$

13.
$$\begin{array}{l} -a^5 + a^4b - a^3b^2 \\ -a - b \\ \hline a^6 - a^5b + a^4b^2 \\ \quad a^5b - a^4b^2 + a^3b^3 \\ \hline a^6 + a^3b^3 \end{array}$$

14.
$$\begin{array}{l} x^3 - 7x + 5 \\ x^2 - 2x + 3 \\ \hline x^5 - 7x^3 + 5x^2 \\ \quad - 2x^4 + 14x^2 - 10x \\ \qquad + 3x^3 - 21x + 15 \\ \hline x^5 - 2x^4 - 4x^3 + 19x^2 - 31x + 15 \end{array}$$

15.
$$\begin{array}{l} a^3 + 2a^2b + 2ab^2 \\ a^2 - 2ab + 2b^2 \\ \hline a^5 + 2a^4b + 2a^3b^2 \\ \quad - 2a^4b - 4a^3b^2 - 4a^2b^3 \\ \qquad + 2a^3b^2 + 4a^2b^3 + 4ab^4 \\ \hline a^5 + 4ab^4 \end{array}$$

16.
$$\begin{array}{l} 4x^2 + 6xy + 9y^2 \\ 2x - 3y \\ \hline 8x^3 + 12x^2y + 18xy^2 \\ \quad - 12x^2y - 18xy^2 - 27y^3 \\ \hline 8x^3 - 27y^3 \end{array}$$

17.
$$\begin{array}{l} x^2 - 3xy - y^2 \\ -x^2 + xy + y^2 \\ \hline -x^4 + 3x^3y + x^2y^2 \\ \quad + x^3y - 3x^2y^2 - xy^3 \\ \qquad + x^2y^2 - 3xy^3 - y^4 \\ \hline -x^4 + 4x^3y - x^2y^2 - 4xy^3 - y^4 \end{array}$$

18.
$$\begin{array}{l} a^3 - a^2b^2 + b^3 \\ a^3 + a^2b^2 + b^3 \\ \hline a^6 - a^5b^2 + a^3b^3 \\ \quad + a^5b^2 - a^4b^4 + a^2b^5 \\ \qquad + a^3b^3 - a^2b^5 + b^6 \\ \hline a^6 - a^4b^4 + 2a^3b^3 + b^6 \end{array}$$

19.
$$\begin{array}{l}
x^2-2xy+y^2\\
x^2+2xy+y^2\\
\hline
x^4-2x^3y+x^2y^2\\
\quad+2x^3y-4x^2y^2+2xy^3\\
\qquad\quad+x^2y^2-2xy^3+y^4\\
\hline
x^4-2x^2y^2+y^4
\end{array}$$

20.
$$\begin{array}{l}
ab+cd+ac+bd\\
ab+cd-ac-bd\\
\hline
a^2b^2+abcd+a^2bc+ab^2d\\
\qquad+abcd+c^2d^2+ac^2d+bcd^2\\
\qquad-abcd-a^2bc-ac^2d-a^2c^2\\
\qquad-abcd-ab^2d-bcd^2-b^2d^2\\
\hline
a^2b^2+c^2d^2-a^2c^2-b^2d^2
\end{array}$$

21.
$$\begin{array}{l}
15a^3b-3a^2b^2+4ab^3\\
5a^2b^2+ab^3-3b^4\\
\hline
75a^5b^3-15a^4b^4+20a^3b^5\\
\qquad+15a^4b^4-3a^3b^5+4a^2b^6\\
\qquad\qquad-45a^3b^5+9a^2b^6-12ab^7\\
\hline
75a^5b^3-28a^3b^5+13a^2b^6-12ab^7
\end{array}$$

22.
$$\begin{array}{l}
27x^3-36ax^2+48a^2x-64a^3\\
3x+4a\\
\hline
81x^4-108ax^3+144a^2x^2-192a^3x\\
\qquad+108ax^3-144a^2x^2+192a^3x-256a^4\\
\hline
81x^4-256a^4
\end{array}$$

23.
$$\begin{array}{l}
a^2-5ab-b^2\\
a^2+5ab+b^2\\
\hline
a^4-5a^3b-a^2b^2\\
\quad+5a^3b-25a^2b^2-5ab^3\\
\qquad\qquad+a^2b^2-5ab^3-b^4\\
\hline
a^4-25a^2b^2-10ab^3-b^4
\end{array}$$

24.
$$\begin{array}{l}
x^2-xy+x+y^2+y+1\\
x+y-1\\
\hline
x^3-x^2y+x^2+xy^2+xy+x\\
\quad+x^2y-xy^2+xy+y^3+y^2+y\\
\qquad-x^2+xy-x-y^2-y-1\\
\hline
x^3+3xy+y^3-1
\end{array}$$

25.
$$\begin{array}{l}
a^2-ab-ac+b^2+c^2-bc\\
a+b+c\\
\hline
a^3-a^2b-a^2c+ab^2+ac^2-abc\\
\quad+a^2b-ab^2-abc+b^3+bc^2-b^2c\\
\qquad+a^2c-ac^2-abc-bc^2+b^2c+c^3\\
\hline
a^3-3abc+b^3+c^3
\end{array}$$

26.
$$\begin{array}{l}
x^4-x^3y+x^2y^2-xy^3+y^4\\
x+y\\
\hline
x^5-x^4y+x^3y^2-x^2y^3+xy^4\\
\quad+x^4y-x^3y^2+x^2y^3-xy^4+y^5\\
\hline
x^5+y^5
\end{array}$$

27.
$$\begin{array}{l}
x^{12}-x^9y^2+x^6y^4-x^3y^6+y^8\\
x^3+y^2\\
\hline
x^{15}-x^{12}y^2+x^9y^4-x^6y^6+x^3y^8\\
\quad+x^{12}y^2-x^9y^4-x^6y^6-x^3y^8+y^{10}\\
\hline
x^{15}+y^{10}
\end{array}$$

28.
$$\begin{array}{l} a^4+2a^3+3a^2+2a+1 \\ a^2-2a+1 \\ \hline a^6+2a^5+3a^4+2a^3+a^2 \\ \quad -2a^5-4a^4-6a^3-4a^2-2a \\ \qquad +a^4+2a^3+3a^2+2a+1 \\ \hline a^6 \qquad -2a^3 \qquad +1 \end{array}$$

29.
$$\begin{array}{l} -ax^2+3axy^2-9ay^4 \\ -ax-3ay^2 \\ \hline a^2x^3-3a^2x^2y^2+9a^2xy^4 \\ \quad +3a^2x^2y^2-9a^2xy^4+27a^2y^6 \\ \hline a^2x^3 \qquad +27a^2y^6 \end{array}$$

30.
$$\begin{array}{l} x^4-2x^3y+3x^2y^2-2xy^3+y^4 \\ x^2+2xy+y^2 \\ \hline x^6-2x^5y+3x^4y^2-2x^3y^3+x^2y^4 \\ \quad +2x^5y-4x^4y^2+6x^3y^3-4x^2y^4+2xy^5 \\ \qquad +x^4y^2-2x^3y^3+3x^2y^4-2xy^5+y^6 \\ \hline x^6 \qquad +2x^3y^3 \qquad +y^6 \end{array}$$

31.
$$\begin{array}{l} \frac{1}{2}a^2+\frac{1}{3}a+\frac{1}{4} \\ \frac{1}{2}a-\frac{1}{3} \\ \hline \frac{1}{4}a^3+\frac{1}{6}a^2+\frac{1}{8}a \\ \quad -\frac{1}{6}a^2-\frac{1}{9}a-\frac{1}{12} \\ \hline \frac{1}{4}a^3 \qquad +\frac{1}{72}a-\frac{1}{12} \end{array}$$

32.
$$\begin{array}{l} \frac{1}{2}x^2-2x+\frac{3}{2} \\ \frac{1}{2}x+\frac{1}{3} \\ \hline \frac{1}{4}x^3-x^2+\frac{3}{4}x \\ \quad +\frac{1}{6}x^2-\frac{2}{3}x+\frac{1}{2} \\ \hline \frac{1}{4}x^3-\frac{5}{6}x^2+\frac{1}{12}x+\frac{1}{2} \end{array}$$

33.
$$\begin{array}{l} \frac{2}{3}x^2+xy+\frac{3}{2}y^2 \\ \frac{1}{3}x-\frac{1}{2}y \\ \hline \frac{2}{9}x^3+\frac{1}{3}x^2y+\frac{1}{2}xy^2 \\ \quad -\frac{1}{3}x^2y-\frac{1}{2}xy^2-\frac{3}{4}y^3 \\ \hline \frac{2}{9}x^3 \qquad -\frac{3}{4}y^3 \end{array}$$

34.
$$\begin{array}{l} \frac{3}{2}x^2-ax-\frac{2}{3}a^2 \\ \frac{3}{4}x^2-\frac{1}{2}ax+\frac{1}{3}a^2 \\ \hline \frac{9}{8}x^4-\frac{3}{4}ax^3-\frac{1}{2}a^2x^2 \\ \quad -\frac{3}{4}ax^3+\frac{1}{2}a^2x^2+\frac{1}{3}a^3x \\ \qquad +\frac{1}{2}a^2x^2-\frac{1}{3}a^3x-\frac{2}{9}a^4 \\ \hline \frac{9}{8}x^4-\frac{3}{2}ax^3+\frac{1}{2}a^2x^2 \qquad -\frac{2}{9}a^4 \end{array}$$

35.
$$\begin{array}{l} \frac{1}{2}x^2-\frac{2}{3}x-\frac{3}{4} \\ \frac{1}{2}x^2+\frac{2}{3}x-\frac{3}{4} \\ \hline \frac{1}{4}x^4-\frac{1}{3}x^3-\frac{3}{8}x^2 \\ \quad +\frac{1}{3}x^3-\frac{4}{9}x^2-\frac{1}{2}x \\ \qquad -\frac{3}{8}x^2+\frac{1}{2}x+\frac{9}{16} \\ \hline \frac{1}{4}x^4 \qquad -\frac{43}{36}x^2 \qquad +\frac{9}{16} \end{array}$$

36.
$$\begin{array}{l} \frac{2}{3}x^2+\frac{2}{3}ax+\frac{1}{3}a^2 \\ \frac{3}{2}x^2-\frac{3}{2}ax+\frac{3}{4}a^2 \\ \hline x^4+ax^3+\frac{1}{2}a^2x^2 \\ \quad -ax^3-a^2x^2-\frac{1}{2}a^3x \\ \qquad +\frac{1}{2}a^2x^2+\frac{1}{2}a^3x+\frac{1}{4}a^4 \\ \hline x^4 \qquad +\frac{1}{4}a^4 \end{array}$$

EXAMPLES. VI. b. PAGE 42.

25.
$$\begin{array}{l} 3x-2\,)\,27x^3+9x^2-3x-10\,(\,9x^2+9x+5 \\ \qquad 27x^3-18x^2 \\ \hline \qquad\qquad 27x^2-3x \\ \qquad\qquad 27x^2-18x \\ \hline \qquad\qquad\qquad 15x-10 \\ \qquad\qquad\qquad 15x-10 \\ \hline \end{array}$$

26.

$$8a-3)\ 16a^3-46a^2+39a-9\ (\ 2a^2-5a+3$$
$$16a^3-\ 6a^2$$
$$-40a^2+39a$$
$$-40a^2+15a$$
$$24a-9$$
$$14a-9$$

27.

$$5-4a)\ 15+\ 3a-\ 7a^2-4a^3\ (\ 3+3a+a^2$$
$$15-12a$$
$$15a-\ 7a^2$$
$$15a-12a^2$$
$$5a^2-4a^3$$
$$5a^2-4a^3$$

28.

$$2-3x)\ 16-96x+216x^2-216x^3+81x^4\ (\ 8-36x+54x^2-27x^3$$
$$16-24x$$
$$-72x+216x^2$$
$$-72x+108x^2$$
$$108x^2-216x^3$$
$$108x^2-162x^3$$
$$-\ 54x^3+81x^4$$
$$-\ 54x^3+81x^4$$

EXAMPLES. VI. c. PAGE 44

1.

$$x^2+3x+3)\ x^3-\ x^2-\ 9x-12\ (\ x-4$$
$$x^3+3x^2+\ 3x$$
$$-4x^2-12x-12$$
$$-4x^2-12x-12$$

2.

$$2y^2-5y-1)\ 2y^3-3y^2-6y-1\ (\ y+1$$
$$2y^3-5y^2-\ y$$
$$2y^2-5y-1$$
$$2y^2-5y-1$$

3.

$$3m^2+4m-1)\ 6m^3-\ m^2-14m+\ 3\ (\ 2m-3$$
$$6m^3+8m^2-\ 2m$$
$$-9m^2-12m+3$$
$$-9m^2-12m+3$$

4.

$$3a^3-2a^2-a)\ 6a^5-13a^4+4a^3+3a^2\ (\ 2a^2-3a$$
$$6a^5-\ 4a^4-2a^3$$
$$-\ 9a^4+6a^3+3a^2$$
$$-\ 9a^4+6a^3+3a^2$$

5.
$$\begin{array}{l}
x^2+2x+8\ )\ x^4+x^3+7x^2-6x+8\ (\ x^2-x+1 \\
\qquad x^4+2x^3+8x^2 \\
\qquad\quad -x^3-x^2-6x \\
\qquad\quad -x^3-2x^2-8x \\
\qquad\qquad x^2+2x+8 \\
\qquad\qquad x^2+2x+8
\end{array}$$

6.
$$\begin{array}{l}
a^2+2a-3\ )\ a^4-a^3-8a^2+12a-9\ (\ a^2-3a+1 \\
\qquad a^4+2a^3-3a^2 \\
\qquad\quad -3a^3-5a^2+12a \\
\qquad\quad -3a^3-6a^2+9a \\
\qquad\qquad a^2+3a-9 \\
\qquad\qquad a^2+2a-3 \\
\qquad\qquad\quad a-6
\end{array}$$

7.
$$\begin{array}{l}
a^2+3a+2\ )\ a^4+6a^3+13a^2+12a+4\ (\ a^2+3a+2 \\
\qquad a^4+3a^3+2a^2 \\
\qquad\quad 3a^3+11a^2+12a \\
\qquad\quad 3a^3+9a^2+6a \\
\qquad\qquad 2a^2+6a+4 \\
\qquad\qquad 2a^2+6a+4
\end{array}$$

8.
$$\begin{array}{l}
x^2-x+3\ )\ 2x^4-x^3+4x^2+7x+1\ (\ 2x^2+x-1 \\
\qquad 2x^4-2x^3+6x^2 \\
\qquad\quad x^3-2x^2+7x \\
\qquad\quad x^3-x^2+3x \\
\qquad\qquad -x^2+4x+1 \\
\qquad\qquad -x^2+x-3 \\
\qquad\qquad\quad 3x+4
\end{array}$$

9.
$$\begin{array}{l}
x^2-3x+2\ )\ x^5-5x^4+9x^3-6x^2-x+2\ (\ x^3-2x^2+x+1 \\
\qquad x^5-3x^4+2x^3 \\
\qquad\quad -2x^4+7x^3-6x^2 \\
\qquad\quad -2x^4+6x^3-4x^2 \\
\qquad\qquad x^3-2x^2-x \\
\qquad\qquad x^3-3x^2+2x \\
\qquad\qquad\quad x^2-3x+2 \\
\qquad\qquad\quad x^2-3x+2
\end{array}$$

10.
$$\begin{array}{l}
x^2-x-2\ )\ x^5-4x^4+3x^3+3x^2-3x+2\ (\ x^3-3x^2+2x-1 \\
\qquad x^5-x^4-2x^3 \\
\qquad\quad -3x^4+5x^3+3x^2 \\
\qquad\quad -3x^4+3x^3+6x^2 \\
\qquad\qquad 2x^3-3x^2-3x \\
\qquad\qquad 2x^3-2x^2-4x \\
\qquad\qquad\quad -x^2+x+2 \\
\qquad\qquad\quad -x^2+x+2
\end{array}$$

11.
$$3x^2+2x-4\ )\ 30x^4+11x^3-82x^2-5x+3\ (\ 10x^2-3x-12$$
$$30x^4+20x^3-40x^2$$
$$-9x^3-42x^2-5x$$
$$-9x^3-6x^2+12x$$
$$-36x^2-17x+3$$
$$-36x^2-24x+48$$
$$7x-45$$

12.
$$4y^2-13y+6\ )\ 28y^4-71y^3-35y^2+30y+9\ (\ 7y^2+5y-3$$
$$28y^4-91y^3+42y^2$$
$$20y^3-77y^2+30y$$
$$20y^3-65y^2+30y$$
$$-12y^2\qquad+9$$
$$-12y^2+39y-18$$
$$-39y+27$$

13.
$$3k^3-k+1\ )\ 6k^5-15k^4+4k^3+7k^2-7k+2\ (\ 2k^2-5k+2$$
$$6k^5\qquad-2k^3+2k^2$$
$$-15k^4+6k^3+5k^2-7k$$
$$-15k^4\qquad+5k^2-5k$$
$$6k^3\qquad-2k+2$$
$$6k^3\qquad-2k+2$$

14.
$$3-2m-m^2\ )\ 15-31m+9m^2+4m^3+2m^4+m^5\ (\ 5-7m-m^3$$
$$15-10m-5m^2$$
$$-21m+14m^2+4m^3$$
$$-21m+14m^2+7m^3$$
$$-3m^3+2m^4+m^5$$
$$-3m^3+2m^4+m^5$$

15.
$$x^2-3x+2\ )\ x^4+2x^3-7x^2-8x+12\ (\ x^2+5x+6$$
$$x^4-3x^3+2x^2$$
$$5x^3-9x^2-8x$$
$$5x^3-15x^2+10x$$
$$6x^2-18x+12$$
$$6x^2-18x+12$$

16.
$$x^3-7x+5\ )\ x^5-2x^4-4x^3+19x^2\ (\ x^2-2x+3$$
$$x^5\qquad-7x^3+5x^2$$
$$-2x^4+3x^3+14x^2$$
$$-2x^4\qquad+14x^2-10x$$
$$3x^3\qquad+10x$$
$$3x^3\qquad-21x+15$$
$$31x-15$$

17.

$$\begin{array}{l}
16 - x^2 \,)\, 192 + 128x + 4x^2 - 8x^3 - x^4 \,(\, 12 + 8x + x^2 \\
192 \quad - 12x^2 \\
\quad 128x + 16x^2 - 8x^3 \\
\quad 128x \quad - 8x^3 \\
\qquad 16x^2 \quad - x^4 \\
\qquad 16x^2 \quad - x^4
\end{array}$$

18.

$$\begin{array}{l}
2x^2 + 5xy + 7y^2 \,)\, 14x^4 + 45x^3y + 78x^2y^2 + 45xy^3 + 14y^4 \,(\, 7x^2 + 5xy + 2y^2 \\
14x^4 + 35x^3y + 49x^2y^2 \\
\quad 10x^3y + 29x^2y^2 + 45xy^3 \\
\quad 10x^3y + 25x^2y^2 + 35xy^3 \\
\qquad 4x^2y^2 + 10xy^3 + 14y^4 \\
\qquad 4x^2y^2 + 10xy^3 + 14y^4
\end{array}$$

19.

$$\begin{array}{l}
x^3 - x - y \,)\, x^5 - x^4y + x^3y^2 - x^3 + x^2 - y^3 \,(\, x^2 - xy + y^2 \\
x^5 \quad - x^3 - x^2y \\
- x^4y + x^3y^2 + x^2y + x^2 \\
- x^4y \quad + x^2y + xy^2 \\
\quad x^3y^2 + x^2 - xy^2 - y^3 \\
\quad x^3y^2 \quad - xy^2 - y^3 \\
\qquad x^2
\end{array}$$

20.

$$\begin{array}{l}
x^2 + xy - y^2 \,)\, x^5 + x^4y - x^3y^2 + x^3 - 2xy^2 + y^3 \,(\, x^3 + x - y \\
x^5 + x^4y - x^3y^2 \\
\quad x^3 - 2xy^2 + y^3 \\
\quad x^3 + x^2y - xy^2 \\
\qquad - x^2y - xy^2 + y^3 \\
\qquad - x^2y - xy^2 + y^3
\end{array}$$

21.

$$\begin{array}{l}
a^3 - b^3 \,)\, a^9 - b^9 \quad (\, a^6 + a^3b^3 + b^6 \\
a^6 - a^6b^3 \\
\quad a^6b^3 - b^9 \\
\quad a^6b^3 - a^3b^6 \\
\qquad a^3b^6 - b^9 \\
\qquad a^3b^6 - b^9
\end{array}$$

22.

$$\begin{array}{l}
x^2 + xy + y^2 \,)\, x^9 - y^9 \quad (\, x^7 - x^6y + x^4y^3 - x^3y^4 + xy^6 - y^7 \\
x^9 + x^8y + x^7y^2 \\
- x^8y - x^7y^2 - y^9 \\
- x^8y - x^7y^2 - x^6y^3 \\
\quad x^6y^3 - y^9 \\
\quad x^6y^3 + x^5y^4 + x^4y^5 \\
\qquad - x^5y^4 - x^4y^5 - y^9 \\
\qquad - x^5y^4 - x^4y^5 - x^3y^6 \\
\qquad\quad x^3y^6 - y^9 \\
\qquad\quad x^3y^6 + x^2y^7 + xy^8 \\
\qquad\qquad - x^2y^7 - xy^8 - y^9 \\
\qquad\qquad - x^2y^7 - xy^8 - y^9
\end{array}$$

23.

$$\begin{array}{l}
x-2y^2)\,x^7-7x^5y^4+14x^3y^8-7xy^{12}-2y^{14}\,(x^6+2x^5y^2-3x^4y^4-6x^3y^6+2x^2y^8+4xy^{10}+y^{12}\\
\quad x^7-2x^6y^2\\
\qquad 2x^6y^2-7x^5y^4\\
\qquad 2x^6y^2-4x^5y^4\\
\qquad\quad -3x^5y^4\\
\qquad\quad -3x^5y^4+6x^4y^6\\
\qquad\qquad -6x^4y^6+14x^3y^8\\
\qquad\qquad -6x^4y^6+12x^3y^8\\
\qquad\qquad\quad 2x^3y^8-7xy^{12}\\
\qquad\qquad\quad 2x^3y^8-4x^2y^{10}\\
\qquad\qquad\qquad 4x^2y^{10}-7xy^{12}\\
\qquad\qquad\qquad 4x^2y^{10}-8xy^{12}\\
\qquad\qquad\qquad\quad xy^{12}-2y^{14}\\
\qquad\qquad\qquad\quad xy^{12}-2y^{14}
\end{array}$$

24.

$$\begin{array}{l}
a+b-1)\,a^3+3a^2b+3ab^2+b^3-1\,(a^2+2ab+a+b^2+b+1\\
\quad a^3+a^2b-a^2\\
\qquad 2a^2b+a^2+3ab^2\\
\qquad 2a^2b+2ab^2-2ab\\
\qquad\quad a^2+ab^2+2ab\\
\qquad\quad a^2+ab-a\\
\qquad\qquad ab^2+ab+a+b^3\\
\qquad\qquad ab^2+b^3-b^2\\
\qquad\qquad\quad ab+a+b^2-1\\
\qquad\qquad\quad ab+b^2-b\\
\qquad\qquad\qquad a+b-1\\
\qquad\qquad\qquad a+b-1
\end{array}$$

25.

$$\begin{array}{l}
x^3+x^2y+xy^2+y^3)\,x^8-y^8\,(x^5-x^4y+xy^4-y^5\\
\quad x^8-x^7y-x^6y^2+x^5y^3\\
\qquad x^7y+x^6y^2+x^5y^3-y^8\\
\qquad x^7y+x^6y^2+x^5y^3-x^4y^4\\
\qquad\qquad x^4y^4-y^8\\
\qquad\qquad x^4y^4+x^3y^5+x^2y^6+xy^7\\
\qquad\qquad\quad -x^3y^5-x^2y^6-xy^7-y^8\\
\qquad\qquad\quad -x^3y^5-x^2y^6-xy^7-y^8
\end{array}$$

26.

$$\begin{array}{l}
a^2-b^2)\,a^{12}-b^{12}\,(a^{10}+a^8b^2+a^6b^4+a^4b^6+a^2b^8+b^{10}\\
\quad a^{12}-a^{10}b^2\\
\qquad a^{10}b^2-b^{12}\\
\qquad a^{10}b^2-a^8b^4\\
\qquad\quad a^8b^4-b^{12}\\
\qquad\quad a^8b^4-a^6b^6\\
\qquad\qquad a^6b^6-b^{12}\\
\qquad\qquad a^6b^6-a^4b^8\\
\qquad\qquad\quad a^4b^8-b^{12}\\
\qquad\qquad\quad a^4b^8-a^2b^{10}\\
\qquad\qquad\qquad a^2b^{10}-b^{12}\\
\qquad\qquad\qquad a^2b^{10}-b^{12}
\end{array}$$

27.

$$\begin{array}{l}
a^4+2a^2b^2+b^4\,)\,a^{12}+2a^6b^6+b^{12}\,(\,a^8-2a^6b^2+3a^4b^4-2a^2b^6+b^8\\
a^{12}+2a^{10}b^2+a^8b^4\\
\hline
-2a^{10}b^2-a^8b^4+2a^6b^6\\
-2a^{10}b^2-4a^8b^4-2a^6b^6\\
\hline
3a^8b^4+4a^6b^6\\
3a^8b^4+6a^6b^6+3a^4b^8\\
\hline
-2a^6b^6-3a^4b^8\\
-2a^6b^6-4a^4b^8-2a^2b^{10}\\
\hline
a^4b^8+2a^2b^{10}+b^{12}\\
a^4b^8+2a^2b^{10}+b^{12}
\end{array}$$

28.

$$\begin{array}{l}
1-a-2x\,)\,1-6ax-a^3-8x^3\,(\,1+a+2x-2ax+a^2+4x^2\\
1-a-2x\\
\hline
a+2x-6ax\\
a-a^2-2ax\\
\hline
2x-4ax+a^2\\
2x-2ax-4x^2\\
\hline
-2ax+a^2+4x^2\\
-2ax+2a^2x+4ax^2\\
\hline
a^2-2a^2x+4x^2-4ax^2-a^3\\
a^2-2a^2x-a^3\\
\hline
4x^2-4ax^2-8x^3\\
4x^2-4ax^2-8x^3
\end{array}$$

29.

$$\begin{array}{l}
\frac{1}{2}a-3x\,)\,\frac{1}{8}a^3-\frac{9}{4}a^2x+\frac{27}{2}ax^2-27x^3\,(\,\frac{1}{4}a^2-3ax+9x^2\\
\frac{1}{8}a^3-\frac{3}{4}a^2x\\
\hline
-\frac{3}{2}a^2x+\frac{27}{2}ax^2\\
-\frac{3}{2}a^2x+9ax^2\\
\hline
\frac{9}{2}ax^2-27x^3\\
\frac{9}{2}ax^2-27x^3
\end{array}$$

30.

$$\begin{array}{l}
\frac{1}{3}a-\frac{1}{4}\,)\,\frac{1}{27}a^3-\frac{1}{12}a^2+\frac{1}{16}a-\frac{1}{64}\,(\,\frac{1}{9}a^2-\frac{1}{6}a+\frac{1}{16}\\
\frac{1}{27}a^3-\frac{1}{36}a^2\\
\hline
-\frac{1}{18}a^2+\frac{1}{16}a\\
-\frac{1}{18}a^2+\frac{1}{24}a\\
\hline
\frac{1}{48}a-\frac{1}{64}\\
\frac{1}{48}a-\frac{1}{64}
\end{array}$$

31.

$$\begin{array}{l}
\frac{1}{5}a^2+\frac{1}{2}ac\,)\,\frac{6}{125}a^5+\frac{3}{4}a^2c^3\,(\,\frac{6}{25}a^3-\frac{3}{5}a^2c+\frac{3}{2}ac^2\\
\frac{6}{125}a^5+\frac{3}{25}a^4c\\
\hline
-\frac{3}{25}a^4c+\frac{3}{4}a^2c^3\\
-\frac{3}{25}a^4c-\frac{3}{10}a^3c^2\\
\hline
\frac{3}{10}a^3c^2+\frac{3}{4}a^2c^3\\
\frac{3}{10}a^3c^2+\frac{3}{4}a^2c^3
\end{array}$$

32.

$$\begin{array}{l}
\frac{3}{2}a^2-a-\frac{8}{3}\,)\,\frac{9}{16}a^4-\frac{3}{4}a^3-\frac{7}{4}a^2+\frac{4}{3}a+\frac{16}{9}\,(\,\frac{3}{8}a^2-\frac{1}{4}a-\frac{2}{3}\\
\frac{9}{16}a^4-\frac{3}{8}a^3-a^2\\
\hline
-\frac{3}{8}a^3-\frac{3}{4}a^2+\frac{4}{3}a\\
-\frac{3}{8}a^3+\frac{1}{4}a^2+\frac{2}{3}a\\
\hline
-a^2+\frac{2}{3}a+\frac{16}{9}\\
-a^2+\frac{2}{3}a+\frac{16}{9}
\end{array}$$

33.
$$
\begin{array}{l}
6x - \tfrac{1}{3}y - \tfrac{1}{2}\ )\ 36x^2 - 4xy - 6x + \tfrac{1}{9}y^2 + \tfrac{1}{3}y + \tfrac{1}{4}\ (\ 6x - \tfrac{1}{3}y - \tfrac{1}{2} \\
\qquad\qquad\qquad\ \underline{36x^2 - 2xy - 3x} \\
\qquad\qquad\qquad\qquad\quad - 2xy - 3x + \tfrac{1}{9}y^2 + \tfrac{1}{3}y \\
\qquad\qquad\qquad\qquad\quad \underline{- 2xy \qquad\ + \tfrac{1}{9}y^2 + \tfrac{1}{6}y} \\
\qquad\qquad\qquad\qquad\qquad\qquad - 3x \qquad + \tfrac{1}{6}y + \tfrac{1}{4} \\
\qquad\qquad\qquad\qquad\qquad\qquad \underline{- 3x \qquad + \tfrac{1}{6}y + \tfrac{1}{4}}
\end{array}
$$

34.
$$
\begin{array}{l}
\tfrac{2}{3}a - \tfrac{3}{4}x\ )\ \tfrac{8}{27}a^5 - \tfrac{243}{512}ax^4\ (\ \tfrac{4}{9}a^4 + \tfrac{1}{2}a^3x + \tfrac{9}{16}a^2x^2 + \tfrac{81}{128}ax^3 \\
\qquad\qquad\ \underline{\tfrac{8}{27}a^5 - \quad \tfrac{1}{3}a^4x} \\
\qquad\qquad\qquad\quad \tfrac{1}{3}a^4x \\
\qquad\qquad\qquad\quad \underline{\tfrac{1}{3}a^4x - \tfrac{3}{8}a^3x^2} \\
\qquad\qquad\qquad\qquad\qquad \tfrac{3}{8}a^3x^2 \\
\qquad\qquad\qquad\qquad\qquad \underline{\tfrac{3}{8}a^3x^2 - \tfrac{27}{64}a^2x^3} \\
\qquad\qquad\qquad\qquad\qquad\qquad\quad \tfrac{27}{64}a^2x^3 - \tfrac{243}{512}ax^4 \\
\qquad\qquad\qquad\qquad\qquad\qquad\quad \underline{\tfrac{27}{64}a^2x^3 - \tfrac{243}{512}ax^4}
\end{array}
$$

EXAMPLES. VII. a. PAGE 47.

1. $a-(b-c)+a+(b-c)+b-(c+a)=a-b+c+a+b-c+b-c-a$
$$=a+b-c.$$

2. $a-[b+\{a-(b+a)\}]=a-[b+\{a-b-a\}]=a-[b+a-b-a]=a.$

3. $a-[2a-\{3b-(4c-2a)\}]=a-[2a-\{3b-4c+2a\}]$
$$=a-[2a-3b+4c-2a]=a-2a+3b-4c+2a=a+3b-4c.$$

4. $\{a-(b-c)\}+\{b-(c-a)\}-\{c-(a-b)\}=\{a-b+c\}+\{b-c+a\}-\{c-a+b\}$
$$=a-b+c+b-c+a-c+a-b=3a-b-c.$$

5. $2a-(5b+[3c-a])-(5a-[b+c])=2a-(5b+3c-a)-(5a-b-c)$
$$=2a-5b-3c+a-5a+b+c=-2a-4b-2c.$$

6. $-\{-[-(a-\overline{b-c})]\}=-\{-[-(a-b+c)]\}=-\{-[-a+b-c]\}$
$$=-\{a-b+c\}=-a+b-c.$$

7. $-[a-\{b-(c-a)\}]-[b-\{c-(a-b)\}]$
$$=-[a-\{b-c+a\}]-[b-\{c-a+b\}]=-[-b+c]-[-c+a]$$
$$=-a+b.$$

8. $-(-(-(-x)))-(-(-y))=-(-(x))-(y)=-(-x)-y=x-y.$

9. $-[-\{-(b+c-a)\}]+[-\{-(c+a-b)\}]$
$$=-[-\{-b-c+a\}]+[-\{-c-a+b\}]=-[b+c-a]+[c+a-b]$$
$$=-b-c+a+c+a-b=2a-2b.$$

10. $-5x-[3y-\{2x-(2y-x)\}]=-5x-[3y-\{2x-2y+x\}]$
$$=-5x-[3y-2x+2y-x]=-5x-3y+2x-2y+x=-2x-5y.$$

11. $-(-(-a))-(-(-(-x)))=-(a)-(-(x))=-a-(-x)=-a+x.$

12. $3a-[a+b-\{a+b+c-(a+b+c+d)\}]$
$$=3a-[a+b-\{-d\}]=3a-[a+b-a-b-c+a+b+c+d]$$
$$=3a-a-b+a+b+c-a-b-c-d=2a-b-d.$$

13. $-2a-[3x+\{3c-(4y+3x+2a)\}]=-2a-[3x+\{3c-4y-3x-2a\}]$
$$=-2a-[+3c-4y-2a]=-2a-3c+4y+2a=-3c+4y.$$

14. $3x-[5y-\{6z-(4x-7y)\}]=3x-[5y-\{6z-4x+7y\}]$
$$=3x-[5y-6z+4x-7y]=3x-5y+6z-4x+7y=-x+2y+6z.$$

15. $-[5x-(11y-3x)]-[5y-(3x-6y)]=-[5x-11y+3x]-[5y-3x+6y]$
$$=-5x+11y-3x-5y+3x-6y=-5x.$$

16. $-[15x-\{14y-(15z+12y)-(10x-15z)\}]$
$$=-[15x-\{14y-15z-12y-10x+15z\}]$$
$$=-[15x-14y+15z+12y+10x-15z]$$
$$=-15x+14y-15z-12y-10x+15z=-25x+2y.$$

17. $8x-\{16y-[3x-(12y-x)-8y]+x\}=8x-\{16y-[3x-12y+x-8y]+x\}$
$$=8x-\{16y-3x+12y-x+8y+x\}=8x-16y+3x-12y+x-8y-x$$
$$=11x-36y.$$

18. $-[x-\{z+(x-z)-(z-x)-z\}-x]=-[-\{z+x-z-z+x-z\}]$
$$=-[-(2x-2z)]=2x-2z.$$

19. $-[a+\{a-(a-x)-(a+x)-a\}-a]=-[a+\{a-a+x-a-x-a\}-a]$
$$=-(-2a)=2a.$$

20. $-[a-\{a+(x-a)-(x-a)-a\}-2a]=-[a-\{a+x-a-x+a-a\}-2a]$
$$=-[a-a-x+a+x-a+a-2a]=-a+a+x-a-x+a-a+2a=a.$$

EXAMPLES. VII. b. PAGE 48.

1. $a-[2b+\{3c-3a-(a+b)\}+2a-(b+3c)]$
$$=a-[2b+\{3c-3a-a-b\}+2a-b-3c]$$
$$=a-[2b+3c-4a-b+2a-b-3c]=a-(-2a)=3a.$$

2. $a+b-(c+a-[b+c-(a+b-\{c+a-(b+c-a)\})])$
$$=a+b-(c+a-[b+c-(a+b-\{c+a-b-c+a\})])$$
$$=a+b-(c+a-[b+c-(a+b-\{2a-b\})])$$
$$=a+b-(c+a-[b+c-(a+b-2a+b)])$$
$$=a+b-(c+a-[b+c-(-a+2b)])=a+b-(c+a-[b+c+a-2b])$$
$$=a+b-(c+a-b-c-a+2b)=a+b-(b)=a+b-b=a.$$

3. $a-(b-c)-[a-b-c-2\{b+c-3(c-a)-d\}]$
$=a-b+c-[a-b-c-2\{b+c-3c+3a-d\}]$
$=a-b+c-[a-b-c-2b-2c+6c-6a+2d]$
$=a-b+c-a+b+c+2b+2c-6c+6a-2d=6a+2b-2c-2d.$

4. $2x-(3y-4z)-\{2x-(3y+4z)\}-\{3y-(4z+2x)\}$
$=2x-3y+4z-\{2x-3y-4z\}-\{3y-4z-2x\}$
$=2x-3y+4z-2x+3y+4z-3y+4z+2x=2x-3y+12z.$

5. $b+c-(a+b-[c+a-(b+c-\{a+b-(c+a-b)\})])$
$=b+c-(a+b-[c+a-(b+c-\{a+b-c-a+b\})])$
$=b+c-(a+b-[c+a-(b+c-a-b+c+a-b)])$
$=b+c-(a+b-[c+a-(2c-b)])=b+c-(a+b-[c+a-2c+b])$
$=b+c-(a+b-c-a+2c-b)=b+c-a-b+c+a-2c+b=+b.$

6. $3b-\{5a-[6a+2(10a-b)]\}=3b-\{5a-[6a+20a-2b]\}$
$=3b-\{5a-26a+2b\}=3b-5a+26a-2b=21a+b.$

7. $a-(b-c)-[a-b-c-2\{b+c\}]=a-b+c-[a-b-c-2b-2c]$
$=a-b+c-a+b+c+2b+2c=2b+4c.$

8. $3a^2-[6a^2-\{8b^2-(9c^2-2a^2)\}]=3a^2-[6a^2-\{8b^2-9c^2+2a^2\}]$
$=3a^2-[6a^2-8b^2+9c^2-2a^2]=3a^2-6a^2+8b^2-9c^2+2a^2$
$=-a^2+8b^2-9c^2.$

9. $b-(c-a)-[b-a-c-2\{c+a-3(a-b)-d\}]$
$=b-c+a-[b-a-c-2\{c+a-3a+3b-d\}]$
$=b-c+a-[b-a-c-2c-2a+6a-6b+2d]$
$=b-c+a-b+a+c+2c+2a-6a+6b-2d=-2a+6b+2c-2d.$

10. $-20(a-d)+3(b-c)-2[b+c+d-3\{c+d-4(d-a)\}]$
$=-20a+20d+3b-3c-2[b+c+d-3\{c+d-4d+4a\}]$
$=-20a+20d+3b-3c-2[b+c+d-3c-3d+12d-12a]$
$=-20a+20d+3b-3c-2b-2c-2d+6c+6d-24d+24a$
$=4a+b+c.$

11. $-4(a+d)+24(b-c)-2[c+d+a-3\{d+a-4(b+c)\}]$
$=-4a-4d+24b-24c-2[c+d+a-3\{d+a-4b-4c\}]$
$=-4a-4d+24b-24c-2[c+d+a-3d-3a+12b+12c]$
$=-4a-4d+24b-24c-2c-2d-2a+6d+6a-24b-24c=-50c.$

12. $-10(a+b)-[c+a+b-3\{a+2b-(c+a-b)\}]+4c$
$=-10a-10b-[c+a+b-3\{a+2b-c-a+b\}]+4c$
$=-10a-10b-[c+a+b-3\{3b-c\}]+4c$
$=-10a-10b-[c+a+b-9b+3c]+4c$
$=-10a-10b-c-a-b+9b-3c+4c=-11a-2b.$

13. $a-2(b-c)-[-\{-(4a-b-c-2\{a+b+c\})\}]$
$=a-2b+2c-[-\{-(4a-b-c-2a-2b-2c)\}]$
$=a-2b+2c-[-\{-(2a-3b-3c)\}]=a-2b+2c-[-\{-2a+3b+3c\}]$
$=a-2b+2c-[2a-3b-3c]=a-2b+2c-2a+3b+3c=-a+b+5c.$

14. $8(b-c)-[-\{a-b-3(c-b+a)\}]=8b-8c-[-\{a-b-3c+3b-3a\}]$
$=8b-8c-[-a+b+3c-3b+3a]=8b-8c+a-b-3c+3b-3a$
$=-2a+10b-11c.$

15. $2(3b-5a)-7[a-6\{2-5(a-b)\}]=6b-10a-7[a-6\{2-5a+5b\}]$
$=6b-10a-7[a-12+30a-30b]=6b-10a-7a+84-210a+210b$
$=84-227a+216b.$

16. $6\{a-2[b-3(c+d)]\}-4\{a-3[b-4(c-d)]\}$
$=6\{a-2[b-3c-3d]\}-4\{a-3[b-4c+4d]\}$
$=6\{a-2b+6c+6d\}-4\{a-3b+12c-12d\}$
$=6a-12b+36c+36d-4a+12b-48c+48d=2a-12c+84d.$

17. $5\{a-2[a-2(a+x)]\}-4\{a-2[a-2(a+x)]\}$
$=5\{a-2[a-2a-2x]\}-4\{a-2[a-2a-2x]\}$
$=5\{a-2a+4a+4x\}-4\{a-2a+4a+4x\}$
$=5\{3a+4x\}-4\{3a+4x\}=3a+4x.$

18. $-10\{a-6[a-(b-c)]\}+60\{b-(c+a)\}$
$=-10\{a-6[a-b+c]\}+60\{b-c-a\}$
$=-10\{a-6a+6b-6c\}+60b-60c-60a$
$=-10a+60a-60b+60c+60b-60c-60a=-10a.$

19. $-3\{-2[-4(-a)]\}+5\{-2[-2(-a)]\}=-3\{-2[4a]\}+5\{-2[2a]\}$
$=-3\{-8a\}+5\{-4a\}=24a-20a=4a.$

20. $-2\{-[-(x-y)]\}+\{-2[-(x-y)]\}=-2\{-[-x+y]\}+\{-2[-x+y]\}$
$=-2\{x-y\}+\{2x-2y\}=-2x+2y+2x-2y=0.$

21. $\frac{1}{4}\{a-5(b-a)\}-\frac{3}{2}\left\{\frac{1}{3}\left(b-\frac{a}{3}\right)-\frac{2}{9}\left[a-\frac{3}{4}\left(b-\frac{4a}{5}\right)\right]\right\}$
$=\frac{1}{4}\{a-5b+5a\}-\frac{3}{2}\left\{\frac{1}{3}b-\frac{a}{9}-\frac{2}{9}\left[a-\frac{3}{4}b+\frac{3a}{5}\right]\right\}$
$=\frac{1}{4}(6a-5b)-\frac{3}{2}\{\frac{1}{3}b-\frac{1}{9}a-\frac{2}{9}a+\frac{1}{6}b-\frac{6}{45}a\}=\frac{3}{2}a-\frac{5}{4}b-\frac{3}{2}\{\frac{1}{2}b-\frac{7}{15}a\}$
$=\frac{3}{2}a-\frac{5}{4}b-\frac{3}{4}b+\frac{7}{10}a=\frac{11}{5}a-2b.$

22. $35\left[\frac{3x-4y}{5}-\frac{1}{15}\{3x-\frac{5}{7}(7x-4y)\}\right]+8(y-2x)$
$=35[\frac{3}{5}x-\frac{4}{5}y-\frac{1}{15}\{3x-5x+\frac{20}{7}y\}]+8y-16x$
$=35[\frac{3}{5}x-\frac{4}{5}y-\frac{3}{15}x+\frac{1}{3}x-\frac{4}{21}y]+8y-16x=35[\frac{4}{5}x-\frac{38}{35}y]+8y-16x$
$=28x-38y+8y-16x=12x-30y.$

23. $\frac{3}{8}\left\{\frac{4}{3}(a-b)-8(b-c)\right\}-\left\{\frac{b-c}{2}-\frac{c-a}{3}\right\}-\frac{1}{2}\left\{c-a-\frac{2}{3}(a-b)\right\}$

$$=\frac{3}{8}\left\{\frac{4}{3}a-\frac{4}{3}b-8b+8c\right\}-\left\{\frac{1}{2}b-\frac{1}{2}c-\frac{1}{3}c+\frac{1}{3}a\right\}-\frac{1}{2}\left\{c-a-\frac{2}{3}a+\frac{2}{3}b\right\}$$
$$=\frac{1}{2}a-\frac{1}{2}b-3b+3c-\frac{1}{2}b+\frac{1}{2}c+\frac{1}{3}c-\frac{1}{3}a-\frac{1}{2}c+\frac{1}{2}a+\frac{1}{3}a-\frac{1}{3}b$$
$$=a-\frac{13}{3}b+\frac{10}{3}c.$$

24. $\frac{1}{2}x-\frac{1}{2}\left(\frac{2}{3}y-\frac{1}{2}z\right)-\left[x-\left\{\frac{1}{2}x-\left(\frac{1}{3}y-\frac{1}{4}z\right)\right\}-\left(\frac{2}{3}y-\frac{1}{2}z\right)\right]$

$$=\frac{1}{2}x-\frac{1}{3}y+\frac{1}{4}z-\left[x-\left\{\frac{1}{2}x-\frac{1}{3}y+\frac{1}{4}z\right\}-\frac{2}{3}y+\frac{1}{2}z\right]$$
$$=\frac{1}{2}x-\frac{1}{3}y+\frac{1}{4}z-\left[x-\frac{1}{2}x+\frac{1}{3}y-\frac{1}{4}z-\frac{2}{3}y+\frac{1}{2}z\right]$$
$$=\frac{1}{2}x-\frac{1}{3}y+\frac{1}{4}z-\left[\frac{1}{2}x-\frac{1}{3}y+\frac{1}{4}z\right]=0.$$

EXAMPLES. VII. c. PAGE 50.

9. The expression $=ax^3-2cx-[bx^2-\{cx-dx-bx^3-3cx^2\}-cx^2+bx]$

$$=ax^3-2cx-[bx^2-cx+dx+bx^3+3cx^2-cx^2+bx]$$
$$=ax^3-2cx-bx^2+cx-dx-bx^3-2cx^2-bx$$
$$=(a-b)x^3-(b+2c)x^2-(b+c+d)x.$$

10. The expression $=5ax^3-7bx+7cx^2-\{6bx^2-3ax^2-2ax-4cx^3\}$

$$=5ax^3-7bx+7cx^2-6bx^2+3ax^2+2ax+4cx^3$$
$$=(5a+4c)x^3+(3a-6b+7c)x^2+(2a-7b)x.$$

11. The expression $=ax^2-3\left\{-ax^3+3bx-4\left[\frac{1}{6}cx^3-\frac{2}{3}ax+\frac{2}{3}bx^2\right]\right\}$

$$=ax^2-3\left\{-ax^3+3bx-\frac{2}{3}cx^3+\frac{8}{3}ax-\frac{8}{3}bx^2\right\}$$
$$=ax^2+3ax^3-9bx+2cx^3-8ax+8bx^2$$
$$=(3a+2c)x^3+(a+8b)x^2-(8a+9b)x.$$

12. The expression $=x^5-4bx^4-\frac{1}{6}\left[12ax-4\left\{3bx^4-\frac{9}{2}cx+9bx^5-\frac{3}{2}ax^4\right\}\right]$

$$=x^5-4bx^4-\frac{1}{6}[12ax-12bx^4+18cx-36bx^5+6ax^4]$$
$$=x^5-4bx^4-2ax+2bx^4-3cx+6bx^5-ax^4$$
$$=(6b+1)x^5-(a+2b)x^4-(2a+3c)x.$$

13. The expression $=x\{x-b-x(a-bx)\}+ax-x\{x-x(ax-b)\}$

$$=x\{x-b-ax+bx^2\}+ax-x\{x-ax^2+bx\}$$
$$=x^2-bx-ax^2+bx^3+ax-x^2+ax^3-bx^2$$
$$=(a+b)x^3-(a+b)x^2+(a-b)x.$$

EXAMPLES. VII. d. PAGE 51.

6. The product $=(ax^2+bx+1)(cx+2)=acx^3+bcx^2+cx+2ax^2+2bx+2$

$$=acx^3+(2a+bc)x^2+(2b+c)x+2.$$

7. The product $=(cx^2-2x+3)(ax-b)=acx^3-2ax^2+3ax-bcx^2+2bx-3b$

$$=acx^3-(2a+bc)x^2+(3a+2b)x-3b.$$

8. The product $=(ax^2-bx-c)(px+q)=apx^3-bpx^2-cpx+aqx^2-bqx-cq$

$$=apx^3+(aq-bp)x^2-(bq+cp)x-cq.$$

9. The product $=(2x^2-3x-1)(bx+c)=2bx^3-3bx^2-bx+2cx^2-3cx-c$
$=2bx^3-(3b-2c)x^2-(b+3c)x-c.$

10. The product $=(ax^2-2bx+3c)(x-1)=ax^3-2bx^2+3cx-ax^2+2bx-3c$
$=ax^3-(a+2b)x^2+(2b+3c)x-3c.$

11. The product $=(px^2-2x-q)(ax-3)=apx^3-2ax^2-aqx-3px^2+6x+3q$
$=apx^3-(2a+3p)x^2+(6-aq)x+3q.$

12. The product $=(x^3+ax^2-bx-c)(x^3-ax^2-bx+c)$
$=x^6+ax^5-bx^4-cx^3-ax^5-a^2x^4+abx^3+acx^2-bx^4-abx^3$
$+b^2x^2+bcx+cx^3+acx^2-bcx-c^2$
$=x^6-(a^2+2b)x^4+(b^2+2ac)x^2-c^2.$

13. The product $=(ax^3-x^2+3x-b)(ax^3+x^2+3x+b)$
$=a^2x^6-ax^5+3ax^4-abx^3+ax^5-x^4+3x^3-bx^2+3ax^4-3x^3$
$+9x^2-3bx+abx^3-bx^2+3bx-b^2$
$=a^2x^6+(6a-1)x^4+(9-2b)x^2-b^2.$

14. The product $=(x^4-ax^3-bx^2+cx+d)(x^4+ax^3-bx^2-cx+d)$
$=x^8-ax^7-bx^6+cx^5+dx^4+ax^7-a^2x^6-abx^5+acx^4+adx^3$
$-bx^6+abx^5+b^2x^4-bcx^3-bdx^2-cx^5+acx^4+bcx^3$
$-c^2x^2-cdx+dx^4-adx^3-bdx^2+cdx+d^2$
$=x^8-(a^2+2b)x^6+(2ac+b^2+2d)x^4-(2bd+c^2)x^2+d^2.$

EXAMPLES. VIII. a. Page 55.

8. $$8(x-3)-(6-2x)=2(x+2)-5(5-x).$$
$$8x-24-6+2x=2x+4-25+5x;$$
$$8x+2x-2x-5x=4-25+24+6;$$
whence we obtain $$x=3.$$

9. $$7(25-x)-2x=2(3x-25).$$
$$175-7x-2x=6x-50;$$
$$-7x-2x-6x=-50-175;$$
whence $$x=15.$$

10. $$3(169-x)-(78+x)=29x.$$
$$507-3x-78-x=29x;$$
$$\therefore\ -3x-x-29x=-507+78;$$
whence $$x=13.$$

11. $$5x-17+3x-5=6x-7-8x+115.$$
$$5x+3x-6x+8x=-7+115+17+5;$$
whence $$x=13.$$

12. $7x - 39 - 10x + 15 = 100 - 33x + 26.$

$7x - 10x + 33x = 100 + 26 + 39 - 15;$

whence $x = 5.$

13. $118 - 65x - 123 = 15x + 35 - 120x.$

$-65x - 15x + 120x = 35 - 118 + 123;$

whence $x = 1.$

14. $157 - 21(x + 3) = 163 - 15(2x - 5).$

$157 - 21x - 63 = 163 - 30x + 75;$

$\therefore \quad -21x + 30x = 163 + 75 - 157 + 63;$

whence $x = 16.$

15. $179 - 18(x - 10) = 158 - 3(x - 17).$

$179 - 18x + 180 = 158 - 3x + 51;$

$\therefore \quad -18x - 3x = 158 + 51 - 179 - 180;$

whence $x = 10.$

16. $97 - 5(x + 20) = 111 - 8(x + 3).$

$97 - 5x - 100 = 111 - 8x - 24;$

$\therefore \quad -5x + 8x = 111 - 24 - 97 + 100;$

whence $x = 30.$

17. $x - [3 + \{x - (3 + x)\}] = 5.$

$x - [3 + \{x - 3 - x\}] = 5;$

$\therefore \quad x - [3 - 3] = 5;$

or $x = 5.$

18. $5x - (3x - 7) - \{4 - 2x - (6x - 3)\} = 10.$

$5x - 3x + 7 - \{4 - 2x - 6x + 3\} = 10;$

$\therefore \quad 5x - 3x + 7 - 7 + 8x = 10;$

whence $x = 1.$

19. $14x - (5x - 9) - \{4 - 3x - (2x - 3)\} = 30.$

$14x - 5x + 9 - \{4 - 3x - 2x + 3\} = 30;$

$\therefore \quad 14x - 5x + 9 - 7 + 5x = 30;$

whence $x = 2.$

20. $25x - 19 - [3 - \{4x - 5\}] = 3x - (6x - 5).$

$25x - 19 - [3 - 4x + 5] = 3x - 6x + 5;$

$\therefore \quad 25x - 19 - 8 + 4x = 3x - 6x + 5;$

whence $x = 1.$

21.
$$(x+1)(2x+1)=(x+3)(2x+3)-14.$$
$$2x^2+3x+1=2x^2+9x+9-14;$$
$$\therefore\ 3x-9x=9-14-1;$$
whence $$x=1.$$

22.
$$(x+1)^2-(x^2-1)=x(2x+1)-2(x+2)(x+1)+20.$$
$$x^2+2x+1-x^2+1=2x^2+x-2(x^2+3x+2)+20;$$
that is, $$2x+2=2x^2+x-2x^2-6x-4+20;$$
$$\therefore\ 2x-x+6x=-4+20-2;$$
whence $$x=2.$$

23.
$$2(x+1)(x+3)+8=(2x+1)(x+5).$$
$$2(x^2+4x+3)+8=2x^2+11x+5;$$
that is, $$2x^2+8x+6+8=2x^2+11x+5;$$
whence $$x=3.$$

24.
$$6(x^2-3x+2)-2(x^2-1)=4(x+1)(x+2)-24.$$
$$6x^2-18x+12-2x^2+2=4(x^2+3x+2)-24;$$
that is, $$4x^2-18x+14=4x^2+12x+8-24;$$
whence $$x=1.$$

25.
$$2(x-4)-(x^2+x-20)=4x^2-(5x+3)(x-4)-64.$$
$$2x-8-x^2-x+20=4x^2-(5x^2-17x-12)-64\cdot$$
that is, $$-x^2+x+12=-x^2+17x+12-64;$$
whence $$x=4.$$

26.
$$(x+15)(x-3)-(x^2-6x+9)=30-15(x-1).$$
$$x^2+12x-45-x^2+6x-9=30-15x+15;$$
whence $$x=3.$$

27.
$$2x-5\{3x-7(4x-9)\}=66.$$
$$2x-5\{3x-28x+63\}=66;$$
that is, $$2x+125x-315=66;$$
whence $$x=3.$$

28.
$$20(2-x)+3(x-7)-2[x+9-3\{9-4(2-x)\}]=22.$$
$$40-20x+3x-21-2[x+9-3\{9-8+4x\}]=22;$$
that is, $$40-20x+3x-21-2[6-11x]=22;$$
or $$40-20x+3x-21-12+22x=22;$$
whence $$x=3.$$

29. $$x+2-[x-8-2\{8-3(5-x)-x\}]=0.$$

$$x+2-[x-8-2\{8-15+3x-x\}]=0;$$

that is, $$x+2-[x-8-4x+14]=0;$$

$$\therefore\ x+2+3x-6=0;$$

whence $$x=1.$$

30. $$3(5-6x)-5[x-5\{1-3(x-5)\}]=23.$$

$$15-18x-5[x-5\{1-3x+15\}]=23;$$

$$15-18x-5[x-80+15x]=23;$$

$$15-18x-5x+400-75x=23;$$

whence $$x=4.$$

31. $$(x+1)(2x+3)=2(x+1)^2+8.$$

$$2x^2+5x+3=2x^2+4x+2+8;$$

$$\therefore\ 5x-4x=10-3;$$

whence $$x=7.$$

32. $$3(x-1)^2-3(x^2-1)=x-15.$$

$$3(x^2-2x+1)-3x^2+3=x-15;$$

$$\therefore\ 3x^2-6x+3-3x^2+3=x-15;$$

whence $$x=3.$$

33. $$(3x+1)(2x-7)=6(x-3)^2+7.$$

$$6x^2-19x-7=6(x^2-6x+9)+7;$$

$$\therefore\ 6x^2-19x-7=6x^2-36x+54+7;$$

whence $$x=4.$$

34. $$x^2-8x+25=x(x-4)-25(x-5)-16.$$

$$x^2-8x+25=x^2-4x-25x+125-16;$$

whence $$x=4.$$

35. $$x(x+1)+(x+1)(x+2)=(x+2)(x+3)+x(x+4)-9.$$

$$x^2+x+x^2+3x+2=x^2+5x+6+x^2+4x-9;$$

$$\therefore\ x+3x-5x-4x=3-2;$$

whence $$x=1.$$

36. $$2(x+2)(x-4)=x(2x+1)-21.$$

$$2(x^2-2x-8)=2x^2+x-21;$$

that is, $$2x^2-4x-16=2x^2+x-21;$$

whence $$x=1.$$

37. $$(x+1)^2+2(x+3)^2=3x(x+2)+35.$$

$$x^2+2x+1+2(x^2+6x+9)=3x^2+6x+35;$$

that is, $$x^2+2x+1+2x^2+12x+18=3x^2+6x+35;$$

whence $$x=2.$$

38. $$4(x+5)^2-(2x+1)^2=3(x-5)+180.$$

$$4(x^2+10x+25)-(4x^2+4x+1)=3x-15+180;$$

that is, $$4x^2+40x+100-4x^2-4x-1=3x-15+180;$$

whence $$x=2.$$

39. $$84+(x+4)(x-3)(x+5)=(x+1)(x+2)(x+3).$$

$$84+(x^2+x-12)(x+5)=(x^2+3x+2)(x+3);$$

that is $$84+x^3+x^2-12x+5x^2+5x-60=x^3+3x^2+2x+3x^2+9x+6;$$

$$\therefore\ 24+6x^2-7x=6x^2+2x+9x+6;$$

whence $$x=1.$$

40. $$(x+1)(x+2)(x+6)=x^3+9x^2+4(7x-1).$$

$$(x^2+3x+2)(x+6)=x^3+9x^2+28x-4;$$

that is, $$x^3+3x^2+2x+6x^2+18x+12=x^3+9x^2+28x-4;$$

$$\therefore\ 9x^2+2x+18x+12=9x^2+28x-4;$$

whence $$x=2.$$

EXAMPLES. VIII. b. PAGE 58.

6. $$\frac{x-1}{8}=1+\frac{x+1}{18}.$$

Multiply by 72; thus $9x-9=72+4x+4$; whence $x=17$.

7. $$\frac{4(x+2)}{5}=7+\frac{5x}{13}.$$

Multiply by 65; thus $52(x+2)=455+25x$; whence $x=13$.

8. $$\frac{x+4}{14}+\frac{x-4}{6}=2.$$

Multiply by 42; thus $3x+12+7x-28=84$; whence $x=10$.

9. $$\frac{x+20}{9}+\frac{3x}{7}=6.$$

Multiply by 63; thus $7x+140+27x=378$; whence $x=7$.

10. $$\frac{x-8}{7}+\frac{x-3}{3}+\frac{5}{21}=0.$$

Multiply by 21; thus $3x-24+7x-21+5=0$; whence $x=4$.

11. $$\frac{x+5}{6} - \frac{x+1}{9} = \frac{x+3}{4}.$$

Multiply by 36; thus $6x + 30 - 4x - 4 = 9x + 27$; whence $x = -\frac{1}{7}$.

12. $$\frac{4-5x}{6} - \frac{1-2x}{3} = \frac{13}{42}.$$

Multiply by 42; thus $28 - 35x - 14 + 28x = 13$; whence $x = \frac{1}{7}$.

13. $$\frac{5(x+5)}{8} - \frac{2(x-3)}{7} = 5\frac{19}{28}.$$

Multiply by 56; thus $35(x+5) - 16(x-3) = \frac{159}{28} \times 56$;

or $35x + 175 - 16x + 48 = 318$; whence $x = 5$.

14. $$\frac{4(x+2)}{3} - \frac{6(x-7)}{7} = 12.$$

Multiply by 21; thus $28(x+2) - 18(x-7) = 252$; whence $x = 7$.

15. $$1 + \frac{x}{2} - \frac{2x}{3} = \frac{3x}{4} - 4\frac{1}{2}.$$

Multiply by 12; thus $12 + 6x - 8x = 9x - 54$; whence $x = 6$.

16. $$\frac{x}{2} + \frac{x}{3} - \frac{x}{4} + \frac{x}{5} = 7\frac{5}{6}.$$

Multiply by 60; thus $30x + 20x - 15x + 12x = 470$; whence $x = 10$.

17. $$\frac{3}{16}(x-1) - \frac{5}{12}(x-4) = \frac{2}{5}(x-6) + \frac{5}{48}.$$

Multiply by 240; thus $45(x-1) - 100(x-4) = 96(x-6) + 25$;

or $45x - 45 - 100x + 400 = 96x - 576 + 25$;

whence $x = 6$.

18. $$x + \frac{5}{3}(x-7) - \frac{6}{7}(x-8) = 3x - 14\frac{1}{3}.$$

Multiply by 21; thus $21x + 35(x-7) - 18(x-8) = 63x - 301$;

or $21x + 35x - 245 - 18x + 144 = 63x - 301$;

whence $x = 8$.

19. $$\frac{3x}{4} - \frac{6}{17}(x+10) - (x-3) = \frac{x-7}{51} - 4\frac{3}{4}.$$

Multiply by 204; thus

$$153x - 72(x+10) - 204(x-3) = 4(x-7) - 969;$$

or $153x - 72x - 720 - 204x + 612 = 4x - 28 - 969$;

whence $x = 7$.

Or we may proceed as in Art. 76.

20. $$\frac{7x}{5} - \frac{1}{14}(x - 11) = \frac{3}{7}(x - 25) + 34.$$

Multiply by 70; thus $98x - 5(x - 11) = 30(x - 25) + 2380,$

or $98x - 5x + 55 = 30x - 750 + 2380;$

whence $x = 25.$

21. $$3 + \frac{x}{4} = \frac{1}{2}\left(4 - \frac{x}{3}\right) - \frac{5}{6} + \frac{1}{3}\left(11 - \frac{x}{2}\right).$$

Multiply by 12; thus $36 + 3x = 6\left(4 - \frac{x}{3}\right) - 10 + 4\left(11 - \frac{x}{2}\right);$

that is, $36 + 3x = 24 - 2x - 10 + 44 - 2x;$ whence $x = 3\frac{1}{7}.$

22. $$\frac{1}{5}(x - 8) + \frac{4 + x}{4} + \frac{x - 1}{7} = 7 - \frac{23 - x}{5}.$$

Multiply by 140; thus $28(x - 8) + 35(4 + x) + 20(x - 1) = 980 - 28(23 - x);$

that is, $28x - 224 + 140 + 35x + 20x - 20 = 980 - 644 + 28x;$

whence $x = 8.$

23. $$\frac{1}{3}\left(\frac{x}{4} - 3\right) + \frac{5x}{6} - \frac{5x}{4} = \frac{x - 12}{5} - \frac{x + 3}{3}.$$

Multiply by 60; thus $20\left(\frac{x}{4} - 3\right) + 50x - 75x = 12(x - 12) - 20(x + 3);$

that is, $5x - 60 + 50x - 75x = 12x - 144 - 20x - 60;$

whence $x = 12.$

24. $$x - 3x + \frac{2x - 5}{10} = \frac{1}{6}(2x - 57) - \frac{5}{3}.$$

Multiply by 30; thus $30x - 90x + 3(2x - 5) = 5(2x - 57) - 50;$

or $30x - 90x + 6x - 15 = 10x - 285 - 50;$

whence $x = 5.$

25. $$\frac{x}{4} - \frac{x + 10}{5} + 4\tfrac{3}{4} = x - 1 - \frac{x - 2}{3}.$$

Multiply by 60; thus $15x - 12(x + 10) + 285 = 60x - 60 - 20(x - 2);$

or $15x - 12x - 120 + 285 = 60x - 60 - 20x + 40;$

whence $x = 5.$

26. $$\cdot 5x - \cdot\dot{3}x = \cdot 25x - 1.$$

Express the decimals as vulgar fractions; thus

$$\tfrac{1}{2}x - \tfrac{1}{3}x = \tfrac{1}{4}x - 1;$$

$$6x - 4x = 3x - 12;$$

whence $x = 12.$

27. $$3+\frac{x}{\cdot 5}=7-\frac{x}{\cdot 2}.$$

$$3+\frac{10x}{5}=7-\frac{10x}{2};$$

$$\therefore\ 3+2x=7-5x;$$

whence $$x=\tfrac{4}{7}.$$

28. $$2\cdot 25x-\cdot 125=3x+3\cdot 75.$$

$$2\cdot 25x-3x=3\cdot 75+\cdot 125;$$

that is, $$-\cdot 75x=3\cdot 875;$$

whence $$x=-5\tfrac{1}{6}.$$

29. $$\cdot 2x-\cdot 1\dot{6}x=\cdot 6-\cdot\dot{3}.$$

Express the decimals as vulgar fractions; thus

$$\tfrac{2}{10}x-\tfrac{15}{90}x=\tfrac{6}{10}-\tfrac{1}{3};$$

$$18x-15x=54-30;$$

whence $$x=8.$$

30. $$\cdot 6x-\cdot 7x+\cdot 75x-\cdot 875x+15=0.$$

$$\cdot 6x-\cdot 7x+\cdot 75x-\cdot 875x=-15;$$

that is, $$(\cdot 6-\cdot 7+\cdot 75-\cdot 875)\,x=-15;$$

$$\therefore\ (1\cdot 35-1\cdot 575)\,x=-15;$$

whence $$x=66\tfrac{2}{3}.$$

31. $$12\{3x-\cdot 25\,(x-4)-\cdot\dot{3}\,(5x+14)\}=47.$$

$$36x-3\,(x-4)-4\,(5x+14)=47;$$ whence $x=7$.

32. $$\frac{\cdot 25\,(x-3)+\dot{3}\,(x-4)}{\cdot 125}=5x-19.$$

Express the decimals as fractions; thus

$$\frac{\frac{1}{4}(x-3)+\frac{1}{3}(x-4)}{\frac{1}{8}}=5x-19;$$

that is, $$2\,(x-3)+\tfrac{8}{3}\,(x-4)=5x-19;$$

or $$6\,(x-3)+8\,(x-4)=15x-57;$$ whence $x=7$.

33. $$\frac{x+\cdot 75}{\cdot 125}-\frac{x-\cdot 25}{\cdot 25}=15.$$

Express the decimals as fractions; thus

$$\frac{x+\frac{3}{4}}{\frac{1}{8}}-\frac{x-\frac{1}{4}}{\frac{1}{4}}=15;$$

$$\therefore\ 8x+6-(4x-1)=15;$$ whence $x=2$

34. $$\tfrac{5}{6}x + \cdot 25x - \cdot\dot{3}x = x - 3.$$

$$\tfrac{5}{6}x + \tfrac{1}{4}x - \tfrac{1}{3}x = x - 3;$$

$$\therefore\ 10x + 3x - 4x = 12x - 36;\ \text{whence } x = 12.$$

35. $$\cdot 5x - \dot{2}x = \cdot\dot{3}x - 1\cdot 5.$$

$$\tfrac{1}{2}x - \tfrac{2}{9}x = \tfrac{1}{3}x - 1\tfrac{1}{2};$$

$$9x - 4x = 6x - 27;$$

whence $$x = 27.$$

36. $$1\cdot 5 = \frac{\cdot 36}{\cdot 2} - \frac{\cdot 09x - 18}{\cdot 9}.$$

$$1\cdot 5 = \frac{36}{20} - \frac{9x - 18}{90};$$

that is, $$1\cdot 5 = \frac{9}{5} - \frac{x - 2}{10};$$

$$\therefore\ 15 = 18 - x + 2;\ \text{or } x = 5.$$

EXAMPLES. IX. a. PAGE 62.

28. The train travels 30 miles in x hours.

,, ,, $30 \times 1760 \times 3$ ft. in $x \times 60 \times 60$ seconds.

,, ,, $\dfrac{30 \times 1760 \times 3}{x \times 60 \times 60} = \dfrac{44}{x}$ ft. in 1 second.

29. a crowns + b florins = $(5a + 2b)$ shillings.

30. £20 = 400*s*.; hence after spending x shillings there are left $400 - x$ shillings.

31. £a + b shillings = $(20a + b)$ shillings = $(240a + 12b)$ pence, so that after spending c pence there are $(240a + 12b - c)$ pence left.

32. $2x - 5 - (x + 1) = 2x - 5 - x - 1 = x - 6.$

33. The number $= a - 2b - (a - 3b) = a - 2b - a + 3b = b.$

34. One person pays 3*s*. 4*d*., or 40*d*., and therefore x persons pay $40x$ pence.

35. The contents of the purse = £a + b florins = $20a$ shillings + $2b$ shillings. There are left $(20a + 2b - c)$ shillings.

36. One horse eats y bushels a week; therefore x horses eat xy bushels a week. Hence the number of weeks $= \dfrac{100}{xy}$.

37. I spend x shillings a week; that is, $52x$ shillings, or £$\frac{52x}{20}$ a year. Therefore I save £$\left(y - \frac{13x}{5}\right)$

39. In the purse are £x, or $240x$ pence, in the pocket y shillings, or $12y$ pence, and in the other pocket z pence. After giving away 2s. 6d., or 30d., there remain in all $(240x + 12y + z - 30)$ pence.

40. The number of workers is $y + z$; the number of idlers is $x - y - z$. Hence the excess of workers over idlers

$$= y + z - (x - y - z) = y + z - x + y + z = 2y + 2z - x.$$

EXAMPLES. IX. b. PAGE 65.

9. Five years ago he was x years old, therefore now he is $x + 5$ years old, and in y years he will be $x + 5 + y$ years old.

10. Five years hence the boy's age will be $x + 5$; then his father's age will be double his age, that is, $2x + 10$. Therefore his father's age now is $2x + 10 - 5 = 2x + 5$.

11. y years ago a man's age was m times that of a child aged x years, that is, it was mx years. Therefore now his age is $mx + y$ years.

12. C's age is x years, B's age is three times C's or $3x$, A's age is double B's or $2 \times 3x = 6x$.

13. Interest on £100 for 1 year is £c.
,, £1000 ,, £$10c$.
,, £1000 for b years is £$10c \times b$ = £$10bc$.

14. Interest on £100 for 1 year is £5.

,, £1 ,, £$\frac{5}{100}$ = £$\frac{1}{20}$.

,, £x ,, £$\frac{x}{20}$.

,, £x for a years is £$\frac{x}{20} \times a$ = £$\frac{ax}{20}$.

15. Interest on £100 for 1 year is £a.

,, £50 ,, £$\frac{a}{2}$.

,, £$50a$,, £$\frac{a}{2} \times a$ = £$\frac{a^2}{2}$.

,, £$50a$ for a years is £$\frac{a^2}{2} \times a$ = £$\frac{a^3}{2}$.

16. Interest on £100 for 1 year is £y.

,, £100 for 1 month is £$\frac{y}{12}$.

,, £1 ,, $\frac{y}{1200}$.

,, £$24xy$,, £$\frac{y}{1200} \times 24xy = £\frac{xy^2}{50}$.

,, £$24xy$ for x months is £$\frac{x^2y^2}{50}$.

17. The room is x yds., or $3x$ ft. long, and y ft. broad, and therefore the area $= 3xy$ sq. ft.

18. The area is x^2 sq. ft., or $\frac{x^2}{9}$ sq. yds.; $\therefore$ $\frac{x^2}{9}$ sq. yds. of carpet are required.

19. Length $= p$ ft.; breadth $= x$ yds. $= 3x$ ft. Area $= p \times 3x$ sq. ft.; $\therefore$ length of carpet required $= \frac{3px}{2}$ ft. $= \frac{3px}{2 \times 3}$ yds. $= \frac{px}{2}$ yds.

20. Length $= a$ yds. $= 3a$ ft.; breadth $= b$ ft. Area $= 3ab$ sq. ft. $= \frac{3ab}{9}$ sq. yds. $= \frac{ab}{3}$ sq. yds.; $\therefore$ cost of carpeting at c shillings a sq. yd. $= \frac{abc}{3}$ shillings $=$ £$\frac{abc}{60}$.

21. Length of room $= y$ ft. $= 12y$ in.; breadth $= z$ ft. $= 12z$ in. Area $= 12y \times 12z = 144yz$ sq. in.; $\therefore$ length of carpet required $= \frac{144yz}{x}$ in. $= \frac{12yz}{x}$ ft. $= \frac{4yz}{x}$ yds.

22. Area of room $= ab$ sq. yds., and the area of the carpet in the middle $= c^2$ sq. ft. $= \frac{c^2}{9}$ sq. yds.; $\therefore$ the number of sq. yds. of oilcloth required $= ab - \frac{c^2}{9}$.

23. In x hours he walks a miles,

in 1 hour ,, $\frac{a}{x}$ miles,

in 1 minute ,, $\frac{a}{60x}$ miles;

$\therefore$ in 45 minutes ,, $\frac{45a}{60x}$, or $\frac{3a}{4x}$ miles.

24. To walk 20 miles he takes c hours,

.. 1 mile .. $\frac{c}{20}$ hours,

,, b miles ,, $\frac{bc}{20}$ hours.

25. The train travels $1760 \times 3a$ ft. in b hours, that is, in $60 \times 60b$ seconds.

$\therefore$,, $\frac{1760 \times 3a}{60 \times 60b}$ ft. in 1 second,

$= \frac{22a}{15b}$ ft. ,,

26. The train travels x ft., or $\frac{x}{1760 \times 3}$ miles in 1 second.

$\therefore$ it travels $\frac{x}{1760 \times 3} \times 60 \times 60$ miles in 60×60 secs. or 1 hour.

$\therefore$,, $\frac{60 \times 60x}{1760 \times 3} \times y$ miles, or $\frac{15xy}{22}$ miles in y hours.

27. 1 man mows z acres in 1 day.

,, ,, 1 acre in $\frac{1}{z}$ day.

x men mow 1 acre in $\frac{1}{xz}$ day.

x men ,, y acres in $\frac{y}{xz}$ days.

28. To do a piece of work in xz hours requires y men.

,, ,, x ,, ,, yz ,,

29. A rate of 1 per cent. produces £10 in one year or £$10r$ in r years.

$\therefore$ rate per cent. $= \frac{y}{10r}$.

30. £100 produces in one year £r interest, or £a produces £$\frac{ar}{100}$ interest.

$\therefore$ number of years $= p \div \frac{ar}{100} = \frac{100p}{ar}$.

[*For Examples* 31–37 *and Examples IX. c., see Appendix, p.* 334.]

EXAMPLES. X. a. PAGE 71.

3. Let x be the number; then $x - 50 = (89 - x) + 11$.

4. Let x = number of miles he travels by train,

then $2x =$,, ,, coach;

then $10 + x + 2x = 70$.

5. Let x be one number, then $58-x$ is the other number;
then $$x-(58-x)=28.$$

6. Let x be the number; then $288+x=3(x-12)$.

7. Let x be the number; then $23x-14=16-7x$.

8. Let x = one part, then $105-x$ = the other part;
then $$x-20=(105-x)-15.$$

9. Let x be the least number; the other two consecutive numbers are $x+1$ and $x+2$;
then $$x+(x+1)+(x+2)=84.$$

10. Let x be one number, then $8-x$ is the other number;
then $$x+22=5(8-x).$$

11. Let x be the less number, then $x+10$ is the greater.
Hence $$x+x+10=20.$$

12. Let £x be what A wins. After playing, A has £$(60+x)$, B has £$(60-x)$. Hence $60+x=2(60-x)$.

13. Let x be the number; then $(x+5)(x+35)=(x+15)^2$.

14. Let x be the less number, then $x+1$ is the greater number.
Hence $$(x+1)^2-x^2=121.$$

15. Let x be the less number, then $x+3$ is the other number.
Hence $$(x+3)^2-x^2=27.$$

16. Let £x be A's share, then B's share is £$(x+30)$, and C's share is £$(x+30+20)$.
Hence $$x+(x+30)+(x+30+20)=380;$$
whence $$x=100.$$
∴ A's share is £100, B's £$(100+30)$ = £130, C's £$(130+20)$ = £150.

17. Let x = number of florins, then $124-x$ = number of shillings.
But x florins + $(124-x)$ shillings = £8. 17*s*.,
or, expressing all in shillings, $2x+124-x=177$.
Whence $x=53$, and the number of shillings is $124-53$, or 71.

18. Let x shillings be the price of linen per yard; then $6x$ shillings is the price of silk per yard,
and $23\times 6x$ shillings + $50x$ shillings = £9. 8*s*. = 188*s*.,
that is, $138x+50x=188$; whence $x=1$;
∴ linen is 1*s*. per yard, silk 6*s*. a yard.

19. Let x = the age of son, then the age of the father is $4x$. In 24 years they will be respectively $x+24$ and $4x+24$ years old.

$$\therefore\ 4x+24=2(x+24);\ \text{whence } x=12.$$

Thus the son's age is 12, and the father's age is 48.

20. Let x = the age of B, then $25+x$ = the age of A.

$$\therefore\ 25+x-20=85-x;\ \text{whence } x=40.$$

Thus B's age is 40, and A's age is 25+40, or 65.

21. Let x = the age of B, then $6x$ = the age of A. In 15 years their respective ages will be $x+15$ and $6x+15$.

$$\therefore\ 6x+15=3(x+15);\ \text{whence } x=10.$$

Thus B's age is 10, and A's age = 60.

22. Let x = number of crowns, then $4x$ = number of half-crowns, and $2x$ = number of shillings. Now amount paid in crowns = $5x$ shillings, in half-crowns = $4x \times 2\frac{1}{2} = 10x$ shillings, and in shillings = $2x$ shillings.

$$\therefore\ 5x+10x+2x=85;\ \text{whence } x=5.$$

Thus there were 5 crowns, 20 half-crowns, 10 shillings.

23. Let x = the age of A, then $30-x$ = the age of B. In 5 years their respective ages will be $x+5$ and $30-x+5$.

$$\therefore\ x+5=3(30-x+5);\ \text{whence } x=25.$$

Thus A's age is 25, and B's age = 30 − 25 = 5.

24. Let x = number of wides, then $2x$ = number of byes, and the remainder of the score = $12\times 2x+3=24x+3$.

$$\therefore\ 24x+3+2x+x=138;\ \text{whence } x=5.$$

Thus the score was made up of 5 wides, 10 byes, 123 runs.

25. Let x ft. = the breadth of room, then the length = $(x+3)$ ft., and the area = $x(x+3)$ sq. ft. The length of second room = $(x+6)$ ft., the breadth = $(x-2)$ ft., and the area = $(x+6)(x-2)$ sq. ft.

$$\therefore\ x(x+3)=(x+6)(x-2);\ \text{whence } x=12.$$

Thus the room is 15 ft. long and 12 ft. broad.

26. Let x ft. = the breadth of room, then the length = $(x+8)$ ft., and the area = $x(x+8)$ sq. ft.

$$\therefore\ (x+2)(x+10)=x(x+8)+60;\ \text{whence } x=10.$$

Thus the room is 18 ft. long and 10 ft. broad.

EXAMPLES. X. b. PAGE 73.

8. Let x = the less number, and $x+1$ the greater number;

$$\text{then } \tfrac{1}{5}(x+1)-\tfrac{1}{7}x=3;\ \text{whence } x=49.$$

9. Let x, $x+1$, $x+2$ be the three consecutive numbers;

then $$\frac{x}{10}+\frac{x+1}{17}+\frac{x+2}{26}=10,$$

or $$221x+130(x+1)+85(x+2)=22100; \text{ whence } x=50.$$

Thus the numbers are 50, 51, and 52.

10. Let £x be what A and B had to start with. When B has lost £$\frac{5}{11}x$, A has gained £$6+\frac{1}{2}$ of £$\frac{6}{11}x$; but B's loss $=A$'s gain;

$$\therefore \tfrac{5}{11}x=6+\tfrac{1}{2}\times\tfrac{6}{11}x, \text{ whence } x=33.$$

11. Let $x=$ the number;

then $$\left(\frac{x-3}{4}+4\right)\div 5=2; \text{ or } \frac{1}{5}\left(\frac{x-3}{4}+4\right)=2;$$

whence $$x=27.$$

12. Let $x=$ the number of bottles of wine in the cellar, of which $\frac{x}{5}$ are port and $\frac{x}{3}$ are claret;

then $$\frac{x}{5}+\frac{x}{3}+180+30=x; \text{ whence } x=450.$$

Thus the number of bottles of port $=\frac{1}{5}\times 450=90$; and the number of bottles of claret $=\frac{1}{3}\times 450=150$.

13. Let £$x=A$'s money; then B's money $=$ £$\frac{2}{5}x$, and C's money $=\frac{7}{9}$ of $\frac{2}{5}x$;

then $$x+\tfrac{2}{5}x+\tfrac{14}{45}x=770; \text{ whence } x=450.$$

Thus A's money $=$ £450, B's money $=\frac{2}{5}\times$ £450 $=$ £180,

C's money $=\frac{14}{45}\times$ £450 $=$ £140.

14. Let £$x=B$'s share, then A's share $=$ £25 $+$ £$\frac{5}{6}x$, and C's share $=$ £$\frac{4}{15}x$;

then $$25+\tfrac{5}{6}x+x+\tfrac{4}{15}x=1285; \text{ whence } x=600.$$

$\therefore$ A's share $=$ £$(25+\frac{5}{6}\times 600)=$ £525; B's share $=$ £600;

C's share $=\frac{4}{15}\times$ £600 $=$ £160.

15. Let £x be what he paid for the horse. He sold it for £$35+$£$\frac{x}{2}$, gaining £$10\frac{1}{2}$.

$$\therefore 35+\frac{x}{2}-x=10\tfrac{1}{2}; \text{ whence } x=49.$$

16. Let x ft. $=$ the length of room, then the breadth $=\frac{2}{3}x$ ft. If the room is square, its length $=$ breadth;

$$\therefore x-3=\tfrac{2}{3}x+3; \text{ whence } x=18.$$

$\therefore$ the dimensions are length 18 ft., breadth 12 ft.

17. Let £x be the person's property.

£$\frac{2}{3}x$ invested at 4 per cent. will yield $\frac{4}{100} \times$ £$\frac{2}{3}x =$ £$\frac{2}{75}x$,

£$\frac{1}{4}x$,, 3 ,, ,, $\frac{3}{100} \times$ £$\frac{1}{4}x =$ £$\frac{3}{400}x$,

the remainder £$\frac{1}{12}x$,, 2 ,, ,, $\frac{2}{100} \times$ £$\frac{1}{12}x =$ £$\frac{1}{600}x$;

$$\therefore \frac{2}{75}x + \frac{3}{400}x + \frac{1}{600}x = 430, \text{ whence } x = 12000.$$

18. Let $x =$ the number of apples bought at 3 a penny, which cost $\frac{x}{3}$ pence.

Then $\frac{5}{6}x =$ number of apples bought at 4 a penny which cost $\frac{1}{4} \times \frac{5}{6}x$ pence.

The total number bought $= x + \frac{5}{6}x = \frac{11}{6}x$, which sold at 16 for sixpence, and would therefore realise $\frac{6}{16} \times \frac{11}{6}x = \frac{11}{16}x$ pence;

$$\therefore \frac{11}{16}x - \left(\frac{x}{3} + \frac{1}{4} \times \frac{5}{6}x\right) = 3\tfrac{1}{2}; \text{ whence } x = 24.$$

Thus the number of apples bought $= \frac{11}{6} \times 24 = 44$.

[*For Miscellaneous Examples II., see Appendix, p.* 337.]

EXAMPLES. XIII. a. PAGE 86.

5. By addition, we have $11x + 11y = 33$; that is, $x + y = 3$.
By subtraction, we obtain $x - y = -1$.

8. $$15x + 7y = 29, \quad 9x + 15y = 39.$$
Multiply the first equation by 3, and the second by 5; thus
$$45x + 21y = 87, \text{ and } 45x + 75y = 195.$$
By subtraction, $54y = 108$; whence $y = 2$.

9. $$14x - 3y = 39, \quad 6x + 17y = 35.$$
Multiply the first equation by 3, and the second by 7; thus
$$42x - 9y = 117, \text{ and } 42x + 119y = 245.$$
By subtraction, $128y = 128$; whence $y = 1$.

10. $$28x - 23y = 33, \quad 63x - 25y = 101.$$
Multiply the first and second equations by 9 and 4 respectively; thus
$$252x - 207y = 297, \text{ and } 252x - 100y = 404.$$
By subtraction, $107y = 107$; whence $y = 1$.

11. $$35x + 17y = 86, \quad 56x - 13y = 17.$$

Multiply the first equation by 8, and the second by 5; thus

$$280x + 136y = 688, \text{ and } 280x - 65y = 85.$$

By subtraction, $201y = 603$; whence $y = 3$.

12. From the second equation, we have $5x - 3y = 2$; whence $15x - 9y = 6$.

From the first equation, $15x + 77y = 92$.

By subtraction, $86y = 86$; whence $y = 1$.

13. From the first equation, we have $x = \frac{7}{5}y$.

By substitution in the second equation, we find

$$\frac{49}{5}y + 5y = 74; \text{ whence } y = 5.$$

14. $$21x - 50y = 60, \quad 28x - 27y = 199.$$

By subtraction, we find

$$7x + 23y = 139; \text{ whence } 21x + 69y = 417.$$

From this subtract the first equation; thus

$$119y = 357; \text{ so that } y = 3.$$

15. $$39x - 8y = 99, \quad 52x - 15y = 80.$$

By subtraction, we have

$$13x - 7y = -19; \text{ whence } 39x - 21y = -57.$$

Subtract this from the first equation; thus

$$13y = 156; \text{ so that } y = 12.$$

16. $$5x - 7y = -21, \quad 21x - 9y = 75.$$

Multiply the first equation by 9 and the second by 7; thus

$$45x - 63y = -189, \quad 147x - 63y = 525.$$

By subtraction, $102x = 714$; whence $x = 7$.

17. From the second equation, we find $x = \frac{3}{4}y$.

On substitution in the first equation, we have

$$6y - \frac{15}{4}y = 18; \text{ whence } y = 8$$

18. From the first equation, $x = \frac{5}{8}y$.

On substitution in the second equation, we have

$$\frac{65}{8}y = 8y + 1; \text{ whence } y = 8.$$

19. Substitute $y=\frac{3}{7}x$ in the second equation; thus

$$\frac{36}{7}x=5x-1; \text{ whence } x=-7.$$

20. $$19x+17y=0, \quad 2x-y=53.$$

Multiply the second equation by 17; thus

$$34x-17y=901.$$

Add this last equation to the first; then

$$53x=901; \text{ whence } x=17.$$

21. $$93x+15y=123, \quad 15x+93y=201.$$

By addition, $108x+108y=324$; that is, $x+y=3$.
By subtraction, $78x-78y=-78$; that is, $x-y=-1$.

EXAMPLES. XIII. b. PAGE 87.

1. $$\frac{2}{3}x+y=16, \quad x+\frac{y}{4}=14.$$

Multiply the second equation by 4; thus $4x+y=56$.
On subtracting the first from this last equation, we have

$$\frac{10}{3}x=40; \text{ whence } x=12.$$

2. $$\frac{x}{5}+\frac{y}{2}=5, \quad x-y=4.$$

Multiply the first equation by 2; thus $\frac{2}{5}x+y=10$.

On adding this last equation to the second, we have

$$\frac{7}{5}x=14; \text{ whence } x=10.$$

3. $$\frac{5}{6}x-y=3, \quad x-\frac{5}{6}y=8.$$

Multiply the first equation by 6, and the second by 5; thus

$$5x-6y=18, \text{ and } 5x-\frac{25}{6}y=40.$$

By subtraction, $\frac{11}{6}y=22$; whence $y=12$.

4. $$x - y = 5, \quad \frac{x}{4} - \frac{y}{5} = 2.$$

Multiply the second equation by 4; thus $x - \frac{4}{5}y = 8$.

On subtracting the first from this last equation, we have

$$\frac{1}{5}y = 3; \text{ whence } y = 15.$$

5. $$\frac{x}{9} + \frac{y}{7} = 10, \quad \frac{x}{3} + y = 50.$$

Multiply the first equation by 3; thus $\frac{x}{3} + \frac{3}{7}y = 30$.

On subtracting this last equation from the second, we have

$$\frac{4}{7}y = 20; \text{ whence } y = 35.$$

6. Substitute from the first equation in the second; thus

$$\frac{1}{3}(3y) + y = 34; \text{ whence } y = 17.$$

7. $$\frac{2}{5}x - \frac{1}{12}y = 3, \quad 4x - y = 20.$$

Multiply the first equation by 12; thus $\frac{24}{5}x - y = 36$.

On subtracting the second from this last equation, we have

$$\frac{4}{5}x = 16; \text{ whence } x = 20.$$

8. $$\frac{1}{2}x - \frac{1}{5}y = 4, \quad \frac{1}{7}x + \frac{1}{15}y = 3.$$

Multiply the second equation by 3; thus $\frac{3}{7}x + \frac{1}{5}y = 9$.

On adding this last equation to the first, we have

$$\frac{13}{14}x = 13; \text{ whence } x = 14.$$

9. From the first equation $y = -2x$; hence by substitution in the second equation, we have

$$\frac{1}{2}(-2x) - 3x = 8; \text{ whence } x = -2.$$

10. $$\frac{1}{7}x+\frac{1}{5}y=\frac{10}{7}, \quad x+\frac{1}{3}y=\frac{14}{3}.$$

Multiply the first equation by 7; thus $x+\frac{7}{5}y=10$.

On subtracting the second from this last equation, we have

$$\frac{16}{15}y=\frac{16}{3}; \text{ whence } y=5.$$

11. From the first equation, we have $x=\frac{7}{3}y$.

Substitute in the second equation; thus

$$\frac{2}{7}\left(\frac{7}{3}y\right)+\frac{5}{3}y=7; \text{ that is, } \frac{7}{3}y=7, \text{ and } y=3.$$

12. From the first equation, we have $x=\frac{5}{4}y$.

Substitute in the second equation; thus

$$\frac{15}{4}y+\frac{1}{2}y=17; \text{ whence } y=4.$$

13. From the equation $\frac{x}{3}+\frac{y}{4}=0$, we have $y=-\frac{4}{3}x$.

On substituting in $3x-7y-37=0$, we have

$$3x+\frac{28}{3}x-37=0; \text{ whence } x=3.$$

14. From $\frac{x+1}{10}=\frac{3y-5}{2}$, we have $15y-x=26$; and from $\frac{3y-5}{2}=\frac{x-y}{8}$, we have $13y-x=20$.

By subtraction, $2y=6$; whence $y=3$.

15. From $\frac{x+3}{5}=\frac{8-y}{4}$, we have $4x+5y=28$; and from $\frac{8-y}{4}=\frac{3(x+y)}{8}$ we have $3x+5y=16$.

By subtraction, $x=12$.

16. From the equation $\frac{x}{13}-\frac{y}{7}=0$, we have $x=\frac{13}{7}y$.

On substituting in $6x-10y-8=0$, we obtain

$$\frac{78}{7}y-10y-8=0; \text{ whence } y=7.$$

EXAMPLES. XIII. c. PAGE 90.

1. $x+2y+2z=11,\quad 2x+y+z=7,\quad 3x+4y+z=14.$

Multiply the second equation by 2; thus $4x+2y+2z=14$; on subtracting the first from this last equation, we have

$$3x=3;\ \text{whence } x=1.$$

Substitute this value of x in the second and third equations; then

$$y+z=5,\ \text{and } 4y+z=11;\ \text{whence } y=2,\ z=3.$$

2. $x+3y+4z=14,\quad x+2y+z=7,\quad 2x+y+2z=2.$

Multiply the second equation by 2, and subtract the third; thus

$$3y=12;\ \text{whence } y=4.$$

Substitute this value of y in the first and second equations; then

$$x+4z=2,\ \text{and } x+z=-1;\ \text{whence } z=1,\ x=-2.$$

3. $x+4y+3z=17,\quad 3x+3y+z=16,\quad 2x+2y+z=11.$

On multiplying the second equation by 2 and the third equation by 3 and subtracting, we obtain $z=1$.

Substitute this value of z in the first and second equations; thus

$$x+4y=14,\ \text{and } 3x+3y=15,\ \text{or } x+y=5;\ \text{whence } y=3,\ x=2.$$

4. $3x-2y+z=2,\quad 2x+3y-z=5,\quad x+y+z=6.$

Add the first and second equations; thus $5x+y=7$.

Add the second and third equations; thus $3x+4y=11$.

From these last two equations, we have $x=1,\ y=2$.

5. $2x+y+z=16,\quad x+2y+z=9,\quad x+y+2z=3.$

On adding together the first and second equations and subtracting the third, we have $\qquad 2x+2y=22$; that is $x+y=11$.

Subtract this last equation from the third; thus

$$2z=-8;\ \text{whence } z=-4.$$

6. $x-2y+3z=2,\quad 2x-3y+z=1,\quad 3x-y+2z=9.$

Multiply the second equation by 3, and subtract the first; thus

$$5x-7y=1.$$

Multiply the second equation by 2, and subtract from the third; thus

$$-x+5y=7.$$

From the last two equations, we obtain $x=3,\ y=2$.

7. $3x+2y-z=20,\quad 2x+3y+6z=70,\quad x-y+6z=41.$

On subtracting the third equation from the second, we have

$$x+4y=29.$$

Multiply the first equation by 6 and add to the second; thus

$$20x+15y=190;\ \text{that is, } 4x+3y=38.$$

From these two equations in x and y, we obtain $x=5$, $y=6$.

8. $2x+3y+4z=20, \quad 3x+4y+5z=26, \quad 3x+5y+6z=31.$

Subtract the first equation from the second; thus

$$x+y+z=6.$$

Subtract the second equation from the third; thus

$$y+z=5.$$

By subtraction, we obtain $x=1$.

Hence from the first equation, we have $3y+4z=18$.

But $y+z=5$, and therefore $y=2$, $z=3$.

9. $3x-4y-6z=-16, \quad 4x-y-z=5, \quad x-3y-2z=-2.$

Multiply the third equation by 3 and subtract from the first equation; thus

$$5y=-10;\ \text{whence } y=-2.$$

On substituting in the second and third equations, we have

$$4x-z=3, \text{ and } x-2z=-8;\ \text{whence } x=2,\ z=5.$$

10. $5x+2y=14, \quad y-6z=-15, \quad x+2y+z=0.$

Multiply the third equation by 6, and add to the second; thus

$$6x+13y=-15.$$

From this equation and $5x+2y=14$, we obtain $x=4$, $y=-3$.

11. $x-\frac{y}{5}=6, \quad y-\frac{z}{7}=8, \quad z-\frac{x}{2}=10.$

On multiplying the third equation by 2 and adding to the first, we have

$$-\frac{1}{5}y+2z=26.$$

Multiply this equation by 5; thus $-y+10z=130$.

On adding this to the second of the given equations, we have

$$\frac{69}{7}z=138;\ \text{whence } z=14.$$

12. From $\frac{y+z}{4}=\frac{x+y}{2}$, we have

$$2x+y-z=0 \quad\text{..............(1).}$$

From $\frac{z+x}{3}=\frac{x+y}{2}$, we have

$$x+3y-2z=0 \quad\text{..............(2).}$$

Also $$x+y+z=27 \quad\text{..............(3).}$$

From (1) and (3) by addition, $3x+2y=27$.

Multiply (3) by 2 and add to (2); thus $3x+5y=54$.

From these last two equations, we have $y=9$, $x=3$.

13. From $\frac{y-z}{3}=\frac{y-x}{2}$, we have

$$3x-y-2z=0 \quad\text{......(1)}.$$

From $\frac{y-x}{2}=5z-4x$, we have

$$7x+y-10z=0 \quad\text{......(2)}.$$

Also $$-2x+y+z=1.$$

From (1) and (2) by addition,

$$10x-12z=0 \text{ or } 5x-6z=0.$$

From (1) and (3) by addition, $x-z=1$.

From these equations in x and z, we have $x=6$, $z=5$.

14. $$2x+3y=5,\quad 2z-y=1,\quad 7x-9z=3.$$

Multiply the second equation by 3, and add to the first; thus

$$2x+6z=8, \text{ or } x+3z=4.$$

From the two equations $7x-9z=3$ and $x+3z=4$, we obtain $x=\frac{3}{2}$, $z=\frac{5}{6}$.

15. From $\frac{1}{2}(x+z-5)=y-z$, we have

$$x-2y+3z=5 \quad\text{......(1)}.$$

From $y-z=2x-11$, by transposition

$$2x-y+z=11 \quad\text{......(2)}.$$

From $2x-11=9-(x+2z)$, we have

$$3x+2z=20 \quad\text{......(3)}.$$

Multiply the second equation by 2, and subtract the first; thus

$$3x-z=17.$$

By combining this last equation with (3), we obtain $z=1$, $x=6$.

16. From the given equations, we easily obtain

$$2x-3y=-20,\quad x-2z=-15,\quad x+y+z=90.$$

Multiply the third equation by 2, and add to the second; thus

$$3x+2y=165.$$

On combining this last equation with $2x-3y=-20$, we obtain

$$x=35,\ y=30.$$

EXAMPLES. XIII. d. Page 92.

1. $$\frac{5}{x}+\frac{6}{y}=3,\qquad \frac{15}{x}+\frac{3}{y}=4.$$

Multiply the second equation by 2, and subtract the first equation; thus

$$\frac{25}{x}=5;\ \text{whence } x=5.$$

2. $$\frac{6}{x}-\frac{7}{y}=2, \quad \frac{2}{x}+\frac{14}{y}=3.$$

Multiply the first equation by 2, and add the second equation; thus

$$\frac{14}{x}=7; \text{ whence } x=2.$$

3. $$\frac{12}{x}-\frac{4}{y}=2, \quad \frac{3}{x}-\frac{2}{y}=0.$$

Multiply the second equation by 2 and subtract from the first; thus

$$\frac{6}{x}=2; \text{ whence } x=3.$$

4. $$\frac{5}{x}+\frac{16}{y}=79, \quad \frac{16}{x}-\frac{1}{y}=44.$$

Multiply the second equation by 16; thus

$$\frac{256}{x}-\frac{16}{y}=704.$$

On adding the first equation, we have

$$\frac{261}{x}=783; \text{ whence } x=\frac{1}{3}.$$

5. $$\frac{21}{x}+\frac{12}{y}=5, \quad \frac{1}{y}-\frac{1}{x}=\frac{1}{42}.$$

Multiply the second equation by 21 and add the first equation; thus

$$\frac{33}{y}=5\tfrac{1}{2}; \text{ whence } y=6.$$

6. $$\frac{5}{x}+\frac{3}{y}=30, \quad \frac{9}{x}-\frac{5}{y}=2.$$

Multiply the first equation by 5 and the second by 3; thus

$$\frac{25}{x}+\frac{15}{y}=150, \quad \frac{27}{x}-\frac{15}{y}=6.$$

By addition, $$\frac{52}{x}=156; \text{ whence } x=\frac{1}{3}.$$

7. $$\frac{8}{x}-\frac{9}{y}=7, \quad \frac{6}{x}+\frac{6}{y}=1.$$

Multiply the first equation by 2 and the second by 3; thus

$$\frac{16}{x}-\frac{18}{y}=14, \quad \frac{18}{x}+\frac{18}{y}=3.$$

By addition, $$\frac{34}{x}=17; \text{ whence } x=2.$$

8. $$\frac{25}{x}+\frac{24}{y}=1, \quad \frac{40}{x}+\frac{60}{y}=7.$$

Multiply the first equation by 8 and the second by 5; thus

$$\frac{200}{x}+\frac{192}{y}=8, \text{ and } \frac{200}{x}+\frac{300}{y}=35.$$

By subtraction, $$\frac{108}{y}=27; \text{ whence } y=4.$$

9. $$\frac{4}{x}+\frac{27}{y}=42, \quad \frac{14}{x}-\frac{15}{y}=1.$$

Multiply the first equation by 7, and the second by 2; thus

$$\frac{28}{x}+\frac{189}{y}=294, \text{ and } \frac{28}{x}-\frac{30}{y}=2.$$

By subtraction, $$\frac{219}{y}=292; \text{ whence } y=\frac{219}{292}=\frac{3}{4}.$$

10. $$\frac{3}{x}+\frac{5}{y}=\frac{8}{15}, \quad \frac{9}{x}-\frac{22}{y}=\frac{3}{25}.$$

Multiply the first equation by 3, and subtract the second; thus

$$\frac{37}{y}=\frac{8}{5}-\frac{3}{25}=\frac{37}{25}; \text{ whence } y=25.$$

11. By clearing of fractional coefficients, we have

$$\frac{3}{x}+\frac{4}{y}=24, \quad \frac{2}{y}-\frac{1}{x}=2.$$

Multiply the second equation by 3 and add the first; thus

$$\frac{10}{y}=30; \text{ whence } y=\frac{1}{3}.$$

12. $$\frac{2}{x}-\frac{1}{y}=4, \quad \frac{4}{y}-\frac{3}{x}=9.$$

Multiply the first equation by 4 and add the second; thus

$$\frac{5}{x}=25; \text{ whence } x=\frac{1}{5}.$$

13. $$\frac{1}{x}-\frac{2}{y}=-4, \quad \frac{1}{y}-\frac{1}{z}=-1, \quad \frac{2}{z}+\frac{3}{x}=14.$$

Multiply the second equation by 2 and add the first; thus

$$\frac{1}{x}-\frac{2}{z}=-6.$$

By adding this equation to the third, we have

$$\frac{4}{x}=8; \text{ whence } x=\frac{1}{2}.$$

14. By clearing the third equation of fractions, we have

$$\frac{1}{x}+\frac{1}{y}+\frac{1}{z}=36, \quad \frac{1}{x}+\frac{3}{y}-\frac{1}{z}=28, \quad \frac{6}{x}+\frac{2}{y}+\frac{3}{z}=120.$$

By adding the first and second equations, we obtain

$$\frac{2}{x}+\frac{4}{y}=64, \text{ or } \frac{1}{x}+\frac{2}{y}=32 \quad\quad (1).$$

Multiply the second equation by 3 and add the third; thus

$$\frac{9}{x}+\frac{11}{y}=204 \quad\quad (2).$$

Multiply (1) by 9 and subtract (2); thus

$$\frac{7}{y}=84; \text{ whence } y=\frac{1}{12}.$$

15. $$\frac{9}{x}-\frac{2}{y}=4, \quad \frac{5}{z}-\frac{3}{x}=4, \quad \frac{14}{y}+\frac{15}{z}=8.$$

Multiply the second equation by 3 and add the first; thus

$$\frac{15}{z}-\frac{2}{y}=16.$$

By subtracting this equation from the third, we have

$$\frac{16}{y}=-8; \text{ whence } y=-2.$$

EXAMPLES. XIV. Page 95.

3. Let x be the greater number and y the less;

then $$\frac{1}{3}(x+y)=14, \text{ and } \frac{1}{2}(x-y)=4.$$

Thus $$x+y=42, \text{ and } x-y=8.$$

Whence $$x=25, \text{ and } y=17.$$

4. Let x be the greater number and y the less;

then $$\frac{1}{19}(x+y)=4, \text{ or } x+y=76.$$

Also $$x-y=30.$$

Whence $$x=53, \ y=23.$$

5. Let x be the greater number and y the less;

then $$\frac{1}{2}(x+y)=20, \text{ and } 3(x-y)=18.$$

Thus $$x+y=40, \text{ and } x-y=6.$$

Whence $$x=23, \ y=17.$$

6. Let the price of tea be x pence per lb. and of sugar y pence per lb.

Now $\quad$ £1. 3s. 8d. = 284d., and £1. 18s. 8d. = 464d.

We have $\quad 6x+11y=284$, and $11x+6y=464$.

By addition, $\quad 17x+17y=748$, or $x+y=44$.

By subtraction, $\quad 5x-5y=180$, or $x-y=36$.

Whence $\quad x=40,\ y=4$.

7. Let £x be the price of a horse, £y the price of a cow;

then $\quad 6x+7y=250$, and $11x+13y=461$.

Multiply the first equation by 11 and the second by 6; thus

$$66x+77y=2750, \text{ and } 66x+78y=2766.$$

By subtraction, $\quad y=16$.

8. Let £x be C's share, then A's share is £$2x$;

" £y " D's " B's " £$3y$.

Thus $\quad 2x+3y+x+y=290$, or $3x+4y=290$.

Also $\quad x+y=2x-50$, or $-x+y=-50$.

By eliminating x, we obtain $y=20$.

9. Let C's share be £x, then A's share is £$3x$;

" D's " £y, " B's " £$5y$.

Thus $\quad 3x+5y+x+y=270$, or $4x+6y=270$.

Also $\quad 3x+5y=8x-50$, or $-x+y=-10$.

By eliminating x, we obtain $y=23$.

10. Let A's age be x years, and B's age y years;

then $\quad 4y-x=20$, and $y-\frac{1}{3}x=2$, or $3y-x=6$.

By subtraction, $\quad y=14$.

11. Let A's age be x years, and B's age y years;

then $$\frac{1}{11}x-\frac{1}{7}y=2 \quad\text{..........(1).}$$

Also $$2y=x-13, \text{ or } x-2y=13 \quad\text{..........(2).}$$

Multiply (1) by 11 and subtract (2); thus

$$\left(2-\frac{11}{7}\right)y=9; \text{ that is, } \frac{3}{7}y=9, \text{ or } y=21.$$

12. Let x = number of miles A can walk an hour.

" y " B "

In 8 hours A walks $8x$ miles, and in 7 hours B walks $7y$ miles;

$$\therefore 8x-7y=12 \quad\text{..........(1).}$$

Again, in 13 hours B walks $13y$ miles, and in 9 hours A walks $9x$ miles;

$$\therefore\ 13y - 9x = 7 \quad \ldots\ldots(2).$$

Multiply (1) by 9 and (2) by 8; thus

$$72x - 63y = 108, \text{ and } 104y - 72x = 56.$$

By addition, $41y = 164$, or $y = 4$.

13. Let x = number of miles C can walk an hour.

" y " D "

In 11 hours C walks $11x$ miles, and in 12 hours D walks $12y$ miles,

$$\therefore\ 12y - 11x = 12\tfrac{1}{2} \quad \ldots\ldots(1).$$

Again, in 5 hours D walks $5y$ miles, and in 7 hours C walks $7x$ miles,

$$\therefore\ 7x - 5y = 3\tfrac{1}{4} \ldots\ldots(2).$$

Multiply (1) by 5 and (2) by 12; thus

$$60y - 55x = 62\tfrac{1}{2}, \text{ and } 84x - 60y = 39.$$

By addition, $29x = 101\frac{1}{2}$, or $x = \frac{7}{2} = 3\frac{1}{2}$.

14. Let x be the numerator and y the denominator;

then the fraction is $\frac{x}{y}$.

On the first hypothesis,

$$\frac{x}{y+1} = \frac{1}{2}; \text{ that is, } 2x - y = 1 \ldots\ldots(1).$$

On the second hypothesis,

$$\frac{x+2}{y} = \frac{3}{5}; \text{ that is, } 3y - 5x = 10 \quad \ldots\ldots(2).$$

On multiplying (1) by 3 and adding to (2), we obtain $x = 13$.

15. As in the preceding Example, we have

$$\frac{x-1}{y+2} = \frac{1}{2}; \text{ that is, } 2x - y = 4;$$

and $$\frac{x-7}{y-2} = \frac{1}{3}; \text{ that is, } 3x - y = 19.$$

By subtraction, we obtain $x = 15$.

16. As in Example 14, we have

$$\frac{x+1}{y} = \frac{1}{5}; \text{ that is, } y - 5x = 5;$$

and $$\frac{x}{y-1} = \frac{1}{7}; \text{ that is, } y - 7x = 1.$$

By subtraction, we obtain $2x = 4$, or $x = 2$.

17. With the notation of Example 14, we have

$$\frac{x+\frac{2}{3}}{y}=\frac{x}{y}+\frac{1}{21}, \text{ and } \frac{x}{y-\frac{1}{2}}=\frac{2}{9}.$$

From the first of these equations, we obtain

$$\frac{x}{y}+\frac{2}{3y}=\frac{x}{y}+\frac{1}{21}; \text{ that is, } \frac{2}{3y}=\frac{1}{21}, \text{ or } y=14.$$

18. Let x be the digit in the tens' place,

,, y ,, ,, units' ,, .

The number is $10x+y$, and the number formed by reversing the digits is

$$10y+x;$$

thus $$10x+y+10y+x=110;$$

that is $$11x+11y=110, \text{ or } x+y=10.$$

Also $$x-y=6.$$

Whence $$x=8, \text{ and } y=2.$$

19. Let x be the digit in the tens' place,

,, y ,, ,, units' ,, .

The number is $10x+y$, and the number formed by reversing the digits is

$$10y+x.$$

Thus $$10x+y-(10y+x)=27;$$

that is, $$9x-9y=27, \text{ or } x-y=3.$$

Also $$x+y=13.$$

Whence $$x=8, \text{ and } y=5.$$

20. Let x be the digit in the tens' place,

,, y ,, ,, units' ,, .

Then the number is $10x+y$, and the number formed by reversing the digits is $10y+x$.

Now $$10x+y=3(x+y), \text{ or } 7x-2y=0.$$

Also $$10x+y+45=10y+x;$$

that is, $$9x-9y=-45, \text{ or } x-y=-5.$$

Whence $$x=2, \text{ and } y=7.$$

21. Let x be the digit in the tens' place,

,, y ,, ,, units' ,, .

The number is $10x+y$, and the number formed by reversing the digits is

$$10y+x.$$

Now $$10x+y=8(x+y), \text{ or } 2x-7y=0.$$

Also $$10x + y - 45 = 10y + x;$$
that is, $$9x - 9y = 45, \text{ or } x - y = 5.$$
Whence $$x = 7, \text{ and } y = 2.$$

22. Let x be the number of pounds he has, and y the number of shillings. The amount he has is $20x + y$ shillings.

On the first supposition, $x + 20y = 20x + y + 114$;

that is, $$-19x + 19y = 114, \text{ or } -x + y = 6.$$

On the second supposition, $10x + \frac{5}{2}y = 20x + y - 33\frac{1}{2}$;

that is, $$20x + 5y = 40x + 2y - 67, \text{ or } -20x + 3y = -67.$$

By eliminating y, we obtain $x = 5$; and therefore $y = 11$.

23. Let x = number of black balls, y the number of white balls:

then $$\frac{y}{2} = \frac{x}{3}, \text{ or } 3y - 2x = 0.$$

Also $$2(x + y) - 3x = 4, \text{ or } -x + 2y = 4.$$
Whence $$x = 12, \text{ and } y = 8.$$

24. Let x be the digits in the hundreds' place,
,, y ,, ,, tens' ,,

Then the number is $100x + 10y$, the right-hand digit being 0.

If the left-hand and middle digits are interchanged the number is
$$100y + 10x.$$

Thus $$100y + 10x = 100x + 10y - 180;$$
that is, $$-x + y = -2 \quad\text{.......(1)}.$$

If the left-hand digit be halved, and the middle and right-hand digits be interchanged, the number formed is $\left(100 \times \frac{x}{2}\right) + y$ or $50x + y$, for the digit in the tens' place is now zero.

Hence $$50x + y = 100x + 10y - 454;$$
that is, $$50x + 9y = 454 \quad\text{.......(2)}.$$

From (1) and (2), we obtain $x = 8$, and $y = 6$.

25. Let x shillings be the wages of a man,
,, y ,, ,, boy.

We have $$10x + 8y = 37, \text{ and } 4x - 6y = 1$$

From these equations, we obtain $x = 2\frac{1}{2}$, $y = 1\frac{1}{2}$.

26. Let x = number of lbs. of spice at 8s. a lb.,
and y = „ „ 5s. „ ;
then $$x+y=60.$$

Again the number of shillings paid is $8x+5y$;
$$\therefore\ 8x+5y=60\times 6=360.$$

From these equations, we obtain $x=20$, $y=40$.

27. Let x miles = distance he walks,
and y miles an hour = pace he walks at.

The time he takes to walk x miles is $\frac{x}{y}$ hours.

On the first supposition, $$\frac{x}{y+\frac{1}{2}}=\frac{4}{5}\times\frac{x}{y}.$$

Dividing by x, $$\frac{1}{y+\frac{1}{2}}=\frac{4}{5y};$$
$$\therefore\ 5y=4y+2,\ \text{or } y=2.$$

On the second supposition, $$\frac{x}{y-\frac{1}{2}}=\frac{x}{y}+2\tfrac{1}{2}.$$

By substituting $y=2$, we obtain $x=15$.

28. Let x hours = time he was walking at 4 miles an hour,
and y „ „ 5 „ „ ;
then $$4x+5y=35.$$
On the second supposition, $5x+4y=37$.
By addition, $9x+9y=72$; that is, $x+y=8$.
By subtraction, $x-y=2$.
Thus $x=5$, and $y=3$.

29. Let the person who walks the faster travel x miles an hour, and let the other person travel y „ .

In 9 hours the first walks $9x$ miles, the second $9y$ miles;
thus $$9x-9y=27,\ \text{or } x-y=3.$$
Again, in 3 hours they walk $3x$ and $3y$ miles respectively;
thus $$3x+3y=27,\ \text{or } x+y=9.$$
Whence $x=6$, and $y=3$.

30. Let x shillings = cost of food of an adult per week,
„ y = „ child „ ;
then $$3x+5y=37\tfrac{1}{2}\ \text{..............................(1).}$$

On the second supposition the weekly cost of food of an adult is $\frac{x}{2}$ shillings, and that of a child is $\frac{2y}{3}$ shillings.

Hence $$\frac{3x}{2}+\frac{10y}{3}=20;$$

and therefore $$3x+\frac{20y}{3}=40.$$

By subtracting (1) from the above equation, we have

$$\frac{5y}{3}=2\tfrac{1}{2};\text{ whence } y=1\tfrac{1}{2}.$$

31. Let £x be the sum lent, and y years the time for which it is lent.

The interest for y years at 6 p. c. $=£\frac{x}{100}\times y\times 6=£\frac{3xy}{50}$.

$$\therefore \frac{3xy}{50}-x=100 \quad\text{.................................(1).}$$

The interest for $\frac{1}{4}y$ years at 3 p. c. $=£\frac{x}{100}\times\frac{y}{4}\times 3=£\frac{3xy}{400}$.

$$\therefore x-\frac{3xy}{400}=425 \quad\text{.................................(2).}$$

Multiply (2) by 8; thus $$8x-\frac{3xy}{50}=3400;$$

by adding (1) to this last equation, we obtain $7x=3500$; or $x=500$.

32. Let x miles an hour be A's rate of walking,

„ y „ „ B's „ „ .

A takes $\frac{30}{x}$ hours, and B $\frac{30}{y}$ hours to walk 30 miles;

$$\therefore \frac{30}{x}-\frac{30}{y}=3.$$

If A walks $2x$ miles an hour, he takes $\frac{15}{x}$ hours to walk the distance.

$$\therefore \frac{30}{y}-\frac{15}{x}=2.$$

By addition, $$\frac{15}{x}=5;\text{ whence } x=3.$$

EXAMPLES. XVI. b. PAGE 105

9.

$$\begin{array}{r|l}
 & a^4 - 2a^3 + 3a^2 - 2a + 1 \;(\, a^2 - a + 1 \\
 & a^4 \\ \hline
2a^2 - a & -2a^3 + 3a^2 \\
 & -2a^3 + a^2 \\ \hline
2a^2 - 2a + 1 & 2a^2 - 2a + 1 \\
 & 2a^2 - 2a + 1 \\ \hline
\end{array}$$

10.

$$\begin{array}{r|l}
 & 4x^4 - 12x^3 + 29x^2 - 30x + 25 \;(\, 2x^2 - 3x + 5 \\
 & 4x^4 \\ \hline
4x^2 - 3x & -12x^3 + 29x^2 \\
 & -12x^3 + 9x^2 \\ \hline
4x^2 - 6x + 5 & 20x^2 - 30x + 25 \\
 & 20x^2 - 30x + 25 \\ \hline
\end{array}$$

11.

$$\begin{array}{r|l}
 & 9x^4 - 12x^3 - 2x^2 + 4x + 1 \;(\, 3x^2 - 2x - 1 \\
 & 9x^4 \\ \hline
6x^2 - 2x & -12x^3 - 2x^2 \\
 & -12x^3 + 4x^2 \\ \hline
6x^2 - 4x - 1 & -6x^2 + 4x + 1 \\
 & -6x^2 + 4x + 1 \\ \hline
\end{array}$$

12.

$$\begin{array}{r|l}
 & x^4 - 4x^3 + 6x^2 - 4x + 1 \;(\, x^2 - 2x + 1 \\
 & x^4 \\ \hline
2x^2 - 2x & -4x^3 + 6x^2 \\
 & -4x^3 + 4x^2 \\ \hline
2x^2 - 4x + 1 & 2x^2 - 4x + 1 \\
 & 2x^2 - 4x + 1 \\ \hline
\end{array}$$

13.

$$\begin{array}{r|l}
 & 4a^4 + 4a^3 - 7a^2 - 4a + 4 \;(\, 2a^2 + a - 2 \\
 & 4a^4 \\ \hline
4a^2 + a & 4a^3 - 7a^2 \\
 & 4a^3 + a^2 \\ \hline
4a^2 + 2a - 2 & -8a^2 - 4a + 4 \\
 & -8a^2 - 4a + 4 \\ \hline
\end{array}$$

14.

$$\begin{array}{r|l}
 & 1 - 10x + 27x^2 - 10x^3 + x^4 \;(\, 1 - 5x + x^2 \\
 & 1 \\ \hline
2 - 5x & -10x + 27x^2 \\
 & -10x + 25x^2 \\ \hline
2 - 10x + x^2 & 2x^2 - 10x^3 + x^4 \\
 & 2x^2 - 10x^3 + x^4 \\ \hline
\end{array}$$

15.

$$\begin{array}{r|l}
 & 4x^2+12xy-20xz+9y^2-30yz+25z^2\ (\,2x+3y-5z \\
 & 4x^2 \\
\hline
4x+3y & 12xy-20xz+9y^2 \\
 & 12xy \qquad\quad +9y^2 \\
\hline
4x+6y-5z & -20xz \qquad -30yz+25z^2 \\
 & -20xz \qquad -30yz+25z^2 \\
\hline
\end{array}$$

16.

$$\begin{array}{r|l}
 & 16x^6+16x^7-4x^8-4x^9+x^{10}\ (\,4x^3+2x^4-x^5 \\
 & 16x^6 \\
\hline
8x^3+2x^4 & 16x^7-4x^8 \\
 & 16x^7+4x^8 \\
\hline
8x^3+4x^4-x^5 & -8x^8-4x^9+x^{10} \\
 & -8x^8-4x^9+x^{10} \\
\hline
\end{array}$$

17.

$$\begin{array}{r|l}
 & x^6-22x^4+34x^3+121x^2-374x+289\ (\,x^3-11x+17 \\
 & x^6 \\
\hline
2x^3-11x & -22x^4+34x^3+121x^2 \\
 & -22x^4 \qquad\quad +121x^2 \\
\hline
2x^3-22x+17 & 34x^3 \qquad -374x+289 \\
 & 34x^3 \qquad -374x+289 \\
\hline
\end{array}$$

18.

$$\begin{array}{r|l}
 & 25x^4-30ax^3+49a^2x^2-24a^3x+16a^4\ (\,5x^2-3ax+4a^2 \\
 & 25x^4 \\
\hline
10x^2-3ax & -30ax^3+49a^2x^2 \\
 & -30ax^3+\ \ 9a^2x^2 \\
\hline
10x^2-6ax+4a^2 & 40a^2x^2-24a^3x+16a^4 \\
 & 40a^2x^2-24a^3x+14a^4 \\
\hline
\end{array}$$

19.

$$\begin{array}{r|l}
 & 4x^4+4x^2y^2-12x^2z^2+y^4-6y^2z^2+9z^4\ (\,2x^2+y^2-3z^2 \\
 & 4x^4 \\
\hline
4x^2+y^2 & 4x^2y^2-12x^2z^2+y^4 \\
 & 4x^2y^2 \qquad\quad +y^4 \\
\hline
4x^2+2y^2-3z^2 & -12x^2z^2 \qquad -6y^2z^2+9z^4 \\
 & -12x^2z^2 \qquad -6y^2z^2+9z^4 \\
\hline
\end{array}$$

20.

$$\begin{array}{r|l}
 & a^2b^2-4a^2bc+4a^2c^2+6ab^2c-12abc^2+9b^2c^2\ (\,ab-2ac+3bc \\
 & a^2b^2 \\
\hline
2ab-2ac & -4a^2bc+4a^2c^2 \\
 & -4a^2bc+4a^2c^2 \\
\hline
2ab-4ac+3bc & 6ab^2c-12abc^2+9b^2c^2 \\
 & 6ab^2c-12abc^2+9b^2c^2 \\
\hline
\end{array}$$

21.

$$\begin{array}{r|l}
 & 4a^4+4a^2b^2-12a^2c^2+b^4-6b^2c^2+9c^4\ (\,2a^2+b^2-3c^2 \\
 & 4a^4 \\
\hline
4a^2+b^2 & 4a^2b^2-12a^2c^2+b^4 \\
 & 4a^2b^2 \qquad\quad +b^4 \\
\hline
4a^2+2b^2-3c^2 & -12a^2c^2 \qquad -6b^2c^2+9c^4 \\
 & -12a^2c^2 \qquad -6b^2c^2+9c^4 \\
\hline
\end{array}$$

22.

$$\begin{array}{r|l}
 & 4x^4 - 4x^3y + 13x^2y^2 - 6xy^3 + 9y^4 \;(\; 2x^2 - xy + 3y^2 \\
 & 4x^4 \\
4x^2 - xy & -4x^3y + 13x^2y^2 \\
 & -4x^3y + x^2y^2 \\
4x^2 - 2xy + 3y^2 & 12x^2y^2 - 6xy^3 + 9y^4 \\
 & 12x^2y^2 - 6xy^3 + 9y^4
\end{array}$$

23.

$$\begin{array}{r|l}
 & 9x^4 - 30x^3 + 67x^2 - 70x + 49 \;(\; 3x^2 - 5x + 7 \\
 & 9x^4 \\
6x^2 - 5x & -30x^3 + 67x^2 \\
 & -30x^3 + 25x^2 \\
6x^2 - 10x + 7 & 42x^2 - 70x + 49 \\
 & 42x^2 - 70x + 49
\end{array}$$

24.

$$\begin{array}{r|l}
 & 1 - 4x + 10x^2 - 20x^3 + 25x^4 - 24x^5 + 16x^6 \;(\; 1 - 2x + 3x^2 - 4x^3 \\
 & 1 \\
2 - 2x & -4x + 10x^2 \\
 & -4x + 4x^2 \\
2 - 4x + 3x^2 & 6x^2 - 20x^3 + 25x^4 \\
 & 6x^2 - 12x^3 + 9x^4 \\
2 - 4x + 6x^2 - 4x^3 & -8x^3 + 16x^4 - 24x^5 + 16x^6 \\
 & -8x^3 + 16x^4 - 24x^5 + 16x^6
\end{array}$$

25.

$$\begin{array}{r|l}
 & a^2x^{10} - 4abx^7 + 6acx^5 + 4b^2x^4 - 12bcx^2 + 9c^2 \;(\; ax^5 - 2bx^2 + 3c \\
 & a^2x^{10} \\
2ax^5 - 2bx^2 & -4abx^7 + 6acx^5 + 4b^2x^4 \\
 & -4abx^7 \qquad\qquad + 4b^2x^4 \\
2ax^5 - 4bx^2 + 3c & 6acx^5 \qquad\qquad - 12bcx^2 + 9c^2 \\
 & 6acx^5 \qquad\qquad - 12bcx^2 + 9c^2
\end{array}$$

EXAMPLES. XVI. c. PAGE 106.

9.

$$\begin{array}{r|l}
 & \dfrac{a^4}{64} + \dfrac{a^3}{8} - a + 1 \;\Big(\; \dfrac{a^2}{8} + \dfrac{a}{2} - 1 \\
 & \dfrac{a^4}{64} \\
\dfrac{a^2}{4} + \dfrac{a}{2} & \dfrac{a^3}{8} - a \\
 & \dfrac{a^3}{8} + \dfrac{a^2}{4} \\
\dfrac{a^2}{4} + a - 1 & -\dfrac{a^2}{4} - a + 1 \\
 & -\dfrac{a^2}{4} - a + 1
\end{array}$$

10.

$$\begin{array}{r|l}
 & x^4 + 2x^3 - x + \tfrac{1}{4} \;(\; x^2 + x - \tfrac{1}{2} \\
 & x^4 \\
2x^2 + x & 2x^3 \\
 & 2x^3 + x^2 \\
2x^2 + 2x - \tfrac{1}{2} & -x^2 - x + \tfrac{1}{4} \\
 & -x^2 - x + \tfrac{1}{4}
\end{array}$$

11.

$$\begin{array}{r|l}
 & a^4 - 3a^3 + \tfrac{67}{12}a^2 - 5a + \tfrac{25}{9} \;(\; a^2 - \tfrac{3}{2}a + \tfrac{5}{3} \\
 & a^4 \\
2a^2 - \tfrac{3}{2}a & -3a^3 + \tfrac{67}{12}a^2 \\
 & -3a^3 + \tfrac{9}{4}a^2 \\
2a^2 - 3a + \tfrac{5}{3} & \tfrac{10}{3}a^2 - 5a + \tfrac{25}{9} \\
 & \tfrac{10}{3}a^2 - 5a + \tfrac{25}{9}
\end{array}$$

12.

$$x^4 - 6x^3 + \tfrac{29}{3}x^2 - 2x + \tfrac{1}{9} \left(x^2 - 3x + \tfrac{1}{3} \right.$$
$$x^4$$
$$2x^2 - 3x \;\Big|\; -6x^3 + \tfrac{29}{3}x^2$$
$$-6x^3 + 9x^2$$
$$2x^2 - 6x + \tfrac{1}{3} \;\Big|\; \tfrac{2}{3}x^2 - 2x + \tfrac{1}{9}$$
$$\tfrac{2}{3}x^2 - 2x + \tfrac{1}{9}$$

13.

$$\frac{a^4}{4} + \frac{a^3}{x} + \frac{a^2}{x^2} - ax - 2 + \frac{x^2}{a^2} \left(\frac{a^2}{2} + \frac{a}{x} - \frac{x}{a} \right.$$
$$\frac{a^4}{4}$$
$$a^2 + \frac{a}{x} \;\Big|\; \frac{a^3}{x} + \frac{a^2}{x^2}$$
$$\frac{a^3}{x} + \frac{a^2}{x^2}$$
$$a^2 + \frac{2a}{x} - \frac{x}{a} \;\Big|\; -ax - 2 + \frac{x^2}{a^2}$$
$$-ax - 2 + \frac{x^2}{a^2}$$

14.

$$x^4 - 2x^3 + \frac{3x^2}{2} - \frac{x}{2} + \frac{1}{16} \left(x^2 - x + \tfrac{1}{4} \right.$$
$$x^4$$
$$2x^2 - x \;\Big|\; -2x^3 + \frac{3x^2}{2}$$
$$-2x^3 + x^2$$
$$2x^2 - 2x + \tfrac{1}{4} \;\Big|\; \frac{x^2}{2} - \frac{x}{2} + \frac{1}{16}$$
$$\frac{x^2}{2} - \frac{x}{2} + \frac{1}{16}$$

15.

$$\frac{x^4}{4} - 2x^3 + 4x^2 + \frac{ax^2}{3} - \frac{4ax}{3} + \frac{a^2}{9} \left(\frac{x^2}{2} - 2x + \frac{a}{3} \right.$$
$$\frac{x^4}{4}$$
$$x^2 - 2x \;\Big|\; -2x^3 + 4x^2$$
$$-2x^3 + 4x^2$$
$$x^2 - 4x + \frac{a}{3} \;\Big|\; \frac{ax^2}{3} - \frac{4ax}{3} + \frac{a^2}{9}$$
$$\frac{ax^2}{3} - \frac{4ax}{3} + \frac{a^2}{9}$$

16.

$$\frac{9a^2}{x^2}-\frac{6a}{5x}+\frac{101}{25}-\frac{4x}{15a}+\frac{4x^2}{9a^2}\left(\frac{3a}{x}-\frac{1}{5}+\frac{2x}{3a}\right.$$

$$\frac{9a^2}{x^2}$$

$$\frac{6a}{x}-\frac{1}{5}\ \Big|\ -\frac{6a}{5x}+\frac{101}{25}$$

$$-\frac{6a}{5x}+\frac{1}{25}$$

$$\frac{6a}{x}-\frac{2}{5}+\frac{2x}{3a}\ \Big|\ 4-\frac{4x}{15a}+\frac{4x^2}{9a^2}$$

$$4-\frac{4x}{15a}+\frac{4x^2}{9a^2}$$

17.

$$16m^4+\tfrac{16}{3}m^2n+8m^2+\tfrac{4}{9}n^2+\tfrac{4}{3}n+1\ (4m^2+\tfrac{2}{3}n+1$$

$$16m^4$$

$$8m^2+\tfrac{2}{3}n\ \Big|\ \tfrac{16}{3}m^2n+8m^2+\tfrac{4}{9}n^2$$

$$\tfrac{16}{3}m^2n+\tfrac{4}{9}n^2$$

$$8m^2+\tfrac{4}{3}n+1\ \Big|\ 8m^2+\tfrac{4}{3}n+1$$

$$8m^2+\tfrac{4}{3}n+1$$

18.

$$4x^4+32x^2+96+\frac{128}{x^2}+\frac{64}{x^4}\left(2x^2+8+\frac{8}{x^2}\right.$$

$$4x^4$$

$$4x^2+8\ \Big|\ 32x^2+96$$

$$32x^2+64$$

$$4x^2+16+\frac{8}{x^2}\ \Big|\ 32+\frac{128}{x^2}+\frac{64}{x^4}$$

$$32+\frac{128}{x^2}+\frac{64}{x^4}$$

EXAMPLES. XVI. d. Page 109.

6.

$$1 + 3x + 6x^2 + 7x^3 + 6x^4 + 3x^5 + x^6 \,(\, 1 + x + x^2$$

$$1$$

$3 \times 1^2 = 3$	$3x + 6x^2 + 7x^3$
$3 \times 1 \times x = \quad + 3x$	
$x^2 = \quad\quad + x^2$	
$3 + 3x + x^2$	$3x + 3x^2 + x^3$
	$3x^2 + 6x^3 + 6x^4 + 3x^5 + x^6$
$3 \times (1 + x)^2 = 3 + 6x + 3x^2$	
$3 \times (1 + x) \times x^2 = \quad\quad 3x^2 + 3x^3$	
$(x^2)^2 = \quad\quad + x^4$	
$3 + 6x + 6x^2 + 3x^3 + x^4$	$3x^2 + 6x^3 + 6x^4 + 3x^5 + x^6$

$$1 - 6x + 21x^2 - 44x^3 + 63x^4 - 54x^5 + 27x^6 \,(\, 1 - 2x + 3x^2$$

$$1$$

$3 \times 1^2 = 3$	$-6x + 21x^2 - 44x^3$
$3 \times 1 \times (-2x) = \quad -6x$	
$(-2x)^2 = \quad\quad + 4x^3$	
$3 - 6x + 4x^2$	$-6x + 12x^2 - 8x^3$
	$9x^2 - 36x^3 + 63x^4 - 54x^5 + 27x^6$
$3 \times (1 - 2x)^2 = 3 - 12x + 12x^2$	
$3 \times (1 - 2x) \times 3x^2 = \quad\quad + 9x^2 - 18x^3$	
$(3x^2)^2 = \quad\quad + 9x^4$	
$3 - 12x + 21x^2 - 18x^3 + 9x^4$	$9x^2 - 36x^3 + 63x^4 - 54x^5 + 27x^6$

8.

$$
\begin{array}{rl|l}
 & & a^3+6a^2b-3a^2c+12ab^2-12abc+3ac^2+8b^3-12b^2c+6bc^2-c^3 \;(\; a+2b-c \\
 & & a^3 \\ \hline
3\times(a)^2= & 3a^2 & 6a^2b-3a^2c+12ab^2-12abc+3ac^2+8b^3 \\
3\times a\times 2b= & \quad +6ab & \\
(2b)^2= & \qquad +4b^2 & \\
 & 3a^2+6ab+4b^2 & 6a^2b \qquad +12ab^2 \qquad\qquad +8b^3 \\ \hline
3\times(a+2b)^2= & 3a^2+12ab+12b^2 & -3a^2c \qquad -12abc+3ac^2 \qquad -12b^2c+6bc^2-c^3 \\
3\times(a+2b)\times(-c)= & -3ac-6bc & \\
(-c)^2= & c^2 & \\
 & 3a^2+12ab-3ac+12b^2-6bc+c^2 & -3a^2c \qquad -12abc+3ac^2 \qquad -12b^2c+6bc^2-c^3
\end{array}
$$

9.

$$
\begin{array}{rl|l}
 & & 8a^6-36a^5+66a^4-63a^3+33a^2-9a+1 \;(\; 2a^2-3a+1 \\
 & & 8a^6 \\ \hline
3\times(2a^2)^2= & 12a^4 & -36a^5+66a^4-63a^3 \\
3\times(2a^2)\times(-3a)= & \quad -18a^3 & \\
(-3a)^2= & \qquad +9a^2 & \\
 & 12a^4-18a^3+9a^2 & -36a^5+54a^4-27a^3 \\ \hline
3\times(2a^2-3a)^2= & 12a^4-36a^3+27a^2 & 12a^4-36a^3+33a^2-9a+1 \\
3\times(2a^2-3a)\times 1= & \qquad +6a^2-9a & \\
1^2= & \qquad\qquad +1 & \\
 & 12a^4-36a^3+33a^2-9a+1 & 12a^4-36a^3+33a^2-9a+1
\end{array}
$$

10.

$$
\begin{array}{rl}
 & y^6 - 3y^5 + 6y^4 - 7y^3 + 6y^2 - 3y + 1 \,(\, y^2 - y + 1 \\
 & y^6 \\
3 \times (y^2)^2 = 3y^4 & \quad -3y^5 + 6y^4 - 7y^3 \\
3 \times y^2 \times (-y) = \quad -3y^3 & \\
(-y)^2 = \quad +y^2 & \\
3y^4 - 3y^3 + y^2 & \quad -3y^5 + 3y^4 - y^3 \\
3(y^2 - y)^2 = 3y^4 - 6y^3 + 3y^2 & \quad 3y^4 - 6y^3 + 6y^2 - 3y + 1 \\
3(y^2 - y) \times 1 = \quad +3y^2 - 3y & \\
1^2 = \quad +1 & \\
3y^4 - 6y^3 + 6y^2 - 3y + 1 & \quad 3y^4 - 6y^3 + 6y^2 - 3y + 1
\end{array}
$$

11.

$$
\begin{array}{rl}
 & 8x^6 + 12x^5 - 30x^4 - 35x^3 + 45x^2 + 27x - 27 \,(\, 2x^2 + x - 3 \\
 & 8x^6 \\
3 \times (2x^2)^2 = 12x^4 & \quad 12x^5 - 30x^4 - 35x^3 \\
3 \times 2x^2 \times x = \quad +6x^3 & \\
(x)^2 = \quad +x^2 & \\
12x^5 + 6x^3 + x^2 & \quad 12x^5 + 6x^4 + x^3 \\
3 \times (2x^2 + x)^2 = 12x^4 + 12x^3 + 3x^2 & \quad -36x^4 - 36x^3 + 45x^2 + 27x - 27 \\
3 \times (2x^2 + x) \times (-3) = \quad -18x^2 - 9x & \\
(-3)^2 = \quad +9 & \\
12x^4 + 12x^3 - 15x^2 - 9x + 9 & \quad -36x^4 - 36x^3 + 45x^2 + 27x - 27
\end{array}
$$

12.

$$27x^6 - 54x^5a + 117x^4a^2 - 116x^3a^3 + 117x^2a^4 - 54xa^5 + 27a^6 \; (\; 3x^2 - 2xa + 3a^2$$
$$27x^6$$

$$3 \times (3x^2)^2 = 27x^4 \qquad\qquad - 54x^5a + 117x^4a^2 - 116x^3a^3$$
$$3 \times 3x^2 \times (-2xa) = \quad - 18x^3a$$
$$(-2xa)^2 = \quad + 4x^2a^2$$
$$\overline{27x^4 - 18x^3a + 4x^2a^2} \qquad - 54x^5a + 36x^4a^2 - 8x^3a^3$$

$$3 \times (3x^2 - 2xa)^2 = 27x^4 - 36x^3a + 12x^2a^2 \qquad 81x^4a^2 - 108x^3a^3 + 117x^2a^4 - 54xa^5 + 27a^6$$
$$3 \times (3x^2 - 2xa) \times 3a^2 = \quad + 27x^2a^2 - 18xa^3$$
$$(3a^2)^2 = \quad + 9a^4$$
$$\overline{27x^4 - 36x^3a + 39x^2a^2 - 18xa^3 + 9a^4} \qquad 81x^4a^2 - 108x^3a^3 + 117x^2a^4 - 54xa^5 + 27a^6$$

13.

$$27x^6 - 27x^5 - 18x^4 + 17x^3 + 6x^2 - 3x - 1 \; (\; 3x^2 - x - 1$$
$$27x^6$$

$$3 \times (3x^2)^2 = 27x^4 \qquad\qquad - 27x^5 - 18x^4 + 17x^3$$
$$3 \times (3x^2) \times (-x) = \quad - 9x^3$$
$$(-x)^2 = \quad + x^2$$
$$\overline{27x^4 - 9x^3 + x^2} \qquad - 27x^5 + 9x^4 - x^3$$

$$3 \times (3x^2 - x)^2 = 27x^4 - 18x^3 + 3x^2 \qquad - 27x^4 + 18x^3 + 6x^2 - 3x - 1$$
$$3 \times (3x^2 - x) \times (-1) = \quad - 9x^2 + 3x$$
$$(-)^2 = \quad + 1$$
$$\overline{27x^4 - 18x^3 - 6x^2 + 3x + 1} \qquad - 27x^4 + 18x^3 + 6x^2 - 3x - 1$$

14.

$$x^6 - 6x^5y + 24x^4y^2 - 56x^3y^3 + 96x^2y^4 - 96xy^5 + 64y^6 \,(\, x^2 - 2xy + 4y^2$$
$$x^6$$

$$\begin{array}{rl}
3 \times (x^2)^2 = 3x^4 & \\
3 \times x^2 \times (-2xy) = \quad -6x^3y & \\
(-2xy)^2 = \quad\quad +4x^2y^2 & -6x^5y + 24x^4y^2 - 56x^3y^3 \\
\hline
3x^4 - 6x^3y + 4x^2y^2 & -6x^5y + 12x^4y^2 - 8x^3y^3 \\
\end{array}$$

$$\begin{array}{rl}
3 \times (x^2 - 2xy)^2 = 3x^4 - 12x^3y + 12x^2y^2 & 12x^4y^2 - 48x^3y^3 + 96x^2y^4 - 96xy^5 + 64y^6 \\
3 \times (x^2 - 2xy) \times 4y^2 = \quad 12x^2y^2 - 24xy^3 & \\
(4y^2)^2 = \quad\quad +16y^4 & \\
\hline
3x^4 - 12x^3y + 24x^2y^2 - 24xy^3 + 16y^4 & 12x^4y^2 - 48x^3y^3 + 96x^2y^4 - 96xy^5 + 64y^6 \\
\end{array}$$

15.

$$216 - 108x + 342x^2 - 109x^3 + 171x^4 - 27x^5 + 27x^6 \,(\, 6 - x + 3x^2$$
$$216$$

$$\begin{array}{rl}
3 \times 6^2 = 108 & -108x + 342x^2 - 109x^3 \\
3 \times 6 \times (-x) = \quad -18x & \\
(-x)^2 = \quad\quad +x^2 & \\
\hline
108 - 18x + x^2 & -108x + 18x^2 - x^3 \\
\end{array}$$

$$\begin{array}{rl}
3 \times (6 - x)^2 = 108 - 36x + 3x^2 & 324x^2 - 108x^3 + 171x^4 - 27x^5 + 27x^6 \\
3 \times (6 - x) \times 3x^2 = \quad 54x^2 - 9x^3 & \\
(3x^2)^2 = \quad\quad +9x^4 & \\
\hline
108 - 36x + 57x^2 - 9x^3 + 9x^4 & 324x^2 - 108x^3 + 171x^4 - 27x^5 + 27x^6 \\
\end{array}$$

EXAMPLES. XVI. e. PAGE 110.

7.

$$\frac{x^3}{y^3}+\frac{6x^2}{y^2}+\frac{9x}{y}-4-\frac{9y}{x}+\frac{6y^2}{x^2}-\frac{y^3}{x^3}\left(\frac{x}{y}+2-\frac{y}{x}\right.$$

$$\frac{x^3}{y^3}$$

$$3\times\left(\frac{x}{y}\right)^2=\frac{3x^2}{y^2} \qquad \frac{6x^2}{y^2}+\frac{9x}{y}-4$$

$$3\times\frac{x}{y}\times 2=\quad +\frac{6x}{y}$$

$$2^2=\quad +4$$

$$\frac{3x^2}{y^2}+\frac{6x}{y}+4 \qquad \frac{6x^2}{y^2}+\frac{12x}{y}+8$$

$$3\times\left(\frac{x}{y}+2\right)^2=\frac{3x^2}{y^2}+\frac{12x}{y}+12 \qquad -\frac{3x}{y}-12-\frac{9y}{x}+\frac{6y^2}{x^2}-\frac{y^3}{x^3}$$

$$3\times\left(\frac{x}{y}+2\right)\times\left(-\frac{y}{x}\right)=\quad -3-\frac{6y}{x}$$

$$\left(-\frac{y}{x}\right)^2=\quad +\frac{y^2}{x^2}$$

$$\frac{3x^2}{y^2}+\frac{12x}{y}+9-\frac{6y}{x}+\frac{y^2}{x^2} \qquad -\frac{3x}{y}-12-\frac{9y}{x}+\frac{6y^2}{x^2}-\frac{y^3}{x^3}$$

8.

$$\frac{x^3}{27}-\frac{x^2}{3}+2x-7+\frac{18}{x}-\frac{27}{x^2}+\frac{27}{x^3}\left(\frac{x}{3}-1+\frac{3}{x}\right.$$

$$\frac{x^3}{27}$$

$$3\times\left(\frac{x}{3}\right)^2=\frac{x^2}{3} \qquad -\frac{x^2}{3}+2x-7$$

$$3\times\left(\frac{x}{3}\right)\times(-1)=\quad -x$$

$$(-1)^2=\quad +1$$

$$\frac{x^2}{3}-x+1 \qquad -\frac{x^2}{3}+x-1$$

$$3\times\left(\frac{x}{3}-1\right)^2=\frac{x^2}{3}-2x+3 \qquad x-6+\frac{18}{x}-\frac{27}{x^2}+\frac{27}{x^3}$$

$$3\times\left(\frac{x}{3}-1\right)\times\frac{3}{x}=\quad 3-\frac{9}{x}$$

$$\left(\frac{3}{x}\right)^2=\quad +\frac{9}{x^2}$$

$$\frac{x^2}{3}-2x+6-\frac{9}{x}+\frac{9}{x^2} \qquad x-6+\frac{18}{x}-\frac{27}{x^2}+\frac{27}{x^3}$$

9.

$$\frac{x^3}{a^3} - \frac{12x^2}{a^2} + \frac{54x}{a} - 112 + \frac{108a}{x} - \frac{48a^2}{x^2} + \frac{8a^3}{x^3} \Big(\frac{x}{a} - 4 + \frac{2a}{x}$$

$$\frac{x^3}{a^3}$$

$$3 \times \left(\frac{x}{a}\right)^2 = \frac{3x^2}{a^2} \qquad - \frac{12x^2}{a^2} + \frac{54x}{a} - 112$$

$$3 \times \frac{x}{a} \times (-4) = \qquad - \frac{12x}{a}$$

$$(-4)^2 = \qquad + 16$$

$$\frac{3x^2}{a^2} - \frac{12x}{a} + 16 \qquad - \frac{12x^2}{a^2} + \frac{48x}{a} - 64$$

$$3 \times \left(\frac{x}{a} - 4\right)^2 = \frac{3x^2}{a^2} - \frac{24x}{a} + 48 \qquad \frac{6x}{a} - 48 + \frac{108a}{x} - \frac{48a^2}{x^2} + \frac{8a^3}{x^3}$$

$$3 \times \left(\frac{x}{a} - 4\right) \times \frac{2a}{x} = \qquad + 6 - \frac{24a}{x}$$

$$\left(\frac{2a}{x}\right)^2 = \qquad + \frac{4a^2}{x^2}$$

$$\frac{3x^2}{a^2} - \frac{24x}{a} + 54 - \frac{24a}{x} + \frac{4a^2}{x^2} \qquad \frac{6x}{a} - 48 + \frac{108a}{x} - \frac{48a^2}{x^2} + \frac{8a^3}{x^3}$$

10.

$$\frac{64a^3}{x^3}-\frac{192a^2}{x^2}+\frac{240a}{x}-160+\frac{60x}{a}-\frac{12x^2}{a^2}+\frac{x^3}{a^3}\left(\frac{4a}{x}-4+\frac{x}{a}\right.$$

$$\frac{64a^3}{x^3}$$

$$3\times\left(\frac{4a}{x}\right)^2=\frac{48a^2}{x^2} \qquad -\frac{192a^2}{x^2}+\frac{240a}{x}-160$$

$$3\times\frac{4a}{x}\times(-4)=\quad -\frac{48a}{x}$$

$$(-4)^2=\quad +16$$

$$\frac{48a^2}{x^2}-\frac{48a}{x}+16 \qquad -\frac{192a^2}{x^2}+\frac{192a}{x}-64$$

$$3\times\left(\frac{4a}{x}-4\right)^2=\frac{48a^2}{x^2}-\frac{96a}{x}+48 \qquad \frac{48a}{x}-96+\frac{60x}{a}-\frac{12x^2}{a^2}+\frac{x^3}{a^3}$$

$$3\times\left(\frac{4a}{x}-4\right)\times\frac{x}{a}=\quad +12-\frac{12x}{a}$$

$$\left(\frac{x}{a}\right)^2=\quad +\frac{x^2}{a^2}$$

$$\frac{48a^2}{x^2}-\frac{96a}{x}+60-\frac{12x}{a}+\frac{x^2}{a^2} \qquad \frac{48a}{x}-96+\frac{60x}{a}-\frac{12x^2}{a^2}+\frac{x^3}{a^3}$$

11.

$$\frac{a^3}{b^3}-\frac{3a^2}{b^2}+\frac{6a}{b}-7+\frac{6b}{a}-\frac{3b^2}{a^2}+\frac{b^3}{a^3}\left(\frac{a}{b}-1+\frac{b}{a}\right.$$

$$\frac{a^3}{b^3}$$

$$3\times\left(\frac{a}{b}\right)^2=\frac{3a^2}{b^2} \qquad -\frac{3a^2}{b^2}+\frac{6a}{b}-7$$

$$3\times\left(\frac{a}{b}\right)\times(-1)=\quad -\frac{3a}{b}$$

$$(-1)^2=\quad +1$$

$$\frac{3a^2}{b^2}-\frac{3a}{b}+1 \qquad -\frac{3a^2}{b^2}+\frac{3a}{b}-1$$

$$3\times\left(\frac{a}{b}-1\right)^2=\frac{3a^2}{b^2}-\frac{6a}{b}+3 \qquad \frac{3a}{b}-6+\frac{6b}{a}-\frac{3b^2}{a^2}+\frac{b^3}{a^3}$$

$$3\times\left(\frac{a}{b}-1\right)\times\frac{b}{a}=\quad +3-\frac{3b}{a}$$

$$\left(\frac{b}{a}\right)^2=\quad +\frac{b^2}{a^2}$$

$$\frac{3a^2}{b^2}-\frac{6a}{b}+6-\frac{3b}{a}+\frac{b^2}{a^2} \qquad \frac{3a}{b}-6+\frac{6b}{a}-\frac{3b^2}{a^2}+\frac{b^3}{a^3}$$

12.

$$\frac{8x^6}{y^6}+\frac{48x^5}{y^5}+\frac{60x^4}{y^4}-\frac{80x^3}{y^3}-\frac{90x^2}{y^2}+\frac{108x}{y}-27\left(\frac{2x^2}{y^2}+\frac{4x}{y}-3\right.$$

$$\frac{8x^6}{y^6}$$

$$\frac{48x^5}{y^5}+\frac{60x^4}{y^4}-\frac{80x^3}{y^3}$$

$$3\times\left(\frac{2x^2}{y^2}\right)^2=\frac{12x^4}{y^4}$$

$$3\times\frac{2x^2}{y^2}\times\frac{4x}{y}=\quad+\frac{24x^3}{y^3}$$

$$\left(\frac{4x}{y}\right)^2=\quad+\frac{16x^2}{y^2}$$

$$\frac{12x^4}{y^4}+\frac{24x^3}{y^3}+\frac{16x^2}{y^2}\qquad\frac{48x^5}{y^5}+\frac{96x^4}{y^4}+\frac{64x^3}{y^3}$$

$$-\frac{36x^4}{y^4}-\frac{144x^3}{y^3}-\frac{90x^2}{y^2}+\frac{108x}{y}-27$$

$$3\times\left(\frac{2x^2}{y^2}+\frac{4x}{y}\right)^2=\frac{12x^4}{y^4}+\frac{48x^3}{y^3}+\frac{48x^2}{y^2}$$

$$3\times\left(\frac{2x^2}{y^2}+\frac{4x}{y}\right)\times(-3)=\quad-\frac{18x^2}{y^2}-\frac{36x}{y}$$

$$(-3)^2=\quad+9$$

$$\frac{12x^4}{y^4}+\frac{48x^3}{y^3}+\frac{30x^2}{y^2}-\frac{36x}{y}+9\qquad-\frac{36x^4}{y^4}-\frac{144x^3}{y^3}-\frac{90x^2}{y^2}+\frac{108x}{y}-27$$

EXAMPLES. XVII. h. PAGE 122.

17. $x^2-2x+1-a^2-4ab-4b^2=(x-1)^2-(a^2+4ab+4b^2)$
$=(x-1)^2-(a+2b)^2.$

18. $9a^2-6a+1-x^2-8dx-16d^2=(3a-1)^2-(x^2+8dx+16d^2)$
$=(3a-1)^2-(x+4d)^2.$

19. $x^2-2xy+y^2-(a^2-2ab+b^2)=(x-y)^2-(a-b)^2.$

21. $4x^2-12ax+9a^2-(c^2+2ck+k^2)=(2x-3a)^2-(c+k)^2.$

22. $a^2-10ab+25b^2-(9b^2x^2-6bx+1)=(a-5b)^2-(3bx-1)^2.$

23. $a^4+8a^2x^2+16x^4-(25x^6-30x^3+9)=(a^2+4x^2)^2-(5x^3-3)^2.$

24. $x^4-2a^2x^2+a^4-(x^2-6x+9)=(x^2-a^2)^2-(x-3)^2.$

25. See Art. 135. Ex. 4.

26. $x^4+4x^2y^2+16y^4=(x^4+8x^2y^2+16y^4)-4x^2y^2$
$=(x^2+4y^2)^2-(2xy)^2$
$=(x^2+4y^2+2xy)(x^2+4y^2-2xy).$

27. $p^4+9p^2q^2+81q^4=(p^4+18p^2q^2+81q^4)-9p^2q^2$
$=(p^2+9q^2)^2-(3pq)^2$
$=(p^2+9q^2+3pq)(p^2+9q^2-3pq).$

28. $c^4+3c^2d^2+4d^4=c^4+4c^2d^2+4d^4-c^2d^2$
$=(c^2+2d^2)^2-c^2d^2.$

29. $x^4+y^4-11x^2y^2=(x^4-2x^2y^2+y^4)-9x^2y^2$
$=(x^2-y^2)^2-9x^2y^2.$

30. $4m^4-5m^2n^2+n^4=(4m^4+4m^2n^2+n^4)-9m^2n^2$
$=(2m^2+n^2)^2-9m^2n^2.$

EXAMPLES. XVII. l. PAGE 124_A.

74. $250p^3+2=2(125p^3+1)=2(5p+1)(25p^2-5p+1).$

79. $9x^3-4xy^2=x(9x^2-4y^2)=x(3x+2y)(3x-2y).$

84. $x^2+2x-323=(x^2+2x+1)-324=(x+1)^2-18^2$
$=(x+1+18)(x+1-18)=(x+19)(x-17).$

86. $l^2+l-272=\left(l^2+l+\frac{1}{4}\right)-272-\frac{1}{4}$

$$=\left(l+\frac{1}{2}\right)^2-\frac{1089}{4}=\left(l+\frac{1}{2}\right)^2-\left(\frac{33}{2}\right)^2$$

$$=\left(l+\frac{1}{2}+\frac{33}{2}\right)\left(l+\frac{1}{2}-\frac{33}{2}\right)$$

$$=(l+17)(l-16).$$

88. $a^2+10a-299=(a^2+10a+25)-324$

$$=(a+5)^2-18^2.$$

91. $x^4-7x^2y^2+y^4=(x^4+2x^2y^2+y^4)-9x^2y^2$

$$=(x^2+y^2)^2-(3xy)^2.$$

93. $b^2-2b-783=(b^2-2b+1)-784=(b-1)^2-28^2.$

96. $x^8-1=(x^4+1)(x^4-1)=(x^4+1)(x^2+1)(x^2-1).$

97. $729a^7b-ab^7=ab(729a^6-b^6)$

$=ab(27a^3+b^3)(27a^3-b^3)$. [Art. 137. Ex. 2.]

98. $a^8x^6-64a^2y^6=a^2(a^6x^6-64y^6)$. [Art. 137. Ex. 2.]

99. $a^{12}-b^{12}=(a^6+b^6)(a^6-b^6)$. [Art. 137. Ex. 2.]

104. $(a+b)^4-1=\{(a+b)^2+1\}\{(a+b)^2-1\}$

$$=\{(a+b)^2+1\}(a+b+1)(a+b-1).$$

107 and 108. See Art 136A.

109. $250(a-b)^3+2=2\{125(a-b)^3+1\}$

$$=2\{5(a-b)+1\}\{25(a-b)^2-5(a-b)+1\}.$$

110. $(c+d)^3+(c-d)^3=\{(c+d)+(c-d)\}\{(c+d)^2-(c+d)(c-d)+(c-d)^2\}$

$$=2c(c^2+3d^2).$$

111. $8(x+y)^3-(2x-y)^3$

$$=\{2(x+y)-(2x-y)\}\{4(x+y)^2+2(x+y)(2x-y)+(2x-y)^2\}$$

$$=3y(4x^2+8xy+4y^2+4x^2+2xy-2y^2+4x^2-4xy+y^2)$$

$$=3y(12x^2+6xy+3y^2)$$

$$=9y(4x^2+2xy+y^2).$$

112. $x^2-4y^2+x-2y=(x+2y)(x-2y)+(x-2y)=(x-2y)(x+2y+1).$

113. $a^2-b^2+a-b=(a+b)(a-b)+(a-b)=(a-b)(a+b+1).$

115. $a^3+b^3+a+b=(a+b)(a^2-ab+b^2)+(a+b)=(a+b)(a^2-ab+b^2+1).$

116. $a^2-9b^2+a+3b=(a+3b)(a-3b)+(a+3b)=(a+3b)(a-3b+1).$

117. $4(x-y)^3-(x-y)=(x-y)\{4(x-y)^2-1\}$
$=(x-y)\{2(x-y)+1\}\{2(x-y)-1\}.$

118. $x^4y-x^2y^3-x^3y^2+xy^4=xy(x^3-xy^2-x^2y+y^3)$
$=xy\{x(x^2-y^2)-y(x^2-y^2)\}$
$=xy(x^2-y^2)(x-y)$
$=xy(x+y)(x-y)(x-y).$

[*For Miscellaneous Examples III., see Appendix, p.* 340.]

EXAMPLES. XVIII. a. PAGE 129.

3. $2x^2-2xy=2x(x-y).$
$x^3-x^2y=x^2(x-y).$
$\therefore$ H. C. F. is $x(x-y)$.

4. $6x^2-9xy=3x(2x-3y).$
$4x^2-9y^2=(2x+3y)(2x-3y).$
$\therefore$ H. C. F. is $2x-3y$.

5. $x^3+x^2y=x^2(x+y).$
$x^3+y^3=(x+y)(x^2-xy+y^2).$
$\therefore$ H. C. F. is $x+y$.

6. $a^3b-ab^3=ab(a^2-b^2)=ab(a+b)(a-b).$
$a^5b^2-a^2b^5=a^2b^2(a^3-b^3)=a^2b^2(a-b)(a^2+ab+b^2).$
$\therefore$ H. C. F. is $ab(a-b)$.

7. $a^3-a^2x=a^2(a-x).$
$a^3-ax^2=a(a^2-x^2)=a(a+x)(a-x).$
$a^4-ax^3=a(a^3-x^3)=a(a-x)(a^2+ax+x^2).$
$\therefore$ H. C. F. is $a(a-x)$.

8. $a^2-4x^2=(a+2x)(a-2x).$
$a^2+2ax=a(a+2x).$
$\therefore$ H. C. F. is $a+2x$.

9. $a^2bx+ab^2x=abx(a+b).$
$a^2b-b^3=b(a^2-b^2)=b(a+b)(a-b).$
$\therefore$ H. C. F. is $b(a+b)$.

10. $2x^2y-6xy^2=2xy(x-3y).$
$x^2-9y^2=(x+3y)(x-3y).$
$\therefore$ H. C. F. is $x-3y$.

11. $a^2-x^2=(a+x)(a-x).$
$a^2-ax=a(a-x).$
$a^2x-ax^2=ax(a-x).$
$\therefore$ H. C. F. is $a-x$.

12. $4x^2+2xy=2x(2x+y).$
$12x^2-3y^2=3(4x^2-y^2)=3(2x+y)(2x-y).$
$\therefore$ H. C. F. is $2x+y$.

13. $20x-4=4(5x-1).$
$50x^2-2=2(25x^2-1)=2(5x+1)(5x-1).$
$\therefore$ H. C. F. is $2(5x-1)$.

14. $6bx+4by=2b(3x+2y)$.
$9cx+6cy=3c(3x+2y)$.
$\therefore$ H. C. F. is $3x+2y$.

15. $x^2+x=x(x+1)$.
$(x+1)^2=(x+1)^2$.
$x^3+1=(x+1)(x^2-x+1)$.
$\therefore$ H. C. F. is $x+1$.

16. $xy-y=y(x-1)$.
$x^4y-xy=xy(x^3-1)=xy(x-1)(x^2+x+1)$.
$\therefore$ H. C. F. is $y(x-1)$.

17. $x^2-2xy+y^2=(x-y)^2$.
$(x-y)^3=(x-y)^3$.
$\therefore$ H. C. F. is $(x-y)^2$.

18. $x^3+a^2x=x(x^2+a^2)$.
$x^4-a^4=(x^2+a^2)(x^2-a^2)$.
$\therefore$ H. C. F. is x^2+a^2.

19. $x^3+8y^3=(x+2y)(x^2-2xy+4y^2)$.
$x^2+xy-2y^2=(x+2y)(x-y)$.
$\therefore$ H. C. F. is $x+2y$.

20. $x^4-27a^3x=x(x^3-27a^3)=x(x-3a)(x^2+3ax+9a^2)$.
$(x-3a)^2=(x-3a)^2$.
$\therefore$ H. C. F. is $x-3a$.

21. $x^2+3x+2=(x+2)(x+1)$.
$x^2-4=(x+2)(x-2)$.
$\therefore$ H. C. F. is $x+2$.

22. $x^2-x-20=(x+4)(x-5)$.
$x^2-9x+20=(x-4)(x-5)$.
$\therefore$ H. C. F. is $x-5$.

23. $x^2-18x+45=(x-15)(x-3)$.
$x^2-9=(x+3)(x-3)$.
$\therefore$ H. C. F. is $x-3$.

24. $2x^2-7x+3=(2x-1)(x-3)$.
$3x^2-7x-6=(3x+2)(x-3)$.
$\therefore$ H. C. F. is $x-3$.

25. $12x^2+x-1=(4x-1)(3x+1)$.
$15x^2+8x+1=(5x+1)(3x+1)$.
$\therefore$ H. C. F. is $3x+1$.

26. $2x^2-x-1=(2x+1)(x-1)$.
$3x^2-x-2=(3x+2)(x-1)$.
$\therefore$ H. C. F. is $x-1$.

27. $c^2x^2-d^2=(cx+d)(cx-d)$.
$acx^2-bcx+adx-bd=(ax-b)(cx+d)$.
$\therefore$ H. C. F. is $cx+d$.

28. $x^5-xy^2=x(x^4-y^2)=x(x^2+y)(x^2-y)$.
$x^3+x^2y+xy+y^2=(x+y)(x^2+y)$.
$\therefore$ H. C. F. is x^2+y.

29. $a^3x-a^2bx-6ab^2x=ax(a^2-ab-6b^2)=ax(a-3b)(a+2b)$.
$a^2bx^2-4ab^2x^2+3b^3x^2=bx^2(a^2-4ab+3b^2)=bx^2(a-b)(a-3b)$.
$\therefore$ H. C. F. is $x(a-3b)$.

30. $2x^2+9x+4=(2x+1)(x+4)$.
$2x^2+11x+5=(x+5)(2x+1)$.
$2x^2-3x-2=(2x+1)(x-2)$.
$\therefore$ H. C. F. is $2x+1$.

31. $3x^4+8x^3+4x^2=x^2(3x^2+8x+4)=x^2(x+2)(3x+2)$.
$3x^5+11x^4+6x^3=x^3(3x^2+11x+6)=x^3(x+3)(3x+2)$.
$3x^4-16x^3-12x^2=x^2(3x^2-16x-12)=x^2(x-6)(3x+2)$.
$\therefore$ H. C. F. is $x^2(3x+2)$.

EXAMPLES. XVIII. b. PAGE 133.

1.
$$\begin{array}{r|r|r|l} 1 & x^3+2x^2-13x+10 & x^3+x^2-10x+8 & \\ & x^3+x^2-10x+8 & x^3-3x^2+2x & \\ & x^2-3x+2 & 4x^2-12x+8 & 4 \\ & & 4x^2-12x+8 & \end{array}$$
$\therefore$ H. C. F. is x^2-3x+2.

2.
$$\begin{array}{r|r|r|l} -x & x^3-5x^2-99x+40 & x^3-6x^2-86x+35 & 1 \\ & x^3-13x^2+5x & x^3-5x^2-99x+40 & \\ -8 & 8x^2-104x+40 & -x^2+13x-5 & \\ & 8x^2-104x+40 & & \end{array}$$
$\therefore$ H. C. F. is $x^2-13x+5$.

3.
$$\begin{array}{r|r|r|l} x & x^3+2x^2-8x-16 & x^3+3x^2-8x-24 & 1 \\ & x^3-8x & x^3+2x^2-8x-16 & \\ 2 & 2x^2-16 & x^2-8 & \\ & 2x^2-16 & & \end{array}$$
$\therefore$ H. C. F. is x^2-8.

4.
$$\begin{array}{r|r|r|l} x & x^3+4x^2-5x-20 & x^3+6x^2-5x-30 & 1 \\ & x^3-5x & x^3+4x^2-5x-20 & \\ 4 & 4x^2-20 & 2\,)\,2x^2-10 & \\ & 4x^2-20 & x^2-5 & \end{array}$$
$\therefore$ H. C. F. is x^2-5.

5.
$$\begin{array}{r|r|r|l} x & x^3-x^2-5x-3 & x^3-4x^2-11x-6 & 1 \\ & x^3+2x^2+x & x^3-x^2-5x-3 & \\ -3 & -3x^2-6x-3 & -3\,)\,-3x^2-6x-3 & \\ & -3x^2-6x-3 & x^2+2x+1 & \end{array}$$
$\therefore$ H. C. F. is x^2+2x+1.

6.
$$\begin{array}{r|r|r|l} 1 & x^3+3x^2-8x-24 & x^3+3x^2-3x-9 & x^2 \\ & x^3+3x^2-3x-9 & x^3+3x^2 & \\ & -5\,)\,-5x-15 & -3x-9 & -3 \\ & x+3 & -3x-9 & \end{array}$$
$\therefore$ H. C. F. is $x+3$.

7.
$$\begin{array}{r|r|r|l} 1 & a^3-5a^2x+7ax^2-3x^3 & a^3-3ax^2+2x^3 & a \\ & a^3-3ax^2+2x^3 & a^3-2a^2x+ax^2 & \\ & -5x\,)\,-5a^2x+10ax^2-5x^3 & 2a^2x-4ax^2+2x^3 & 2x \\ & a^2-2ax+x^2 & 2a^2x-4ax^2+2x^3 & \end{array}$$
$\therefore$ H. C. F. is $a^2-2ax+x^2$.

8.

x	$x^4-2x^3 \quad -4x-7$	$x^4+x^3-3x^2-x+2$	1
	$x^4-x^3+x^2+3x$	$x^4-2x^3 \quad -4x-7$	
-1	$-x^3-x^2-7x-7$	$3\,)\,3x^3-3x^2+3x+9$	
	$-x^3+x^2-x-3$	x^3-x^2+x+3	x
	$-2\,)\,-2x^2-6x-4$	x^3+3x^2+2x	
x	x^2+3x+2	$-4x^2-x+3$	-4
	x^2+x	$-4x^2-12x-8$	
2	$2x+2$	$11\,)\,11x+11$	
	$2x+2$	$x+1$	

$\therefore$ H. C. F. is $x+1$.

9.

$-2x$	$2x^3-5x^2+11x+7$	$4x^3-11x^2+25x+7$	2
	$2x^3-6x^2+14x$	$4x^3-10x^2+22x+14$	
	x^2-3x+7	$-x^2+3x-7$	-1
		$-x^2+3x-7$	

$\therefore$ H. C. F. is x^2-3x+7.

10.

x	$2x^3+4x^2-7x-14$	$6x^3-10x^2-21x+35$	3
	$2x^3 \quad -7x$	$6x^3+12x^2-21x-42$	
2	$4x^2 \quad -14$	$-11\,)\,-22x^2 \quad +77$	
	$4x^2 \quad -14$	$2x^2 \quad -7$	

$\therefore$ H. C. F. is $2x^2-7$.

11.

x	$3x^4-3x^3-2x^2-x-1$	$9x^4-3x^3 \quad -x-1$	3
	$3x^4+3x^3+x^2+x$	$9x^4-9x^3-6x^2-3x-3$	
-2	$-6x^3-3x^2-2x-1$	$2\,)\,6x^3+6x^2+2x+2$	4
	$-6x^3-6x^2-2x-2$	$3x^3+3x^2+x+1$	x
	$3x^2 \quad +1$	$3x^3 \quad +x$	
		$3x^2 \quad +1$	1
		$3x^2 \quad +1$	

$\therefore$ H. C. F. is $3x^2+1$.

12.

x	$2x^4-2x^3+x^2+3x-6$	$4x^4-2x^3 \quad +3x-9$	2
	$2x^4-2x^3-3x^2+3x$	$4x^4-4x^3+2x^2+6x-12$	
	$2\,)\,4x^2 \quad -6$	$2x^3-2x^2-3x+3$	x
	$2x^2 \quad -3$	$2x^3 \quad -3x$	
		$-2x^2 \quad +3$	-1
		$-2x^2 \quad +3$	

$\therefore$ H. C. F. is $2x^2-3$.

13.

x	$3x^3-3ax^2+2a^2x-2a^3$	$3x^3+12ax^2+2a^2x+8a^3$	1
	$3x^3 \quad +2a^2x$	$3x^3-3ax^2+2a^2x-2a^3$	
$-a$	$-3ax^2 \quad -2a^3$	$5a\,)\,15ax^2 \quad +10a^3$	
	$-3ax^2 \quad -2a^3$	$3x^2 \quad +2a^2$	

$\therefore$ H. C. F. is $3x^2+2a^2$.

14.

$$\begin{array}{r|r|r|l}
2x & 2x^3-9ax^2+9a^2x-7a^3 & 4x^3-20ax^2+20a^2x-16a^3 & 2\\
 & 2x^3-2ax^2+2a^2x & 4x^3-18ax^2+18a^2x-14a^3 & \\
-7a & -7ax^2+7a^2x-7a^3 & -2a)\ -2ax^2+2a^2x-2a^3 & \\
 & -7ax^2+7a^2x-7a^3 & x^2-ax+a^2 &
\end{array}$$

$\therefore$ H. C. F. is x^2-ax+a^2.

15.

$$\begin{array}{r|r|r|l}
 & 5)\ 10x^3+25ax^2-5a^3 & 4x^3+9ax^2-2a^2x-a^3 & 2\\
2x & 2x^3+5ax^2-a^3 & 4x^3+10ax^2-2a^3 & \\
 & 2x^3+4ax^2-2a^2x & -a)\ -ax^2-2a^2x+a^3 & \\
a & ax^2+2a^2x-a^3 & x^2+2ax-a^2 & \\
 & ax^2+2a^2x-a^3 & &
\end{array}$$

$\therefore$ H. C. F. is $x^2+2ax-a^2$.

16.

$$\begin{array}{r|r|r|l}
2a & 6a^3+13a^2x-9ax^2-10x^3 & 9a^3+12a^2x-11ax^2-10x^3 & \\
 & 6a^3-2a^2x-4ax^2 & 2 & \\
5x & 15a^2x-5ax^2-10x^3 & 18a^3+24a^2x-22ax^2-20x^3 & 3\\
 & 15a^2x-5ax^2-10x^3 & 18a^3+39a^2x-27ax^2-30x^3 & \\
 & & -5x)\ -15a^2x+5ax^2+10x^3 & \\
 & & 3a^2-ax-2x^2 &
\end{array}$$

$\therefore$ H. C. F. is $3a^2-ax-2x^2$.

17.

$$24x^4y+72x^3y^2-6x^2y^3-90xy^4=6xy\,(4x^3+12x^2y-xy^2-15y^3).$$
$$6x^4y^2+13x^3y^3-4x^2y^4-15xy^5=xy^2\,(6x^3+13x^2y-4xy^2-15y^3).$$

$\therefore$ xy is a common factor.

$$\begin{array}{r|r|r|l}
2x & 4x^3+12x^2y-xy^2-15y^3 & 6x^3+13x^2y-4xy^2-15y^3 & \\
 & 4x^3+2x^2y-6xy^2 & 2 & \\
5y & 10x^2y+5xy^2-15y^3 & 12x^3+26x^2y-8xy^2-30y^3 & 3\\
 & 10x^2y+5xy^2-15y^3 & 12x^3+36x^2y-3xy^2-45y^3 & \\
 & & -5y)\ -10x^2y-5xy^2+15y^3 & \\
 & & 2x^2+xy-3y^2 &
\end{array}$$

$\therefore$ H. C. F. is $xy\,(2x^2+xy-3y^2)$.

18.

$$4x^5a^2+10x^4a^3-60x^3a^4+54x^2a^5=2x^2a^2\,(2x^3+5x^2a-30xa^2+27a^3).$$
$$24x^5a^3+30x^3a^5-126x^2a^6=6x^2a^3\,(4x^3+5xa^2-21a^3).$$

$\therefore$ $2x^2a^2$ is a common factor.

$$\begin{array}{r|r|r|l}
2 & 4x^3+5xa^2-21a^3 & 2x^3+5x^2a-30xa^2+27a^3 & x\\
 & 4x^3+10x^2a-60xa^2+54a^3 & 2x^3-13x^2a+15xa^2 & \\
 & -5a)\ -10x^2a+65xa^2-75a^3 & 18x^2a-45xa^2+27a^3 & 9a\\
x & 2x^2-13xa+15a^2 & 18x^2a-117xa^2+135a^3 & \\
 & 2x^2-3xa & 36a^2)\ 72xa^2-108a^3 & \\
-5a & -10xa+15a^2 & 2x-3a & \\
 & -10xa+15a^2 & &
\end{array}$$

$\therefore$ H. C. F. is $2x^2a^2\,(2x-3a)$.

19.
$$4x^5+14x^4+20x^3+70x^2=2x^2(2x^3+7x^2+10x+35).$$
$$8x^7+28x^6-8x^5-12x^4+56x^3=4x^3(2x^4+7x^3-2x^2-3x+14).$$
$\therefore$ $2x^2$ is a common factor.

$$\begin{array}{r|r|r|l}
 & 2x^3+7x^2+10x+35 & 2x^4+7x^3-2x^2-3x+14 & x \\
 & 3 & 2x^4+7x^3+10x^2+35x & \\
x & 6x^3+21x^2+30x+105 & -2)\ -12x^2-38x+14 & \\
 & 6x^3+19x^2-7x & 6x^2+19x-7 & 3 \\
x & 2x^2+37x+105 & 6x^2+111x+315 & \\
 & 2x^2+7x & -46)\ -92x-322 & \\
15 & 30x+105 & 2x+7 & \\
 & 30x+105 & &
\end{array}$$

$\therefore$ H. C. F. is $2x^2(2x+7)$.

20.
$$72x^3-12ax^2+72a^2x-420a^3=12(6x^3-ax^2+6a^2x-35a^3).$$
$$18x^3+42ax^2-282a^2x+270a^3=6(3x^3+7ax^2-47a^2x+45a^3).$$
$\therefore$ 6 is a common factor.

$$\begin{array}{r|r|r|l}
2 & 6x^3-ax^2+6a^2x-35a^3 & 3x^3+7ax^2-47a^2x+45a^3 & x \\
 & 6x^3+14ax^2-94a^2x+90a^3 & 3x^3-20ax^2+25a^2x & \\
 & -5a)\ -15ax^2+100a^2x-125a^3 & 27ax^2-72a^2x+45a^3 & 9a \\
x & 3x^2-20ax+25a^2 & 27ax^2-180a^2x+225a^3 & \\
 & 3x^2-5ax & 36a^2)\ 108a^2x-180a^3 & \\
-5a & -15ax+25a^2 & 3x-5a & \\
 & -15ax+25a^2 & &
\end{array}$$

$\therefore$ H. C. F. is $6(3x-5a)$.

21.
$$\begin{array}{r|r|r|l}
3x & 9x^4+2x^2y^2+y^4 & x)\ 3x^4-8x^3y+5x^2y^2-2xy^3 & \\
 & 9x^4-24x^3y+15x^2y^2-6xy^3 & 3x^3-8x^2y+5xy^2-2y^3 & x \\
8y & 24x^3y-13x^2y^2+6xy^3+y^4 & 3x^3-2x^2y+xy^2 & \\
 & 24x^3y-64x^2y^2+40xy^3-16y^4 & -6x^2y+4xy^2-2y^3 & -2y \\
 & 17y^2)\ 51x^2y^2-34xy^3+17y^4 & -6x^2y+4xy^2-2y^3 & \\
 & 3x^2-2xy+y^2 & &
\end{array}$$

$\therefore$ H. C. F. is $3x^2-2xy+y^2$.

22.
$$\begin{array}{r|r|r|l}
x & x^5-x^3-x+1 & x^7+x^6+x^4-1 & x^2 \\
 & x^5+x^4-x & x^7-x^5-x^3+x^2 & \\
-1 & -x^4-x^3+1 & x^6+x^5+x^4+x^3-x^2-1 & x \\
 & -x^4-x^3+1 & x^6-x^4-x^2+x & \\
 & & x^5+2x^4+x^3-x-1 & 1 \\
 & & x^5-x^3-x+1 & \\
 & & 2)\ 2x^4+2x^3-2 & \\
 & & x^4+x^3-1 &
\end{array}$$

$\therefore$ H. C. F. is x^4+x^3-1.

23.

$$\begin{array}{r|l|l|l}
1 & 1+x+x^3-x^5 & 1-x^4-x^6+x^7 & 1 \\
 & 1+x^3-x^4 & 1+x+x^3-x^5 & \\
\hline
x & x+x^4-x^5 & -x-x^3-x^4+x^5-x^6+x^7 & -x \\
 & x+x^4-x^5 & -x-x^2-x^4+x^6 & \\
\hline
 & & x^2-x^3+x^5-2x^6+x^7 & x^2 \\
 & & x^2+x^3+x^5-x^7 & \\
\hline
 & & -2x^3)\,-2x^3-2x^6+2x^7 & \\
 & & 1+x^3-x^4 &
\end{array}$$

$\therefore$ H. C. F. is $1+x^3-x^4$.

[*For Examples 24–27, see Appendix, p.* 343.]

EXAMPLES. XIX. a PAGE 137.

1. $\dfrac{3a^2-6ab}{2a^2b-4ab^2}=\dfrac{3a(a-2b)}{2ab(a-2b)}=\dfrac{3}{2b}$.

2. $\dfrac{abx+bx^2}{acx+cx^2}=\dfrac{bx(a+x)}{cx(a+x)}=\dfrac{b}{c}$.

3. $\dfrac{ax}{a^2x^2-ax}=\dfrac{ax}{ax(ax-1)}=\dfrac{1}{ax-1}$.

4. $\dfrac{15a^2b^2c}{100(a^3-a^2b)}=\dfrac{15a^2b^2c}{100a^2(a-b)}=\dfrac{3b^2c}{20(a-b)}$.

5. $\dfrac{4x^2-9y^2}{4x^2+6xy}=\dfrac{(2x+3y)(2x-3y)}{2x(2x+3y)}=\dfrac{2x-3y}{2x}$.

6. $\dfrac{20(x^3-y^3)}{5x^2+5xy+5y^2}=\dfrac{20(x-y)(x^2+xy+y^2)}{5(x^2+xy+y^2)}=4(x-y)$.

7. $\dfrac{x(2a^2-3ax)}{a(4a^2x-9x^3)}=\dfrac{xa(2a-3x)}{ax(4a^2-9x^2)}=\dfrac{2a-3x}{(2a+3x)(2a-3x)}=\dfrac{1}{2a+3x}$.

8. $\dfrac{x^3-2xy^2}{x^4-4x^2y^2+4y^4}=\dfrac{x(x^2-2y^2)}{(x^2-2y^2)^2}=\dfrac{x}{x^2-2y^2}$.

9. $\dfrac{(xy-3y^2)^2}{x^3y^2-27y^5}=\dfrac{\{y(x-3y)\}^2}{y^2(x^3-27y^3)}=\dfrac{y^2(x-3y)^2}{y^2(x-3y)(x^2+3xy+9y^2)}=\dfrac{x-3y}{x^2+3xy+9y^2}$.

10. $\dfrac{x^2-5x}{x^2-4x-5}=\dfrac{x(x-5)}{(x+1)(x-5)}=\dfrac{x}{x+1}$.

11. $\dfrac{3x^2+6x}{x^2+4x+4}=\dfrac{3x(x+2)}{(x+2)^2}=\dfrac{3x}{x+2}$.

12. $\dfrac{5a^3b+10a^2b^2}{3a^2b^2+6ab^3}=\dfrac{5a^2b(a+2b)}{3ab^2(a+2b)}=\dfrac{5a}{3b}$.

13. $\frac{x^3y+2x^2y+4xy}{x^3-8}=\frac{xy(x^2+2x+4)}{(x-2)(x^2+2x+4)}=\frac{xy}{x-2}.$

14. $\frac{3a^4+9a^3b+6a^2b^2}{a^4+a^3b-2a^2b^2}=\frac{3a^2(a^2+3ab+2b^2)}{a^2(a^2+ab-2b^2)}=\frac{3(a+b)(a+2b)}{(a-b)(a+2b)}=\frac{3(a+b)}{a-b}.$

15. $\frac{x^4-14x^2-51}{x^4-2x^2-15}=\frac{(x^2+3)(x^2-17)}{(x^2+3)(x^2-5)}=\frac{x^2-17}{x^2-5}.$

16. $\frac{x^2+xy-2y^2}{x^3-y^3}=\frac{(x+2y)(x-y)}{(x-y)(x^2+xy+y^2)}=\frac{x+2y}{x^2+xy+y^2}.$

17. $\frac{2x^2+17x+21}{3x^2+26x+35}=\frac{(2x+3)(x+7)}{(3x+5)(x+7)}=\frac{2x+3}{3x+5}.$

18. $\frac{a^2x^2-16a^2}{ax^2+9ax+20a}=\frac{a^2(x^2-16)}{a(x^2+9x+20)}=\frac{a(x+4)(x-4)}{(x+4)(x+5)}=\frac{a(x-4)}{x+5}.$

19. $\frac{3x^2+23x+14}{3x^2+41x+26}=\frac{(3x+2)(x+7)}{(3x+2)(x+13)}=\frac{x+7}{x+13}.$

20. $\frac{27a+a^4}{18a-6a^2+2a^3}=\frac{a(27+a^3)}{2a(9-3a+a^2)}=\frac{(3+a)(9-3a+a^2)}{2(9-3a+a^2)}=\frac{3+a}{2}.$

EXAMPLES. XIX. b. PAGE 139.

1. The sum of the numerator and denominator is $2a^3+2a^2b+2ab^2$ or $2a(a^2+ab+b^2)$. Hence if the numerator and denominator have a common factor it must be a^2+ab+b^2. [See Art. 131, Second Method.]

$$\frac{a^3-a^2b-ab^2-2b^3}{a^3+3a^2b+3ab^2+2b^3}=\frac{(a^2+ab+b^2)(a-2b)}{(a^2+ab+b^2)(a+2b)}=\frac{a-2b}{a+2b}.$$

2. The difference of numerator and denominator is

$$5x^2-10x+5=5(x-1)^2=5(x-1)(x-1).$$

We therefore try whether $x-1$ is a factor common to both.

$$\frac{x^3-5x^2+7x-3}{x^3-3x+2}=\frac{(x-1)(x^2-4x+3)}{(x-1)(x^2+x-2)}=\frac{x^2-4x+3}{x^2+x-2}=\frac{x-3}{x+2}.$$

3. The difference of numerator and denominator is

$$a^2-3a+2=(a-2)(a-1).$$

We therefore try whether either of the factors $a-2$ or $a-1$ is common to both.

$$\frac{a^3+2a^2-13a+10}{a^3+a^2-10a+8}=\frac{(a-1)(a^2+3a-10)}{(a-1)(a^2+2a-8)}=\frac{a^2+3a-10}{a^2+2a-8}=\frac{a+5}{a+4}.$$

4. On multiplying the numerator by 2 and subtracting the denominator from the product, we obtain

$$10x^2y - 65xy^2 + 75y^3 = 5y\,(2x^2 - 13xy + 15y^2) = 5y\,(2x - 3y)(x - 5y).$$

If there be a common divisor it must clearly be $2x - 3y$.

$$\frac{2x^3 + 5x^2y - 30xy^2 + 27y^3}{4x^3 + 5xy^2 - 21y^3} = \frac{(2x - 3y)(x^2 + 4xy - 9y^2)}{(2x - 3y)(2x^2 + 3xy + 7y^2)} = \frac{x^2 + 4xy - 9y^2}{2x^2 + 3xy + 7y^2}.$$

5. The difference of numerator and denominator is

$$2a^3 + a^2b - 3ab^2 = a\,(2a + 3b)(a - b).$$

$$\frac{4a^3 + 12a^2b - ab^2 - 15b^3}{6a^3 + 13a^2b - 4ab^2 - 15b^3} = \frac{(a - b)(4a^2 + 16ab + 15b^2)}{(a - b)(6a^2 + 19ab + 15b^2)}$$

$$= \frac{4a^2 + 16ab + 15b^2}{6a^2 + 19ab + 15b^2} = \frac{(2a + 3b)(2a + 5b)}{(2a + 3b)(3a + 5b)} = \frac{2a + 5b}{3a + 5b}.$$

6. The difference between numerator and denominator is

$$x^2 + x^3 + x^4 = x^2\,(1 + x + x^2),$$

and

$$\frac{1 + 2x^2 + x^3 + 2x^4}{1 + 3x^2 + 2x^3 + 3x^4} = \frac{(1 + x + x^2)(1 - x + 2x^2)}{(1 + x + x^2)(1 - x + 3x^2)} = \frac{1 - x + 2x^2}{1 - x + 3x^2}.$$

7.
$$\frac{x^2 - 2x + 1}{3x^3 + 7x - 10} = \frac{(x - 1)^2}{(x - 1)(3x^2 + 3x + 10)} = \frac{x - 1}{3x^2 + 3x + 10}.$$

8. The factors of the denominator are $4a - b$, and $a - b$, of which the former evidently is not a factor of the numerator. Thus

$$\frac{3a^3 - 3a^2b + ab^2 - b^3}{4a^2 - 5ab + b^2} = \frac{(a - b)(3a^2 + b^2)}{(4a - b)(a - b)} = \frac{3a^2 + b^2}{4a - b}.$$

9. The denominator $= x^3\,(x + a) + a^3\,(x + a) = (x + a)\,(x^3 + a^3)$.

$$\frac{4x^3 + 3ax^2 + a^3}{x^4 + ax^3 + a^3x + a^4} = \frac{(x + a)(4x^2 - ax + a^2)}{(x + a)(x^3 + a^3)} = \frac{4x^2 - ax + a^2}{x^3 + a^3}$$

10. The denominator $= 3x\,(x^3 - 1) - 2\,(x^3 - 1) = (x^3 - 1)(3x - 2)$.

Now $3x - 2$ is not a common divisor, nor is $x^3 - 1$; hence if there be a common divisor it must be either $x - 1$ or $x^2 + x + 1$, which are the factors of $x^3 - 1$.

$$\frac{4x^3 - 10x^2 + 4x + 2}{3x^4 - 2x^3 - 3x + 2} = \frac{(x - 1)(4x^2 - 6x - 2)}{(x - 1)(3x^3 + x^2 + x - 2)} = \frac{2\,(2x^2 - 3x - 1)}{3x^3 + x^2 + x - 2}.$$

11.
$$\frac{16x^4 - 72x^2a^2 + 81a^4}{4x^2 + 12ax + 9a^2} = \frac{(4x^2 - 9a^2)^2}{(2x + 3a)^2}$$

$$= \frac{\{(2x + 3a)(2x - 3a)\}^2}{(2x + 3a)^2}$$

$$= (2x - 3a)^2.$$

12. The difference between the numerator and denominator is

$$4x^2+2x, \text{ or } 2x(2x+1);$$

thus $$\frac{6x^3+x^2-5x-2}{6x^3+5x^2-3x-2}=\frac{(2x+1)(3x^2-x-2)}{(2x+1)(3x^2+x-2)}=\frac{3x^2-x-2}{3x^2+x-2}.$$

13. On multiplying the numerator by 2 and adding to the denominator, we obtain $$17x^3-51x=17x(x^2-3).$$

$$\frac{5x^3+2x^2-15x-6}{7x^3-4x^2-21x+12}=\frac{(x^2-3)(5x+2)}{(x^2-3)(7x-4)}=\frac{5x+2}{7x-4}.$$

14. The difference of numerator and denominator is

$$20x^2-30x+50=10(2x^2-3x+5).$$

$$\frac{4x^4+11x^2+25}{4x^4-9x^2+30x-25}=\frac{(2x^2-3x+5)(2x^2+3x+5)}{(2x^2-3x+5)(2x^2+3x-5)}=\frac{2x^2+3x+5}{2x^2+3x-5}.$$

15. The expression $=\dfrac{3(x^3-9ax^2+26a^2x-24a^3)}{2(x^3+5ax^2-2a^2x-24a^3)}$.

The difference between the two expressions in brackets is

$$14ax^2-28a^2x \text{ or } 14ax(x-2a).$$

It will easily be found that $x-2a$ is a common divisor and that the expression reduces to

$$\frac{3(x^2-7ax+12a^2)}{2(x^2+7ax+12a^2)} \text{ or } \frac{3(x-4a)(x-3a)}{2(x+4a)(x+3a)}.$$

16. The expression $=\dfrac{a(x^3-5ax^2-99a^2x+40a^3)}{x(x^3-6ax^2-86a^2x+35a^3)}$.

The difference between the two expressions in brackets is

$$ax^2-13a^2x+5a^3 \text{ or } a(x^2-13ax+5a^2).$$

It will be found that $x^2-13ax+5a^2$ is a common divisor and that the expression reduces to $$\frac{a(x+8a)}{x(x+7a)}.$$

EXAMPLES. XIX. c. PAGE 142.

1. $$\frac{14x^2-7x}{12x^3+24x^2}\div\frac{2x-1}{x^2+2x}=\frac{14x^2-7x}{12x^3+24x^2}\times\frac{x^2+2x}{2x-1}$$

$$=\frac{7x(2x-1)}{12x^2(x+2)}\times\frac{x(x+2)}{2x-1}=\frac{7}{12}.$$

2. $$\frac{a^2b^2+3ab}{4a^2-1} \div \frac{ab+3}{2a+1} = \frac{a^2b^2+3ab}{4a^2-1} \times \frac{2a+1}{ab+3}$$
$$= \frac{ab(ab+3)}{(2a+1)(2a-1)} \times \frac{2a+1}{ab+3}$$
$$= \frac{ab}{2a-1}.$$

3. $$\frac{x^2-4a^2}{ax+2a^2} \times \frac{2a}{x-2a} = \frac{(x+2a)(x-2a)}{a(x+2a)} \times \frac{2a}{x-2a} = 2.$$

4. $$\frac{a^2-121}{a^2-4} \div \frac{a+11}{a+2} = \frac{(a+11)(a-11)}{(a+2)(a-2)} \times \frac{a+2}{a+11} = \frac{a-11}{a-2}.$$

5. $$\frac{16x^2-9a^2}{x^2-4} \times \frac{x-2}{4x-3a} = \frac{(4x+3a)(4x-3a)}{(x+2)(x-2)} \times \frac{x-2}{4x-3a} = \frac{4x+3a}{x+2}.$$

6. $$\frac{25a^2-b^2}{9a^2x^2-4x^2} \times \frac{x(3a+2)}{5a+b} = \frac{(5a+b)(5a-b)}{x^2(3a+2)(3a-2)} \times \frac{x(3a+2)}{5a+b} = \frac{5a-b}{x(3a-2)}.$$

7. $$\frac{x^2+5x+6}{x^2-1} \times \frac{x^2-2x-3}{x^2-9} = \frac{(x+2)(x+3)}{(x+1)(x-1)} \times \frac{(x+1)(x-3)}{(x+3)(x-3)}$$
$$= \frac{(x+2)}{(x-1)}.$$

8. $$\frac{x^2+3x+2}{x^2+9x+20} \times \frac{x^2+7x+12}{x^2+5x+6} = \frac{(x+1)(x+2)}{(x+4)(x+5)} \times \frac{(x+3)(x+4)}{(x+2)(x+3)}$$
$$= \frac{x+1}{x+5}.$$

9. $$\frac{2x^2+5x+2}{x^2-4} \times \frac{x^2+4x}{2x^2+9x+4} = \frac{(2x+1)(x+2)}{(x+2)(x-2)} \times \frac{x(x+4)}{(2x+1)(x+4)}$$
$$= \frac{x}{x-2}.$$

10. $$\frac{2x^2+13x+15}{4x^2-9} \div \frac{2x^2+11x+5}{4x^2-1} = \frac{2x^2+13x+15}{4x^2-9} \times \frac{4x^2-1}{2x^2+11x+5}$$
$$= \frac{(2x+3)(x+5)}{(2x+3)(2x-3)} \times \frac{(2x+1)(2x-1)}{(2x+1)(x+5)}$$
$$= \frac{2x-1}{2x-3}.$$

11. $$\frac{x^2-14x-15}{x^2-4x-45} \div \frac{x^2-12x-45}{x^2-6x-27} = \frac{x^2-14x-15}{x^2-4x-45} \times \frac{x^2-6x-27}{x^2-12x-45}$$
$$= \frac{(x+1)(x-15)}{(x+5)(x-9)} \times \frac{(x+3)(x-9)}{(x+3)(x-15)} = \frac{x+1}{x+5}.$$

12. $\dfrac{2x^2-x-1}{2x^2+5x+2}\times\dfrac{4x^2+x-14}{16x^2-49}=\dfrac{(2x+1)(x-1)}{(2x+1)(x+2)}\times\dfrac{(4x-7)(x+2)}{(4x+7)(4x-7)}=\dfrac{x-1}{4x+7}.$

13. $\dfrac{b^4-27b}{2b^2+5b}\times\dfrac{4b^2-25}{2b^2-11b+15}=\dfrac{b(b-3)(b^2+3b+9)}{b(2b+5)}\times\dfrac{(2b+5)(2b-5)}{(2b-5)(b-3)}$

$=b^2+3b+9.$

14. $\dfrac{x^3-6x^2+36x}{x^2-49}\div\dfrac{x^4+216x}{x^2-x-42}=\dfrac{x^3-6x^2+36x}{x^2-49}\times\dfrac{x^2-x-42}{x^4+216x}$

$=\dfrac{x(x^2-6x+36)}{(x+7)(x-7)}\times\dfrac{(x+6)(x-7)}{x(x+6)(x^2-6x+36)}$

$=\dfrac{1}{x+7}.$

15. $\dfrac{64p^2q^2-z^4}{x^2-4}\times\dfrac{(x-2)^2}{8pq+z^2}\div\dfrac{x^2-4}{(x+2)^2}$

$=\dfrac{(8pq+z^2)(8pq-z^2)}{(x+2)(x-2)}\times\dfrac{(x-2)^2}{8pq+z^2}\times\dfrac{(x+2)^2}{(x+2)(x-2)}=8pq-z^2.$

16. $\dfrac{x^2-x-20}{x^2-25}\times\dfrac{x^2-x-2}{x^2+2x-8}\div\dfrac{x+1}{x^2+5x}$

$=\dfrac{(x-5)(x+4)}{(x-5)(x+5)}\times\dfrac{(x+1)(x-2)}{(x+4)(x-2)}\times\dfrac{x(x+5)}{x+1}=x.$

17. $\dfrac{x^2-18x+80}{x^2-5x-50}\times\dfrac{x^2-6x-7}{x^2-15x+56}\times\dfrac{x+5}{x-1}$

$=\dfrac{(x-8)(x-10)}{(x+5)(x-10)}\times\dfrac{(x+1)(x-7)}{(x-7)(x-8)}\times\dfrac{x+5}{x-1}=\dfrac{x+1}{x-1}.$

18. $\dfrac{x^2-8x-9}{x^2-17x+72}\times\dfrac{x^2-25}{x^2-1}\div\dfrac{x^2+4x-5}{x^2-9x+8}=\dfrac{x^2-8x-9}{x^2-17x+72}\times\dfrac{x^2-25}{x^2-1}\times\dfrac{x^2-9x+8}{x^2+4x-5}$

$=\dfrac{(x+1)(x-9)}{(x-8)(x-9)}\times\dfrac{(x+5)(x-5)}{(x+1)(x-1)}\times\dfrac{(x-1)(x-8)}{(x-1)(x+5)}$

$=\dfrac{x-5}{x-1}.$

19. The expression $=\dfrac{4x^2+x-14}{6xy-14y}\times\dfrac{4x^2}{x^2-4}\times\dfrac{x-2}{4x-7}\times\dfrac{3x^2-x-14}{2x^2+4x}$

$=\dfrac{(4x-7)(x+2)}{2y(3x-7)}\times\dfrac{4x^2}{(x+2)(x-2)}\times\dfrac{x-2}{4x-7}\times\dfrac{(3x-7)(x+2)}{2x(x+2)}$

$=\dfrac{x}{y}.$

20. The expression $=\dfrac{x^2+x-2}{x^2-x-20}\times\dfrac{x^2+5x+4}{x^2-x}\times\left(\dfrac{x^2-2x-15}{x^2+3x+2}\times\dfrac{x^2}{x+3}\right)$

$$=\frac{(x+2)(x-1)}{(x+4)(x-5)}\times\frac{(x+1)(x+4)}{x(x-1)}\times\frac{(x+3)(x-5)}{(x+1)(x+2)}\times\frac{x^2}{x+3}$$

$$=x.$$

21. $\dfrac{4x^2-16x+15}{2x^2+3x+1}\times\dfrac{x^2-6x-7}{2x^2-17x+21}\times\dfrac{4x^2-1}{4x^2-20x+25}$

$$=\frac{(2x-3)(2x-5)}{(2x+1)(x+1)}\times\frac{(x+1)(x-7)}{(2x-3)(x-7)}\times\frac{(2x+1)(2x-1)}{(2x-5)^2}=\frac{2x-1}{2x-5}.$$

22. The expression $=\dfrac{x^4-8x}{x^2-4x-5}\times\dfrac{x^2+2x+1}{x^3-x^2-2x}\times\dfrac{x-5}{x^2+2x+4}$

$$=\frac{x(x-2)(x^2+2x+4)}{(x+1)(x-5)}\times\frac{(x+1)^2}{x(x+1)(x-2)}\times\frac{x-5}{x^2+2x+4}=1.$$

23. $\dfrac{(a+b)^2-c^2}{a^2+ab-ac}\times\dfrac{a}{(a+c)^2-b^2}\times\dfrac{(a-b)^2-c^2}{ab-b^2-bc}$

$$=\frac{(a+b+c)(a+b-c)}{a(a+b-c)}\times\frac{a}{(a+c+b)(a+c-b)}\times\frac{(a-b+c)(a-b-c)}{b(a-b-c)}=\frac{1}{b}.$$

24. $\dfrac{a^2+2ab+b^2-c^2}{a^2-b^2-c^2-2bc}\times\dfrac{a^2-2ac+c^2-b^2}{b^2-2bc+c^2-a^2}=\dfrac{(a+b)^2-c^2}{a^2-(b+c)^2}\times\dfrac{(a-c)^2-b^2}{(b-c)^2-a^2}$

$$=\frac{(a+b+c)(a+b-c)}{(a+b+c)(a-b-c)}\times\frac{(a-c+b)(a-c-b)}{(b-c+a)(b-c-a)}$$

$$=\frac{a+b-c}{b-c-a}.$$

25. The expression $=\dfrac{x^2-64}{x^2+24x+128}\times\dfrac{x^2+12x-64}{x^3-64}\times\dfrac{x^2+4x+16}{x^2-16x+64}$

$$=\frac{(x+8)(x-8)}{(x+8)(x+16)}\times\frac{(x-4)(x+16)}{(x-4)(x^2+4x+16)}\times\frac{x^2+4x+16}{(x-8)^2}$$

$$=\frac{1}{x+8}.$$

[*For Examples 26–32, see Appendix, p.* 345.]

EXAMPLES. XX. a. PAGE 144.

5. $x^2-1=(x+1)(x-1)$.
$x^2+x=x(x+1)$.
$\therefore$ L. C. M. is $x(x+1)(x-1)$.

6. $a^2+ab=a(a+b)$.
$ab+b^2=b(a+b)$.
$\therefore$ L. C. M. is $ab(a+b)$.

7. $4x^2y - y = y(2x+1)(2x-1)$.
$2x^2 + x = x(2x+1)$.
$\therefore$ L. C. M. is $xy(2x+1)(2x-1)$.

8. $6x^2 - 2x = 2x(3x-1)$.
$9x^2 - 3x = 3x(3x-1)$.
$\therefore$ L. C. M. is $6x(3x-1)$.

9. $x^2 + 2x = x(x+2)$.
$x^2 + 3x + 2 = (x+1)(x+2)$.
$\therefore$ L. C. M. is $x(x+1)(x+2)$.

10. $x^2 - 3x + 2 = (x-1)(x-2)$.
$x^2 - 1 = (x+1)(x-1)$.
$\therefore$ L. C. M. is $(x+1)(x-1)(x-2)$.

11. $x^2 + 4x + 4 = (x+2)^2$.
$x^2 + 5x + 6 = (x+2)(x+3)$.
$\therefore$ L. C. M. is $(x+2)^2(x+3)$.

12. $x^2 - 5x + 4 = (x-1)(x-4)$.
$x^2 - 6x + 8 = (x-2)(x-4)$.
$\therefore$ L. C. M. is $(x-1)(x-2)(x-4)$.

13. $x^2 - x - 6 = (x+2)(x-3)$.
$x^2 + x - 2 = (x-1)(x+2)$.
$x^2 - 4x + 3 = (x-1)(x-3)$.
$\therefore$ L. C. M. is $(x-1)(x+2)(x-3)$.

14. $x^2 + x - 20 = (x-4)(x+5)$.
$x^2 - 10x + 24 = (x-4)(x-6)$.
$x^2 - x - 30 = (x+5)(x-6)$.
$\therefore$ L. C. M. is $(x-4)(x+5)(x-6)$.

15. $x^2 + x - 42 = (x-6)(x+7)$.
$x^2 - 11x + 30 = (x-5)(x-6)$.
$x^2 + 2x - 35 = (x-5)(x+7)$.
$\therefore$ L. C. M. is $(x-5)(x-6)(x+7)$.

16. $2x^2 + 3x + 1 = (2x+1)(x+1)$.
$2x^2 + 5x + 2 = (2x+1)(x+2)$.
$x^2 + 3x + 2 = (x+1)(x+2)$.
$\therefore$ L. C. M. is $(x+1)(x+2)(2x+1)$.

17. $3x^2 + 11x + 6 = (3x+2)(x+3)$.
$3x^2 + 8x + 4 = (3x+2)(x+2)$.
$x^2 + 5x + 6 = (x+3)(x+2)$.
$\therefore$ L. C. M. is $(x+2)(x+3)(3x+2)$.

18. $5x^2 + 11x + 2 = (5x+1)(x+2)$.
$5x^2 + 16x + 3 = (5x+1)(x+3)$.
$x^2 + 5x + 6 = (x+2)(x+3)$.
$\therefore$ L. C. M. is $(x+2)(x+3)(5x+1)$.

19. $2x^2 + 3x - 2 = (2x-1)(x+2)$.
$2x^2 + 15x - 8 = (2x-1)(x+8)$.
$x^2 + 10x + 16 = (x+2)(x+8)$.
$\therefore$ L. C. M. is $(x+2)(x+8)(2x-1)$.

20. $3x^2 - x - 14 = (3x-7)(x+2)$.
$3x^2 - 13x + 14 = (3x-7)(x-2)$.
$x^2 - 4 = (x+2)(x-2)$.
$\therefore$ L. C. M. is $(x+2)(x-2)(3x-7)$.

21. $12x^2 + 3x - 42 = 3(4x^2 + x - 14) = 3(x+2)(4x-7)$.
$12x^3 + 30x^2 + 12x = 6x(2x^2 + 5x + 2) = 6x(2x+1)(x+2)$.
$32x^2 - 40x - 28 = 4(8x^2 - 10x - 7) = 4(2x+1)(4x-7)$.
$\therefore$ L. C. M. is $12x(x+2)(2x+1)(4x-7)$.

22. $3x^4 + 26x^3 + 35x^2 = x^2(3x^2 + 26x + 35) = x^2(3x+5)(x+7)$.
$6x^2 + 38x - 28 = 2(3x^2 + 19x - 14) = 2(3x-2)(x+7)$.
$27x^3 + 27x^2 - 30x = 3x(9x^2 + 9x - 10) = 3x(3x-2)(3x+5)$.
$\therefore$ L. C. M. is $6x^2(x+7)(3x+5)(3x-2)$.

23. $$60x^4+5x^3-5x^2=5x^2(12x^2+x-1)=5x^2(4x-1)(3x+1).$$
$$60x^2y+32xy+4y=4y(15x^2+8x+1)=4y(5x+1)(3x+1).$$
$$40x^3y-2x^2y-2xy=2xy(20x^2-x-1)=2xy(5x+1)(4x-1).$$
$$\therefore \text{ L. C. M.}=20x^2y(4x-1)(3x+1)(5x+1).$$

24. $$8x^2-38xy+35y^2=(2x-7y)(4x-5y).$$
$$4x^2-xy-5y^2=(4x-5y)(x+y).$$
$$2x^2-5xy-7y^2=(2x-7y)(x+y).$$
$$\therefore \text{ L. C. M.}=(x+y)(2x-7y)(4x-5y).$$

25. $$12x^2-23xy+10y^2=(3x-2y)(4x-5y).$$
$$4x^2-9xy+5y^2=(x-y)(4x-5y).$$
$$3x^2-5xy+2y^2=(3x-2y)(x-y).$$
$$\therefore \text{ L. C. M. is } (x-y)(3x-2y)(4x-5y).$$

26. $$6ax^3+7a^2x^2-3a^3x=ax(6x^2+7ax-3a^2)=ax(2x+3a)(3x-a).$$
$$3a^2x^2+14a^3x-5a^4=a^2(3x^2+14ax-5a^2)=a^2(x+5a)(3x-a).$$
$$6x^2+39ax+45a^2=3(2x^2+13ax+15a^2)=3(2x+3a)(x+5a).$$
$$\therefore \text{ L. C. M. is } 3a^2x(2x+3a)(3x-a)(x+5a).$$

27. $$4ax^2y^2+11axy^2-3ay^2=ay^2(4x^2+11x-3)=ay^2(4x-1)(x+3).$$
$$3x^3y^3+7x^2y^3-6xy^3=xy^3(3x^2+7x-6)=xy^3(x+3)(3x-2).$$
$$24ax^2-22ax+4a=2a(12x^2-11x+2)=2a(4x-1)(3x-2).$$
$$\therefore \text{ L. C. M. is } 2axy^3(x+3)(4x-1)(3x-2).$$

[*For Examples* 28-31, *see Appendix, p.* 346.]

EXAMPLES. XX. b. PAGE 146.

1. $$x^2-5x+6=(x-2)(x-3).$$
$$x^2-4=(x+2)(x-2).$$
$$x^3-3x-2=(x-2)(x^2+2x+1)=(x-2)(x+1)^2.$$
$$\therefore \text{ H. C. F. is } x-2, \text{ and L. C. M. is } (x+1)^2(x+2)(x-2)(x-3).$$

2. $$ab(x^2+1)+x(a^2+b^2)=abx^2+ab+a^2x+b^2x$$
$$=(bx+a)(ax+b).$$
$$ab(x^2-1)+x(a^2-b^2)=abx^2-ab+a^2x-b^2x$$
$$=(bx+a)(ax-b).$$
$$\therefore \text{ L. C. M. is } (bx+a)(ax+b)(ax-b).$$

3.
$$xy - bx = x(y - b).$$
$$xy - ay = y(x - a).$$
$$y^2 - 3by + 2b^2 = (y - b)(y - 2b).$$
$$xy - 2bx - ay + 2ab = (x - a)(y - 2b).$$
$$xy - bx - ay + ab = (x - a)(y - b).$$
$$\therefore \text{ L. C. M. is } xy(x - a)(y - b)(y - 2b).$$

4.
$$x^3 + 2x^2 - 3x = x(x^2 + 2x - 3) = x(x+3)(x - 1).$$
$$2x^3 + 5x^2 - 3x = x(2x^2 + 5x - 3) = x(2x - 1)(x + 3).$$
$$\therefore \text{ H. C. F. is } x(x+3), \text{ and L. C. M. is } x(x+3)(x-1)(2x-1).$$

5.
$$(1 - x^2)^2 = (1 - x)^2(1 + x)^2.$$
L. C. M. of $1 - x$, $(1 - x)^2(1 + x)^2$, and $(1 + x)^3$ is $(1 + x)^3(1 - x)^2$.

6.
$$x^2 - 10x + 24 = (x - 4)(x - 6).$$
$$x^2 - 8x + 12 = (x - 2)(x - 6).$$
$$x^2 - 6x + 8 = (x - 2)(x - 4).$$
$$\therefore \text{ L. C. M. is } (x - 2)(x - 4)(x - 6).$$

7. The difference of these two expressions is $4x^2 + 2x$ or $2x(2x + 1)$. It will easily be found that $2x + 1$ divides both expressions, and that

$$6x^3 + x^2 - 5x - 2 = (2x + 1)(3x^2 - x - 2) = (2x + 1)(3x + 2)(x - 1).$$
$$6x^3 + 5x^2 - 3x - 2 = (2x + 1)(3x^2 + x - 2) = (2x + 1)(3x - 2)(x + 1).$$
$$\therefore \text{ H. C. F. is } 2x + 1, \text{ and L. C. M. is } (2x + 1)(x + 1)(x - 1)(3x + 2)(3x - 2).$$

8.
$$(bc^2 - abc)^2 = b^2c^2(c - a)^2.$$
$$b^2(ac^2 - a^3) = b^2a(c^2 - a^2) = ab^2(c + a)(c - a).$$
$$a^2c^2 + 2ac^3 + c^4 = c^2(a^2 + 2ac + c^2) = c^2(a + c)^2.$$
$$\therefore \text{ L. C. M. is } ab^2c^2(c + a)^2(c - a)^2.$$

9.
$$x^3 - y^3 = (x - y)(x^2 + xy + y^2).$$
$$x^3y - y^4 = y(x^3 - y^3) = y(x - y)(x^2 + xy + y^2).$$

Hence the L. C. M. of the four given expressions is $y^2(x - y)^2(x^2 + xy + y^2)$; while the H. C. F. of the first three expressions is $y(x - y)$.

10.
$$6x^2 - 13x + 6 = (3x - 2)(2x - 3).$$
$$2x^2 + 5x - 12 = (2x - 3)(x + 4).$$
$$6x^2 - x - 12 = (3x + 4)(2x - 3).$$
$$\therefore \text{ H. C. F. is } 2x - 3.$$

$$\text{L. C. M.} = (3x - 2)(2x - 3)(x + 4)(3x + 4)$$
$$= \frac{(3x - 2)(2x - 3)(2x - 3)(x + 4)(2x - 3)(3x + 4)}{(2x - 3)^2}$$
$$= \frac{(6x^2 - 13x + 6)(2x^2 + 5x - 12)(6x^2 - x - 12)}{(2x - 3)^2}$$
= product of the three quantities divided by square of H. C. F.

11. The difference of the two expressions is

$$ax^3 - a^2x^2 + a^3x \text{ or } ax(x^2 - ax + a^2).$$

Now $$x^4 + ax^3 + a^3x + a^4 = (x^2 - ax + a^2)(x^2 + 2ax + a^2).$$
$$x^4 + a^2x^2 + a^4 = (x^2 - ax + a^2)(x^2 + ax + a^2).$$
$$\therefore \text{ L. C. M. is } (x^2 - ax + a^2)(x^2 + ax + a^2)(x + a)^2.$$

12. The difference of the first two expressions is

$$6x^3 - 8x^2y + 2xy^2 \text{ or } 2x(x - y)(3x - y).$$

Now

$$3x^3 - 7x^2y + 5xy^2 - y^3 = (x - y)(3x^2 - 4xy + y^2) = (x - y)^2(3x - y).$$
$$-3x^3 + x^2y + 3xy^2 - y^3 = (x - y)(-3x^2 - 2xy + y^2) = -(x - y)(x + y)(3x - y).$$

The difference of the first and last of the given expressions is

$$12x^2y - 4xy^2 \text{ or } 4xy(3x - y).$$

But $3x^3 + 5x^2y + xy^2 - y^3 = (3x - y)(x^2 + 2xy + y^2) = (3x - y)(x + y)^2$.

Thus the H. C. F. is $3x - y$; and the L. C. M. is $(3x - y)(x - y)^2(x + y)^2$.

13. The H. C. F. of the two given expressions is $x - 1$ (see XIX. b. Ex. 10.).

14.
$$a^2 - b^2 = (a + b)(a - b).$$
$$a^3 - b^3 = (a - b)(a^2 + ab + b^2).$$
$$a^3 - a^2b - ab^2 - 2b^3 = (a^2 + ab + b^2)(a - 2b).$$
$$\therefore \text{ L. C. M. is } (a + b)(a - b)(a - 2b)(a^2 + ab + b^2).$$

15. The first expression $= 2x^2y - 3a^2y + 2a^2x - 3xy^2$

$$= 2x(xy + a^2) - 3y(a^2 + xy)$$
$$= (a^2 + xy)(2x - 3y).$$

Similarly the second expression $= 2x(a^2 + xy) + 3y(xy + a^2)$

$$= (a^2 + xy)(2x + 3y).$$

$\therefore$ H. C. F. is $a^2 + xy$, and L. C. M. is $(a^2 + xy)(2x - 3y)(2x + 3y)$.

16. The difference of the two expressions is

$$3x^2 - 21x + 36 \text{ or } 3(x - 4)(x - 3).$$

Now $x^3 - 9x^2 + 26x - 24 = (x - 3)(x^2 - 6x + 8) = (x - 3)(x - 4)(x - 2)$;

and $x^3 - 12x^2 + 47x - 60 = (x - 3)(x^2 - 9x + 20) = (x - 3)(x - 4)(x - 5)$.

$\therefore$ H. C. F. is $(x - 3)(x - 4)$, and L. C. M. is $(x - 2)(x - 3)(x - 4)(x - 5)$.

17. $x^2 - 10ax + 16a^2 = (x - 2a)(x - 8a)$.

$$x^3 - 15ax^2 + 48a^2x + 64a^3 = (x - 8a)(x^2 - 7ax - 8a^2) = (x - 8a)(x - 8a)(x + a).$$
$$\therefore \text{ H. C. F. is } x - 8a.$$

18.
$$21x(xy - y^2)^2 = 21xy^2(x - y)^2.$$
$$35(x^4y^2 - x^2y^4) = 35x^2y^2(x^2 - y^2) = 35x^2y^2(x + y)(x - y).$$
$$15y(x^2 + xy)^2 = 15yx^2(x + y)^2.$$
$$\therefore \text{ L. C. M. is } 105x^2y^2(x + y)^2(x - y)^2.$$

EXAMPLES. XXI. b. PAGE 151.

21. The expression $=\frac{(x+2a)(x-2a)}{x(x-2a)}-\frac{x+4a}{x+2a}$

$$=\frac{x+2a}{x}-\frac{x+4a}{x+2a}$$

$$=\frac{(x+2a)^2-x(x+4a)}{x(x+2a)}$$

$$=\frac{x^2+4ax+4a^2-x^2-4ax}{x(x+2a)}$$

$$=\frac{4a^2}{x(x+2a)}.$$

22. The expression

$$=\frac{(x-y)(x^2+xy+y^2)+(x+y)(x^2-xy+y^2)}{(x+y)(x-y)}=\frac{x^3-y^3+x^3+y^3}{x^2-y^2}=\frac{2x^3}{x^2-y^2}$$

23. The expression

$$=\frac{a^2+2ax+4x^2-(a+2x)^2}{(a^3-8x^3)}=\frac{a^2+2ax+4x^2-(a^2+4ax+4x^2)}{a^3-8x^3}=-\frac{2ax}{a^3-8x^3}.$$

24. The expression $=\frac{(a+b)(a^2-ab+b^2)}{a^2-ab+b^2}-\frac{(a-b)(a^2+ab+b^2)}{a^2+ab+b^2}$

$$=(a+b)-(a-b)=2b.$$

[*For Examples 25, 26, see Appendix, p. 346.*]

EXAMPLES. XXI. c. PAGE 153.

1. The expression $=\frac{(x-y)-(x+y)+2x}{x^2-y^2}=\frac{2(x-y)}{x^2-y^2}=\frac{2}{x+y}.$

2. The expression $=\frac{(2x-y)+(2x+y)-3x}{4x^2-y^2}=\frac{x}{4x^2-y^2}.$

3. The expression $=\frac{5(1-2x)-3x(1+2x)-(4-13x)}{1-4x^2}$

$$=\frac{5-10x-3x-6x^2-4+13x}{1-4x^2}$$

$$=\frac{1-6x^2}{1-4x^2}.$$

4. The expression

$$=\frac{2a(2a-3b)+3b(2a+3b)-8b^2}{4a^2-9b^2}=\frac{4a^2-6ab+6ab+9b^2-8b^2}{4a^2-9b^2}=\frac{4a^2+b^2}{4a^2-9b^2}.$$

5. The expression $=\dfrac{10-2(3-a)-(3+a)}{9-a^2}=\dfrac{10-6+2a-3-a}{9-a^2}=\dfrac{1+a}{9-a^2}.$

6. The expression

$$=\frac{5x-3(x+1)+2(x-1)}{6(x^2-1)}=\frac{5x-3x-3+2x-2}{6(x^2-1)}=\frac{4x-5}{6(x^2-1)}.$$

7. The expression $=\dfrac{(a+b)-(a-b)-2b}{2(a^2-b^2)}=\dfrac{0}{2(a^2-b^2)}=0.$

8. The expression $=\dfrac{2a}{2a-3}-\dfrac{5}{3(2a+3)}-\dfrac{4(3a+2)}{3(4a^2-9)}$

$$=\frac{6a(2a+3)-5(2a-3)-4(3a+2)}{3(4a^2-9)}$$

$$=\frac{12a^2+18a-10a+15-12a-8}{3(4a^2-9)}$$

$$=\frac{12a^2-4a+7}{3(4a^2-9)}.$$

9. The expression $=\dfrac{3}{x-2}+\dfrac{2}{3(x+2)}+\dfrac{5x}{x^2-4}=\dfrac{9(x+2)+2(x-2)+15x}{3(x^2-4)}$

$$=\frac{9x+18+2x-4+15x}{3(x^2-4)}=\frac{2(13x+7)}{3(x^2-4)}.$$

10. The expression $=\dfrac{x(x^3-y^3)-y(x^3+y^3)+x^3y+xy^3}{x^6-y^6}=\dfrac{x^4-y^4}{x^6-y^6}$

$$=\frac{(x^2+y^2)(x^2-y^2)}{(x^2-y^2)(x^4+x^2y^2+y^4)}$$

$$=\frac{x^2+y^2}{x^4+x^2y^2+y^4}.$$

11. The expression

$$=\frac{1}{(x-4)(x-5)}+\frac{1}{(x-5)(x-6)}=\frac{(x-6)+(x-4)}{(x-4)(x-5)(x-6)}$$

$$=\frac{2(x-5)}{(x-4)(x-5)(x-6)}=\frac{2}{(x-4)(x-6)}.$$

12. The expression

$$=\frac{1}{(x-3)(x-4)}-\frac{1}{(x-2)(x-3)}=\frac{(x-2)-(x-4)}{(x-2)(x-3)(x-4)}=\frac{2}{(x-2)(x-3)(x-4)}.$$

13. The expression

$$=\frac{1}{(2x+1)(x-1)}-\frac{1}{(2x+3)(x-1)}=\frac{(2x+3)-(2x+1)}{(x-1)(2x+1)(2x+3)}$$

$$=\frac{2}{(x-1)(2x+1)(2x+3)}.$$

14. The expression

$$=\frac{1}{(2x+1)(x-1)}-\frac{3}{(2x+1)(3x-2)}=\frac{(3x-2)-3(x-1)}{(2x+1)(x-1)(3x-2)}$$

$$=\frac{1}{(2x+1)(x-1)(3x-2)}.$$

15. The expression

$$=\frac{4}{(4+a)(1-2a)}-\frac{3}{(3+5a)(1-2a)}=\frac{4(3+5a)-3(4+a)}{(4+a)(1-2a)(3+5a)}$$

$$=\frac{17a}{(4+a)(1-2a)(3+5a)}.$$

16. The expression

$$=\frac{5}{(1+2x)(5-9x)}-\frac{2}{(1+2x)(2+x)}=\frac{5(2+x)-2(5-9x)}{(1+2x)(5-9x)(2+x)}$$

$$=\frac{23x}{(1+2x)(5-9x)(2+x)}.$$

17. The expression

$$=\frac{(x+2)(x+3)-(x+3)+1}{(x+1)(x+2)(x+3)}=\frac{x^2+5x+6-x-3+1}{(x+1)(x+2)(x+3)}=\frac{x^2+4x+4}{(x+1)(x+2)(x+3)}$$

$$=\frac{x+2}{(x+1)(x+3)}.$$

18. The expression $=\dfrac{40x(x-2)-15(x+1)(x-1)-9(x-3)(x+3)}{16(x+1)(x-2)(x-3)}$

$$=\frac{40x^2-80x-15x^2+15-9x^2+81}{16(x+1)(x-2)(x-3)}$$

$$=\frac{16x^2-80x+96}{16(x+1)(x-2)(x-3)}=\frac{x^2-5x+6}{(x+1)(x-2)(x-3)}$$

$$=\frac{(x-2)(x-3)}{(x+1)(x-2)(x-3)}$$

$$=\frac{1}{x+1}.$$

19. The expression $=\dfrac{(a+3b)^2+4(a+2b)^2-(a+b)^2}{4(a+b)(a+2b)(a+3b)}$

$$=\frac{a^2+6ab+9b^2+4a^2+16ab+16b^2-a^2-2ab-b^2}{4(a+b)(a+2b)(a+3b)}$$

$$=\frac{4a^2+20ab+24b^2}{4(a+b)(a+2b)(a+3b)}=\frac{a^2+5ab+6b^2}{(a+b)(a+2b)(a+3b)}$$

$$=\frac{(a+2b)(a+3b)}{(a+b)(a+2b)(a+3b)}$$

$$=\frac{1}{a+b}.$$

20. The expression

$$=\frac{2}{(x-1)(x-2)}+\frac{2}{(x+1)(x-2)}-\frac{1}{(x+1)(x-1)}$$

$$=\frac{2(x+1)+2(x-1)-(x-2)}{(x+1)(x-1)(x-2)}=\frac{2x+2+2x-2-x+2}{(x+1)(x-1)(x-2)}=\frac{3x+2}{(x+1)(x-1)(x-2)}.$$

21. The expression $=\dfrac{x}{(x+2)(x+3)}+\dfrac{15}{(x+2)(x+7)}-\dfrac{12}{(x+3)(x+7)}$

$$=\frac{x(x+7)+15(x+3)-12(x+2)}{(x+2)(x+3)(x+7)}$$

$$=\frac{x^2+10x+21}{(x+2)(x+3)(x+7)}$$

$$=\frac{(x+3)(x+7)}{(x+2)(x+3)(x+7)}$$

$$=\frac{1}{x+2}.$$

22. The expression $=\dfrac{3}{(x+1)(x-1)}+\dfrac{4}{2x+1}+\dfrac{4x+2}{(x+1)(2x+1)}$

$$=\frac{3(2x+1)+4(x^2-1)+(4x+2)(x-1)}{(x+1)(x-1)(2x+1)}$$

$$=\frac{6x+3+4x^2-4+4x^2-2x-2}{(x+1)(x-1)(2x+1)}=\frac{8x^2+4x-3}{(x+1)(x-1)(2x+1)}.$$

23. The expression

$$=\frac{5(2x-3)}{11(3x-1)(2x+1)}+\frac{7x}{(3x-1)(2x+3)}-\frac{12(3x+1)}{11(2x+1)(2x+3)}$$

$$=\frac{5(2x+3)(2x-3)+77x(2x+1)-12(3x-1)(3x+1)}{11(3x-1)(2x+1)(2x+3)}$$

$$=\frac{20x^2-45+154x^2+77x-108x^2+12}{11(3x-1)(2x+1)(2x+3)}$$

$$=\frac{66x^2+77x-33}{11(3x-1)(2x+1)(2x+3)}=\frac{6x^2+7x-3}{(3x-1)(2x+1)(2x+3)}$$

$$=\frac{(3x-1)(2x+3)}{(3x-1)(2x+1)(2x+3)}$$

$$=\frac{1}{2x+1}.$$

24. The expression $=\frac{x^2-9-(x^2-4)}{(x+2)(x+3)}+\frac{1}{x-1}$

$$=-\frac{5}{(x+2)(x+3)}+\frac{1}{x-1}=\frac{-5(x-1)+x^2+5x+6}{(x+2)(x+3)(x-1)}$$

$$=\frac{x^2+11}{(x+2)(x+3)(x-1)}.$$

25. The expression $=\frac{x^2-9-(x^2-16)}{(x-4)(x+3)}-\frac{5}{(x+4)(x-4)}$

$$=\frac{7}{(x-4)(x+3)}-\frac{5}{(x+4)(x-4)}$$

$$=\frac{7(x+4)-5(x+3)}{(x+3)(x+4)(x-4)}=\frac{2x+13}{(x+3)(x+4)(x-4)}.$$

26. The expression $=\frac{(1+2a)^2-(1-2a)^2}{(1-2a)(1+2a)}-\frac{8a}{(1-2a)^2}$

$$=\frac{8a}{(1-2a)(1+2a)}-\frac{8a}{(1-2a)^2}$$

$$=\frac{8a(1-2a)-8a(1+2a)}{(1-2a)^2(1+2a)}=-\frac{32a^2}{(1-2a)^2(1+2a)}.$$

27. The expression $=\frac{24x}{(3-2x)^2}-\frac{(3+2x)^2-(3-2x)^2}{(3-2x)(3+2x)}$

$$=\frac{24x}{(3-2x)^2}-\frac{24x}{(3-2x)(3+2x)}$$

$$=\frac{24x(3+2x)-24x(3-2x)}{(3-2x)^2(3+2x)}$$

$$=\frac{96x^2}{(3-2x)^2(3+2x)}.$$

28. $$\frac{1}{3-x}-\frac{1}{3+x}-\frac{2x}{9+x^2}=\frac{(3+x)-(3-x)}{9-x^2}-\frac{2x}{9+x^2}$$

$$=\frac{2x}{9-x^2}-\frac{2x}{9+x^2}$$

$$=\frac{18x+2x^3-18x+2x^3}{81-x^4}$$

$$=\frac{4x^3}{81-x^4}.$$

29. $$\frac{1}{2a+3}+\frac{1}{2a-3}-\frac{4a}{4a^2+9}=\frac{2a-3+2a+3}{4a^2-9}-\frac{4a}{4a^2+9}$$

$$=\frac{4a}{4a^2-9}-\frac{4a}{4a^2+9}$$

$$=\frac{16a^3+36a-16a^3+36a}{16a^4-81}$$

$$=\frac{72a}{16a^4-81}.$$

30. $$\frac{1}{4(1+x)}+\frac{1}{4(1-x)}+\frac{1}{2(1+x^2)}=\frac{1-x+1+x}{4(1-x^2)}+\frac{1}{2(1+x^2)}$$

$$=\frac{2}{4(1-x^2)}+\frac{1}{2(1+x^2)}$$

$$=\frac{1}{2(1-x^2)}+\frac{1}{2(1+x^2)}$$

$$=\frac{1+x^2+1-x^2}{2(1-x^4)}$$

$$=\frac{2}{2(1-x^4)}$$

$$=\frac{1}{1-x^4}.$$

31. $\frac{3}{8(a-x)}+\frac{1}{8(a+x)}-\frac{a-x}{4(a^2+x^2)}=\frac{3a+3x+a-x}{8(a^2-x^2)}-\frac{a-x}{4(a^2+x^2)}$

$$=\frac{4a+2x}{8(a^2-x^2)}-\frac{a-x}{4(a^2+x^2)}$$

$$=\frac{2a+x}{4(a^2-x^2)}-\frac{a-x}{4(a^2+x^2)}$$

$$=\frac{(2a+x)(a^2+x^2)-(a-x)(a^2-x^2)}{4(a^4-x^4)}$$

$$=\frac{2a^3+2ax^2+a^2x+x^3-a^3+ax^2+a^2x-x^3}{4(a^4-x^4)}$$

$$=\frac{a^3+2a^2x+3ax^2}{4(a^4-x^4)}$$

$$=\frac{a(a^2+2ax+3x^2)}{4(a^4-x^4)}$$

32. $\frac{2x}{4+x^2}+\frac{1}{2-x}-\frac{1}{2+x}=\frac{2x}{4+x^2}+\frac{2+x-2+x}{4-x^2}$

$$=\frac{2x}{4+x^2}+\frac{2x}{4-x^2}$$

$$=\frac{8x-2x^3+8x-2x^3}{16-x^4}$$

$$=\frac{16x}{16-x^4}.$$

33. The expression $=\frac{5(1+2x)-5(1-2x)}{3(1-2x)(1+2x)}-\frac{x}{2(1+4x^2)}$

$$=\frac{20x}{3(1-4x^2)}-\frac{x}{2(1+4x^2)}$$

$$=\frac{40x(1+4x^2)-3x(1-4x^2)}{6(1-16x^4)}$$

$$=\frac{40x+160x^3-3x+12x^3}{6(1-16x^4)}$$

$$=\frac{x(37+172x^2)}{6(1-16x^4)}.$$

34. $\frac{1}{2a-8x}+\frac{1}{2a+8x}-\frac{a}{3a^2+48x^2}=\frac{a+4x+a-4x}{2(a-4x)(a+4x)}-\frac{a}{3(a^2+16x^2)}$

$$=\frac{a}{a^2-16x^2}-\frac{a}{3(a^2+16x^2)}$$

$$=\frac{3a^3+48ax^2-a^3+16ax^2}{3(a^4-256x^4)}$$

$$=\frac{2a(a^2+32x^2)}{3(a^4-256x^4)}.$$

35. The expression $= \dfrac{1}{6(a^2+9)} + \dfrac{1}{3(a-3)} - \dfrac{a}{3(a^2-9)}$

$$= \frac{1}{6(a^2+9)} + \frac{a+3-a}{3(a^2-9)}$$

$$= \frac{1}{6(a^2+9)} + \frac{1}{a^2-9}$$

$$= \frac{a^2-9+6a^2+54}{6(a^4-81)}$$

$$= \frac{7a^2+45}{6(a^4-81)}$$

36. The expression $= \dfrac{1}{8(1-x)} - \dfrac{1}{8(1+x)} + \dfrac{x}{4(1+x^2)} - \dfrac{x}{2(1+x^4)}$

$$= \frac{1+x-1+x}{8(1-x^2)} + \frac{x}{4(1+x^2)} - \frac{x}{2(1+x^4)}$$

$$= \frac{x}{4(1-x^2)} + \frac{x}{4(1+x^2)} - \frac{x}{2(1+x^4)}$$

$$= \frac{x}{2(1-x^4)} - \frac{x}{2(1+x^4)}$$

$$= \frac{x^5}{1-x^8}.$$

37. The expression $= \dfrac{1}{6(a-3)} - \dfrac{1}{6(a+3)} - \dfrac{1}{a^2+9} + \dfrac{18}{a^4+81}$

$$= \frac{a+3-a+3}{6(a^2-9)} - \frac{1}{a^2+9} + \frac{18}{a^4+81}$$

$$= \frac{1}{a^2-9} - \frac{1}{a^2+9} + \frac{18}{a^4+81}$$

$$= \frac{18}{a^4-81} + \frac{18}{a^4+81}$$

$$= \frac{36a^4}{a^8-6561}.$$

[*For Examples* 38–43, *see Appendix, p.* 346.]

EXAMPLES. XXI. d. PAGE 156.

1. The expression $= \dfrac{1}{4(x-1)} - \dfrac{1}{5(x+1)} - \dfrac{1}{x^2-1}$

$$= \frac{5x+5-4x+4-20}{20(x^2-1)}$$

$$= \frac{x-11}{20(x^2-1)}.$$

2. The expression $= \frac{3}{1+a} - \frac{2}{1-a} + \frac{5a}{1-a^2} = \frac{3-3a-2-2a+5a}{1-a^2} = \frac{1}{1-a^2}$.

3. The expression $= \frac{x-2a}{x+a} - \frac{2(a^2-4ax)}{x^2-a^2} - \frac{3a}{x-a}$

$$= \frac{(x-2a)(x-a) - 2(a^2-4ax) - 3a(x+a)}{x^2-a^2}$$

$$= \frac{x^2-3ax+2a^2-2a^2+8ax-3ax-3a^2}{x^2-a^2}$$

$$= \frac{x^2+2ax-3a^2}{x^2-a^2} = \frac{(x-a)(x+3a)}{x^2-a^2}$$

$$= \frac{x+3a}{x+a}.$$

4. The expression $= \frac{x-a}{x+a} - \frac{a^2+3ax}{x^2-a^2} + \frac{x+a}{x-a}$

$$= \frac{(x-a)^2 - (a^2+3ax) + (x+a)^2}{x^2-a^2}$$

$$= \frac{x^2-2ax+a^2-a^2-3ax+x^2+2ax+a^2}{x^2-a^2}$$

$$= \frac{2x^2-3ax+a^2}{x^2-a^2} = \frac{(2x-a)(x-a)}{x^2-a^2}$$

$$= \frac{2x-a}{x+a}.$$

5. The expression $= \frac{1}{2x+1} + \frac{1}{2x-1} - \frac{4x}{4x^2-1}$

$$= \frac{2x-1+2x+1}{4x^2-1} - \frac{4x}{4x^2-1}$$

$$= \frac{4x}{4x^2-1} - \frac{4x}{4x^2-1} = 0.$$

6. The expression $= \frac{3x}{1-x^2} + \frac{2}{1-x} - \frac{2}{1+x}$

$$= \frac{3x+2+2x-2+2x}{1-x^2} = \frac{7x}{1-x^2}.$$

7. The expression $= \frac{2-5x}{x+3} + \frac{3+x}{x-3} + \frac{2x(2x-11)}{x^2-9}$

$$= \frac{(2-5x)(x-3) + (x+3)^2 + 2x(2x-11)}{x^2-9}$$

$$= \frac{-5x^2+17x-6+x^2+6x+9+4x^2-22x}{x^2-9}$$

$$= \frac{x+3}{x^2-9} = \frac{1}{x-3}.$$

8. The expression $= -\dfrac{2x-3}{2x+3} + \dfrac{2x+3}{2x-3} + \dfrac{12}{4x^2-9}$

$$= \frac{-(2x-3)^2 + (2x+3)^2 + 12}{4x^2-9}$$

$$= \frac{-4x^2 + 12x - 9 + 4x^2 + 12x + 9 + 12}{4x^2-9}$$

$$= \frac{12(2x+1)}{4x^2-9}.$$

9. The expression $= \dfrac{5}{2(b+1)} + \dfrac{3}{4(1-b)} + \dfrac{11}{6(1-b^2)}$

$$= \frac{30(1-b) + 9(1+b) + 22}{12(1-b^2)}$$

$$= \frac{30 - 30b + 9 + 9b + 22}{12(1-b^2)}$$

$$= \frac{61-21b}{12(1-b^2)}.$$

10. The expression $= \dfrac{1}{6(1+a)} + \dfrac{1}{6(1-a)} + \dfrac{1}{3(1-a^2)}$

$$= \frac{1-a+1+a+2}{6(1-a^2)} = \frac{2}{3(1-a^2)}.$$

11. The expression $= \dfrac{y^2}{x^3-y^3} - \dfrac{x^3y^2}{x^6-y^6} = \dfrac{y^2(x^3+y^3) - x^3y^2}{x^6-y^6}$

$$= \frac{y^5}{x^6-y^6}.$$

12. The expression $= \dfrac{x^2-y^2}{xy} - \dfrac{y(x-y)}{x(y-x)} = \dfrac{x^2-y^2}{xy} + \dfrac{y}{x}$

$$= \frac{x^2-y^2+y^2}{xy} = \frac{x}{y}.$$

13. The expression $= \dfrac{x^2+y^2}{x^2-y^2} + \dfrac{x}{x+y} - \dfrac{y}{x-y}$

$$= \frac{x^2+y^2+x(x-y) - y(x+y)}{x^2-y^2}$$

$$= \frac{x^2+y^2+x^2-xy-xy-y^2}{x^2-y^2}$$

$$= \frac{2x^2-2xy}{x^2-y^2} = \frac{2x}{x+y}$$

14. The expression $=\dfrac{x^2+2x+4}{x+2}+\dfrac{x^2-2x+4}{x-2}$

$$=\frac{(x-2)(x^2+2x+4)+(x+2)(x^2-2x+4)}{x^2-4}$$

$$=\frac{x^3-8+x^3+8}{x^2-4}=\frac{2x^3}{x^2-4}.$$

15. The expression $=\dfrac{1}{2a+5b}-\dfrac{3a}{4a^2-25b^2}+\dfrac{1}{2a-5b}$

$$=\frac{2a-5b-3a+2a+5b}{4a^2-25b^2}=\frac{a}{4a^2-25b^2}.$$

16. The expression $=\dfrac{2b-a}{x-b}+\dfrac{3x(a-b)}{x^2-b^2}+\dfrac{b-2a}{b+x}$

$$=\frac{(2b-a)(x+b)+3x(a-b)+(b-2a)(x-b)}{x^2-b^2}$$

$$=\frac{2bx-ax+2b^2-ab+3ax-3bx+bx-2ax-b^2+2ab}{x^2-b^2}$$

$$=\frac{b(a+b)}{x^2-b^2}.$$

17. The expression $=\dfrac{ax^2+b}{2x-1}-\dfrac{2(bx+ax^2)}{4x^2-1}-\dfrac{ax^2-b}{2x+1}$

$$=\frac{(ax^2+b)(2x+1)-2(bx+ax^2)-(ax^2-b)(2x-1)}{4x^2-1}$$

$$=\frac{2ax^3+2bx+ax^2+b-2bx-2ax^2-2ax^3+2bx+ax^2-b}{4x^2-1}$$

$$=\frac{2bx}{4x^2-1}.$$

18. The expression $=\dfrac{a+c}{(a-b)(x-a)}-\dfrac{b+c}{(a-b)(x-b)}$

$$=\frac{(a+c)(x-b)-(b+c)(x-a)}{(a-b)(x-a)(x-b)}$$

$$=\frac{ax+cx-ab-bc-bx-cx+ab+ac}{(a-b)(x-a)(x-b)}$$

$$=\frac{ax-bx+ac-bc}{(a-b)(x-a)(x-b)}$$

$$=\frac{(a-b)(x+c)}{(a-b)(x-a)(x-b)}$$

$$=\frac{x+c}{(x-a)(x-b)}.$$

19. The expression $=\dfrac{a-c}{(a-b)(x-a)}-\dfrac{b-c}{(a-b)(x-b)}$

$$=\frac{(a-c)(x-b)-(b-c)(x-a)}{(a-b)(x-a)(x-b)}$$

$$=\frac{ax-ab-cx+bc-bx+ab+cx-ac}{(a-b)(x-a)(x-b)}$$

$$=\frac{ax-bx-ac+bc}{(a-b)(x-a)(x-b)}$$

$$=\frac{(a-b)(x-c)}{(a-b)(x-a)(x-b)}$$

$$=\frac{x-c}{(x-a)(x-b)}.$$

20. The expression $=\dfrac{2a+y}{(x-a)(a-b)}-\dfrac{a+b+y}{(x-b)(a-b)}-\dfrac{x+y-a}{(x-a)(x-b)}$

$$=\frac{(2a+y)(x-b)-(a+b+y)(x-a)-(x+y-a)(a-b)}{(a-b)(x-a)(x-b)}$$

$$=\frac{2ax+xy-2ab-by-ax-bx-xy+a^2+ab+ay-ax-ay+a^2+bx+by-ab}{(a-b)(x-a)(x-b)}$$

$$=\frac{2a^2-2ab}{(a-b)(x-a)(x-b)}=\frac{2a}{(x-a)(x-b)}.$$

21. The expression

$$=\frac{1}{(a^2-b^2)(x^2+b^2)}-\frac{1}{(a^2-b^2)(x^2+a^2)}-\frac{1}{(x^2+a^2)(x^2+b^2)}$$

$$=\frac{x^2+a^2-x^2-b^2-a^2+b^2}{(a^2-b^2)(x^2+a^2)(x^2+b^2)}=0.$$

22. The expression $=\dfrac{1}{x+a}-\dfrac{1}{x-a}+\dfrac{4a}{x^2-a^2}-\dfrac{2a}{x^2+a^2}$

$$=\frac{x-a-x-a+4a}{x^2-a^2}-\frac{2a}{x^2+a^2}$$

$$=\frac{2a}{x^2-a^2}-\frac{2a}{x^2+a^2}=\frac{4a^3}{x^4-a^4}.$$

23. The expression $=\dfrac{3}{x+a}-\dfrac{3}{x-a}+\dfrac{1}{x-3a}-\dfrac{1}{x+3a}$

$$=\frac{3x-3a-3x-3a}{x^2-a^2}+\frac{x+3a-x+3a}{x^2-9a^2}$$

$$=\frac{-6a}{x^2-a^2}+\frac{6a}{x^2-9a^2}$$

$$=\frac{-6ax^2+54a^3+6ax^2-6a^3}{(x^2-a^2)(x^2-9a^2)}=\frac{48a^3}{(x^2-a^2)(x^2-9a^2)}.$$

24. The expression $=\frac{1}{4a^3(a+x)}+\frac{1}{4a^3(a-x)}+\frac{1}{2a^2(a^2+x^2)}-\frac{a^4}{a^8-x^8}$

$$=\frac{a-x+a+x}{4a^3(a^2-x^2)}+\frac{1}{2a^2(a^2+x^2)}-\frac{a^4}{a^8-x^8}$$

$$=\frac{1}{2a^2(a^2-x^2)}+\frac{1}{2a^2(a^2+x^2)}-\frac{a^4}{a^8-x^8}$$

$$=\frac{a^2+x^2+a^2-x^2}{2a^2(a^4-x^4)}-\frac{a^4}{a^8-x^8}$$

$$=\frac{1}{a^4-x^4}-\frac{a^4}{a^8-x^8}$$

$$=\frac{x^4}{a^8-x^8}.$$

25. The expression $=\frac{x}{x^2-y^2}-\frac{y}{x^2+y^2}-\frac{x^3+y^3}{x^4-y^4}+\frac{xy}{(x+y)(x^2+y^2)}$

$$=\frac{x^3+xy^2-x^2y+y^3-x^3-y^3}{x^4-y^4}+\frac{xy}{(x+y)(x^2+y^2)}$$

$$=\frac{-xy(x-y)}{x^4-y^4}-\frac{xy}{(x+y)(x^2+y^2)}$$

$$=-\frac{xy}{(x+y)(x^2+y^2)}+\frac{xy}{(x+y)(x^2+y^2)}$$

$$=0.$$

26. The expression $=\frac{b}{a(a^2-b^2)}+\frac{a}{b(a^2+b^2)}-\frac{a^4+b^4}{ab(a^4-b^4)}+\frac{a^6}{a^8-b^8}$

$$=\frac{b^2(a^2+b^2)+a^2(a^2-b^2)-a^4-b^4}{ab(a^4-b^4)}+\frac{a^6}{a^8-b^8}$$

$$=\frac{a^2b^2+b^4+a^4-a^2b^2-a^4-b^4}{ab(a^4-b^4)}+\frac{a^6}{a^8-b^8}$$

$$=\frac{a^6}{a^8-b^8}.$$

[*For Examples* 27-32, *see Appendix, p.* 348.]

EXAMPLES. XXI. e. PAGE 159.

1. The expression $=-\frac{a}{(a-b)(c-a)}-\frac{b}{(b-c)(a-b)}-\frac{c}{(c-a)(b-c)}$

$$=\frac{-a(b-c)-b(c-a)-c(a-b)}{(a-b)(b-c)(c-a)}$$

$$=\frac{-ab+ac-bc+ab-ac+bc}{(a-b)(b-c)(c-a)}$$

$$=0.$$

2. The expression $= -\dfrac{b}{(a-b)(c-a)} - \dfrac{c}{(b-c)(a-b)} - \dfrac{a}{(c-a)(b-c)}$

$$= \frac{-b(b-c)-c(c-a)-a(a-b)}{(a-b)(b-c)(c-a)}$$

$$= \frac{bc+ca+ab-a^2-b^2-c^2}{(a-b)(b-c)(c-a)}.$$

3. The expression $= -\dfrac{z}{(x-y)(z-x)} - \dfrac{x}{(y-z)(x-y)} - \dfrac{y}{(z-x)(y-z)}$

$$= \frac{-z(y-z)-x(z-x)-y(x-y)}{(x-y)(y-z)(z-x)}$$

$$= \frac{x^2+y^2+z^2-yz-zx-xy}{(x-y)(y-z)(z-x)}.$$

4. The expression $= -\dfrac{y+z}{(x-y)(z-x)} - \dfrac{z+x}{(y-z)(x-y)} - \dfrac{x+y}{(z-x)(y-z)}$

$$= \frac{-(y+z)(y-z)-(z+x)(z-x)-(x+y)(x-y)}{(x-y)(y-z)(z-x)}$$

$$= \frac{-y^2+z^2-z^2+x^2-x^2+y^2}{(x-y)(y-z)(z-x)}$$

$$=0.$$

5. The expression $= \dfrac{b-c}{(a-b)(c-a)} - \dfrac{c-a}{(b-c)(a-b)} - \dfrac{a-b}{(c-a)(b-c)}$

$$= \frac{-(b-c)^2-(c-a)^2-(a-b)^2}{(a-b)(b-c)(c-a)}$$

$$= \frac{-b^2+2bc-c^2-c^2+2ac-a^2-a^2+2ab-b^2}{(a-b)(b-c)-(c-a)}$$

$$= \frac{2(bc+ca+ab-a^2-b^2-c^2)}{(a-b)(b-c)(c-a)}.$$

6. The expression $= -\dfrac{x^2yz}{(x-y)(z-x)} - \dfrac{y^2zx}{(y-z)(x-y)} - \dfrac{z^2xy}{(z-x)(y-z)}$

$$= \frac{-x^2yz(y-z)-y^2zx(z-x)-z^2xy(x-y)}{(x-y)(y-z)(z-x)}$$

$$= \frac{-x^2y^2z+x^2yz^2-xy^2z^2+x^2y^2z-x^2yz^2+xy^2z^2}{(x-y)(y-z)(z-x)}$$

$$=0.$$

7. The expression $= -\dfrac{1+a}{(a-b)(c-a)} - \dfrac{1+b}{(b-c)(a-b)} - \dfrac{1+c}{(c-a)(b-c)}$

$$= \frac{-(1+a)(b-c)-(1+b)(c-a)-(1+c)(a-b)}{(a-b)(b-c)(c-a)}$$

$$= \frac{-b+c-ab+ac-c+a-bc+ab-a+b-ac+bc}{(a-b)(b-c)(c-a)}$$

$$=0.$$

8. The expression

$$= -\frac{p-a}{(p-q)(r-p)} - \frac{q-a}{(q-r)(p-q)} - \frac{r-a}{(r-p)(q-r)}$$

$$= \frac{-(p-a)(q-r)-(q-a)(r-p)-(r-a)(p-q)}{(p-q)(q-r)(r-p)}$$

$$= \frac{-pq+aq+pr-ar-qr+ar+pq-ap-pr+ap+qr-aq}{(p-q)(q-r)(r-p)}$$

$$= 0.$$

9. The expression

$$= -\frac{p+q-r}{(p-q)(r-p)} - \frac{q+r-p}{(q-r)(p-q)} - \frac{r+p-q}{(r-p)(q-r)}$$

$$= \frac{-(p+q-r)(q-r)-(q+r-p)(r-p)-(r+p-q)(p-q)}{(p-q)(q-r)(r-p)}$$

$$= \frac{-p(q-r)-q(r-p)-r(p-q)-(q-r)^2-(r-p)^2-(p-q)^2}{(p-q)(q-r)(r-p)}$$

$$= \frac{2(qr+rp+pq-p^2-q^2-r^2)}{(p-q)(q-r)(r-p)}.$$

10. The expression

$$= -\frac{a^2}{(a^2-b^2)(c^2-a^2)} - \frac{b^2}{(b^2-c^2)(a^2-b^2)} - \frac{c^2}{(c^2-a^2)(b^2-c^2)}$$

$$= \frac{-a^2(b^2-c^2)-b^2(c^2-a^2)-c^2(a^2-b^2)}{(a^2-b^2)(b^2-c^2)(c^2-a^2)}$$

$$= \frac{-a^2b^2+a^2c^2-b^2c^2+a^2b^2-a^2c^2+b^2c^2}{(a^2-b^2)(b^2-c^2)(c^2-a^2)}$$

$$= 0.$$

11. The expression $= -\dfrac{x+y}{(p-q)(r-p)} - \dfrac{x+y}{(q-r)(p-q)} - \dfrac{x+y}{(r-p)(q-r)}$

$$= \frac{-(x+y)\{q-r+r-p+p-q\}}{(p-q)(q-r)(r-p)}$$

$$= 0.$$

12. The expression

$$= -\frac{q+r}{(x-y)(z-x)} - \frac{r+p}{(y-z)(x-y)} - \frac{p+q}{(z-x)(y-z)}$$

$$= \frac{-(q+r)(y-z)-(r+p)(z-x)-(p+q)(x-y)}{(x-y)(y-z)(z-x)}$$

$$= \frac{-qy+qz-ry+rz-rz+rx-pz+px-px+py-qx+qy}{(x-y)(y-z)(z-x)}$$

$$= \frac{rx-qx+py-ry+qz-pz}{(x-y)(y-z)(z-x)}$$

$$= \frac{p(y-z)+q(z-x)+r(x-y)}{(x-y)(y-z)(z-x)}.$$

EXAMPLES. XXII. a. PAGE 163.

1. $$\frac{\frac{m}{n}-\frac{l}{m}}{\frac{a}{m}-\frac{b}{n}}=\left(\frac{m}{n}-\frac{l}{m}\right)\div\left(\frac{a}{m}-\frac{b}{n}\right)=\frac{m^2-ln}{mn}\div\frac{na-mb}{mn}$$
$$=\frac{m^2-ln}{mn}\times\frac{mn}{na-mb}=\frac{m^2-ln}{na-mb}.$$

2. $$\frac{\frac{1}{x}+\frac{1}{y}}{\frac{1}{x}-\frac{1}{y}}=\left(\frac{1}{x}+\frac{1}{y}\right)\div\left(\frac{1}{x}-\frac{1}{y}\right)=\frac{y+x}{xy}\div\frac{y-x}{xy}$$
$$=\frac{y+x}{xy}\times\frac{xy}{y-x}=\frac{y+x}{y-x}.$$

3. $$\frac{a+\frac{b}{d}}{x-\frac{y}{d}}=\left(a+\frac{b}{d}\right)\div\left(x-\frac{y}{d}\right)=\frac{ad+b}{d}\div\frac{dx-y}{d}=\frac{ad+b}{dx-y}.$$

4. $$\frac{1+\frac{c}{x}}{\frac{b}{x}-1}=\left(1+\frac{c}{x}\right)\div\left(\frac{b}{x}-1\right)=\frac{x+c}{x}\div\frac{b-x}{x}=\frac{x+c}{x}\times\frac{x}{b-x}=\frac{x+c}{b-x}.$$

5. $$\frac{2+\frac{3a}{4b}}{a+\frac{8b}{3}}=\left(2+\frac{3a}{4b}\right)\div\left(a+\frac{8b}{3}\right)=\frac{8b+3a}{4b}\div\frac{3a+8b}{3}=\frac{3}{4b}.$$

6. $$\frac{3a+\frac{7b}{8c}}{3c+\frac{7b}{8a}}=\left(3a+\frac{7b}{8c}\right)\div\left(3c+\frac{7b}{8a}\right)=\frac{24ac+7b}{8c}\div\frac{24ac+7b}{8a}=\frac{a}{c}.$$

7. $$\frac{1-\frac{y^2}{x^2}}{1+\frac{y^2}{x^2}}=\left(1-\frac{y^2}{x^2}\right)\div\left(1+\frac{y^2}{x^2}\right)=\frac{x^2-y^2}{x^2}\div\frac{x^2+y^2}{x^2}=\frac{x^2-y^2}{x^2+y^2}.$$

8. $$\frac{1}{a+\frac{b}{c}}=1\div\left(a+\frac{b}{c}\right)=1\div\frac{ac+b}{c}=\frac{c}{ac+b}.$$

9. $$\frac{a}{b+\frac{c}{d}}=a\div\left(b+\frac{c}{d}\right)=a\div\frac{bd+c}{d}=\frac{ad}{bd+c}.$$

10. $\dfrac{x}{x-\dfrac{m}{n}}=x\div\left(x-\dfrac{m}{n}\right)=x\div\dfrac{nx-m}{n}=\dfrac{nx}{nx-m}$.

11. $\dfrac{\left(\dfrac{a}{b}+\dfrac{c}{d}\right)}{\dfrac{m}{n}+\dfrac{k}{p}}=\left(\dfrac{a}{b}+\dfrac{c}{d}\right)\div\left(\dfrac{m}{n}+\dfrac{k}{p}\right)=\dfrac{ad+bc}{bd}\div\dfrac{pm+kn}{pn}=\dfrac{pn(ad+bc)}{bd(pm+kn)}$.

12. $\dfrac{x-\dfrac{1}{x}}{1+\dfrac{1}{x}}=\left(x-\dfrac{1}{x}\right)\div\left(1+\dfrac{1}{x}\right)=\dfrac{x^2-1}{x}\div\dfrac{x+1}{x}=\dfrac{x^2-1}{x+1}=x-1$.

13. $\dfrac{x+5+\dfrac{6}{x}}{1+\dfrac{6}{x}+\dfrac{8}{x^2}}=\left(x+5+\dfrac{6}{x}\right)\div\left(1+\dfrac{6}{x}+\dfrac{8}{x^2}\right)=\dfrac{x^2+5x+6}{x}\div\dfrac{x^2+6x+8}{x^2}$

$$=\frac{(x+2)(x+3)}{x}\times\frac{x^2}{(x+2)(x+4)}=\frac{x(x+3)}{x+4}.$$

14. $\dfrac{\dfrac{1}{x}-\dfrac{2}{x^2}-\dfrac{3}{x^3}}{\dfrac{9}{x}-x}=\left(\dfrac{1}{x}-\dfrac{2}{x^2}-\dfrac{3}{x^3}\right)\div\left(\dfrac{9}{x}-x\right)=\dfrac{x^2-2x-3}{x^3}\div\dfrac{9-x^2}{x}$

$$=\frac{(x-3)(x+1)}{x^3}\times\frac{x}{(3-x)(3+x)}=-\frac{x+1}{x^2(x+3)}.$$

15. $\dfrac{2x^2-x-6}{\dfrac{4}{x^2}-1}=(2x^2-x-6)\div\left(\dfrac{4}{x^2}-1\right)=(2x-3)(x-2)\div\dfrac{4-x^2}{x^2}$

$$=(2x+3)(x-2)\times\frac{x^2}{(2-x)(2+x)}=-\frac{x^2(2x+3)}{x+2}.$$

16. $\dfrac{2}{1-x^2}\div\left(\dfrac{1}{1-x}-\dfrac{1}{1+x}\right)=\dfrac{2}{1-x^2}\div\dfrac{1+x-1+x}{1-x^2}$

$$=\frac{2}{1-x^2}\times\frac{1-x^2}{2x}=\frac{1}{x}.$$

17. $\dfrac{a^3-b^3}{a-b}-\dfrac{a^3+b^3}{a+b}=(a^2+ab+b^2)-(a^2-ab+b^2)=2ab$;

$\therefore$ the fraction $=2ab\div\dfrac{4ab}{a^2-b^2}=2ab\times\dfrac{a^2-b^2}{4ab}=\dfrac{a^2-b^2}{2}$

18. $\left(\frac{a^2-ax+x^2}{a-x}-\frac{a^2+ax+x^2}{a+x}\right)\div\frac{x^3}{a^2-x^2}$

$$=\frac{(a+x)(a^2-ax+x^2)-(a-x)(a^2+ax+x^2)}{a^2-x^2}\times\frac{a^2-x^2}{x^3}$$

$$=\frac{(a^3+x^3)-(a^3-x^3)}{a^2-x^2}\times\frac{a^2-x^2}{x^3}=2.$$

19. $\left(y+\frac{xy}{y-x}\right)\left(y-\frac{xy}{x+y}\right)\times\frac{y^2-x^2}{y^2+x^2}$

$$=\frac{y^2-xy+xy}{y-x}\times\frac{xy+y^2-xy}{x+y}\times\frac{y^2-x^2}{y^2+x^2}$$

$$=\frac{y^2}{y-x}\times\frac{y^2}{x+y}\times\frac{(y-x)(y+x)}{y^2+x^2}$$

$$=\frac{y^4}{y^2+x^2}.$$

20. $\left(\frac{x}{1+x}+\frac{1-x}{x}\right)\div\left(\frac{x}{1+x}-\frac{1-x}{x}\right)=\frac{x^2+1-x^2}{x(1+x)}\div\frac{x^2-(1-x^2)}{x(1+x)}$

$$=\frac{1}{x(1+x)}\times\frac{x(1+x)}{2x^2-1}$$

$$=\frac{1}{2x^2-1}.$$

21. The numerator $=\frac{(a+b)^2-(a-b)^2}{a^2-b^2}=\frac{4ab}{a^2-b^2}$,

and the denominator $=\frac{(a+b)^2-(a^2+b^2)}{(a+b)^2}=\frac{2ab}{(a+b)^2}$;

$\therefore$ the fraction $=\frac{4ab}{a^2-b^2}\div\frac{2ab}{(a+b)^2}=\frac{4ab}{(a+b)(a-b)}\times\frac{(a+b)^2}{2ab}=\frac{2(a+b)}{a-b}$.

22. $\dfrac{\frac{a}{x^2}+\frac{x}{a^2}}{\frac{1}{a^2}-\frac{1}{ax}+\frac{1}{x^2}}=\left(\frac{a}{x^2}+\frac{x}{a^2}\right)\div\left(\frac{1}{a^2}-\frac{1}{ax}+\frac{1}{x^2}\right)=\frac{a^3+x^3}{a^2x^2}\div\frac{x^2-ax+a^2}{a^2x^2}$

$$=\frac{a^3+x^3}{a^2x^2}\times\frac{a^2x^2}{x^2-ax+a^2}=a+x.$$

23. The numerator $=\frac{3x+2-3x+2}{9x^2-4}=\frac{4}{9x^2-4}$,

and the denominator $=1\div\left(9-\frac{4}{x^2}\right)=1\div\frac{9x^2-4}{x^2}=\frac{x^2}{9x^2-4}$;

$\therefore$ the fraction $=\frac{4}{9x^2-4}\div\frac{x^2}{9x^2-4}=\frac{4}{9x^2-4}\times\frac{9x^2-4}{x^2}=\frac{4}{x^2}$.

24. $$1+\dfrac{x}{1+x+\dfrac{2x^2}{1-x}}=1+\dfrac{x}{\dfrac{1-x^2+2x^2}{1-x}}=1+\dfrac{x(1-x)}{1+x^2}$$

$$=\frac{1+x^2+x-x^2}{1+x^2}=\frac{1+x}{1+x^2}.$$

25. $$\dfrac{1}{a-\dfrac{a^2-1}{a+\dfrac{1}{a-1}}}=\dfrac{1}{a-\dfrac{a^2-1}{\dfrac{a^2-a+1}{a-1}}}=\dfrac{1}{a-\dfrac{(a-1)(a^2-1)}{a^2-a+1}}$$

$$=\dfrac{1}{\dfrac{a(a^2-a+1)-(a-1)(a^2-1)}{a^2-a+1}}$$

$$=\frac{a^2-a+1}{a^3-a^2+a-a^3+a^2+a-1}=\frac{a^2-a+1}{2a-1}.$$

26. $$\dfrac{1}{4x+\dfrac{4x}{1+\dfrac{2(x+y)}{6-x}}}=\dfrac{1}{4x+\dfrac{4x}{\dfrac{6-x+2x+2y}{6-x}}}=\dfrac{1}{4x+\dfrac{4x(6-x)}{6+x+2y}}$$

$$=\dfrac{1}{\dfrac{4x(6+x+2y)+4x(6-x)}{6+x+2y}}$$

$$=\frac{6+x+2y}{24x+4x^2+8xy+24x-4x^2}$$

$$=\frac{6+x+2y}{8x(y+6)}.$$

27. $$\dfrac{a}{x+\dfrac{m}{y+\dfrac{n}{z}}}=\dfrac{a}{x+\dfrac{m}{\dfrac{yz+n}{z}}}=\dfrac{a}{x+\dfrac{mz}{yz+n}}=\dfrac{a(yz+n)}{xyz+nx+mz}.$$

28. $$\dfrac{1}{1-\dfrac{1+x}{x-\dfrac{1}{x}}}=\dfrac{1}{1-\dfrac{1+x}{\dfrac{x^2-1}{x}}}=\dfrac{1}{1-\dfrac{x}{x-1}}=\dfrac{1}{\dfrac{x-1-x}{x-1}}=\dfrac{x-1}{-1}=1-x.$$

29. $$\frac{x-2}{x-2-\dfrac{x}{x-\dfrac{x-1}{x-2}}} = \frac{x-2}{x-2-\dfrac{x}{\dfrac{x^2-2x-x+1}{x-2}}} = \frac{x-2}{x-2-\dfrac{x(x-2)}{x^2-3x+1}}$$

$$= \frac{1}{1-\dfrac{x}{x^2-3x+1}} = \frac{1}{\dfrac{x^2-3x+1-x}{x^2-3x+1}} = \frac{x^2-3x+1}{x^2-4x+1}.$$

30. $$\frac{1}{x-\dfrac{1}{x+\dfrac{1}{x}}} - \frac{1}{x+\dfrac{1}{x-\dfrac{1}{x}}} = \frac{1}{x-\dfrac{x}{x^2+1}} - \frac{1}{x+\dfrac{x}{x^2-1}}$$

$$= \frac{1}{\dfrac{x^3+x-x}{x^2+1}} - \frac{1}{\dfrac{x^3-x+x}{x^2-1}} = \frac{x^2+1}{x^3} - \frac{x^2-1}{x^3} = \frac{2}{x^3}.$$

[*For Examples* 31-36, *see Appendix*, p. 349.]

EXAMPLES. XXII. b. PAGE 167.

7. $$\begin{array}{l} 1+x\,)\,x\,(\,x-x^2+x^3-x^4 \\ \quad x+x^2 \\ \quad -x^2 \\ \quad -x^2-x^3 \\ \quad\quad x^3 \\ \quad\quad x^3+x^4 \\ \quad\quad -x^4 \\ \quad\quad -x^4-x^5 \\ \quad\quad\quad x^5 \end{array}$$

8. $$\begin{array}{l} a-b\,)\,a\,(\,1+\dfrac{b}{a}+\dfrac{b^2}{a^2}+\dfrac{b^3}{a^3} \\ \quad a-b \\ \quad\quad b \\ \quad\quad b-\dfrac{b^2}{a} \\ \quad\quad\quad \dfrac{b^2}{a} \\ \quad\quad\quad \dfrac{b^2}{a}-\dfrac{b^3}{a^2} \\ \quad\quad\quad\quad \dfrac{b^3}{a^2} \\ \quad\quad\quad\quad \dfrac{b^3}{a^2}-\dfrac{b^4}{a^3} \\ \quad\quad\quad\quad\quad \dfrac{b^4}{a^3} \end{array}$$

9. $$\begin{array}{l} 1-x\,)\,1+x\,(\,1+2x+2x^2+2x^3 \\ \quad 1-x \\ \quad\quad 2x \\ \quad\quad 2x-2x^2 \\ \quad\quad\quad 2x^2 \\ \quad\quad\quad 2x^2-2x^3 \\ \quad\quad\quad\quad 2x^3 \\ \quad\quad\quad\quad 2x^3-2x^4 \\ \quad\quad\quad\quad\quad 2x^4 \end{array}$$

10. $$\begin{array}{l} 1-x+x^2\,)\,1\,(\,1+x-x^3-x^4 \\ \quad 1-x+x^2 \\ \quad\quad x-x^2 \\ \quad\quad x-x^2+x^3 \\ \quad\quad\quad -x^3 \\ \quad\quad\quad -x^3+x^4-x^5 \\ \quad\quad\quad\quad -x^4+x^5 \\ \quad\quad\quad\quad -x^4+x^5-x^6 \\ \quad\quad\quad\quad\quad x^6 \end{array}$$

11.

$$
\begin{array}{l}
x+3\;)\;x^2\;\left(x-3+\dfrac{9}{x}-\dfrac{27}{x^2}\right.\\
\quad x^2+3x\\
\quad\quad -3x\\
\quad\quad -3x-9\\
\quad\quad\quad\quad 9\\
\quad\quad\quad\quad 9+\dfrac{27}{x}\\
\quad\quad\quad\quad -\dfrac{27}{x}\\
\quad\quad\quad\quad -\dfrac{27}{x}-\dfrac{81}{x^2}\\
\quad\quad\quad\quad\quad\quad \dfrac{81}{x^2}
\end{array}
$$

12.

$$
\begin{array}{l}
1-2x+x^2\;)\;1\;(1+2x+3x^2+4x^3\\
\quad 1-2x+x^2\\
\quad\quad 2x-x^2\\
\quad\quad 2x-4x^2+2x^3\\
\quad\quad\quad 3x^2-2x^3\\
\quad\quad\quad 3x^2-6x^3+3x^4\\
\quad\quad\quad\quad 4x^3-3x^4\\
\quad\quad\quad\quad 4x^3-8x^4+4x^5\\
\quad\quad\quad\quad\quad 5x^4-4x^5
\end{array}
$$

13. $\dfrac{a^3-b^3}{(a-b)^2}=\dfrac{(a-b)(a^2+ab+b^2)}{(a-b)^2}=\dfrac{a^2+ab+b^2}{a-b}.$

By division

$$
\begin{array}{l}
a-b\;)\;a^2+ab+b^2\;(a+2b\\
\quad a^2-ab\\
\quad\quad 2ab+b^2\\
\quad\quad 2ab-2b^2\\
\quad\quad\quad\quad 3b^2
\end{array}
$$

$$\therefore \text{ the fraction}=a+2b+\frac{3b^2}{a-b}.$$

14. $x^2-xy+y^2-\dfrac{2y^3}{x+y}=\dfrac{(x+y)(x^2-xy+y^2)-2y^3}{x+y}$

$$=\frac{x^3+y^3-2y^3}{x+y}=\frac{x^3-y^3}{x+y}.$$

15.

$$
\begin{array}{l}
5x^2+9x-2\;)\;60x^3-17x^2-4x+1\;(12x-25\\
\quad 60x^3+108x^2-24x\\
\quad\quad -125x^2+20x+1\\
\quad\quad -125x^2-225x+50\\
\quad\quad\quad\quad 245x-49
\end{array}
$$

$$\therefore \text{ the fraction}=12x-25+\frac{245x-49}{5x^2+9x-2}=12x-25+\frac{49(5x-1)}{(5x-1)(x+2)}$$

$$=12x-25+\frac{49}{x+2}.$$

16. $1+\dfrac{a^2+b^2-c^2}{2ab}=\dfrac{2ab+a^2+b^2-c^2}{2ab}=\dfrac{(a+b)^2-c^2}{2ab}=\dfrac{(a+b+c)(a+b-c)}{2ab}.$

17. $\left(x+\dfrac{16x-27}{x^2-16}\right)\div\left(x-1+\dfrac{13}{x+4}\right)=\dfrac{x^3-16x+16x-27}{x^2-16}\div\dfrac{(x-1)(x+4)+13}{x+4}$

$$=\frac{x^3-27}{(x+4)(x-4)}\times\frac{x+4}{x^2+3x+9}=\frac{x-3}{x-4}.$$

18. $\left(a^2 - 2ax + 4x^2 - \dfrac{16x^3}{a+2x}\right) \times \left(3 - \dfrac{6x(a+4x)}{a^2+2ax+4x^2}\right)$

$$= \frac{(a^2-2ax+4x^2)(a+2x) - 16x^3}{a+2x} \times \frac{3a^2+6ax+12x^2-6ax-24x^2}{a^2+2ax+4x^2}$$

$$= \frac{a^3+8x^3-16x^3}{a+2x} \times \frac{3a^2-12x^2}{a^2+2ax+4x^2}$$

$$= \frac{a^3-8x^3}{a+2x} \times \frac{3(a^2-4x^2)}{a^2+2ax+4x^2} = 3(a-2x)^2.$$

19. $\left(b^2+3b-2-\dfrac{12}{b-3}\right) \div \left(3b+6-\dfrac{2b^2}{b-3}\right)$

$$= \frac{(b^2+3b-2)(b-3)-12}{b-3} \div \frac{(3b+6)(b-3)-2b^2}{b-3}$$

$$= \frac{b^3-11b+6-12}{b-3} \times \frac{b-3}{3b^2-3b-18-2b^2} = \frac{b^3-11b-6}{b^2-3b-18}$$

$$= \frac{(b+3)(b^2-3b-2)}{(b-6)(b+3)} = \frac{b^2-3b-2}{b-6}.$$

20. $\left(a^2+9b^2+\dfrac{65b^4}{a^2-9b^2}\right) \div \left(a+3b+\dfrac{13b^2}{a-3b}\right)$

$$= \frac{(a^2+9b^2)(a^2-9b^2)+65b^4}{a^2-9b^2} \div \frac{(a+3b)(a-3b)+13b^2}{a-3b}$$

$$= \frac{a^4-81b^4+65b^4}{a^2-9b^2} \times \frac{a-3b}{a^2-9b^2+13b^2}$$

$$= \frac{a^4-16b^4}{a^2-9b^2} \times \frac{a-3b}{a^2+4b^2} = \frac{a^2-4b^2}{a+3b}.$$

21. $\left(4x^2+14x+\dfrac{98x-27}{2x-7}\right) \times \left(\dfrac{1}{6} - \dfrac{3x+29}{12x^2+18x+27}\right)$

$$= \frac{(4x^2+14x)(2x-7)+98x-27}{2x-7} \times \frac{4x^2+6x+9-2(3x+29)}{2(12x^2+18x+27)}$$

$$= \frac{8x^3-98x+98x-27}{2x-7} \times \frac{4x^2+6x+9-6x-58}{6(4x^2+6x+9)}$$

$$= \frac{8x^3-27}{2x-7} \times \frac{4x^2-49}{6(4x^2+6x+9)} = \frac{(2x-3)(2x+7)}{6}.$$

EXAMPLES. XXII. c. PAGE 168.

1. The fraction $= \dfrac{4a(a^2-x^2)}{3b(c^2-x^2)} \times \dfrac{bc+bx}{a^2-ax} \times \dfrac{c^2-2cx+x^2}{a^2+2ax+x^2}$

$$= \frac{4a(a-x)(a+x)}{3b(c-x)(c+x)} \times \frac{b(c+x)}{a(a-x)} \times \frac{(c-x)^2}{(a+x)^2} = \frac{4(c-x)}{3(a+x)}.$$

2. The fraction $= \dfrac{2x(x+a)(x+2a) - x(x+a)(2x+a)}{6a}$

$$= \frac{x(x+a)\{2(x+2a) - (2x+a)\}}{6a}$$

$$= \frac{x(x+a)\,3a}{6a} = \frac{x(x+a)}{2}.$$

3. The fraction $= \dfrac{1}{b}\left(\dfrac{a+2b-a+b}{(a-b)(a+2b)}\right) - \dfrac{2}{a^2+ab-2b^2}$

$$= \frac{1}{b} \cdot \frac{3b}{a^2+ab-2b^2} - \frac{2}{a^2+ab-2b^2} = \frac{1}{a^2+ab-2b^2}.$$

4. $\left(\dfrac{x+y}{x-y}\right)^2 - \left(\dfrac{x-y}{x+y}\right)^2 = \left(\dfrac{x+y}{x-y} + \dfrac{x-y}{x+y}\right)\left(\dfrac{x+y}{x-y} - \dfrac{x-y}{x+y}\right)$

$$= \frac{x^2+2xy+y^2+x^2-2xy+y^2}{x^2-y^2} \times \frac{x^2+2xy+y^2-x^2+2xy-y^2}{x^2-y^2}$$

$$= \frac{2(x^2+y^2)}{x^2-y^2} \times \frac{4xy}{x^2-y^2} = \frac{8xy(x^2+y^2)}{(x^2-y^2)^2}.$$

5. $\dfrac{2}{x-1} + \dfrac{2}{x+1} - \dfrac{4x}{x^2-x+1} = \dfrac{4x}{(x+1)(x-1)} - \dfrac{4x}{x^2-x+1}$

$$= \frac{4x(x^2-x+1-x^2+1)}{(x+1)(x-1)(x^2-x+1)} = \frac{4x(2-x)}{(x-1)(x^3+1)}.$$

6. $\left(\dfrac{x^2}{1-x^4} + \dfrac{2x^4}{1-x^8}\right) \div \left(\dfrac{x^2+1}{x^2}\right)^2 = \dfrac{x^2(1+x^4)+2x^4}{1-x^8} \times \dfrac{x^2}{(x^2+1)^2}$

$$= \frac{x^2(1+2x^2+x^4)}{1-x^8} \times \frac{x^2}{(x^2+1)^2} = \frac{x^4}{1-x^8}.$$

7. The expression $= \dfrac{1}{x} - \dfrac{2}{x+1} + \dfrac{x}{1+x+x^2} - \dfrac{1}{(x+1)^2}$

$$= \frac{(x+1)^2 - 2x(x+1) - x}{x(x+1)^2} + \frac{x}{1+x+x^2}$$

$$= \frac{1-x-x^2}{x(x+1)^2} + \frac{x}{1+x+x^2}$$

$$= \frac{1-(x+x^2)^2+x^2(x+1)^2}{x(x+1)^2(1+x+x^2)}$$

$$= \frac{1}{x(x+1)^2(1+x+x^2)}.$$

8. $\dfrac{1+x^3}{1+2x+2x^2+x^3} = \dfrac{(1+x)(1-x+x^2)}{1+x^3+2x(1+x)}$

$$= \frac{1-x+x^2}{1-x+x^2+2x} = \frac{1-x+x^2}{1+x+x^2}.$$

9. The denominator $= 3\,(x^3 - 27x + 54)$;

and $2\,(x^3 - 27x + 54) - (2x^3 - 9x^2 + 27) = 9\,(x^2 - 6x + 9) = 9\,(x - 3)^2$.

Thus $$\frac{2x^3 - 9x^2 + 27}{3\,(x^3 - 27x + 54)} = \frac{(x-3)\,(2x^2 - 3x - 9)}{3\,(x-3)\,(x^2 + 3x - 18)} = \frac{2x+3}{3\,(x+6)}.$$

10. The fraction $$= \frac{ab\,(b+ax) - (a^2 - b^2)\,\{x\,(b+ax) - ax^2\}}{b^2\,(b+ax)}$$
$$= \frac{ab^2 + a^2bx - (a^2 - b^2)\,bx}{b^2\,(b+ax)}$$
$$= \frac{ab^2 + b^3x}{b^2\,(b+ax)} = \frac{a+bx}{b+ax}.$$

11. $$\left\{\frac{x^4 - a^4}{x^2 - 2ax + a^2} \div \frac{x^2 + ax}{x-a}\right\} \times \frac{x^5 - a^2x^3}{x^3 + a^3} \div \left(\frac{x}{a} - \frac{a}{x}\right)$$
$$= \left\{\frac{(x+a)\,(x-a)\,(x^2+a^2)}{(x-a)^2} \times \frac{x-a}{x\,(x+a)}\right\} \times \frac{x^3\,(x^2 - a^2)}{x^3 + a^3} \div \left(\frac{x^2 - a^2}{ax}\right)$$
$$= \frac{x^2 + a^2}{x} \times \frac{x^3\,(x^2 - a^2)}{x^3 + a^3} \times \frac{ax}{x^2 - a^2} = \frac{ax^3\,(x^2 + a^2)}{x^3 + a^3}.$$

12. The fraction
$$= \frac{a^2 - x^2}{a^2 + ax + x^2} \div \frac{(a-x)^3\,(a+x)^3}{a^3 \times a^3 \times (a^3 - x^3)}$$
$$= \frac{(a+x)\,(a-x)}{a^2 + ax + x^2} \times \frac{a^6\,(a-x)\,(a^2 + ax + x^2)}{(a-x)^3\,(a+x)^3} = \frac{a^6}{(a-x)\,(a+x)^2}.$$

13. The numerator $= (x^2 - 1)^2 = (x-1)^2\,(x+1)^2$; and the denominator is divisible by $(x-1)^2$, the quotient being $3x^3 + 6x^2 - x - 8$.

Thus the fraction $$= \frac{(x+1)^2}{3x^3 + 6x^2 - x - 8}.$$

14. $$\frac{a^3 + a\,(1+a)\,y + y^2}{a^4 - y^2} = \frac{a^3 + a^2y + ay + y^2}{a^4 - y^2} = \frac{(a^2+y)\,(a+y)}{(a^2+y)\,(a^2-y)} = \frac{a+y}{a^2-y}.$$

15. The expression $$= \frac{1}{a} + \frac{2}{a+1} + \frac{3}{a+2} - \frac{4}{a+1} = \frac{1}{a} - \frac{2}{a+1} + \frac{3}{a+2}$$
$$= \frac{(a+1)\,(a+2) - 2a\,(a+2) + 3a\,(a+1)}{a\,(a+1)\,(a+2)}$$
$$= \frac{a^2 + 3a + 2 - 2a^2 - 4a + 3a^2 + 3a}{a\,(a+1)\,(a+2)} = \frac{2\,(a^2 + a + 1)}{a\,(a+1)\,(a+2)}.$$

16. The expression $= \dfrac{x+3}{(x+3)(2x+3)} + \dfrac{1}{2} \cdot \dfrac{1}{2x-3} - \dfrac{1}{\dfrac{4x^2-9}{4x}}$

$$= \frac{1}{2x+3} + \frac{1}{2(2x-3)} - \frac{4x}{4x^2-9} = \frac{4x-6+2x+3-8x}{2(4x^2-9)}$$

$$\frac{-(2x+3)}{2(4x^2-9)} = \frac{1}{2(3-2x)}.$$

17. $\dfrac{2}{x^3+x^2+x+1} - \dfrac{2}{x^3-x^2+x-1} = \dfrac{2}{(x+1)(x^2+1)} - \dfrac{2}{(x-1)(x^2+1)}$

$$= \frac{2(x-1)-2(x+1)}{(x+1)(x-1)(x^2+1)}$$

$$= \frac{-4}{(x^2-1)(x^2+1)} = \frac{4}{1-x^4}.$$

18. $\dfrac{1-a^2}{(1+ax)^2-(a+x)^2} \div \dfrac{1}{2}\left(\dfrac{1}{1-x} + \dfrac{1}{1+x}\right)$

$$= \frac{1-a^2}{1+2ax+a^2x^2-a^2-2ax-x^2} \div \frac{1}{2} \cdot \frac{2}{1-x^2}$$

$$= \frac{1-a^2}{1-a^2-x^2+a^2x^2} \div \frac{1}{1-x^2}$$

$$= \frac{1-a^2}{(1-a^2)(1-x^2)} \times (1-x^2) = 1.$$

19. $$2x^3-x^2-2x+1 = 2x(x^2-1)-(x^2-1)$$

$$= (x^2-1)(2x-1);$$

and $$x^3-3x+2 = (x-1)(x^2+x-2)$$

$$= (x-1)(x+2)(x-1).$$

Thus the fraction $$= \frac{(x+1)(x-1)(2x-1)}{(x-1)(x+2)(x-1)}$$

$$= \frac{(x+1)(2x-1)}{(x+2)(x-1)}.$$

20. The numerator $= (x-2)(x-4)$;

and the denominator $= (x-4)(4x^2-5x-5)$.

Thus the fraction $= \dfrac{x-2}{4x^2-5x-5}$.

21. The expression $= \dfrac{2a}{(x-2a)^2} - \dfrac{x-a}{(x-2a)(x-3a)} + \dfrac{2}{x-3a}$

$$= \frac{2a(x-3a)-(x-a)(x-2a)+2(x-2a)^2}{(x-2a)^2(x-3a)}$$

$$= \frac{2ax-6a^2-x^2+3ax-2a^2+2x^2-8ax+8a^2}{(x-2a)^2(x-3a)}$$

$$= \frac{x(x-3a)}{(x-2a)^2(x-3a)} = \frac{x}{(x-2a)^2}.$$

22. $\dfrac{1}{2}\left(\dfrac{a^2+x^2}{a^2-x^2}\right) - \dfrac{1}{2}\cdot\dfrac{a+x}{a-x} - \dfrac{a^2}{(a+x)^2}$

$$= \frac{(a^2+x^2)(a+x)-(a+x)^3-2a^2(a-x)}{2(a-x)(a+x)^2}$$

$$= \frac{a^3+a^2x+ax^2+x^3-a^3-3a^2x-3ax^2-x^3-2a^3+2a^2x}{2(a-x)(a+x)^2}$$

$$= -\frac{2a(a^2+x^2)}{2(a-x)(a+x)^2} = \frac{a(a^2+x^2)}{(x-a)(a+x)^2}.$$

23. $\dfrac{x}{2}\left(\dfrac{1}{x-y} - \dfrac{1}{x+y}\right) \times \dfrac{x^2-y^2}{x^2y+xy^2} \div \dfrac{1}{x+y}$

$$= \frac{x}{2} \times \frac{2y}{x^2-y^2} \times \frac{x^2-y^2}{xy(x+y)} \times (x+y) = 1.$$

24. $\dfrac{y}{2}\left(\dfrac{1}{x+y} + \dfrac{1}{x-y}\right) \times \dfrac{x^2-y^2}{x^2y+xy^2} = \dfrac{y}{2} \times \dfrac{2x}{x^2-y^2} \times \dfrac{x^2-y^2}{xy(x+y)} = \dfrac{1}{x+y}.$

Hence the expression $= \dfrac{1}{x+y} \div \dfrac{1}{x+y} = 1.$

25. $\left(3x-5-\dfrac{2}{x}\right)\left(3x+5-\dfrac{2}{x}\right) \div \left(x-\dfrac{4}{x}\right)$

$$= \frac{3x^2-5x-2}{x} \times \frac{3x^2+5x-2}{x} \div \frac{x^2-4}{x}$$

$$= \frac{(3x+1)(x-2)}{x} \times \frac{(3x-1)(x+2)}{x} \times \frac{x}{(x-2)(x+2)}$$

$$= \frac{(3x+1)(3x-1)}{x} = \frac{9x^2-1}{x} = 9x - \frac{1}{x}.$$

26. The dividend $= \dfrac{2(a^2-x^2)-x(a-x)+x(a+x)}{x(a+x)(a-x)}$

$$= \frac{2a^2-2x^2-ax+x^2+ax+x^2}{x(a^2-x^2)} = \frac{2a^2}{x(a^2-x^2)};$$

the divisor $= \dfrac{(a+x)^2-(a-x)^2}{a^2-x^2} = \dfrac{4ax}{a^2-x^2}.$

$\therefore$ the expression $= \dfrac{2a^2}{x(a^2-x^2)} \div \dfrac{4ax}{a^2-x^2} = \dfrac{2a^2}{x(a^2-x^2)} \times \dfrac{a^2-x^2}{4ax} = \dfrac{a}{2x^2}.$

27. $$\frac{1}{2x-1}-\frac{2x-\frac{1}{2x}}{4x^2-1}=\frac{1}{2x-1}-\left\{\frac{4x^2-1}{2x}\div(4x^2-1)\right\}$$

$$=\frac{1}{2x-1}-\frac{1}{2x}=\frac{1}{2x(2x-1)}.$$

28. $$\left(b+\frac{ab}{b-a}\right)\left(b-\frac{ab}{a+b}\right)\left(\frac{b^2-a^2}{b^2+a^2}\right)=\frac{b^2-ab+ab}{b-a}\times\frac{ab+b^2-ab}{a+b}\times\frac{b^2-a^2}{b^2+a^2}$$

$$=\frac{b^2}{b-a}\times\frac{b^2}{a+b}\times\frac{(b-a)(b+a)}{b^2+a^2}$$

$$=\frac{b^4}{b^2+a^2}.$$

29. $$\frac{b+\frac{a-b}{1+ab}}{1-\frac{(a-b)b}{1+ab}}=\frac{b+ab^2+a-b}{1+ab}\div\frac{1+ab-ab+b^2}{1+ab}$$

$$=\frac{a(b^2+1)}{1+ab}\times\frac{1+ab}{1+b^2}=a;$$

and $$\frac{a-\frac{a-b}{1-ab}}{1-\frac{a(a-b)}{1-ab}}=\frac{a-a^2b-a+b}{1-ab}\div\frac{1-ab-a^2+ab}{1-ab}$$

$$=\frac{b(1-a^2)}{1-ab}\times\frac{1-ab}{1-a^2}=b.$$

Thus the expression

$$=(a-b)\div\left(\frac{a}{b}-\frac{b}{a}\right)=(a-b)\div\frac{a^2-b^2}{ab}=(a-b)\times\frac{ab}{a^2-b^2}=\frac{ab}{a+b}.$$

30. $$\frac{\frac{x^2+y^2}{y}-x}{\frac{1}{y}-\frac{1}{x}}=\frac{x^2+y^2-xy}{y}\div\left(\frac{1}{y}-\frac{1}{x}\right)=\frac{x^2-xy+y^2}{y}\div\frac{x-y}{xy}$$

$$=\frac{x^2-xy+y^2}{y}\times\frac{xy}{x-y};$$

hence the expression $$=\frac{x^2-xy+y^2}{y}\times\frac{xy}{x-y}\times\frac{(x+y)(x-y)}{(x+y)(x^2-xy+y^2)}=x.$$

31. $$\frac{2\left(x^2-\frac{1}{4}\right)}{2x+1}+\frac{1}{2}=\frac{x^2-\frac{1}{4}}{x+\frac{1}{2}}+\frac{1}{2}=x-\frac{1}{2}+\frac{1}{2}=x.$$

32. $$\frac{a+\frac{b-a}{1+ab}}{1-\frac{a(b-a)}{1+ab}}=\frac{a+a^2b+b-a}{1+ab}\div\frac{1+ab-ab+a^2}{1+ab}=\frac{b(a^2+1)}{1+ab}\times\frac{1+ab}{1+a^2}=b;$$

and $$\frac{\frac{x+y}{1-xy}-y}{1+\frac{y(x+y)}{1-xy}}=\frac{x+y-y+xy^2}{1-xy}\div\frac{1-xy+xy+y^2}{1-xy}=\frac{x(1+y^2)}{1-xy}\times\frac{1-xy}{1+y^2}=x;$$

$$\therefore \text{ the product}=bx.$$

33. $$\frac{\frac{a+b}{a-b}+\frac{a-b}{a+b}}{\frac{a-b}{a+b}-\frac{a+b}{a-b}}\times\frac{ab^3-a^3b}{a^2+b^2}=\frac{(a+b)^2+(a-b)^2}{(a-b)^2-(a+b)^2}\times\frac{ab(b^2-a^2)}{a^2+b^2}$$

$$=\frac{2(a^2+b^2)}{-4ab}\times\frac{ab(b^2-a^2)}{a^2+b^2}$$

$$=-\frac{b^2-a^2}{2}=\frac{a^2-b^2}{2}.$$

34. $$\frac{(1-x^2)(1-x^3)}{x(1+x)(1-x)^2}-\frac{x^3+\frac{1}{x^3}}{x^2+\frac{1}{x^2}-1}$$

$$=\frac{(1+x)(1-x)(1-x)(1+x+x^2)}{x(1+x)(1-x)^2}-\left\{\frac{x^6+1}{x^3}\div\frac{x^4+1-x^2}{x^2}\right\}$$

$$=\frac{1+x+x^2}{x}-\left\{\frac{x^6+1}{x^3}\times\frac{x^2}{x^4-x^2+1}\right\}$$

$$=\frac{1+x+x^2}{x}-\frac{x^2+1}{x}=1.$$

35. $$\left\{\left(x^3-\frac{1}{x^3}\right)-3\left(x-\frac{1}{x}\right)\right\}\div\left(x-\frac{1}{x}\right)$$

$$=\left\{\left(x-\frac{1}{x}\right)\left(x^2+1+\frac{1}{x^2}\right)-3\left(x-\frac{1}{x}\right)\right\}\div\left(x-\frac{1}{x}\right)$$

$$=x^2+1+\frac{1}{x^2}-3$$

$$=x^2-2+\frac{1}{x^2}=\left(x-\frac{1}{x}\right)^2.$$

36. The expression $$=(m+1)\times\frac{1}{m^3+1}\div\frac{1}{m^2-m+1}$$

$$=(m+1)\times\frac{1}{m^3+1}\times(m^2-m+1)=1.$$

37. The expression $$=\frac{y(x^2-xy+y^2)}{x(x^2+xy+y^2)}\times\frac{x+y}{x(x-y)}\div\frac{y(x^3+y^3)}{x^2(x^3-y^3)}$$

$$=\frac{y(x^3+y^3)}{x^2(x^3-y^3)}\div\frac{y(x^3+y^3)}{x^2(x^3-y^3)}=1.$$

38. The dividend $=\left(\frac{3x+x^3}{1+3x^2}+1\right)\left(\frac{3x+x^3}{1+3x^2}-1\right)\div\left(\frac{3x^2-1}{x^3-3x}+1\right)$

$$=\frac{(x^3+3x+3x^2+1)(x^3+3x-3x^2-1)}{(1+3x^2)^2}\times\frac{x(x^2-3)}{x^3+3x^2-3x-1}$$

$$=\frac{(x+1)^3(x-1)^3}{(1+3x^2)^2}\times\frac{x(x^2-3)}{(x-1)(x^2+4x+1)}$$

$$=\frac{x(x+1)^3(x-1)^2(x^2-3)}{(1+3x^2)^2(x^2+4x+1)}.$$

The divisor $=\frac{27x^2+9-33x^2+x^4}{x^2(3x^2+1)}\div\frac{3(x^2-1)^2-2(x^2+3)}{x^2(x^2-1)^2}$

$$=\frac{x^4-6x^2+9}{x^2(3x^2+1)}\times\frac{x^2(x^2-1)^2}{3x^4-8x^2-3}$$

$$=\frac{(x^2-3)^2}{x^2(3x^2+1)}\times\frac{x^2(x^2-1)^2}{(3x^2+1)(x^2-3)}$$

$$=\frac{(x^2-3)(x^2-1)^2}{(3x^2+1)^2}$$

Thus the expression

$$=\frac{x(x+1)^3(x-1)^2(x^2-3)}{(1+3x^2)^2(x^2+4x+1)}\div\frac{(x^2-3)(x^2-1)^2}{(3x^2+1)^2}$$

$$=\frac{x(x+1)^3(x-1)^2(x^2-3)}{(1+3x^2)^2(x^2+4x+1)}\times\frac{(1+3x^2)^2}{(x^2-3)(x-1)^2(x+1)^2}$$

$$=\frac{x(x+1)}{x^2+4x+1}.$$

39. The expression $=\frac{1}{a^2-2}-\frac{1}{a^2+2}+\frac{2}{a^2+1}-\frac{2}{a^2-1}$

$$=\frac{4}{a^4-4}-\frac{4}{a^4-1}$$

$$=\frac{4a^4-4-4a^4+16}{(a^4-4)(a^4-1)}$$

$$=\frac{12}{(a^4-4)(a^4-1)}.$$

40. The expression $= \dfrac{1}{2(3m-n)} + \dfrac{1}{3m+2n} - \dfrac{3}{2(3m+n)}$

$$= \frac{3m+n-3(3m-n)}{2(3m-n)(3m+n)} + \frac{1}{3m+2n}$$

$$= \frac{-3m+2n}{(3m-n)(3m+n)} + \frac{1}{3m+2n}$$

$$= \frac{-9m^2+4n^2+(9m^2-n^2)}{2(3m-n)(3m+n)(3m+2n)}$$

$$= \frac{3n^2}{2(3m-n)(3m+n)(3m+2n)}.$$

41. The expression $= \dfrac{3}{4(1-x)^2} + \dfrac{3(1+x)+(1-x)}{8(1-x)(1+x)} + \dfrac{x-1}{4(1+x^2)}$

$$= \frac{3}{4(1-x)^2} + \frac{2+x}{4(1-x)(1+x)} + \frac{x-1}{4(1+x^2)}$$

$$= \frac{3}{4(1-x)^2} + \frac{(2+x+2x^2+x^3)+(x-1-x^3+x^2)}{4(1-x)(1+x)(1+x^2)}$$

$$= \frac{3}{4(1-x)^2} + \frac{1+2x+3x^2}{4(1-x)(1+x)(1+x^2)}$$

$$= \frac{3(1+x+x^2+x^3)+(1+x+x^2-3x^3)}{4(1-x)^2(1+x)(1+x^2)}$$

$$= \frac{1+x+x^2}{(1-x)^2(1+x)(1+x^2)}.$$

42. The expression $= \dfrac{4}{9(x-2)} + \dfrac{5}{9(x+1)} - \dfrac{1}{3(x+1)^2} - \dfrac{x}{(x+1)^2}$

$$= \frac{4(x+1)^2+5(x-2)(x+1)-3(x-2)-9x(x-2)}{9(x-2)(x+1)^2}$$

$$= \frac{4x^2+8x+4+5x^2-5x-10-3x+6-9x^2+18x}{9(x-2)(x+1)^2}$$

$$= \frac{18x}{9(x-2)(x+1)^2} = \frac{2x}{(x-2)(x+1)^2}.$$

43. The expression

$$= \frac{x^3+y^3}{xy(y^2-x^2)} - \frac{y}{x(x+y)} - \frac{x}{y(y-x)}$$

$$= \frac{x^3+y^3-y^2(y-x)-x^2(x+y)}{xy(y-x)(x+y)}$$

$$= \frac{x^3+y^3-y^3+xy^2-x^3-x^2y}{xy(y-x)(x+y)} = \frac{xy(y-x)}{xy(y-x)(x+y)} = \frac{1}{x+y}.$$

44. The expression

$$=\frac{(x+y-z)(x-y+z)}{(x+z+y)(x+z-y)}+\frac{(y+z-x)(y-z+x)}{(x+y+z)(x+y-z)}+\frac{(z+x-y)(z-x+y)}{(y+z+x)(y+z-x)}$$

$$=\frac{x+y-z}{x+y+z}+\frac{y+z-x}{x+y+z}+\frac{z+x-y}{x+y+z}=\frac{x+y+z}{x+y+z}=1.$$

45. The expression

$$=\frac{(x+y-2z)(x-y+2z)}{(2z+x+y)(2z+x-y)}+\frac{(y+2z-x)(y-2z+x)}{(x+y+2z)(x+y-2z)}+\frac{(2z+x-y)(2z-x+y)}{(y+2z+x)(y+2z-x)}$$

$$=\frac{x+y-2z}{x+y+2z}+\frac{y+2z-x}{x+y+2z}+\frac{2z+x-y}{x+y+2z}=\frac{x+y+2z}{x+y+2z}=1.$$

46. The expression $=\dfrac{xy-y^2-xz+yz+yz-z^2-xy+xz+xz-x^2-yz+xy}{xz-x^2+xy-y^2+yz-z^2}$

$$=\frac{yz+xz+xy-x^2-y^2-z^2}{xz+xy+yz-x^2-y^2-z^2}=1.$$

47. The expression

$$=\frac{-a+b+c}{(a-b)(c-a)}+\frac{-b+c+a}{(b-c)(a-b)}+\frac{-c+a+b}{(c-a)(b-c)}$$

$$=\frac{(b-c)(-a+b+c)+(c-a)(-b+c+a)+(a-b)(-c+a+b)}{(a-b)(b-c)(c-a)}$$

$$=\frac{-ab+ac+b^2-c^2-bc+ab+c^2-a^2-ac+bc+a^2-b^2}{(a-b)(b-c)(c-a)}=0.$$

48. The expression

$$=-\frac{c+a}{(a-b)(c-a)}-\frac{a+b}{(b-c)(a-b)}-\frac{b+c}{(c-a)(b-c)}$$

$$=\frac{-(c+a)(b-c)-(a+b)(c-a)-(b+c)(a-b)}{(a-b)(b-c)(c-a)}$$

$$=\frac{-bc-ab+c^2+ac-ac-bc+a^2+ab-ab-ac+b^2+bc}{(a-b)(b-c)(c-a)}$$

$$=\frac{a^2+b^2+c^2-bc-ac-ab}{(a-b)(b-c)(c-a)}.$$

49. The expression

$$=\frac{(x+2y-3z)(x-2y+3z)}{(3z+x+2y)(3z+x-2y)}+\frac{(2y+3z-x)(2y-3z+x)}{(x+2y+3z)(x+2y-3z)}+\frac{(3z+x-2y)(3z-x+2y)}{(2y+3z+x)(2y+3z-x)}$$

$$=\frac{x+2y-3z}{x+2y+3z}+\frac{2y+3z-x}{x+2y+3z}+\frac{3z+x-2y}{2y+3z+x}=\frac{x+2y+3z}{x+2y+3z}=1.$$

50. The expression $=\dfrac{(3y+4z-2x)(3y-4z+2x)}{(2x+3y+4z)(2x+3y-4z)}$

$$+\frac{(4z+2x-3y)(4z-2x+3y)}{(3y+4z+2x)(3y+4z-2x)}+\frac{(2x+3y-4z)(2x-3y+4z)}{(4z+2x+3y)(4z+2x-3y)}$$

$$=\frac{3y+4z-2x}{2x+3y+4z}+\frac{4z+2x-3y}{2x+3y+4z}+\frac{2x+3y-4z}{2x+3y+4z}$$

$$=\frac{2x+3y+4z}{2x+3y+4z}=1.$$

51. $$\frac{\dfrac{1}{x}-\dfrac{x+a}{x^2+a^2}}{\dfrac{1}{a}-\dfrac{a+x}{a^2+x^2}}+\frac{\dfrac{1}{x}-\dfrac{x-a}{x^2+a^2}}{\dfrac{1}{a}-\dfrac{a-x}{x^2+a^2}}$$

$$=\left\{\frac{x^2+a^2-x^2-ax}{x(x^2+a^2)}\div\frac{a^2+x^2-a^2-ax}{a(a^2+x^2)}\right\}+\left\{\frac{x^2+a^2-x^2+ax}{x(x^2+a^2)}\div\frac{x^2+a^2-a^2+ax}{a(x^2+a^2)}\right\}$$

$$=\left\{\frac{a^2-ax}{x(x^2+a^2)}\times\frac{a(a^2+x^2)}{x^2-ax}\right\}+\left\{\frac{a^2+ax}{x(x^2+a^2)}\times\frac{a(x^2+a^2)}{x^2+ax}\right\}$$

$$=\left\{\frac{a(a-x)}{x}\times\frac{a}{x(x-a)}\right\}+\left\{\frac{a(a+x)}{x}\times\frac{a}{x(a+x)}\right\}$$

$$=-\frac{a^2}{x^2}+\frac{a^2}{x^2}$$

$$=0.$$

52. The expression

$$=\frac{x^2+ax+bx+ab-y^2-ay-by-ab}{x-y}-\frac{xy-ay-bx+ab-xy+ax+by-ab}{a-b}$$

$$=\frac{x^2-y^2+a(x-y)+b(x-y)}{x-y}-\frac{x(a-b)-y(a-b)}{a-b}$$

$$=(x+y+a+b)-(x-y)$$

$$=2y+a+b.$$

53. The dividend $=\dfrac{(a+x)(a^2+ax+x^2)-(a-x)(a^2-ax+x^2)}{(a^2-ax+x^2)(a^2+ax+x^2)}$

$$=\frac{a^3+2a^2x+2ax^2+x^3-a^3+2a^2x-2ax^2+x^3}{(a^4+a^2x^2+x^4)}$$

$$=\frac{2x(2a^2+x^2)}{(a^4+a^2x^2+x^4)}.$$

The divisor $=\dfrac{(a^2+x^2)(a^3+x^3)-(a^2-x^2)(a^3-x^3)}{(a^3-x^3)(a^3+x^3)}$

$=\dfrac{2a^2x^2(a+x)}{a^6-x^6}$.

$\therefore$ the expression $=\dfrac{2x(2a^2+x^2)}{a^4+a^2x^2+x^4}\div\dfrac{2a^2x^2(a+x)}{a^6-x^6}$

$=\dfrac{2x(2a^2+x^2)}{a^4+a^2x^2+x^4}\times\dfrac{(a^2-x^2)(a^4+a^2x^2+x^4)}{2a^2x^2(a+x)}$

$=\dfrac{(2a^2+x^2)(a-x)}{a^2x}$.

54. The dividend $=\dfrac{(x-1)(x-2)+3(x-1)}{3(x-2)}\div\dfrac{(x+2)(x-3)+4(x+2)}{4(x-3)}$

$=\dfrac{x^2-1}{3(x-2)}\times\dfrac{4(x-3)}{x^2+3x+2}$

$=\dfrac{4(x-1)(x-3)}{3(x-2)(x+2)}$.

The divisor $=\dfrac{(x+3)(x+4)-7(x+3)}{7(x+4)}\div\dfrac{(x-2)(x-1)+3(x-2)}{3(x-1)}$

$=\dfrac{x^2-9}{7(x+4)}\times\dfrac{3(x-1)}{x^2-4}$.

Thus the expression $=\dfrac{4(x-1)(x-3)}{3(x-2)(x+2)}\times\dfrac{7(x+4)(x+2)(x-2)}{3(x-1)(x+3)(x-3)}=\dfrac{28(x+4)}{9(x+3)}$

[*For Examples* 55–58, *see Appendix, p.* 351.]

[*For Miscellaneous Examples IV., see Appendix, p.* 352.]

EXAMPLES. XXIII. a. PAGE 180.

1. Multiplying up, we have

$$(x+4)(3x-7)=(x+5)(3x-8),$$

$$3x^2+5x-28=3x^2+7x-40;$$

whence $x=6$.

2. Multiplying up, we have

$$(3x+1)(x-1)=3(x-2)^2,$$

$$3x^2-2x-1=3x^2-12x+12;$$

whence $x=\dfrac{13}{10}$

3. Multiplying up, we have

$$(7-5x)(1+3x)=(11-15x)(1+x),$$

$$7+16x-15x^2=11-4x-15x^2;$$

whence $$x=\frac{1}{5}.$$

4. Multiplying up, we have

$$3(7+6x)(9+2x)=(35+4x)(2+9x),$$

$$189+204x+36x^2=70+323x+36x^2;$$

whence $$x=1.$$

5. Multiplying by 15, we have

$$6x+13-\frac{3(3x+5)}{x-5}=6x.$$

By transposition, $$13=\frac{3(3x+5)}{x-5}.$$

Multiplying across, $$13x-65=9x+15;$$

$$\therefore x=20.$$

6. Multiplying up, we have

$$(6x+8)(x+12)-(2x+38)(2x+1)=(2x+1)(x+12),$$

$$6x^2+80x+96-(4x^2+78x+38)=2x^2+25x+12;$$

whence $$-23x=-46; \text{ or } x=2.$$

7. Multiplying up, we have

$$6(3x-1)^2-12(2x-1)^2=(2x-1)(3x-1),$$

$$54x^2-36x+6-48x^2+48x-12=6x^2-5x+1;$$

whence $$17x=7; \text{ or } x=\frac{7}{17}.$$

8. Multiplying up, we have

$$(x+25)(2x-15)=(2x+75)(x-5),$$

$$2x^2+35x-375=2x^2+65x-375;$$

whence $$x=0.$$

9. Multiplying up, we have

$$x(x+6)+4(x+2)=(x+2)(x+6),$$

$$x^2+6x+4x+8=x^2+8x+12;$$

whence $$x=2.$$

10. The equation may be written $\dfrac{6x+7}{3(3x+2)} - \dfrac{5x-5}{4(3x+2)} = \dfrac{1}{12}$;

$$\therefore \frac{24x+28-15x+15}{12(3x+2)} = \frac{1}{12}.$$

Multiplying by 12, we have $\dfrac{9x+43}{3x+2} = 1$;

whence $x = -\dfrac{41}{6}$.

11. Multiplying by 10, we have

$$4x - 10 + \frac{10x-30}{2x-15} = 4x - 3 - 11.$$

By transposition, $\dfrac{10x-30}{2x-15} = -4$;

whence $x = 5$.

12. Multiplying by 18, we have

$$8x + 24 = 8x + 37 - \frac{126x-522}{5x-12}.$$

By transposition, $\dfrac{126x-522}{5x-12} = 13$;

whence $x = 6$.

13. We have $(2x-1)(3x+8) = 6x(x+4)$;

that is, $6x^2 + 13x - 8 = 6x^2 + 24x$;

whence $x = -\dfrac{8}{11}$.

14. We have $(2x+5)(5x+2) = (2x+1)(5x+3)$,

$$10x^2 + 29x + 10 = 10x^2 + 11x + 3;$$

whence $x = -\dfrac{7}{18}$.

15. By transposing, $\dfrac{4}{x+3} - \dfrac{5}{2(x+3)} = \dfrac{2}{x+1} - \dfrac{5}{4(x+1)}$.

Simplifying each side separately, we have

$$\frac{3}{2(x+3)} = \frac{3}{4(x+1)}.$$

Thus, $2(x+3) = 4(x+1)$; and $x = 1$.

16. We have $$\frac{7}{x-4}-\frac{60}{5(x-6)}=\frac{10\frac{1}{2}}{3(x-4)}-\frac{8}{x-6}.$$

Transposing and simplifying the fractions, we have

$$\frac{7}{x-4}-\frac{7}{2(x-4)}=\frac{12}{x-6}-\frac{8}{x-6};$$

that is, $$\frac{7}{2(x-4)}=\frac{4}{x-6}.$$

Multiplying across, $$7x-42=8x-32;$$

$$\therefore\ x=-10.$$

17. By transposition, we have

$$\frac{3}{2(2-x)}-\frac{3}{2-x}=-\frac{15}{4(1-x)}+\frac{5}{2(1-x)};$$

that is, $$\frac{-3}{2(2-x)}=\frac{-5}{4(1-x)}.$$

Multiplying across, $$12-12x=20-10x;$$

$$\therefore\ x=-4.$$

18. By transposing, $$\frac{25-\frac{x}{3}}{x+1}-\frac{23}{x+1}=5-\frac{16x+\frac{21}{5}}{3x+2},$$

$$\frac{25-\frac{x}{3}-23}{x+1}=\frac{15x+10-16x-\frac{21}{5}}{3x+2}$$

$$\frac{2-\frac{x}{3}}{x+1}=\frac{\frac{29}{5}-x}{3x+2}.$$

Multiplying across, $$\left(2-\frac{x}{3}\right)(3x+2)=\left(\frac{29}{5}-x\right)(x+1),$$

$$6x-x^2+4-\frac{2}{3}x=\frac{29}{5}x+\frac{29}{5}-x^2-x;$$

whence $$\frac{1}{5}x+\frac{1}{3}x=\frac{9}{5};\quad\text{or}\quad x=\frac{27}{8}.$$

19. By transposing, $$\frac{30+6x}{x+1}-\frac{48}{x+1}=14-\frac{60+8x}{x+3};$$

$$\therefore\ \frac{30+6x-48}{x+1}=\frac{14x+42-60-8x}{x+3},$$

$$\frac{6x-18}{x+1}=\frac{6x-18}{x+3}.$$

As $6x-18$ is a factor of both sides, the equation is satisfied by

$$6x-18=0,\quad\text{or}\quad x=3.$$

20. Simplifying each side separately, we have

$$\frac{x(x-1)-(x+1)(x-2)}{(x-1)(x-2)}=\frac{(x-8)(x-7)-(x-9)(x-6)}{(x-6)(x-7)}.$$

$$\frac{x^2-x-x^2+x+2}{x^2-3x+2}=\frac{x^2-15x+56-x^2+15x-54}{x^2-13x+42}$$

$$\therefore \frac{2}{x^2-3x+2}=\frac{2}{x^2-13x+42};$$

$$\therefore x^2-3x+2=x^2-13x+42;$$

$$\therefore x=4.$$

21. Simplifying each side separately, we have

$$\frac{(x+5)(x-7)-(x-6)(x+4)}{(x+4)(x-7)}=\frac{(x-4)(x-16)-(x-15)(x-5)}{(x-5)(x-16)},$$

$$\frac{x^2-2x-35-x^2+2x+24}{x^2-3x-28}=\frac{x^2-20x+64-x^2+20x-75}{x^2-21x+80}.$$

$$\therefore \frac{-11}{x^2-3x-28}=\frac{-11}{x^2-21x+80};$$

$$\therefore x^2-3x-28=x^2-21x+80;$$

$$\therefore x=6.$$

22. Simplifying each side separately, we have

$$\frac{(x-7)(x-11)-(x-9)^2}{(x-9)(x-11)}=\frac{(x-13)(x-17)-(x-15)^2}{(x-15)(x-17)},$$

$$\frac{x^2-18x+77-x^2+18x-81}{x^2-20x+99}=\frac{x^2-30x+221-x^2+30x-225}{x^2-32x+255};$$

$$\therefore \frac{-4}{x^2-20x+99}=\frac{-4}{x^2-32x+255};$$

whence $x^2-20x+99=x^2-32x+255$; and $x=13$.

23. Simplifying each side separately, we have

$$\frac{(x+3)(x+9)-(x+6)^2}{(x+6)(x+9)}=\frac{(x+2)(x+8)-(x+5)^2}{(x+5)(x+8)},$$

$$\frac{x^2+12x+27-x^2-12x-36}{x^2+15x+54}=\frac{x^2+10x+16-x^2-10x-25}{x^2+13x+40};$$

$$\therefore \frac{-9}{x^2+15x+54}=\frac{-9}{x^2+13x+40};$$

whence $x^2+15x+54=x^2+13x+40$; and $x=-7$.

24. By transposition, $\dfrac{x+2}{x} - \dfrac{x+3}{x+1} = \dfrac{x-6}{x-4} - \dfrac{x-7}{x-5}$.

Simplifying each side separately, we have

$$\frac{x^2+3x+2-x^2-3x}{x^2+x} = \frac{x^2-11x+30-x^2+11x-28}{x^2-9x+20};$$

$$\therefore \frac{2}{x^2+x} = \frac{2}{x^2-9x+20};$$

whence $x^2+x = x^2-9x+20$; and $x=2$.

25. The given equation may be written

$$4-\frac{1}{x-4}+5+\frac{2}{2x-3} = 4-\frac{2}{2x-7}+5+\frac{1}{x-1};$$

that is, $$-\frac{1}{x-4}+\frac{2}{2x-3} = -\frac{2}{2x-7}+\frac{1}{x-1}.$$

Simplifying each side separately, we have

$$\frac{-2x+3+2x-8}{(x-4)(2x-3)} = \frac{-2x+2+2x-7}{(2x-7)(x-1)},$$

$$\frac{-5}{2x^2-11x+12} = \frac{-5}{2x^2-9x+7};$$

whence $2x^2-11x+12 = 2x^2-9x+7$; and $x=2\frac{1}{2}$.

26. The given equation may be written

$$5+\frac{2}{x-2}+6-\frac{2}{x-7}-\left(10+\frac{2}{x-1}\right) = 1-\frac{2}{x-6};$$

$$\therefore \frac{2}{x-2}-\frac{2}{x-7}-\frac{2}{x-1} = -\frac{2}{x-6};$$

that is, $$\frac{1}{x-2}-\frac{1}{x-7} = \frac{1}{x-1}-\frac{1}{x-6}.$$

Simplifying each side separately, we have

$$\frac{-5}{x^2-9x+14} = \frac{-5}{x^2-7x+6};$$

whence $x^2-9x+14 = x^2-7x+6$; and $x=4$.

27. The equation may be written

$$\frac{20x-30}{3x-4} = \frac{40x-60}{6x-7},$$

or $$\frac{2x-3}{3x-4} = \frac{4x-6}{6x-7}.$$

Multiplying across, we have

$$12x^2-32x+21 = 12x^2-34x+24;$$

whence $x = 1\frac{1}{2}$.

28. Since $\quad 0\dot{5}=\frac{5}{90}=\frac{1}{18}$, and $\cdot 0625=\frac{625}{10000}=\frac{1}{16}$,

the equation may be written

$$18(x-2)-16(x-4)=56;$$

whence $\quad x=14.$

29. Changing the recurring decimals into fractions, we have

$$\frac{1}{12}(x-\cdot 625)=\frac{1}{11}(x-\cdot 59375),$$

$$11x-6\cdot 875=12x-7\cdot 125;$$

$$\therefore\ x=7\cdot 125-6\cdot 875=\cdot 25=\frac{1}{4}.$$

30. Changing the decimals into fractions, we have

$$\left(2x+\frac{3}{2}\right)\left(3x-\frac{9}{4}\right)=\left(2x-\frac{9}{8}\right)\left(3x+\frac{5}{4}\right),$$

$$6x^2+\frac{9}{2}x-\frac{9}{2}x-\frac{27}{8}=6x^2-\frac{27}{8}x+\frac{5}{2}x-\frac{45}{32};$$

whence $\quad -\frac{27}{8}=-\frac{7}{8}x-\frac{45}{32};$

that is, $\quad 108=28x+45, \text{ or } x=\frac{9}{4}.$

31. On multiplying each numerator and denominator by 10, we have

$$\frac{3x-10}{5x-4}=\frac{5+12x}{20x-1};$$

that is, $\quad (3x-10)(20x-1)=(12x+5)(5x-4),$

$$60x^2-203x+10=60x^2-23x-20; \text{ whence } x=\frac{1}{6}.$$

32. On multiplying each numerator and denominator by 10, we have

$$\frac{10-14x}{2+10x}=\frac{7(x-1)}{1-5x};$$

that is, $\quad (10-14x)(1-5x)=7(x-1)(10x+2),$

$$10-64x+70x^2=70x^2-56x-14; \text{ whence } x=3.$$

33. Multiplying through by 10, the equation may be written

$$\frac{(3x-20)(3x-10)}{2x-10}-\frac{1}{6}(3x-20)=4x-20.$$

Multiplying by $6(x-5)$, we have

$$3(3x-20)(3x-10)-(x-5)(3x-20)=6(4x-20)(x-5),$$
$$27x^2-270x+600-3x^2+35x-100=24x^2-240x+600;$$
$$\therefore\ x=20.$$

EXAMPLES. XXIII. b. PAGE 182.

2. Clearing of brackets, we have

$$a^2x-a^3+b^2x-b^3=abx,$$
$$(a^2-ab+b^2)\,x=a^3+b^3;$$
$$\therefore\ x=a+b.$$

4. Clearing of brackets, we have

$$x^2-ax+bx-ab=x^2+a^2+b^2-2ax+2bx-2ab$$
$$ax-bx=a^2-ab+b^2;$$
$$\therefore\ x=\frac{a^2-ab+b^2}{a-b}$$

5. We have
$$ax-2a+2x=6+a,$$
$$(a+2)\,x=6+3a;$$
$$\therefore\ a=3.$$

6. Here
$$m^3-m^2x-mnx=n^3+n^2x,$$
$$(m^2+mn+n^2)\,x=m^3-n^3;$$
$$\therefore\ x=m-n.$$

7. Clearing of brackets, we have

$$ab+ax+bx+x^2=x^2-cx;$$

whence
$$x=-\frac{ab}{a+b+c}.$$

8. On multiplying out, we have

$$ax-bx-a^2+ab=ax-cx-ab+bc;$$
$$\therefore\ (c-b)\,x=a^2-2ab+bc;$$
$$\therefore\ x=\frac{a^2-2ab+bc}{c-b}$$

9. We have
$$(2x+3a)(3x+a)=2(3x+2a)(x+a),$$
$$6x^2+11ax+3a^2=6x^2+10ax+4a^2;$$
$$\therefore\ ax=a^2;\ \text{and}\ x=a.$$

10. We have $6(x-b)(x-c)=(2x+b)(3x-c),$

$$6x^2-6bx-6cx+6bc=6x^2+3bx-2cx-bc;$$

whence $-9bx-4cx=-7bc;$ and $x=\frac{7bc}{9b+4c}.$

11. On transposing, we have

$$\frac{2}{x}=\frac{1}{a}+\frac{1}{b}=\frac{a+b}{ab};$$

$$\therefore x=\frac{2ab}{a+b}.$$

12. On multiplying each side by 12, we have

$$\frac{8x}{a}+8=\frac{9x}{a}-9;$$

whence $\frac{x}{a}=17,$ or $x=17a.$

13. On transposition, we have

$$\frac{a-b}{x}=c(a-b);$$

$$\therefore \frac{1}{x}=c; \text{ and } x=\frac{1}{c}.$$

14. Clearing of fractions, we have

$$9a^2-3ax=4b^2-2bx,$$

$$(2b-3a)x=4b^2-9a^2;$$

$$\therefore x=2b+3a.$$

15. We have $\frac{x-a}{b-x}=\frac{x-b}{a-x}=\frac{b-x}{x-a}$

$$(x-a)^2=(b-x)^2,$$

$$x^2-2ax+a^2=b^2-2bx+x^2,$$

$$2(b-a)x=b^2-a^2;$$

$$\therefore x=\frac{a+b}{2}.$$

16. We have $(x-a)(2x-a)=2(x-b)^2,$

$$2x^2-3ax+a^2=2x^2-4bx+2b^2,$$

$$(3a-4b)x=a^2-2b^2;$$

$$\therefore x=\frac{a^2-2b^2}{3a-4b}$$

17. On multiplication throughout by 12, we obtain

$$3x(x-a)-3(x+a)^2=8a\left(x-\frac{a}{2}\right),$$

$$3x^2-3ax-3x^2-6ax-3a^2=8ax-4a^2;$$

whence $$-17ax=-a^2; \text{ and } x=\frac{a}{17}.$$

18. On removal of brackets, we have

$$ax^2+bx^2-abx-a^3=bx^2-abx+ax^2-abx;$$

whence $$abx=a^3; \text{ and } x=\frac{a^2}{b}.$$

19. By removing brackets, we obtain

$$ab+bx-ab-bx+ax+x^2=x^2+\frac{bc^2}{a};$$

that is, $$ax=\frac{bc^2}{a}; \text{ and } x=\frac{bc^2}{a^2}.$$

20. On multiplying each term by b, we have

$$b^2(a-x)-a(b^2+2bx+x^2)+ab^2\left(\frac{x^2}{b^2}+\frac{2x}{b}+1\right)=0,$$

$$ab^2-b^2x-ab^2-2abx-ax^2+ax^2+2abx+ab^2=0;$$

whence $$ab^2-b^2x=0; \text{ and } x=a.$$

21. On removal of brackets, we obtain

$$x^2+2a^2-ax-\frac{3b^2}{4}=x^2-bx+\frac{b^2}{4}+a^2;$$

$$\therefore (b-a)x=b^2-a^2:$$

$$\therefore x=a+b.$$

22. Multiplying out, we have

$$2x^2-ax+\frac{4ax}{3}-\frac{2a^2}{3}=\frac{4ax}{3}-4x^2-a^2+4ax-\frac{3ax}{2}+6x^2;$$

that is, $$5ax-\frac{3ax}{2}=a^2-\frac{2a^2}{3};$$

whence $$\frac{7ax}{2}=\frac{a^2}{3}, \text{ or } x=\frac{2a}{21}.$$

[*For Examples 23-25, see Appendix, p. 362.*]

EXAMPLES. XXIII. c. PAGE 171.

$$ax + by = l \quad\text{.......}\quad (1),$$
$$bx + ay = m \quad\text{.......}\quad (2).$$

Multiply (1) by b, $\quad abx + b^2y = bl,$

" (2) by a, $\quad abx + a^2y = am.$

Subtract, $\quad (b^2 - a^2)\,y = bl - am;$

$$\therefore\ y = \frac{am - bl}{a^2 - b^2}.$$

In like manner we obtain $\quad x = \dfrac{al - bm}{a^2 - b^2}.$

2. $$lx + my = n \quad\text{.......}\quad (1),$$
$$px + qy = r \quad\text{.......}\quad (2).$$

Multiply (1) by q, $\quad lqx + mqy = nq,$

" (2) by m, $\quad mpx + mqy = mr.$

Subtract, $\quad (lq - mp)\,x = nq - mr;$

$$\therefore\ x = \frac{nq - mr}{lq - mp}.$$

In like manner we obtain $\quad y = \dfrac{lr - np}{lq - mp}.$

3. $$ax - by = 0 \quad\text{.......}\quad (1),$$
$$bx + ay = c \quad\text{.......}\quad (2).$$

Substitute $y = \dfrac{a}{b}x$ in (2),

thus $$bx + \frac{a^2x}{b} = c,$$

or $$a^2x + b^2x = bc;$$

$$\therefore\ x = \frac{bc}{a^2 + b^2},$$

and $$y = \frac{a}{b}x = \frac{ac}{a^2 + b^2}.$$

4. $$ax + by = a^2 \quad\text{.......}\quad (1),$$
$$bx + ay = b^2 \quad\text{.......}\quad (2).$$

Multiply (1) by a, $\quad a^2x + aby = a^3,$

" (2) by b, $\quad b^2x + aby = b^3.$

Subtract, $\quad (a^2 - b^2)\,x = a^3 - b^3;$

$$\therefore\ x = \frac{a^3 - b^3}{a^2 - b^2} = \frac{a^2 + ab + b^2}{a + b}.$$

In like manner we obtain $\quad y = -\dfrac{ab}{a + b}.$

5. $$x + ay = a' \quad\ldots\ldots\ldots\ldots(1),$$
$$ax + a'y = 1 \quad\ldots\ldots\ldots\ldots(2).$$

Multiply (1) by a, $$ax + a^2y = aa';$$
by subtraction, $$(a^2 - a')\,y = aa' - 1;$$
$$\therefore\ y = \frac{aa' - 1}{a^2 - a'}.$$

Also $$x = a' - ay = a' - \frac{a(aa' - 1)}{a^2 - a'} = \frac{a - a'^2}{a^2 - a'}.$$

6. $$px - qy = r \quad\ldots\ldots\ldots\ldots(1),$$
$$rx - py = q \quad\ldots\ldots\ldots\ldots(2).$$

Multiply (1) by p, $$p^2x - pqy = pr,$$
,, (2) by q, $$qrx - pqy = q^2.$$
Subtract, $$(p^2 - qr)\,x = pr - q^2;$$
$$\therefore\ x = \frac{pr - q^2}{p^2 - qr}.$$

In like manner we obtain $$y = \frac{r^2 - pq}{p^2 - qr}.$$

7. $$\frac{x}{a} + \frac{y}{b} = \frac{1}{ab}, \text{ or } bx + ay = 1 \quad\ldots\ldots\ldots\ldots(1),$$
$$\frac{x}{a'} - \frac{y}{b'} = \frac{1}{a'b'}, \text{ or } b'x - a'y = 1 \quad\ldots\ldots\ldots\ldots(2).$$

Multiply (1) by a', $$a'bx + aa'y = a',$$
,, (2) by a, $$ab'x - aa'y = a.$$
Add, $$(a'b + ab')\,x = a' + a;$$
$$\therefore\ x = \frac{a' + a}{a'b + ab'}.$$

In like manner we obtain $$y = \frac{b' - b}{ab' + a'b}.$$

8. From the first equation $bx = ay$; on substituting in the second equation we have
$$2bx = 4ab, \text{ or } x = 2a.$$

Also $$y = \frac{b}{a}x = 2b.$$

9. From the first equation, $\frac{9x}{a}+\frac{6y}{b}=9$;

but $$\frac{9x}{a}-\frac{6y}{b}=3.$$

$$\therefore\ \frac{12y}{b}=6;\ \text{and}\ y=\frac{b}{2}.$$

Substituting in the first equation, we have

$$\frac{3x}{a}+1=3;\ \text{whence}\ x=\frac{2a}{3}.$$

10. $$qx-rb=p(a-y),\ \text{or}\ qx+py=pa+rb\ \ldots\ldots\ldots\ (1),$$
$$\frac{qx}{a}+r=p\left(1+\frac{y}{b}\right),\ \text{or}\ bqx-pay=pab-rab\ \ldots\ldots\ (2).$$

Multiply (1) by b, $\quad bqx+pby=pab+rb^2$.

On subtracting (2), $\quad (pb+pa)y=r(b^2+ab)$;

$$y=\frac{rb(a+b)}{p(a+b)}=\frac{rb}{p}.$$

Hence from (1), $\quad qx+rb=pa+rb$; whence $x=\frac{pa}{q}$.

11. $$\frac{x}{m}+\frac{y}{m'}=1,\ \text{or}\ m'x+my=mm'\ \ldots\ldots\ldots\ (1),$$
$$\frac{x}{m'}-\frac{y}{m}=1,\ \text{or}\ mx-m'y=mm'\ \ldots\ldots\ldots\ (2).$$

Multiply (1) by m', $\quad m'^2x+mm'y=mm'^2$,

„ (2) by m, $\quad m^2x-mm'y=m^2m'$.

Adding, $\quad (m'^2+m^2)x=mm'^2+m^2m'$;

$$\therefore\ x=\frac{mm'(m'+m)}{m'^2+m^2}.$$

In like manner we obtain $$y=\frac{mm'(m-m')}{m^2+m'^2}.$$

12. Substitute $y=-\frac{p}{q}x$ in the second equation; thus

$$lx-\frac{mp}{q}x=n,$$

or $$(lq-mp)x=nq;$$

$$\therefore\ x=\frac{nq}{lq-mp};$$

and $$y=-\frac{p}{q}x=\frac{np}{mp-lq}$$

13. Multiplying the second equation by $a+b$, we have

$$(a-b)x-(a+b)y=0 \text{..........................} (1),$$
$$(a+b)x+(a+b)y=c(a+b) \text{..................} (2);$$

$\therefore$ by addition,
$$2ax=c(a+b);$$
$$\therefore x=\frac{c(a+b)}{2a}.$$

Substituting in (1) we have $y=\dfrac{c(a-b)}{2a}.$

14.
$$(a-b)x+(a+b)y=2(a-b)(a+b) \text{..................} (1),$$
$$(a+b)x-(a-b)y=4ab \text{..................................} (2).$$

Multiply (1) by $a-b$, thus $(a-b)^2x+(a^2-b^2)y=2(a-b)^2(a+b)$,

" (2) by $a+b$, thus $(a+b)^2x-(a^2-b^2)y=4ab(a+b)$.

Add,
$$2(a^2+b^2)x=2(a^2+b^2)(a+b);$$
$$\therefore x=a+b.$$

Substituting in (2), we obtain $y=a-b$.

15.
$$\frac{x}{a}+\frac{y}{b}=1 \text{.................................} (1),$$
$$\frac{x}{3a}+\frac{y}{6b}=\frac{2}{3} \text{.................................} (2).$$

Multiply (2) by 6 and subtract (1); thus

$$\frac{x}{a}=3; \text{ whence } x=3a.$$

Substituting in (1), we obtain $y=-2b$.

16. Substitute $y=\dfrac{b'}{a'}x$ in the first equation;

thus
$$x\left(\frac{1}{a}+\frac{b'}{ba'}\right)=2; \text{ whence } x=\frac{2aa'b}{a'b+ab'}.$$

Also
$$y=\frac{b'}{a'}x=\frac{2abb'}{a'b+ab'}.$$

17.
$$\frac{x}{a}-\frac{y}{b}=1, \text{ or } bx-ay=ab \text{........................} (1),$$
$$\frac{x}{b}+\frac{y}{a}=\frac{a}{b}, \text{ or } ax+by=a^2 \text{........................} (2).$$

Multiply (1) by b, $b^2x-aby=ab^2$,

" (2) by a, $a^2x+aby=a^3$.

Add, $(a^2+b^2)x=ab^2+a^3$;
$$\therefore x=a.$$

Substituting in (1) we obtain $y=0$.

18. From $$\frac{m}{l}x + \frac{l}{m}y = \left(\frac{1}{l} + \frac{1}{m}\right)(m^2 + l^2),$$

we obtain $$m^2x + l^2y = (m + l)(m^2 + l^2).\quad\text{.....................(1)}.$$

From $$(x + y)(m^2 + l^2) = 2(m^3 + l^3) + ml(x + y),$$

we have $$(m^2 - ml + l^2)(x + y) = 2(m^3 + l^3);$$

that is, $$x + y = 2(m + l).\quad\text{.....................(2)}.$$

Multiply (2) by l^2 and subtract from (1); thus

$$(m^2 - l^2)x = (m + l)(m^2 - l^2);$$

$$\therefore\ x = m + l.$$

Substituting in (2), we obtain $y = m + l$.

19. From $$ax\left(\frac{1}{a - b} - \frac{1}{a + b}\right) + cy\left(\frac{1}{b - a} - \frac{1}{b + a}\right) = \frac{2a}{a + b},$$

we obtain $$ax \cdot \frac{2b}{a^2 - b^2} + cy \cdot \frac{2a}{b^2 - a^2} = \frac{2a}{a + b}.$$

Clearing of fractions, $2abx - 2acy = 2a(a - b)$,

or $$bx - cy = a - b.$$

By combination with $bx + cy = a + b$, we have $bx = a$, and $cy = b$.

20. $$(a - b)x + (a + b)y = 2(a^2 - b^2)\quad\text{.....................(1)}.$$

$$ax - by = a^2 + b^2\quad\text{.....................(2)}.$$

Multiply (1) by b, and (2) by $a + b$; thus

$$(ab - b^2)x + b(a + b)y = 2(a^2b - b^3),$$

$$(a^2 + ab)x - b(a + b)y = a^3 + a^2b + ab^2 + b^3;$$

by addition, $$(a^2 + 2ab - b^2)x = a^3 + 3a^2b + ab^2 - b^3;$$

$$\therefore\ x = \frac{a^3 + 3a^2b + ab^2 - b^3}{a^2 + 2ab - b^2} = a + b.$$

Substituting in (2), we obtain $y = a - b$.

EXAMPLES. XXIV. PAGE 188.

1. Let x = the number of persons; then

each receives £ $\frac{10}{x} = \frac{200}{x}$ shillings.

If the number had been $\frac{5x}{4}$, each would have received $\frac{160}{x}$ shillings;

$$\therefore \frac{200}{x} - \frac{160}{x} = 1;$$

whence $x = 40$.

2. Let x = the number of eggs I bought, for which I paid $\frac{x}{4}$ pence.

I then sold $\frac{4x}{5}$ eggs at 3 a penny receiving $\frac{4x}{15}$ pence;

$$\therefore \frac{4x}{15} - \frac{x}{4} = 1;$$

whence $x = 60$.

3. Let x = the number bought, for which I paid $\frac{6x}{5}$ pence;

at the second price I should have paid $\frac{12x}{11}$.

$$\therefore \frac{6x}{5} - \frac{12x}{11} = 6;$$

whence $x = 55$.

4. Let x shillings be the amount he begins with.

He wins $2x$, then loses 16s., leaving $(3x - 16)$ shillings; he then loses $\frac{4}{5}$ths of this, so that he has left $\left(\frac{3x-16}{5}\right)$ shillings. He now wins x shillings;

$$\therefore \frac{3x - 16}{5} + x = 80;$$

whence $x = 52$.

Thus he began with 52s., or £2. 12s.

5. Let the price of silk be x shillings per yard; then the price of calico is $\frac{x}{12}$ shillings a yard.

And $\left(20 \times \frac{x}{12} + 30x\right)$ shillings = £14. 5s.,

or $$\frac{5x}{3} + 30x = 285;$$

whence $x = 9$.

That is, the price of silk is 9s. a yard, and the price of calico is 9d. per yard.

6. Let x be the digit in the tens' place, y the digit in the units' place. Then the number is $10x+y$. Thus we have $10x+y-5(x+y)=9$.

Also $x-y=1$. Whence $x=5$, and $y=4$; thus the number is 54.

7. Let x be the digit in the tens' place, y the digit in the units' place. Then the number is $10x+y$, and the number with digits reversed is $10y+x$.

Hence $$x+y=6, \text{ and } 10x+y-(10y+x)=18;$$

whence $x=4$, and $y=2$; thus the number is 42.

8. Let $x=$ the man's age, $y=$ his wife's age.

Then $$x-2=2y,$$

and $$y-3=\frac{1}{3}(x+12).$$

Whence we obtain $x=48$, $y=23$.

9. Let x denote the required number of minutes after 1 o'clock. Then the hour hand moves $\frac{x}{12}$ minute divisions whilst the minute hand moves x minutes.

At 1 o'clock the minute hand is 5 minute divisions behind the hour hand, and to be at right angles it has to be 15 in front; that is it has to gain 20.

$$\therefore x-\frac{x}{12}=20;$$

$$\therefore x=21\tfrac{9}{11}.$$

Thus the time is $21\frac{9}{11}$ minutes past 1.

10. Let x denote the required number of minutes after 3 o'clock.

At 3 o'clock the minute hand is 15 minute divisions behind the hour hand; therefore to be 1 in front it has to gain 16.

$$\therefore x-\frac{x}{12}=16;$$

$$\therefore x=17\tfrac{5}{11}.$$

Thus the time is $17\frac{5}{11}$ min. past 3.

11. Let x denote the required number of minutes after 6 o'clock.

At 6 o'clock the minute hand is 30 minute divisions behind the hour hand.

$$\therefore x-\frac{x}{12}=30;$$

$$\therefore x=\frac{360}{11}=32\tfrac{8}{11}.$$

Thus the time is $32\frac{8}{11}$ min. past 6.

12. Let x denote the required number of minutes after 2 o'clock; then the hour hand stands at $10+\frac{x}{12}$ minutes; after 10 minutes the minute hand stands at $x+10$ minutes, the hour hand at $10+\frac{x}{12}+\frac{10}{12}$ minutes;

$$\therefore \left(10+\frac{x}{12}\right)-x=(x+10)-\left(10+\frac{x}{12}+\frac{10}{12}\right);$$

whence $$x=\frac{65}{11}=5\tfrac{10}{11}.$$

Thus the time is $5\frac{10}{11}$ min. past 2.

13. Let x = the number of voters in the majority,

then $x-162=$ .... minority,

and $2x-162=$ the whole number of voters.

Thus $$162=\frac{3}{11}(2x-162);$$

whence $$x=378.$$

That is, the voters were 378 and 216.

14. Let x = the number who paid the bill, and let y shillings be the amount each paid; then the bill $=xy$ shillings.

If there had been 10 more they would have paid 2*s.* less,

or the bill $$=(x+10)(y-2)=xy-2x+10y-20.$$

If there had been 5 less they would have paid $2\frac{1}{2}$*s.* more,

or the bill $$=(x-5)(y+2\tfrac{1}{2})=xy-5y+\frac{5}{2}x-\frac{25}{2}.$$

$$\therefore xy-2x+10y-20=xy,$$

and $$xy-5y+\frac{5}{2}x-\frac{25}{2}=xy;$$

that is, $$-2x+10y=20,$$

and $$5x-10y=25;$$

whence we obtain $$x=15,\ y=5;$$

thus there were 15 persons who paid 5*s.* each.

15. Let x = the number of yards he bought at 4*s.* 6*d.* a yd.,

$y=$ .... 4*s.* ......

He buys x yds. of the first and y yds. of the second for 100 shillings and sells the same for $100s.+\frac{2}{100}\times 100s.$ or $102s.$

Thus $$\frac{9}{2}x+4y=100,$$

and $$\frac{17}{4}(x+y)=102;$$

whence $$x=8,\ y=16.$$

That is, he buys 8 yds. at 4s. 6d. a yard, and 16 yds. at 4s. a yd.

16. Let x = the age of father, and y = the age of eldest son; then the age of younger son $=y-2$.

Ten years ago their ages were $x-10$, $y-10$, $y-12$ respectively.

$$\therefore\ y-10+y-12=\frac{1}{3}(x-10),$$

and $$y+y-2=x-14;$$

that is, $$-x+6y=56,$$

$$-x+2y=-12;$$

whence we obtain $$x=46, \text{ and } y=17.$$

Their ages are therefore 46, 17 and 15.

17. Suppose B begins by walking at the rate of x miles an hour. A walking at the rate of 5 miles an hour takes 3 hours to walk 15 miles and in the same time B walks $3x$ miles. Again, B walking $2x$ miles an hour, in 6 hours walks $12x$ miles, whilst A walks 30 miles. Now, the total number of miles walked by A and B being equal,

$$3x+12x=15+30;$$

$$\therefore\ x=3.$$

18. Let x = the number of oranges in the basket.

The first person takes $\frac{x}{2}+1$, leaving $\left(\frac{x}{2}-1\right)$;

the second person takes $\frac{1}{2}\left(\frac{x}{2}-1\right)+1$ leaving $\frac{1}{2}\left(\frac{x}{2}-1\right)-1$;

the third person takes $\frac{1}{2}\left\{\frac{1}{2}\left(\frac{x}{2}-1\right)-1\right\}+6$.

$$\therefore\ \frac{x}{2}+1+\frac{1}{2}\left(\frac{x}{2}-1\right)+1+\frac{1}{2}\left\{\frac{1}{2}\left(\frac{x}{2}-1\right)-1\right\}+6=x;$$

whence $$x=54.$$

19. Let x miles an hour be his rate of swimming.

He goes up stream at the rate of $x-\frac{3}{2}$ miles an hour,

and down stream at the rate of $x+\frac{3}{2}$ miles an hour;

He takes $\dfrac{1}{x-\dfrac{3}{2}}$ hours to swim a mile up stream, and $\dfrac{1}{x+\dfrac{3}{2}}$ hours to swim a mile down stream.

Thus
$$\frac{4}{x+\dfrac{3}{2}}=\frac{1}{x-\dfrac{3}{2}};$$

whence
$$4x-6=x+\frac{3}{2}; \text{ and } x=\frac{5}{2}.$$

That is, he swims at the rate of $2\frac{1}{2}$ miles an hour.

20. Let x denote the required number of minutes past 7 o'clock.

At 7 o'clock the minute hand is 35 minutes behind the hour hand, and to be at right angles the first time it has to gain 20 minutes, and to be at right angles the second time it has to gain 50 minutes.

In the first case $x-\dfrac{x}{12}=20$; whence $x=21\frac{9}{11}$.

And in the second case $x-\dfrac{x}{12}=50$; whence $x=54\frac{6}{11}$.

Thus the hands will be at right angles at $21\frac{9}{11}$ and $54\frac{6}{11}$ minutes past 7.

Again for the hands to be in the same straight line, the minute hand has to gain 5 minutes.

$$\therefore x-\frac{x}{12}=15; \text{ whence } x=5\tfrac{5}{11}.$$

Thus the hands are in the same straight line at $5\frac{5}{11}$ minutes past 7.

21. Let x be the numerator of the fraction; then $x+4=$ the denominator, and the fraction $=\dfrac{x}{x+4}$.

The new fraction is $\dfrac{x-5}{x+4-5}$ or $\dfrac{x-5}{x-1}$.

Hence
$$\frac{x-1}{x-5}+\frac{4x}{x+4}=5.$$

By clearing of fractions, we obtain

$$x^2+3x-4+4x^2-20x=5x^2-5x-100;$$

whence
$$x=8.$$

Thus the fraction is $\dfrac{8}{12}$.

22. Suppose the man walking at the rate of 4 miles an hour to meet the other man x miles from the town he started from.

The first man walks x miles in $\frac{x}{4}$ hours and stops $2\frac{1}{2}$ hours,

the second man walks $60 - x$ miles in $\frac{60-x}{3}$ hours.

Hence $$\frac{x}{4} + 2\tfrac{1}{2} = \frac{60-x}{3};$$

$$\therefore x = 30;$$

that is, they meet half way.

Again the time taken $= \left(\frac{x}{4} + 2\frac{1}{2}\right)$ hours $= 10$ hours; that is, they meet at 10 P. M.

23. Let C start x hours after B, and let A, B, and C meet at a distance y miles from the starting point.

The times taken by A, B, and C from the time when A starts are $\frac{y}{4}$, $\frac{y}{5} + 2$, $\frac{y}{6} + 2 + x$ hours respectively; and these are equal.

$$\therefore \frac{y}{4} = \frac{y}{5} + 2; \text{ whence } y = 40.$$

And $$\frac{y}{4} = \frac{y}{6} + 2 + x; \text{ whence } x = 1\tfrac{1}{3}.$$

$\therefore$ C starts $1\frac{1}{3}$ hours after B.

24. Let £x be the price of the horse.

He expected to sell it at 10 p. c. profit, that is, at £ $\frac{110}{100}x$, but he sold it at 15 p. c. loss; that is, at £ $\frac{85}{100}x$.

$$\therefore \frac{110}{100}x - 50 = \frac{85}{100}x:$$

whence $$x = 200.$$

Thus the price was £200.

25. Let x miles be the distance he walks, y miles the distance he goes by coach, so that the distance from A to $B = x + y$ miles.

The times taken by the man and the coach to go x miles are $\frac{x}{4}$ and $\frac{x}{12}$ respectively.

$$\therefore \frac{x}{4} = \frac{x}{12} + 1; \text{ whence } x = 6.$$

Again the time taken by the coach to go y miles is $\frac{y}{12}$.

Hence $$\frac{y}{12} = 2, \text{ or } y = 24.$$

Thus the total distance $= 6 + 24 = 30$ miles.

26. Let £x be the man's property.

His income is made from £$\frac{x}{12}$ at 2 p. c. yielding £$\frac{2}{100} \times \frac{x}{12}$,

£$\frac{x}{2}$ at 3 p. c............ £$\frac{3}{100} \times \frac{x}{2}$,

£$\frac{x}{3}$ at $4\frac{1}{2}$ p. c............ £$\frac{4\frac{1}{2}}{100} \times \frac{x}{3}$,

the rest yielding nothing.

$$\therefore \left(\frac{2}{100} \times \frac{x}{12}\right) + \left(\frac{3}{100} \times \frac{x}{2}\right) + \left(\frac{9}{200} \times \frac{x}{3}\right) = 1140,$$

$$\frac{x}{600} + \frac{3x}{200} + \frac{3x}{200} = 1140;$$

whence $x = 36000.$

Thus his property is £36000.

27. Let £x be the amount of his debts, and £y the amount of his income.

Now 5 p. c. of his income $= 7\frac{1}{2}$ p. c. of his debts.

$$\therefore \frac{5y}{100} = \frac{7\frac{1}{2}x}{100}, \text{ or } 2y = 3x \text{ ....(1).}$$

Again, $\frac{1}{3}y + \frac{1}{4}y + \frac{5y}{100} + 110 = y;$

whence $y = 300$; and $x = \frac{2}{3}y = 200.$

Thus the amount of his debts $=$ £200.

28. Let x gallons be drawn from the first, and y gallons from the second vessel.

Then $x + y = 7$ ....(1).

From the first vessel $\frac{3x}{4}$ gallons of wine, and $\frac{x}{4}$ gallons of water are drawn;

.... second ... $\frac{y}{6}$ .... $\frac{5y}{6}$ ....

The quantities of wine and water drawn are equal.

$$\therefore \frac{3x}{4} + \frac{y}{6} = \frac{x}{4} + \frac{5y}{6}, \text{ or } 3x = 4y.$$

By combination with (1), we obtain $x = 4$, $y = 3$.

Thus 4 gallons are drawn from the first, 3 gallons from the second vessel.

29. Let x pints be taken from the first, y pints from the second mixture. Then $x+y=1$ .... (1).

From the first vessel $\frac{x}{3}$ pints of wine and $\frac{2x}{3}$ pints of water are taken;

.... second ...... $\frac{3y}{4}$ .... $\frac{y}{4}$ ....

The quantities of wine and water drawn are equal.

$$\therefore \frac{x}{3}+\frac{3y}{4}=\frac{2x}{3}+\frac{y}{4}, \text{ or } 2x=3y.$$

By combination with (1), we obtain $x=\frac{3}{5}$, $y=\frac{2}{5}$.

Thus $\frac{3}{5}$ of a pint is drawn from the first, $\frac{2}{5}$ of a pint from the second vessel.

30. Suppose them to meet x miles from A.

One man has walked x miles at the rate of p miles an hour, thus taking $\frac{x}{p}$ hours; whilst the other man has walked $a-x$ miles at the rate of q miles an hour, thus taking $\frac{a-x}{q}$ hours.

These times must be equal;

$$\therefore \frac{x}{p}=\frac{a-x}{q}; \text{ whence } x=\frac{ap}{p+q}.$$

31. Let x miles be the length of the North Western line,

then $x+15$ miles is .... Great Western line.

The North Western train travels x miles in 3 hours, or $\frac{x}{3}$ miles an hour; the Great Western train travels $x+15$ miles in $3\frac{1}{2}$ hours, or $\frac{x+15}{3\frac{1}{2}}$ miles an hour.

Thus $$\frac{x}{3}-\frac{x+15}{3\frac{1}{2}}=1; \text{ whence } x=111.$$

That is, the North Western route is 111 miles long, the Great Western 126 miles.

32. Let x lbs. of coffee be mixed with each lb. of chicory; then the cost of $(x+1)$ lbs. of the mixture is $(12x+3)$ pence.

To gain 10 p. c. the mixture must be sold for $\frac{11}{10}(12x+3)$ pence.

But the sale price is also $11(x+1)$ pence.

$$\therefore \frac{11}{10}(12x+3)=11(x+1);$$

$$\text{whence } x=\frac{7}{2}.$$

That is, the proportion of coffee to chicory is 7 to 2.

33. Suppose that x lbs. at a pence per lb. are mixed with y lbs. at b pence per lb.

These $(x+y)$ lbs. cost $(ax+by)$ pence.

$$\therefore\ ax+by=(a-b)c.$$

Also $$x+y=a-b.$$

Whence we obtain $$x=c-b,\ y=a-c.$$

34. Let $x=$ the number of yards bought at a shillings a yard,

$y=$ . ... b ... .;

then $$ax+by=\frac{5c}{2}.$$

Also $$3ax+\tfrac{1}{2}by=\frac{5c}{2}.$$

Whence we obtain $$x=\frac{c}{2a},\ y=\frac{2c}{b}.$$

35. Let x miles be the distance from A to B.

He rides $\frac{x}{3}$ miles at the rate of a miles an hour, which takes $\frac{x}{3a}$ hours;

.... $\frac{2x}{3}$ miles .... $2b$ .... $\frac{x}{3b}$ hours.

If he had ridden the whole $2x$ miles at the rate of $3c$ miles an hour, it would have taken him $\frac{2x}{3c}$ hours.

Hence $$\frac{x}{3a}+\frac{x}{3b}=\frac{2x}{3c};$$

that is, $$\frac{1}{a}+\frac{1}{b}=\frac{2}{c}.$$

36. Let the distances between A and B, B and C, C and A, be x, y, z miles respectively, so that the circuit is $x+y+z$ miles.

If he starts from A he walks x miles in ax minutes, rides y miles in by minutes, and drives z miles in cz minutes.

$$\therefore\ ax+by+cz=60\,(c+b-a).$$

Similarly, $$cx+ay+bz=60\,(a+c-b),$$

$$bx+cy+az=60\,(a+b-c).$$

By adding these equations we obtain

$$(x+y+z)(a+b+c)=60\,(a+b+c);$$

$$\therefore\ x+y+z=60.$$

Thus the circuit is 60 miles.

EXAMPLES. XXV. a. PAGE 194.

19. $x^2 - \frac{2}{3}x = 32;$

$x^2 - \frac{2}{3}x + \left(\frac{1}{3}\right)^2 = 32 + \frac{1}{9} = \frac{289}{9};$

that is, $x - \frac{1}{3} = \pm\frac{17}{3}.$

20. $x^2 + \frac{4}{15}x = \frac{1}{5};$

$x^2 + \frac{4}{15}x + \left(\frac{2}{15}\right)^2 = \frac{1}{5} + \frac{4}{225} = \frac{49}{225};$

that is, $x + \frac{2}{15} = \pm\frac{7}{15}.$

21. $x^2 - \frac{7}{6}x = \frac{1}{2};$

$x^2 - \frac{7}{6}x + \left(\frac{7}{12}\right)^2 = \frac{1}{2} + \frac{49}{144} = \frac{121}{144};$

that is, $x - \frac{7}{12} = \pm\frac{11}{12}.$

22. $x^2 + \frac{19}{5}x = \frac{4}{5};$

$x^2 + \frac{19}{5}x + \left(\frac{19}{10}\right)^2 = \frac{4}{5} + \frac{361}{100} = \frac{441}{100};$

that is, $x + \frac{19}{10} = \pm\frac{21}{10}.$

23. We have $\frac{3}{5}(x+6)(x-2) = \frac{2}{3}\left(62\tfrac{1}{10} + \frac{18x}{5}\right).$

Multiply by 15, $9(x^2 + 4x - 12) = 10\left(62\tfrac{1}{10} + \frac{18x}{5}\right);$

$9x^2 + 36x - 108 = 621 + 36x;$

whence $x^2 = 81.$

EXAMPLES. XXV. b. PAGE 197.

1. $x^2 + \frac{14}{5}x = 11.$

$x^2 + \frac{14}{5}x + \left(\frac{7}{5}\right)^2 = 11 + \frac{49}{25} = \frac{324}{25};$

whence $x + \frac{7}{5} = \pm\frac{18}{5}.$

2. $x^2 - \frac{44}{3}x = -\frac{121}{3};$

$x^2 - \frac{44}{3}x + \left(\frac{22}{3}\right)^2 = -\frac{121}{3} + \frac{484}{9} = \frac{121}{9};$

whence $x - \frac{22}{3} = \pm\frac{11}{3}.$

3. $x^2 - \frac{25}{6}x = -\frac{21}{6};$

$x^2 - \frac{25}{6}x + \left(\frac{25}{12}\right)^2 = -\frac{21}{6} + \frac{625}{144} = \frac{121}{144};$

whence $x - \frac{25}{12} = \pm\frac{11}{12}.$

4. $x^2 + \frac{1}{8}x = \frac{30}{8};$

$x^2 + \frac{1}{8}x + \left(\frac{1}{16}\right)^2 = \frac{30}{8} + \frac{1}{256} = \frac{961}{256};$

whence $x + \frac{1}{16} = \pm\frac{31}{16}$

5. $$x^2 - \frac{22}{3}x = -\frac{35}{3};$$

$$x^2 - \frac{22}{3}x + \left(\frac{11}{3}\right)^2 = -\frac{35}{3} + \frac{121}{9} = \frac{16}{9};$$

whence $x - \frac{11}{3} = \pm\frac{4}{3}$.

6. $$x^2 - \frac{1}{6}x = \frac{22}{6};$$

$$x^2 - \frac{1}{6}x + \left(\frac{1}{12}\right)^2 = \frac{22}{6} + \frac{1}{144} = \frac{529}{144};$$

whence $x - \frac{1}{12} = \pm\frac{23}{12}$.

7. $$x^3 + \frac{17}{4}x = \frac{15}{4};$$

$$x^2 + \frac{17}{4}x + \left(\frac{17}{8}\right)^2 = \frac{15}{4} + \frac{289}{64} = \frac{529}{64};$$

whence $x + \frac{17}{8} = \pm\frac{23}{8}$.

8. $$x^2 - \frac{1}{2}x = \frac{21}{2};$$

$$x^2 - \frac{1}{2}x + \left(\frac{1}{4}\right)^2 = \frac{21}{2} + \frac{1}{16} = \frac{169}{16};$$

whence $x - \frac{1}{4} = \pm\frac{13}{4}$.

9. $$x^2 - \frac{2}{3}x = \frac{143}{9};$$

$$x^2 - \frac{2}{3}x + \left(\frac{1}{3}\right)^2 = \frac{143}{9} + \frac{1}{9} = \frac{144}{9};$$

whence $x - \frac{1}{3} = \pm\frac{12}{3}$.

10. $$x^2 - \frac{29}{12}x = -\frac{14}{12};$$

$$x^2 - \frac{29}{12}x + \left(\frac{29}{24}\right)^2 = -\frac{14}{12} + \frac{841}{576} = \frac{169}{576};$$

whence $x - \frac{29}{24} = \pm\frac{13}{24}$.

11. $$x^2 + \frac{1}{20}x = \frac{12}{20};$$

$$x^2 + \frac{1}{20}x + \left(\frac{1}{40}\right)^2 = \frac{12}{20} + \frac{1}{1600} = \frac{961}{1600};$$

whence $x + \frac{1}{40} = \pm\frac{31}{40}$.

12. $$x^2 + \frac{19}{8}x = \frac{15}{8};$$

$$x^2 + \frac{19}{8}x + \left(\frac{19}{16}\right)^2 = \frac{15}{8} + \frac{361}{256} = \frac{841}{256};$$

whence $x + \frac{19}{16} = \pm\frac{29}{16}$.

13. $$x^2 + \frac{22}{21}x = -\frac{5}{21}.$$

$$x^2 + \frac{22}{21}x + \left(\frac{11}{21}\right)^2 = -\frac{5}{21} + \frac{121}{441} = \frac{16}{441};$$

whence $x + \frac{11}{21} = \pm\frac{4}{21}$.

14. $$x^2 - \frac{3}{10}x = \frac{27}{50};$$

$$x^2 - \frac{3}{10}x + \left(\frac{3}{20}\right)^2 = \frac{27}{50} + \frac{9}{400} = \frac{225}{400};$$

whence $x - \frac{3}{20} = \pm\frac{15}{20}$.

15. $$x^2 - \frac{3}{2}x = \frac{13}{9};$$

$$x^2 - \frac{3}{2}x + \left(\frac{3}{4}\right)^2 = \frac{13}{9} + \frac{9}{16} = \frac{289}{144};$$

whence $x - \frac{3}{4} = \pm\frac{17}{12}$.

16. $$x^2 - \frac{8}{5}x = \frac{21}{5};$$

$$x^2 - \frac{8}{5}x + \left(\frac{4}{5}\right)^2 = \frac{21}{5} + \frac{16}{25} = \frac{121}{25};$$

whence $x - \frac{4}{5} = \pm\frac{11}{5}$.

17. $x^2 - \frac{2a}{15}x = \frac{a^2}{15}$;

$$x^2 - \frac{2a}{15}x + \left(\frac{a}{15}\right)^2 = \frac{a^2}{15} + \frac{a^2}{225} = \frac{16a^2}{225};$$

whence $x - \frac{a}{15} = \pm\frac{4a}{15}$.

18. $x^2 - \frac{2a}{21}x = \frac{3a^2}{21}$;

$$x^2 - \frac{2a}{21}x + \left(\frac{a}{21}\right)^2 = \frac{3a^2}{21} + \frac{a^2}{441} = \frac{64a^2}{441};$$

whence $x - \frac{a}{21} = \pm\frac{8a}{21}$.

19. We have $x^2 - \frac{11k}{6}x = \frac{7k^2}{6}$;

$$x^2 - \frac{11k}{6}x + \left(\frac{11k}{12}\right)^2 = \frac{7k^2}{6} + \frac{121k^2}{144} = \frac{289k^2}{144};$$

whence $$x - \frac{11k}{12} = \pm\frac{17k}{12}.$$

20. We have $x^2 + \frac{23k}{12}x = -\frac{10k^2}{12}$;

$$x^2 + \frac{23k}{12}x + \left(\frac{23k}{24}\right)^2 = -\frac{10k^2}{12} + \frac{529k^2}{576} = \frac{49k^2}{576};$$

whence $$x + \frac{23k}{24} = \pm\frac{7k}{24}.$$

21. We have $x^2 - \frac{c}{12}x = \frac{20c^2}{12}$;

$$x^2 - \frac{c}{12}x + \left(\frac{c}{24}\right)^2 = \frac{20c^2}{12} + \frac{c^2}{576} = \frac{961c^2}{576};$$

whence $$x - \frac{c}{24} = \pm\frac{31c}{24}.$$

22. Divide out the factor $x - 3$;

thus $x - 3 = 0$, and $2 = 3(x+2)$. [Compare Art. 201, Ex. 2.]

23. $2x^2 + 5x + 3 = 4x^2 - 22$;

$$2x^2 - 5x = 25;$$

$$x^2 - \frac{5}{2}x + \left(\frac{5}{4}\right)^2 = \frac{25}{2} + \frac{25}{16} = \frac{225}{16};$$

whence $x - \frac{5}{4} = \pm\frac{15}{4}$.

24. $6x^2 - 25x + 25 = x^2 + 2x - 3$;

$$5x^2 - 27x = -28;$$

$$x^2 - \frac{27}{5}x + \left(\frac{27}{10}\right)^2 = -\frac{28}{5} + \frac{729}{100} = \frac{169}{100};$$

whence $x - \frac{27}{10} = \pm\frac{13}{10}$.

25. $5x + 7 = 3x^2 - x - 2$;

$$x^2 - 2x = 3;$$

$$x^2 - 2x + 1 = 3 + 1 = 4;$$

whence $x - 1 = \pm 2$.

26. $3x^2 - 7x = -2$;

$$x^2 - \frac{7}{3}x + \left(\frac{7}{6}\right)^2 = -\frac{2}{3} + \frac{49}{36} = \frac{25}{36};$$

whence $x - \frac{7}{6} = \pm\frac{5}{6}$.

27. $\frac{3x-8}{x-2} = \frac{5x-2}{x+5}$;

$3x^2+7x-40=5x^2-12x+4$;

$2x^2-19x=-44$;

$x^2-\frac{19}{2}x+\left(\frac{19}{4}\right)^2=-22+\frac{361}{16}=\frac{9}{16}$;

whence $x-\frac{19}{4}=\pm\frac{3}{4}$

28. $\frac{3x-1}{4x+7}=1-\frac{6}{x+7}=\frac{x+1}{x+7}$;

$3x^2+20x-7=4x^2+11x+7$;

$x^2-9x=-14$;

$x^2-9x+\left(\frac{9}{2}\right)^2=-14+\frac{81}{4}=\frac{25}{4}$;

whence $x-\frac{9}{2}=\pm\frac{5}{2}$.

29. $\frac{5x-7}{7x-5}=\frac{x-5}{2x-13}$;

$10x^2-79x+91=7x^2-40x+25$;

$x^2-13x=-22$;

$x^2-13x+\left(\frac{13}{2}\right)^2=-22+\frac{169}{4}=\frac{81}{4}$;

whence $x-\frac{13}{2}=\pm\frac{9}{2}$.

30. $\frac{x+3}{2x-7}=\frac{2x-1}{x-3}$;

$x^2-9=4x^2-16x+7$;

$3x^2-16x=-16$;

$x^2-\frac{16}{3}x+\left(\frac{8}{3}\right)^2=-\frac{16}{3}+\frac{64}{9}=\frac{16}{9}$;

whence $x-\frac{8}{3}=\pm\frac{4}{3}$.

31. We have $\frac{(3-x)-(1+x)}{(1+x)(3-x)}=\frac{6}{35}$;

$70-70x=18+12x-6x^2$;

$6x^2-82x=-52$;

$x^2-\frac{41}{3}x+\left(\frac{41}{6}\right)^2=\frac{1681}{36}-\frac{52}{6}=\frac{1369}{36}$;

whence $x-\frac{41}{6}=\pm\frac{37}{6}$.

32. We have $\frac{x+4}{x-4}+\frac{x-2}{x-3}=\frac{19}{3}$;

$3(x+4)(x-3)+3(x-2)(x-4)=19(x-4)(x-3)$;

$3x^2+3x-36+3x^2-18x+24=19x^2-133x+228$;

$13x^2-118x=-240$;

$x^2-\frac{118}{13}x+\left(\frac{59}{13}\right)^2=-\frac{240}{13}+\frac{3481}{169}=\frac{361}{169}$;

whence $x-\frac{59}{13}=\pm\frac{19}{13}$.

33. We have $$\frac{1}{9-2x}=\frac{1}{3-x}-\frac{4}{5}=\frac{-7+4x}{5(3-x)};$$

$$15-5x=-63+14x+36x-8x^2;$$

$$8x^2-55x=-78;$$

$$x^2-\frac{55}{8}x+\left(\frac{55}{16}\right)^2=\frac{3025}{256}-\frac{78}{8}=\frac{529}{256};$$

whence $$x-\frac{55}{16}=\pm\frac{23}{16}.$$

34. $$\frac{4(x+2)-5(x-1)}{(x-1)(x+2)}=\frac{3}{x};$$

$$x(13-x)=3(x^2+x-2);$$

$$2x^2-5x=3;$$

$$x^2-\frac{5}{2}x+\left(\frac{5}{4}\right)^2=\frac{25}{16}+\frac{3}{2}=\frac{49}{16};$$

whence $$x-\frac{5}{4}=\pm\frac{7}{4}.$$

35. $$\frac{5x-4(x-2)}{x(x-2)}=\frac{3}{x+6};$$

$$(x+6)(x+8)=3x(x-2);$$

$$x^2-10x=24.$$

$$x^2-10x+25=49.$$

whence $$x-5=\pm 7.$$

[*For Examples* 36-39, *see Appendix*, *p.* 363.]

EXAMPLES. XXV. c. PAGE 201.

These equations are equivalent to

13. $(x+1)(6x-7)=0.$ 14. $(4x-7)(2x-3)=0.$ 15. $(11x-7)(x+3)=0.$

16. $(5x+6)(x+4)=0.$ 17. $(10x-9)(6x+5)=0.$ 18. $(4x+3)(3x-8)=0.$

19. $(7x-2)(x+14)=0.$ 20. $(12x-5)(8x+3)=0.$ 21. $(5x+2)(5x-3)=0.$

22. $(2x-5)(2x+7)=0.$ 23. $(4x-9a)(3x+4a)=0.$

24. $(4x-9a)(3x-4a)=0.$ 25. $(3x-5b)(3x+7b)=0.$

26. $(6x-7b)(6x+5b)=0.$ 27. $(x-2a)(x-2b)=0.$

28. $(x-2a)(x+8)=0.$ 29. $x(3x-2a-b)=0.$ 30. $x(ax+2-b)=0.$

31. $(x^2-4)(x^2-1)=0.$ 32. $(x^2-4)(x^2-9)=0.$ 33. $(x^3+8)(x^3-1)=0.$

34. $(x^3-27)(x^3+8)=0.$ 35. $16x^4-257x^2+16=0$, or $(16x^2-1)(x^2-16)=0.$

36. $x^4-a^2x^2-b^2x^2+a^2b^2=0$, or $(x^2-a^2)(x^2-b^2)=0.$

37. $x^6+19x^3-216=0$, or $(x^3+27)(x^3-8)=0.$

38. Put $y=x^2+2$; then $y^2-29y+198=0$;

whence $(y-18)(y-11)=0$; that is, $x^2+2=18$ or $11.$

39. Put $y = x^2 - x$; then $y^2 - 18y + 72 = 0$;
whence $(y-12)(y-6) = 0$; that is, $x^2 - x - 12 = 0$, or $x^2 - x - 6 = 0$.
Thus $(x-4)(x+3) = 0$, or $(x-3)(x+2) = 0$.

40. Put $y = x(x - 2a)$; then $y^2 - 7a^2y - 8a^4 = 0$;
whence $(y - 8a^2)(y + a^2) = 0$.
Thus $x^2 - 2ax - 8a^2 = 0$, or $x^2 - 2ax + a^2 = 0$;
that is, $(x - 4a)(x + 2a) = 0$, or $(x - a)^2 = 0$.

[*For Examples XXV. d., see Appendix, p. 364.*]

EXAMPLES. XXVI. a. PAGE 203.

22. Subtract the second equation from the square of the first; thus $xy = 20$. Also $x + y = 9$.

23. Subtract the second equation from the square of the first; thus $xy = 28$. Also $x - y = 3$.

24. Subtract the first equation from the square of the second; thus $3xy = 120$, or $xy = 40$. Also $x + y = 14$.

25. We have $x - y = 10$. Square this equation and from it subtract the second equation; thus $2xy = 48$, or $xy = 24$. Also $x - y = 10$.

26. From the first equation $2xy = x + y = 2$, by using the second equation; thus $xy = 1$. Also $x + y = 2$.

27. We have $x + y = \frac{7}{12} xy = \frac{7}{12} \times 12$;
whence $x + y = 7$. Also $xy = 12$.

28.
$$ax + by = 2 \quad\ldots\ldots(1),$$
$$abxy = 1 \quad\ldots\ldots(2).$$
Multiply (2) by 4 and subtract from the square of (1):
thus $a^2x^2 - 2abxy + b^2y^2 = 0$;
$$\therefore\ ax - by = 0 \quad\ldots\ldots(3).$$
From (1) and (3) we obtain $x = \frac{1}{a}$, $y = \frac{1}{b}$.

29.
$$x^2 + pxy + y^2 = p + 2 \quad\ldots\ldots(1),$$
$$qx^2 + xy + qy^2 = 2q + 1 \quad\ldots\ldots(2).$$
Multiply (1) by q and subtract (2) from the result;
thus $(pq - 1)xy = pq - 1$;
$$\therefore\ xy = 1 \quad\ldots\ldots(3).$$
By substitution in (1), we have $x^2 + y^2 = 2 \quad\ldots\ldots(4)$.
Equations (3) and (4) can now be solved by the method of Art. 204, Ex. 1.

EXAMPLES. XXVI. b. PAGE 205.

11. Subtract the first equation from the square of the second; then

$$\frac{2}{xy}=\left(\frac{29}{24}\right)^2-\frac{481}{576}=\frac{360}{576}.$$

Subtract this result from the first equation; then $\left(\frac{1}{x}-\frac{1}{y}\right)^2=\frac{121}{576}$.

Thus $$\frac{1}{x}+\frac{1}{y}=\frac{29}{24};\quad \frac{1}{x}-\frac{1}{y}=\pm\frac{11}{24}.$$

12. We have $$\frac{1}{xy}=\frac{1}{30}.$$

Multiply this by 2, and add to and then subtract from the first equation;

then $$\left(\frac{1}{x}+\frac{1}{y}\right)^2=\frac{61}{900}+\frac{2}{30}=\frac{121}{900};$$

and $$\left(\frac{1}{x}-\frac{1}{y}\right)^2=\frac{61}{900}-\frac{2}{30}=\frac{1}{900}.$$

Thus $$\frac{1}{x}+\frac{1}{y}=\pm\frac{11}{30},\quad \frac{1}{x}-\frac{1}{y}=\pm\frac{1}{30}.$$

13. From the first equation, $x^2-\frac{5}{2}xy+y^2=0$;

and from the second, $x^2+2xy+y^2=36$;

by subtraction, $\frac{9}{2}xy=36$; whence $xy=8$.

Combine this equation with $x+y=6$.

14. From the first equation, $x^2-\frac{58}{21}xy+y^2=0$;

and from the second, $x^2-2xy+y^2=16$;

by subtraction, $\frac{16}{21}xy=16$; whence $xy=21$.

Combine this equation with $x-y=4$.

15. From the first equation, $x^2-\frac{34}{15}xy+y^2=0$;

and from the second, $x^2+2xy+y^2=64$;

by subtraction, $\frac{64}{15}xy=64$; that is, $xy=15$.

Combine this equation with $x+y=8$.

16. Divide the first equation by the second; thus $x-y=2$. From this and the second equation, we obtain $xy=8$.

17. From the first equation, $x^2+y^2-\frac{17}{4}xy=0$;

and from the second, $$x^2+y^2-2xy=36;$$

by subtraction $$\frac{9}{4}xy=36; \text{ that is, } xy=16.$$

Combine this equation with $$x-y=6.$$

18. Divide the first equation by the second; thus $x+y=6$. From this and the second equation, we obtain $xy=5$.

19. By division, $\frac{1}{x^2}-\frac{1}{xy}+\frac{1}{y^2}=\frac{126}{125}\div\frac{6}{5}=\frac{21}{25}$.

Also $$\frac{1}{x^2}+\frac{2}{xy}+\frac{1}{y^2}=\frac{36}{25};$$

whence $$\frac{3}{xy}=\frac{15}{25}, \text{ and } \frac{4}{xy}=\frac{20}{25}.$$

By subtraction, $\frac{1}{x^2}-\frac{2}{xy}+\frac{1}{y^2}=\frac{16}{25}$; that is, $\frac{1}{x}-\frac{1}{y}=\pm\frac{4}{5}$.

20. By division, $$\frac{1}{x^2}+\frac{1}{xy}+\frac{1}{y^2}=91.$$

But $$\frac{1}{x^2}-\frac{2}{xy}+\frac{1}{y^2}=1;$$

hence $$\frac{3}{xy}=90, \text{ and } \frac{4}{xy}=120.$$

By addition, $\frac{1}{x^2}+\frac{2}{xy}+\frac{1}{y^2}=121$; that is, $\frac{1}{x}+\frac{1}{y}=\pm 11$.

EXAMPLES. XXVI. c. PAGE 208.

1. Since $y=5x-17$, we have $x(5x-17)=12$;
that is, $$(5x+3)(x-4)=0.$$

2. By addition, $x^2+2xy+y^2=0$; that is, $x+y=\pm 5$. Divide each of the given equations by this last equation; thus $x=\pm 3$, $y=\pm 2$.

3. Substitute $x=y+10$ in the second equation;
thus $$(y+10)^2-2y(y+10)-3y^2=84;$$
that is, $$-4y^2=-16; \text{ or } y=\pm 2.$$

4. We have $3x+\frac{20}{x}=16$; whence $(x-2)(3x-10)=0$.

5. Substitute $y=3x-11$ in the second equation; thus

$$3x^2-(3x-11)^2=47;$$

whence we obtain $x^2-11x+28=0$, or $(x-4)(x-7)=0$.

6. Substitute $x=3y+1$ in the second equation;

thus $$(3y+1)^2-2y(3y+1)+9y^2=17;$$

whence $$3y^2+y-4=0: \text{ or } (3y+4)(y-1)=0.$$

7. Since $x=9-2y$, we have

$$3y^2-5(9-2y)^2=43, \text{ or } 17y^2-180y+448=0;$$

whence $$(y-4)(17y-112)=0.$$

8. We have $\frac{x^2+y^2}{2xy-y^2}=\frac{5}{3}$ or $3x^2-10xy+8y^2=0$;

that is, $$(x-2y)(3x-4y)=0;$$

whence $$x=2y, \text{ or } x=\frac{4y}{3}.$$

Combine each of these equations with $x^2+y^2=5$.

9. Since $y=3-5x$, we have

$$2x^2-3x(3-5x)-(3-5x)^2=1;$$

that is, $8x^2-21x+10=0$, or $(x-2)(8x-5)=0$.

10. We have $$\frac{3x^2-5y^2}{3xy-4y^2}=\frac{28}{8}=\frac{7}{2};$$

that is, $2x^2-7xy+6y^2=0$; whence $(2x-3y)(x-2y)=0$.

Combine each of the equations $x=2y$, and $x=\frac{3}{2}y$ with $3x^2-5y^2=28$.

11. We have $$\frac{3x^2-y^2}{2x^2-xy}=\frac{23}{12};$$

that is, $10x^2-23xy+12y^2=0$; whence $(2x-3y)(5x-4y)=0$.

Combine each of the equations $x=\frac{3}{2}y$, and $x=\frac{4}{5}y$ with $2x^2-xy=12$.

12. Multiply the first equation by 2 and subtract the second equation from the product;

then $$5xy = 6\tfrac{1}{2} - 2\tfrac{3}{4} = 3\tfrac{3}{4}\,;\ \text{whence } xy = \frac{3}{4}.$$

From the first equation we now have $x^2 + y^2 = \dfrac{5}{2}$.

13. Multiply the first equation by 3 and subtract from the second equation; thus $8xy = 16$; whence $xy = 2$.

From the first equation, we obtain $x^2 + y^2 = 5$.

14. By division, $$\frac{7xy - 8x^2}{8y^2 - 9xy} = \frac{10}{18} = \frac{5}{9};$$

that is, $$72x^2 - 108xy + 40y^2 = 0, \text{ or } 18x^2 - 27xy + 10y^2 = 0.$$

Thus $$(6x - 5y)(3x - 2y) = 0.$$

Combine each of the equations $x = \dfrac{5}{6}y$, and $x = \dfrac{2}{3}y$ with $8y^2 - 9xy = 18$.

15. By division, $$\frac{x^2 - 2xy}{xy + y^2} = \frac{21}{18} = \frac{7}{6};$$

that is, $6x^2 - 19xy - 7y^2 = 0$; whence $(3x + y)(2x - 7y) = 0$.

Combine each of the equations $x = -\dfrac{1}{3}y$, and $x = \dfrac{7}{2}y$ with $xy + y^2 = 18$.

16. By division, $$\frac{x^2 + 3xy}{xy + 4y^2} = \frac{54}{115};$$

that is, $115x^2 + 291xy - 216y^2 = 0$; whence $(5x - 3y)(23x + 72y) = 0$.

Combine each of the equations $x = \dfrac{3}{5}y$, and $x = -\dfrac{72}{23}y$ with $xy + 4y^2 = 115$.

17. Multiply the second equation by 3 and add to the first; thus $(x + y)^3 = 512$; that is, $x + y = 8$.

But $xy(x + y) = 120$; therefore $xy = 15$.

18. Multiply the second equation by 3 and subtract from the first; thus
$$(x - y)^3 = 1;\ \text{that is, } x - y = 1.$$
But $xy(x - y) = 42$; therefore $xy = 42$.

19. Multiply the second equation by 3 and subtract from the first; thus
$$(x - y)^3 = 64;\ \text{that is, } x - y = 4.$$
But $xy(x - y) = 48$; therefore $xy = 12$.

20. From the first equation $(xy + 12)(xy - 7) = 0$.

Combine each of the equations $xy = 7$, and $xy = -12$ with $x + y = 8$.

21. Multiply the second equation by 4 and add to the first; thus

$$(x+2y)^2-15(x+2y)+56=0.$$

This is a quadratic equation in $x+2y$, of which the solution is

$$x+2y=7, \text{ or } x+2y=8.$$

Combine each of these equations with $xy=6$.

22. Multiply the second equation by 6 and add to the first; thus

$$(3x+y)^2-21(3x+y)+104=0.$$

This is a quadratic equation in $3x+y$, of which the solution is

$$3x+y=13, \text{ or } 3x+y=8.$$

Combine each of these equations with $xy=4$.

EXAMPLES. XXVII. Page 212.

1. Let x = the number:

then $$x^2-119=10(x-8); \text{ whence } x=13, \text{ or } -3.$$

2. Let x years be the age of the son: the age of the father is $5x$ years,

and $$x^2+25x^2=2106; \text{ whence } x=\pm 9.$$

Thus the son's age is 9, the father's age 45.

3. Let x, $x+1$ be the two consecutive numbers;

then $$\frac{1}{x}+\frac{1}{x+1}=\frac{15}{56}.$$

Multiplying up, we have

$$56(x+1)+56x=15x(x+1);$$

whence $$x=7, \text{ or } -\frac{8}{15}.$$

Thus the two numbers are 7 and 8, the negative root being inadmissible.

4. Let x be the given number; then $\frac{1}{x}$ is its reciprocal.

Hence $$x+17=\frac{60}{x}; \text{ whence } x=3, \text{ or } -20.$$

5. Let x and y be the numbers, x being the greater;

then $$x+y=9(x-y),$$

and $$x^2-y^2=81.$$

From these equations we obtain $x=\pm 15$, $y=\pm 12$.

6. Let x be the number; $x+1$ is the next highest number;

then $$x+x^2=9(x+1); \text{ whence } x=9, \text{ or } -1.$$

7. Let x hours be the time the train takes; then it travels $\frac{210}{x}$ miles an hour. If it travelled at $\frac{210}{x}+5$ miles an hour, it would take $x-1$ hours;

therefore $$\left(\frac{210}{x}+5\right)(x-1)=210;$$

whence $$x=7, \text{ or } -6.$$

8. Let x and y be the numbers; then $x^2+y^2=74$; $x+y=12$. These equations may be solved as in Art. 204, Ex. 2.

9. Let x yards and y yards represent the lengths of the two sides. Then the perimeter is $(2x+2y)$ yards, and the area is xy square yards.

Thus $$x+y=250; \quad xy=14400.$$

10. Let x feet and y feet represent the sides of the two squares, x being the greater. Then their perimeters are $4x$, $4y$ feet, and their areas x^2, y^2 square feet respectively.

Thus $$4x-4y=100; \quad x^2-3y^2=325.$$

From the first equation $y=x-25$; substituting this value in the second equation, we obtain $$x^2-75x+1100=0;$$

whence $$x=55, \text{ or } 20; \quad y=30, \text{ or } -5.$$

11. Suppose the larger pipe takes x minutes to fill the cistern; then the smaller pipe takes $x+24$ minutes.

In one minute the first pipe fills $\frac{1}{x}$ of the cistern;

" " " second " $\frac{1}{x+24}$ " "

Hence $$\frac{1}{x}+\frac{1}{x+24}=\frac{1}{22\frac{1}{2}};$$

that is, $$45(x+24)+45x=2x(x+24);$$

or $$x^2-21x-540=0;$$

whence $$x=36, \text{ or } -15.$$

12. Suppose he travels x miles an hour; then he takes $\frac{108}{x}$ hours over the journey. But if he travelled $x+2$ miles an hour he would take $\frac{108}{x+2}$ hours;

$$\therefore \frac{108}{x}=\frac{108}{x+2}+4\tfrac{1}{2};$$

whence $$x=6, \text{ or } -8.$$ [See Art. 209, Ex. 1.]

13. Let x be the price of each ball in shillings; then the number bought is $\frac{100}{x}$. If they had each cost $x-1$ shillings, the number would have been $\frac{100}{x-1}$.

$$\therefore \quad \frac{100}{x-1} - \frac{100}{x} = 5;$$

whence $$x = 5, \text{ or } -4.$$

14. Suppose the master got x eggs; then the boy bought $x+3$. Therefore the price per egg paid by the master was $\frac{12}{x}$ pence, and the price paid by the boy was $\frac{12}{x+3}$ pence.

Also the master paid on each egg $\frac{1}{5}$ of a penny more than the market price;

$$\therefore \quad \frac{12}{x} - \frac{12}{x+3} = \frac{1}{5};$$

whence $$x = 12, \text{ or } -15.$$

15. Let x be the number of eggs bought for a shilling; then the cost of each egg is $\frac{12}{x}$ pence, and that of a dozen eggs is $\frac{144}{x}$ pence.

If $x+2$ eggs had been bought for a shilling, the price of a dozen eggs would have been $\frac{144}{x+2}$ pence.

Therefore $$\frac{144}{x} - \frac{144}{x+2} = 1; \text{ or } x^2 + 2x = 288;$$

whence $$x = 16, \text{ and } \frac{144}{x} = 9.$$

16. Let the width of the path be x feet; then for the path and lawn together the length is $50+2x$ feet and the breadth is $34+2x$ feet. The total area is $(50+2x)(34+2x)$ square feet; therefore the area of the path in square feet is $$(50+2x)(34+2x) - 50 \times 34.$$

Thus we obtain the equation

$$4x^2 + 168x = 540;$$

whence $$x = 3, \text{ or } -45.$$

17. Let x inches be the length of each tile; then its area is x^2 square inches, and the area of the hall is $200x^2$ square inches.

Thus $$200x^2 = 128(x+1)^2, \text{ or } 5x = \pm 4(x+1);$$

whence $$x = 4, \text{ or } -\frac{4}{9}.$$

18. Let x feet be the length of the lawn; then its area is x^2 square feet. The length of the whole garden is $x+8+12$ feet, and the length of the lawn and path is $x+8$ feet. Therefore the area of the flower-bed

$$=(x+20)^2-(x+8)^2 \text{ square feet};$$

$$\therefore\ (x+20)^2-(x+8)^2+x^2=721;$$

whence $$x=11, \text{ or } -35.$$

19. Suppose that at first she sells apples at x pence per dozen. At that price she sells $\frac{720}{x}$ for 5 shillings. At $x-1$ pence a dozen she sells $\frac{720}{x-1}$ for 5 shillings;

$$\therefore\ \frac{720}{x-1}-\frac{720}{x}=60;$$

whence $$x=4, \text{ or } -3.$$

20. Let the first rectangle be x yards long and y yards wide; then the other rectangle is $x-10$ yards long and $y+4$ yards wide;

$$\therefore\ (x-10)(y+4)=480;$$

also $$xy=480;$$

whence $$x=40, \text{ or } -30,$$

and $$y=12, \text{ or } -16.$$

21. Let x, y be the digits in the tens' and units' places respectively, so that the number is $10x+y$.

Then $$x(10x+y)=280;$$

and $$x(x+y)=55;$$

whence $$x=\pm 5.$$

Rejecting the negative value we obtain $y=6$; thus the number is 56.

22. His actual profit is $75-x$ shillings a head, and his outlay x shillings a head. His gain on each sheep is also $\frac{x}{100}\times x$ shillings.

$$\therefore\ 75-x=\frac{x^2}{100};$$

whence $$x=50, \text{ or } -150.$$

23. Let x be the number of yards he bought, so that the price per yard is £$\frac{5}{x}$. The selling price per yard was £$\left(\frac{5}{x}+\frac{1}{10}\right)$;

$$\therefore\ (x-5)\left(\frac{5}{x}+\frac{1}{10}\right)=6, \text{ or } x^2-15x-250=0;$$

whence $$x=25, \text{ or } -10.$$

24. Suppose the carriage to travel x miles an hour; then the number of revolutions made by the wheel per hour is $\dfrac{x \times 1760 \times 3}{14\frac{2}{3}}$, or $360x$. Therefore the time taken to revolve is $\dfrac{1}{360x} \times 60 \times 60$, or $\dfrac{10}{x}$ seconds. If the carriage travels $x - 2\frac{2}{3}$ miles an hour the number of revolutions made per hour is $\dfrac{(x-2\frac{2}{3}) \times 1760 \times 3}{14\frac{2}{3}}$, or $120(3x-8)$. Therefore the time taken to revolve is

$$\frac{1}{120(3x-8)} \times 60 \times 60, \text{ or } \frac{30}{3x-8} \text{ seconds.}$$

Thus $$\frac{30}{3x-8} - \frac{10}{x} = 1, \text{ or } 3x^2 - 8x - 80 = 0;$$

whence $$x = \frac{20}{3}, \text{ or } -4.$$

25. Let x = number of shares he bought, so that he paid £$\dfrac{1875}{x}$ for each. He sold $x-15$ for £1740 and obtained £$\dfrac{1740}{x-15}$ for each.

$$\therefore \frac{1740}{x-15} - \frac{1875}{x} = 4, \text{ or } 4x^2 + 75x - (15 \times 1875) = 0.$$

Put $x = 75y$; then $4y^2 + y - 5 = 0$; whence $y = 1$, or $-\dfrac{5}{4}$.

26. Suppose the trains meet at a point P which is x miles from A and $300 - x$ miles from B. Then the train from A travels $300 - x$ miles in 9 hours, and the train from B travels x miles in 4 hours. That is, the train from A travels the distance AP in $\dfrac{9x}{300-x}$ hours, and the train from B travels the distance BP in $\dfrac{4(300-x)}{x}$ hours, and these two times are equal;

$$\therefore \frac{9x}{300-x} = \frac{4(300-x)}{x}; \text{ that is, } 2(300-x) = \pm 3x;$$

whence $x = 120$, the other value being inadmissible.

The rates of the two trains are $\dfrac{300-x}{9}$ and $\dfrac{x}{4}$ miles an hour. That is, one train travels 20 and the other 30 miles an hour.

27. In 2 hours B travels the same distance that A has travelled in $2\frac{1}{4}$ hours; therefore B's pace is to A's as 9 to 8.

Suppose A travels at $8x$ miles and B at $9x$ miles per hour, and that in 2 hours B has reached a point X; then A is $\dfrac{3}{4} \times 8x$ miles beyond X, and the distance XQ is $240 - 18x$ miles. Hence B travels $240 - 18x$ miles at the rate of $9x+5$ miles an hour in the same time that A travels $240-18x-6x$ miles at $8x$ miles an hour;

$$\frac{240-18x}{9x+5}=\frac{240-24x}{8x}; \text{ that is, } \frac{80-6x}{9x+5}=\frac{10-x}{x};$$

whence $3x^2-5x-50=0$, and $x=5$.

Thus A starts at 40 miles, and B at 45 miles an hour.

28. After one replacement P contains $50-\frac{x}{2}$ gallons of water, and $\frac{x}{2}$ gallons of brandy; Q contains $\frac{x}{2}$ gallons of water, $40-\frac{x}{2}$ gallons of brandy. Now taking x gallons from each, out of P we take $\frac{x}{50}\times\frac{x}{2}$ gallons of brandy, leaving $\left(1-\frac{x}{50}\right)\frac{x}{2}$; out of Q we take $\frac{x}{40}\left(40-\frac{x}{2}\right)$ gallons of brandy. Therefore in the $2x$ gallons there are

$$\left(\frac{x}{50}\times\frac{x}{2}\right)+\frac{x}{40}\left(40-\frac{x}{2}\right) \text{ gallons of brandy,}$$

half of which is returned to P. Thus in P there will be

$$\left(1-\frac{x}{50}\right)\frac{x}{2}+\frac{1}{2}\left\{\frac{x}{50}\cdot\frac{x}{2}+\frac{x}{40}\left(40-\frac{x}{2}\right)\right\} \text{ gallons of brandy.}$$

On reduction this gives $\quad x-\frac{9x^2}{800}$.

$$\therefore\ x-\frac{9x^2}{800}=8\tfrac{7}{8}; \text{ or } 9x^2-800x+7100=$$

whence $x=10$, or $\frac{710}{9}$. The latter value must be rejected, as it is too great for the cask.

29. Suppose that A and B have x and $30-x$ cows respectively, and that each receives a sum of £y.

Then $\frac{y}{x}$ $=A$'s price for each cow in pounds.

$\frac{y}{30-x}=B$'s ,, ,, ,, ,,

Hence by the question

$$\frac{xy}{30-x}=320; \text{ and } \frac{(30-x)\,y}{x}=245.$$

By division, $$\frac{x^2}{(30-x)^2}=\frac{64}{49}; \text{ that is, } \frac{x}{30-x}=\pm\frac{8}{7}.$$

whence $x=16$, the other value being inadmissible.

30. Suppose the distance from the station to the house to be x miles, and that the carriage travels at y miles an hour. By walking $x-8$ miles at 4 miles an hour instead of driving at y miles an hour he loses half-an-hour;

$$\therefore \frac{x-8}{4} - \frac{x-8}{y} = \frac{1}{2}, \text{ or } (x-8)(y-4) = 2y.$$

Again, the carriage only goes 16 miles instead of $2x$ miles, and it reaches home an hour earlier in consequence;

$$\therefore \frac{2x}{y} - \frac{16}{y} = 1, \text{ or } 2(x-8) = y.$$

Hence $\qquad y(y-4) = 4y, \text{ or } y = 8; \text{ and } x = 12.$

[*For Examples* 31-39, *see Appendix, p.* 367.]

EXAMPLES. XXVIII. a. PAGE 216.

1. $x^4+16x^2+256 = (x^4+32x^2+256) - 16x^2 = (x^2+16)^2 - (4x)^2.$

2. $81a^4+9a^2b^2+b^4 = (81a^4+18a^2b^2+b^4) - 9a^2b^2 = (9a^2+b^2)^2 - (3ab)^2.$

3. $x^4+y^4-7x^2y^2 = (x^4+2x^2y^2+y^4) - 9x^2y^2 = (x^2+y^2)^2 - (3xy)^2.$

4. $m^4+n^4-18m^2n^2 = (m^4-2m^2n^2+n^4) - 16m^2n^2 = (m^2-n^2)^2 - (4mn)^2.$

5. $x^4-6x^2y^2+y^4 = (x^4-2x^2y^2+y^4) - 4x^2y^2 = (x^2-y^2)^2 - (2xy)^2.$

6. $4x^4+9y^4-93x^2y^2 = (4x^4-12x^2y^2+9y^4) - 81x^2y^2 = (2x^2-3y^2)^2 - (9xy)^2.$

7. $4m^4+9n^4-24m^2n^2 = (4m^4+12m^2n^2+9n^4) - 36m^2n^2 = (2m^2+3n^2)^2 - (6mn)^2.$

8. $9x^4+4y^4+11x^2y^2 = (9x^4+12x^2y^2+4y^4) - x^2y^2 = (3x^2+2y^2)^2 - (xy)^2.$

9. $x^4-19x^2y^2+25y^4 = (x^2-10x^2y^2+25y^4) - 9x^2y^2 = (x^2-5y^2)^2 - (3xy)^2.$

10. $16a^4+b^4-28a^2b^2 = (16a^4+8a^2b^2+b^4) - 36a^2b^2 = (4a^2+b^2)^2 - (6ab)^2.$

17. $x^2y+3xy^2-3x^3-y^3 = y(x^2-y^2) - 3x(x^2-y^2) = (y-3x)(x^2-y^2).$

18. $4mn^2-20n^3+45nm^2-9m^3 = m(4n^2-9m^2) - 5n(4n^2-9m^2)$
$$= (m-5n)(4n^2-9m^2).$$

19. $ab(x^2+1)+x(a^2+b^2) = abx^2+(a^2+b^2)x+ab = (ax+b)(bx+a).$

20. $y^2z^2(x^4-1)+x^2(y^4-z^4) = x^4y^2z^2 - y^2z^2 + x^2y^4 - x^2z^4$
$$= x^2y^2(x^2z^2+y^2) - z^2(y^2+x^2z^2)$$
$$= (x^2z^2+y^2)(x^2y^2-z^2).$$

22. $pn(m^2+1) - m(p^2+n^2) = pm^2n + pn - p^2m - mn^2$
$$= mn(pm-n) - p(pm-n)$$
$$= (mn-p)(pm-n).$$

23. $6bx(a^2+1)-a(4x^2+9b^2)=6a^2bx^2+6bx-4ax^2-9ab^2$
$$=3ab(2ax-3b)-2x(2ax-3b)$$
$$=(3ab-2x)(2ax-3b).$$

24. $(2a^2+3y^2)x+(2x^2+3a^2)y=2a^2x+3xy^2+2x^2y+3a^2y$
$$=a^2(2x+3y)+xy(3y+2x)=(a^2+xy)(2x+3y).$$

25. $(2x^2-3a^2)y+(2a^2-3y^2)x=2x^2y-3a^2y+2a^2x-3xy^2$
$$=xy(2x-3y)+a^2(2x-3y)$$
$$=(xy+a^2)(2x-3y).$$

Examples **31**—**36** can be put in the following forms and the factors written down by the formula of Art. 213.

31. $b^3+c^3+(-1)^3-3bc(-1).$ **32.** $a^3+(2c)^3+(1)^3-3a(2c)(1).$

33. $a^3+b^3+(2c)^3-3ab(2c).$ **34.** $a^3+(-3b)^3+c^3-3a(-3b)c.$

35. $a^3+(-b)^3+(-c)^3-3a(-b)(-c).$

36. $(2a)^3+(3b)^3+c^3-3(2a)(3b)c.$

37. $x^8+81x^4+6561=(x^4+81)^2-(9x^2)^2$
$$=(x^4+9x^2+81)(x^4-9x^2+81)$$
$$=\{(x^2+9)^2-(3x)^2\}(x^4-9x^2+81)$$
$$=(x^2+3x+9)(x^2-3x+9)(x^4-9x^2+81).$$

38. $(a^4-2a^2b^2-b^4)^2-4a^4b^4=(a^4-2a^2b^2-b^4+2a^2b^2)(a^4-2a^2b^2-b^4-2a^2b^2)$
$$=(a^4-b^4)(a^4-4a^2b^2-b^4)$$
$$=(a^2+b^2)(a+b)(a-b)(a^4-4a^2b^2-b^4).$$

39. $4(ab+cd)^2-(a^2+b^2-c^2-d^2)^2$
$$=(2ab+2cd+a^2+b^2-c^2-d^2)(2ab+2cd-a^2-b^2+c^2+d^2)$$
$$=\{(a+b)^2-(c-d)^2\}\{(c+d)^2-(a-b)^2\}$$
$$=(a+b+c-d)(a+b-c+d)(c+d+a-b)(c+d-a+b).$$

40. $x^8-\dfrac{1}{256}=x^8-\left(\dfrac{1}{2}\right)^8=\left(x^4+\dfrac{1}{2^4}\right)\left(x^4-\dfrac{1}{2^4}\right)$
$$=\left(x^4+\frac{1}{16}\right)\left(x^2+\frac{1}{4}\right)\left(x+\frac{1}{2}\right)\left(x-\frac{1}{2}\right).$$

41. $x^{16}-y^{16}=(x^8+y^8)(x^8-y^8)=(x^8+y^8)(x^4+y^4)(x^4-y^4)$
$$=(x^8+y^8)(x^4+y^4)(x^2+y^2)(x+y)(x-y).$$

42. $x^{18}-y^{18}=(x^9+y^9)(x^9-y^9)$
$$=(x^3+y^3)(x^6-x^3y^3+y^6)(x^3-y^3)(x^6+x^3y^3+y^6)$$
$$=(x+y)(x^2-xy+y^2)(x^6-x^3y^3+y^6)(x-y)(x^2+xy+y^2)(x^6+x^3y^3+y^6).$$

43. $\frac{a^3}{x^2} - 8x - a^3 + 8x^3 = a^3\left(\frac{1}{x^2} - 1\right) - 8x^3\left(\frac{1}{x^2} - 1\right) = \left(\frac{1}{x^2} - 1\right)(a^3 - 8x^3)$

$$= \left(\frac{1}{x} + 1\right)\left(\frac{1}{x} - 1\right)(a - 2x)(a^2 + 2ax + 4x^2).$$

44. $x^9 + x^3y^6 - 8x^6y^3 - 8y^9 = (x^6 + y^6)(x^3 - 8y^3)$

$$= (x^2 + y^2)(x^4 - x^2y^2 + y^4)(x - 2y)(x^2 + 2xy + 4y^2).$$

45. $x^9 + x^6 + 64x^3 + 64 = (x^3 + 1)(x^6 + 64)$

$$= (x + 1)(x^2 - x + 1)(x^2 + 4)(x^4 - 4x^2 + 16).$$

46. $4a - 9b + \frac{4b^3}{a^2} - \frac{9a^3}{b^2} = \frac{4a^3 + 4b^3}{a^2} - \frac{9a^3 + 9b^3}{b^2} = (a^3 + b^3)\left(\frac{4}{a^2} - \frac{9}{b^2}\right)$

$$= (a + b)(a^2 - ab + b^2)\left(\frac{2}{a} + \frac{3}{b}\right)\left(\frac{2}{a} - \frac{3}{b}\right).$$

47. $\frac{xy^3}{72} - \frac{x^3y^5}{32} - \frac{1}{9x^2} + \frac{y^2}{4} = \left(\frac{1}{9x^2} - \frac{y^2}{4}\right)\left(\frac{x^3y^3}{8} - 1\right)$

$$= \left(\frac{1}{3x} + \frac{y}{2}\right)\left(\frac{1}{3x} - \frac{y}{2}\right)\left(\frac{xy}{2} - 1\right)\left(\frac{x^2y^2}{4} + \frac{xy}{2} + 1\right).$$

48. $x^6 - 25x^2 + \frac{25}{4} - \frac{x^4}{4} = \left(x^2 - \frac{1}{4}\right)(x^4 - 25) = \left(x + \frac{1}{2}\right)\left(x - \frac{1}{2}\right)(x^2 + 5)(x^2 - 5).$

49. $x^7 + x^4 - 16x^3 - 16 = (x^3 + 1)(x^4 - 16) = (x^3 + 1)(x^2 + 4)(x^2 - 4)$

$$= (x + 1)(x^2 - x + 1)(x^2 + 4)(x + 2)(x - 2).$$

50. $16x^7 - 81x^3 - 16x^4 + 81 = (16x^4 - 81)(x^3 - 1)$

$$= (4x^2 + 9)(4x^2 - 9)(x^3 - 1)$$

$$= (4x^2 + 9)(2x + 3)(2x - 3)(x - 1)(x^2 + x + 1).$$

EXAMPLES. XXVIII. b. PAGE 220.

1. $(2x - 7y + 3z)(2x + 7y - 3z) = (2x - \overline{7y - 3z})(2x + \overline{7y - 3z})$

$$= 4x^2 - (7y - 3z)^2$$

$$= 4x^2 - 49y^2 + 42yz - 9z^2.$$

2. $(3x^2 - 4xy + 7y^2)(3x^2 + 4xy + 7y^2)$

$$= (\overline{3x^2 + 7y^2} - 4xy)(\overline{3x^2 + 7y^2} + 4xy)$$

$$= (3x^2 + 7y^2)^2 - 16x^2y^2 = 9x^4 + 26x^2y^2 + 49y^4.$$

3. $(5x^2+5xy-9y^2)(5x^2-5xy-9y^2)=(\overline{5x^2-9y^2}+5xy)(5x^2-\overline{9y^2-5xy})$

$$=(5x^2-9y^2)^2-25x^2y^2$$
$$=25x^4-115x^2y^2+81y^4.$$

4. $(7x^2-8xy+3y^2)(7x^2+8xy-3y^2)=(7x^2-\overline{8xy-3y^2})(7x^2+\overline{8xy-3y^2})$

$$=49x^4-(8xy-3y^2)^2$$
$$=49x^4-64x^2y^2+48xy^3-9y^4.$$

5. $(x^3+2x^2y+2xy^2+y^3)(x^3-2x^2y+2xy^2-y^3)$

$$=(\overline{x^3+2xy^2}+\overline{2x^2y+y^3})(\overline{x^3+2xy^2}-\overline{2x^2y+y^3})$$
$$=(x^3+2xy^2)^2-(2x^2y+y^3)^2$$
$$=x^6-y^6.$$

6. $\{\overline{(x+y)^2+4}+2(x+y)\}\{(x+y)^2+4-2(x+y)\}=\{(x+y)^2+4\}^2-4(x+y)^2$

$$=(x+y)^4+4(x+y)^2+16.$$

7. The first expression

$$=\{(1+x+2x^2)+(1-x-2x^2)\}\{(1+x+2x^2)-(1-x-2x^2)\}$$
$$=2(2x+4x^2)=4x(1+2x).$$

The second expression

$$=\{(1+x-2x^2)+(1-x+2x^2)\}\{(1+x-2x^2)-(1-x+2x^2)\}$$
$$=2(2x-4x^2)=4x(1-2x).$$

$\therefore$ the product $=16x^2(1+2x)(1-2x)=16x^2(1-4x^2).$

8. The first expression

$$=\{(a^2+3a-1)+(a^2-3a-1)\}\{(a^2+3a-1)-(a^2-3a-1)\}$$
$$=(2a^2-2)\,.\,6a=12a(a^2-1).$$

The second expression

$$=\{(a^2+a+1)+(a^2-a+1)\}\{(a^2+a+1)-(a^2-a+1)\}$$
$$=(2a^2+2)2a=4a(a^2+1).$$

$\therefore$ the product $=48a^2(a^2-1)(a^2+1)=48a^2(a^4-1).$

9. $\{(x^3+8x)-(4x^2+8)\}\{(x^3+8x)+(4x^2+8)\}$

$$=(x^3+8x)^2-(4x^2+8)^2$$
$$=x^6+16x^4+64x^2-16x^4-64x^2-64$$
$$=x^6-64.$$

10. $\{(x^3+18a^2x)-(6ax^2+27a^3)\}\{(x^3+18a^2x)+(6ax^2+27a^3)\}$

$$=(x^3+18a^2x)^2-(6ax^2+27a^3)^2$$
$$=x^6+36a^2x^4+324a^4x^2-36a^2x^4-324a^4x^2-729a^6$$
$$=x^6-729a^6.$$

11. $\left\{\left(x-\frac{a^2}{x}\right)-\left(a+\frac{x^2}{a}\right)\right\}\left\{\left(x-\frac{a^2}{x}\right)+\left(a+\frac{x^2}{a}\right)\right\}$

$$=\left(x-\frac{a^2}{x}\right)^2-\left(a+\frac{x^2}{a}\right)^2=x^2-2a^2+\frac{a^4}{x^2}-a^2-2x^2-\frac{x^4}{a^2}$$

$$=\frac{a^4}{x^2}-x^2-3a^2-\frac{x^4}{a^2}.$$

12. The first expression

$$=\{(2x^2+3x+1)+(2x^2-3x+1)\}\{(2x^2+3x+1)-(2x^2-3x-1)\}$$
$$=4x^2(6x+2)=8x^2(3x+1).$$

The second expression

$$=\{(x^2+6x-2)+(x^2-6x+2)\}\{(x^2+6x-2)-(x^2-6x+2)\}$$
$$=2x^2(12x-4)=8x^2(3x-1).$$

$\therefore$ the product $=16x^4(3x+1)(3x-1)=64x^4(9x^2-1)$.

13. $(x^2+ax+a^2)(x^2-ax+a^2)(x^4-a^2x^2+a^4)=(x^4+a^2x^2+a^4)(x^4-a^2x^2+a^4)$

$$=x^8+a^4x^4+a^8.$$

14. $(1-x+x^2)(1+x+x^2)(1-x^2+x^4)(1-x^4+x^8)$

$$=(1+x^2+x^4)(1-x^2+x^4)(1-x^4+x^8)$$
$$=(1+x^4+x^8)(1-x^4+x^8)=1+x^8+x^{16}.$$

15. The product $=\{(a-x)(a+x)(a^2+x^2)\}^3=\{(a^2-x^2)(a^2+x^2)\}^3$

$$=(a^4-x^4)^3=a^{12}-3a^8x^4+3a^4x^8-x^{12}.$$

16. The product $=\{(1-x)(1+x)(1+x^2)(1+x^4)\}^2=\{(1-x^2)(1+x^2)(1+x^4)\}^2$

$$=\{(1-x^4)(1+x^4)\}^2=(1-x^8)^2=1-2x^8+x^{16}.$$

17. $(x^2+4x+3)(x^2+x-2)(x^2-5x+6)$

$$=(x+1)(x+3)(x+2)(x-1)(x-2)(x-3)$$
$$=(x+1)(x-1)(x+2)(x-2)(x+3)(x-3)$$
$$=(x^2-1)(x^2-4)(x^2-9)$$
$$=x^6-14x^4+49x^2-36.$$

18. $(x^2+2x-3)(x^2-5x+6)(x^2+3x+2)$

$$=(x+3)(x-1)(x-3)(x-2)(x+2)(x+1)$$
$$=(x+1)(x-1)(x+2)(x-2)(x+3)(x-3)$$
$$=(x^2-1)(x^2-4)(x^2-9)$$
$$=x^6-14x^4+49x^2-36.$$

19. $(x+2)(x^2-2x+4)(x-2)(x^2+2x+4)=(x^3+8)(x^3-8)=x^6-64.$

20. The product $=(a+3b)^2\times(a-3b)^2=(a^2-9b^2)^2=a^4-18a^2b^2+81b^4.$

21. The product $=\frac{1}{2}(a^2-2ab+b^2+b^2-2bc+c^2+c^2-2ac+a^2)(a+b+c)$

$$=(a^2+b^2+c^2-bc-ca-ab)(a+b+c)$$
$$=a^3+b^3+c^3-3abc. \quad \text{[Art. 213].}$$

22. The quotient $=\dfrac{(4x+3y-2z)^2-(3x-2y+3z)^2}{x+5y-5z}$

$$=\frac{(4x+3y-2z+3x-2y+3z)(4x+3y-2z-3x+2y-3z)}{x+5y-5z}$$
$$=\frac{(7x+y+z)(x+5y-5z)}{x+5y-5z}$$
$$=7x+y+z.$$

23. The quotient $=\dfrac{x^8+16a^4x^4+256a^8}{x^2+2ax+4a^2}$

$$=\frac{(x^4+4a^2x^2+16a^4)(x^4-4a^2x^2+16a^4)}{x^2+2ax+4a^2}$$
$$=\frac{(x^2+2ax+4a^2)(x^2-2ax+4a^2)(x^4-4a^2x^2+16a^4)}{x^2+2ax+4a^2}$$
$$=(x^2-2ax+4a^2)(x^4-4a^2x^2+16a^4).$$

24. The quotient $=\dfrac{(3x+4y-2z)^2-(2x+3y-4z)^2}{x+y+2z}$

$$=\frac{(3x+4y-2z+2x+3y-4z)(3x+4y-2z-2x-3y+4z)}{x+y+2z}$$
$$=\frac{(5x+7y-6z)(x+y+2z)}{x+y+2z}$$
$$=5x+7y-6z.$$

25. The quotient $=\dfrac{(x^2+7x+10)(x+3)}{x^2+5x+6}=\dfrac{(x+5)(x+2)(x+3)}{(x+2)(x+3)}=x+5.$

26. The quotient $=\dfrac{2x(x^2-1)(x+2)}{x^2+x-2}=\dfrac{2x(x+1)(x-1)(x+2)}{(x+2)(x-1)}=2x(x+1).$

27. The quotient $=\dfrac{5x(x-11)(x^2-x-156)}{x^3+x^2-132x}=\dfrac{5x(x-11)(x+12)(x-13)}{x(x+12)(x-11)}$

$$=5(x-13).$$

28. The quotient $=\dfrac{x^6+19x^3-216}{(x^2-3x+9)(x-2)}=\dfrac{(x^3+27)(x^3-8)}{(x^2-3x+9)(x-2)}$

$$=(x+3)(x^2+2x+4).$$

29. The quotient $=\dfrac{(5x^2-3x-6)^2-(2x^2-7x+9)^2}{(3x-5)(x+3)}$

$=\dfrac{(7x^2-10x+3)(3x^2+4x-15)}{3x^2+4x-15}$

$=7x^2-10x+3.$

30. The quotient $=\dfrac{a^9-b^9}{(a^2+ab+b^2)(a^6+a^3b^3+b^6)}=\dfrac{(a^3-b^3)(a^6+a^3b^3+b^6)}{(a^2+ab+b^2)(a^6+a^3b^3+b^6)}$

$=a-b.$

31. The quotient $=\dfrac{(x^3-3x^2y)^2-(3xy^2-y^3)^2}{(x-y)^3}$

$=\dfrac{(x^3-3x^2y+3xy^2-y^3)(x^3-3x^2y-3xy^2+y^3)}{x^3-3x^2y+3xy^2-y^3}$

$=x^3-3x^2y-3xy^2+y^3.$

32. The quotient $=\dfrac{(x^2-yz)^3+(2yz)^3}{(x^2-yz)+2yz}$

$=(x^2-yz)^2-2yz(x^2-yz)+4y^2z^2.$

33. The quotient $=\dfrac{27x^3-8y^3+1+18xy}{3x-2y+1}$

$=\dfrac{(3x)^3+(-2y)^3+(1)^3-3(3x)(-2y)(1)}{3x-2y+1}$

$=9x^2+4y^2+1-3x+2y+6xy.$

34. The quotient $=\dfrac{(2x^2+3x-1)^2-(x^2+4x+5)^2}{(3x+4)(x+2)}$

$=\dfrac{(3x^2+7x+4)(x^2-x-6)}{(3x+4)(x+2)}$

$=(x+1)(x-3).$

35. The quotient $=\dfrac{(6a^2-23a+20)(22a^2-81a+14)}{33a^2-50a+8}$

$=\dfrac{(3a-4)(2a-5)(11a-2)(2a-7)}{(3a-4)(11a-2)}$

$=(2a-5)(2a-7).$

36. The quotient $=\dfrac{\{x^2+(a-b)x-ab\}\{x^2-(a-b)x-ab\}}{x^2+(a+b)x+ab}$

$=\dfrac{(x+a)(x-b)(x-a)(x+b)}{(x+a)(x+b)}$

$=(x-a)(x-b).$

37. The quotient $=\dfrac{a^3-27x^3-8y^3-18axy}{a-3x-2y}$

$$=\frac{a^3+(-3x)^3+(-2y)^3-3a(-3x)(-2y)}{a-3x-2y}$$

$$=a^2+9x^2+4y^2+3ax+2ay-6xy.$$

38. The quotient $=\dfrac{27-8x^3-64y^3-72xy}{3-2x-4y}$

$$=\frac{(3)^3+(-2x)^3+(-4y)^3-3(3)(-2x)(-4y)}{3-2x-4y}$$

$$=9+4x^2+16y^2+6x+12y-8xy.$$

39. The expression being of the form A^3-B^3 has a factor of the form $A-B$; therefore $(2x-3y+1)^3-(1-3x+2y)^3$ is divisible by

$$(2x-3y+1)-(1-3x+2y);$$

that is, by $5(x-y)$.

40. The expression being of the form A^3-B^3 has a factor of the form $A-B$; therefore $(x^3+x^2+4)^3-(x^3-2x+3)^3$ is divisible by

$$(x^3+x^2+4)-(x^3-2x+3);$$

that is, by x^2+2x+1.

41. The expression being of the form A^3-B^3 has a factor of the form $A-B$; therefore $(a+b+c+d)^3-(a-b+c-d)^3$ is divisible by

$$(a+b+c+d)-(a-b+c-d);$$

that is, by $2b+2d$.

42. The expression being of the form A^3-B^3 has a factor of the form $A-B$; therefore $(3x^2-7x+2)^3-(x^2-8x+8)^3$ is divisible by

$$(3x^2-7x+2)-(x^2-8x+8);$$

that is, by $2x^2+x-6$, or $(2x-3)(x+2)$.

43. The expression being of the form A^3+B^3 has a factor of the form $A+B$; therefore $(7x^2+3x-3)^3+(5x^2-4x-3)^3$ is divisible by

$(7x^2+3x-3)+(5x^2-4x-3)$; that is, by $12x^2-x-6$, or $(4x-3)(3x+2)$.

44. The sum of the cubes of $2x^2-5x-9$ and x^2+6x-5 is divisible by $(2x^2-5x-9)+(x^2+6x-5)$; that is, by $3x^2+x-14$, or $(3x+7)(x-2)$.

45.
$$x^3+y^3=(x+y)(x^2-xy+y^2)=(x+y)\{(x-y)^2+xy\}$$

$$=m(n^2+xy)=mn^2+\frac{m}{4}\{(x+y)^2-(x-y)^2\}$$

$$=mn^2+\frac{m}{4}(m^2-n^2)=\frac{m(m^2+3n^2)}{4}.$$

46. $$16(x^4 - 7x^2y^2 + y^4) = 16\{(x^2-y^2)^2 - 5x^2y^2\}$$
$$= 16\left\{(x^2-y^2)^2 - 5\left[\frac{(x+y)^2-(x-y)^2}{4}\right]^2\right\}$$
$$= 16(x+y)^2(x-y)^2 - 5[(x+y)^2-(x-y)^2]^2$$
$$= 16m^2n^2 - 5(m^2-n^2)^2$$
$$= -5m^4 + 26m^2n^2 - 5n^4$$
$$= (5m^2-n^2)(5n^2-m^2).$$

47. $$x^4 + x^2y^2 + y^4 = (x^2+xy+y^2)(x^2-xy+y^2)$$
$$= \{(x+y)^2 - xy\}\{(x-y)^2 + xy\}.$$

Now $4xy = (x+y)^2 - (x-y)^2 = 4a^2 - 4b^2$;

$\therefore$ by substitution we have $(4a^2 - a^2 + b^2)(4b^2 + a^2 - b^2)$,

or $(3a^2+b^2)(3b^2+a^2)$.

48. $$x^4 - 23x^2y^2 + y^4 = (x^2-y^2)^2 - 21x^2y^2$$
$$= (x+y)^2(x-y)^2 - 21\left\{\frac{(x+y)^2-(x-y)^2}{4}\right\}^2$$
$$= 4a^2 \,.\, 4b^2 - 21\left(\frac{4a^2-4b^2}{4}\right)^2$$
$$= 16a^2b^2 - 21(a^4 - 2a^2b^2 + b^4)$$
$$= -21a^4 + 58a^2b^2 - 21b^4$$
$$= (7a^2-3b^2)(7b^2-3a^2).$$

49. $$x^4 - 47x^2y^2 + y^4 = (x^2-y^2)^2 - 45x^2y^2$$
$$= (x+y)^2(x-y)^2 - 45\left\{\frac{(x+y)^2-(x-y)^2}{4}\right\}^2$$
$$= p^2q^2 - \frac{45}{16}(p^2-q^2)^2$$
$$= \frac{1}{16}(106p^2q^2 - 45p^4 - 45q^4)$$
$$= \frac{1}{16}(9p^2-5q^2)(9q^2-5p^2).$$

50. $$x^4 - 2x^3y + 2xy^3 - y^4 = x^4 - y^4 - 2xy(x^2-y^2)$$
$$= (x^2-y^2)(x^2+y^2-2xy) = (x-y)^3(x+y)$$
$$= (a+b-a+b)^3(a+b+a-b)$$
$$= (2b)^3(2a) = 16ab^3.$$

EXAMPLES. XXIX. a. PAGE 223.

1. $$\begin{array}{l|l}
x^2+(a+b)x+ab & x^3+(a+b+c)x^2+(bc+ca+ab)x+abc \mid x+c \\
 & x^3+(a+b)\quad x^2+abx \\ \hline
 & \qquad cx^2+(bc+ca)\,x+abc \\
 & \qquad cx^2+(ac+bc)\,x+abc \\ \hline
\end{array}$$

2. $$\begin{array}{l|l}
x^2-5x+4 & x^4-(5+a)x^3+(4+5a+b)x^2-(4a+5b)x+4b \mid x^2-xa+b \\
 & x^4-\quad 5x^3+\quad 4x^2 \\ \hline
 & \quad -ax^3+\quad (5a+b)x^2-(4a+5b)x \\
 & \quad -ax^3+\quad 5ax^2-\quad 4ax \\ \hline
 & \qquad bx^2-\quad 5bx+4b \\
 & \qquad bx^2-\quad 5bx+4b \\ \hline
\end{array}$$

3. $$\begin{array}{l|l}
x-b & x^3-(a-\ b)x^2-(ab+2b^2)x+2ab^2 \mid x^2-(a-2b)x-2ab \\
 & x^3-\quad bx^2 \\ \hline
 & \quad -(a-2b)x^2-(ab+2b^2)x \\
 & \quad -(a-2b)x^2+(ab-2b^2)x \\ \hline
 & \qquad -2abx+2ab^2 \\
 & \qquad -2abx+2ab^2 \\ \hline
\end{array}$$

4. $$\begin{array}{l|l}
x+p+q & x^3-(p^2+3q^2)x\ +2p^2q-2q^3 \mid x^2-(p+q)x+2pq-2q^2 \\
 & x^3+(p+q)x^2 \\ \hline
 & \quad -(p+q)x^2-(p^2+3q^2)x \\
 & \quad -(p+q)x^2-(p^2+2pq+q^2)x \\ \hline
 & \qquad (2pq-2q^2)x+2p^2q-2q^3 \\
 & \qquad (2pq-2q^2)x+2p^2q-2q^3 \\ \hline
\end{array}$$

5. $$\begin{array}{l|l}
x+m+n & x^3-(3mn+n^2)x+m(m^2-n^2) \mid x^2-(m+n)x+m^2-mn \\
 & x^3+(m+n)x^2 \\ \hline
 & \quad -(m+n)x^2-\quad (3mn+n^2)x \\
 & \quad -(m+n)x^2-(m^2+2mn+n^2)x \\ \hline
 & \qquad (m^2-\ mn)x+m^3-mn^2 \\
 & \qquad (m^2-\ mn)x+m^3-mn^2 \\ \hline
\end{array}$$

6. $$\begin{array}{l|l}
(a-1)x+a & a(a-1)x^2+(2a^2-1)x+a(a+1) \mid ax+a+1 \\
 & a(a-1)x^2+\quad a^2x \\ \hline
 & \qquad (a^2-1)x+a(a+1) \\
 & \qquad (a^2-1)x+a(a+1) \\ \hline
\end{array}$$

7. $$\begin{array}{l|l}
x^2+ax+b^2 & x^4+(a+b)x^3+(a^2+ab+b^2)x^2+(a^3+b^3)x+a^2b^2 \mid x^2+bx+a^2 \\
 & x^4+\quad ax^3+\quad b^2x^2 \\ \hline
 & \quad bx^3+\quad (a^2+ab)x^2+(a^3+b^3)x \\
 & \quad bx^3+\quad abx^2+\quad b^3x \\ \hline
 & \qquad a^2x^2+\quad a^3x+a^2b^2 \\
 & \qquad a^2x^2+\quad a^3x+a^2b^2 \\ \hline
\end{array}$$

8.

$$\begin{array}{l|l}
lx+2(m-n)y & 2l^2x^2+lmxy-2(3m-4n)(m-n)y^2 \mid 2lx-(3m-4n)y \\
& \underline{2l^2x^2+4l(m-n)xy} \\
& \qquad -(3m-4n)lxy-2(3m-4n)(m-n)y^2 \\
& \qquad \underline{-(3m-4n)lxy-2(3m-4n)(m-n)y^2}
\end{array}$$

9.

$$\begin{array}{l|l}
(a-1)x-ay & (a^2+a-2)x^2-(2a+1)xy-(a^2+a)y^2 \mid (a+2)x+(a+1)y \\
& \underline{(a^2+a-2)x^2-(a^2+2a)xy} \\
& \qquad (a^2-1)xy-(a^2+a)y^2 \\
& \qquad \underline{(a^2-1)xy-(a^2+a)y^2}
\end{array}$$

10.

$$\begin{array}{l|l}
x^2-(a-2)x-2a & x^3-(a-b-2)x^2-(ab+2a-2b)x-2ab \mid x+b \\
& \underline{x^3-(a-2)x^2-2ax} \\
& \qquad bx^2-(ab-2b)x-2ab \\
& \qquad \underline{bx^2-(ab-2b)x-2ab}
\end{array}$$

11.

$$\begin{array}{l|l}
(x+1)^2+1 & (x+1)^8+4(x+1)^6+6(x+1)^4+4(x+1)^2+1 \mid (x+1)^6+3(x+1)^4+3(x+1)^2+1 \\
& \underline{(x+1)^8+(x+1)^6} \\
& \quad 3(x+1)^6+6(x+1)^4 \\
& \quad \underline{3(x+1)^6+3(x+1)^4} \\
& \qquad 3(x+1)^4+4(x+1)^2 \\
& \qquad \underline{3(x+1)^4+3(x+1)^2} \\
& \qquad\quad (x+1)^2+1 \\
& \qquad\quad \underline{(x+1)^2+1}
\end{array}$$

12. $(m+1)(bx+an)\,b^2x^2-(n+1)(mbx+a)\,a^2=(m+1)\,b^3x^3+n\,(m+1)\,ab^2x^2-(n+1)\,ma^2bx-(n+1)\,a^3$

$$
\begin{array}{r|l|l}
bx-a & (m+1)\,b^3x^3+n\,(m+1)\,ab^2x^2-(n+1)\,ma^2bx-(n+1)\,a^3 & (m+1)\,b^2x^2+(n+1)(m+1)\,abx+(n+1)\,a^2 \\
 & (m+1)\,b^3x^3-\quad(m+1)\,ab^2x^2 & \\ \hline
 & \quad(n+1)(m+1)\,ab^2x^2-(n+1)\,ma^2bx & \\
 & \quad(n+1)(m+1)\,ab^2x^2-(n+1)(m+1)\,a^2bx & \\ \hline
 & \qquad(n+1)\,a^2bx-(n+1)\,a^3 & \\
 & \qquad(n+1)\,a^2bx-(n+1)\,a^3 & \\ \hline
\end{array}
$$

13.

$$
\begin{array}{r|l|l}
1 & (m^2-3m+2)\,x^2+(2m^2-4m+1)\,x+m\,(m-1) & m\,(m-1)\,x^2+(2m^2-1)\,x+m\,(m+1)\ \big|\ m \\
 & \quad m\,(m-1)\,x^2+\quad(2m^2-1)\,x+m\,(m+1) & m\,(m-1)\,x^2+(2m^2-m)\,x+m^2 \\ \hline
-2 & -2\,(m-1)\,x^2-\quad2\,(2m-1)\,x-2m & \qquad(m-1)\,x+m \\ \hline
x & \quad(m-1)\,x^2+\quad(2m-1)\,x+m & \\
 & \quad(m-1)\,x^2+\quad mx & \\ \hline
1 & \qquad(m-1)\,x+m & \therefore\ \text{H. C. F. is } (m-1)\,x+m. \\
 & \qquad(m-1)\,x+m & \\ \hline
\end{array}
$$

14. $mpx^3+(mq-np)\,x^2-(mr+nq)\,x+nr$

$$
\begin{array}{|l}
max^3-\quad(mc+na)\,x^2-\quad(mb-nc)\,x+\ nb \\
\quad p \\ \hline
mapx^3-(mpc+npa)\,x^2-(mpb-npc)\,x+npb\ \big|\ a \\
mapx^3+(mqa-npa)\,x^2-(mra+nqa)\,x+nra \\ \hline
\quad-\ m\,(pc+qa)\,x^2-(mpb-npc-mra-nqa)\,x+npb-nra
\end{array}
$$

The remainder $=-\{(pc+qa)\,x+pb-ra\}\,\{mx-n\}$ of which the former may clearly be rejected. Therefore if there is a common factor it must be $mx-n$, and by division we find it is a factor.

$\therefore$ H. C. F. is $mx-n$.

15.
$$2ap^3+(3a-2b)p^2q+(a-3b)pq^2-bq^3 \left| \begin{array}{l} 3ap^3-(a+3b)p^2q+(2a+b)pq^2-2bq^3 \\ 2 \\ \hline 6ap^3-(2a+6b)p^2q+(4a+2b)pq^2-4bq^3 \quad | \; 3 \\ 6ap^3+(9a-6b)p^2q+(3a-9b)pq^2-3bq^3 \\ \hline \qquad -11ap^2q+(a+11b)pq^2-bq^3 \end{array} \right.$$

The remainder $=-11pq(ap-bq)+q^2(ap-bq)=q(q-11p)(ap-bq)$, of which q and $q-11p$ are clearly not factors, but $ap-bq$ is a factor of the two expressions.

$\therefore$ H. C. F. is $ap-bq$.

16.
$$acx^3+(bc+ad)x^2+(bd+ac)x+bc \left| \begin{array}{l} 2acx^3+(2bc-ad)x^2-(3ac+bd)x-3bc \quad | \; 2 \\ 2acx^3+(2bc+2ad)x^2+(2bd+2ac)x+2bc \\ \hline \qquad -3adx^2-(5ac+3bd)x-5bc \end{array} \right.$$

The remainder $=-3dx(ax+b)-5c(ax+b)=-(3dx+5c)(ax+b)$, of which $3dx+5c$ is clearly not a factor, but $ax+b$ is a factor of the two expressions.

$\therefore$ H. C. F. is $ax+b$.

17.
$$2a^2x^3-(4b+3)ax^2+2(3b-ac)x+3c \left| \begin{array}{l} 2a^2x^3+(2b-3)ax^2-(4ac+3b)x+6c \quad | \; 1 \\ 2a^2x^3-(4b+3)ax^2+(6b-2ac)x+3c \\ \hline \qquad 6abx^2-(2ac+9b)x+3c \end{array} \right.$$

The remainder $=(2ax-3)(3bx-c)$, of which $3bx-c$ is clearly not a factor, but $2ax-3$ is a factor of the two expressions.

$\therefore$ H. C. F. is $2ax-3$.

18.
$$2 \left| \begin{array}{l} 2ax^3+(4a^2-1)bx^2-(2ab^2+3c)x-6abc \\ 2ax^3-(6-4a^2)bx^2+(4c-12ab^2)x+8abc \\ \hline \qquad 5bx^2+(10ab^2-7c)x-14abc \end{array} \right| \; ax^3-(3-2a^2)bx^2+(2c-6ab^2)x+4abc$$

The remainder $=(5bx-7c)(x+2ab)$, of which clearly $5bx-7c$ is not a factor, but $x+2ab$ is a factor of the two expressions.

$\therefore$ H. C. F. is $x+2ab$.

19.

$$
\begin{array}{r|l}
x & x^4 - px^3 + (q-1)x^2 + px - q \\
 & x^4 + x^3 - x^2 - x \\ \hline
-(p+1) & -(p+1)x^3 + qx^2 - (p+1)x - q \\
 & -(p+1)x^3 - (p+1)x^2 + (p+1)x + (p+1) \\ \hline
 & p+q+1 \,\vdots\, (p+q+1)x^2 - (p+q+1) \\
 & \qquad x^2 - 1
\end{array}
\qquad
\begin{array}{r|l|l}
 & x^4 - qx^3 + (p-1)x^2 + qx - p & 1 \\
 & x^4 - px^3 + (q-1)x^2 + px - q & \\ \hline
p-q & (p-q)x^3 + (p-q)x^2 - (p-q)x - (p-q) & x+1 \\
 & x^3 + x^2 - x - 1 & \\
 & x^3 \qquad - x & \\ \hline
 & x^2 - 1 & \\
 & x^2 - 1 & \\ \hline
\end{array}
$$

$\therefore$ H. C. F. is $x^2 - 1$.

Again, $x^4 - px^3 + (q-1)x^2 + px - q = (x^2 - px + q)(x^2 - 1)$;

$x^4 - qx^3 + (p-1)x^2 + qx - p = (x^2 - qx + p)(x^2 - 1)$.

$\therefore$ L. C. M. is $(x^2 - px + q)(x^2 - qx + p)(x^2 - 1)$.

20. Here $p(p+1)x^2 + x - p(p-1) = \{(p+1)x + p\}\{px - (p-1)\}$,

and $p(p+2)x^2 + 2x - (p^2 - 1) = \{(p+2)x + p + 1\}\{px - (p-1)\}$.

$\therefore$ L. C. M. is $\{px - (p-1)\}\{(p+1)x + p\}\{(p+2)x + p + 1\}$.

21. Here $(a^2 - 5a + 6)x^2 + 2(a-1)x - a(a+1) = (a-2)(a-3)x^2 + 2(a-1)x - a(a+1)$

$= \{(a-2)x - a\}\{(a-3)x + a + 1\}$,

and $a(a-3)x^2 + 12x - (a+1)(a+4) = \{(a-3)x + a + 1\}\{ax - (a+4)\}$.

$\therefore$ L. C. M. is $\{(a-3)x + a + 1\}\{(a-2)x - a\}\{ax - (a+4)\}$.

EXAMPLES. XXIX. b. Page 227.

In Examples 1—5 it will be sufficient to find the square root; the remaining step is evident.

1.

$$
\begin{array}{l|l}
 & x^4 - 28x^3 + 294x^2 - 1372x + 2401 \;(\; x^2 - 14x + 49 \\
 & x^4 \\ \hline
2x^2 - 14x & -28x^3 + 294x^2 \\
 & -28x^3 + 196x^2 \\ \hline
2x^2 - 28x + 49 & 98x^2 - 1372x + 2401 \\
 & 98x^2 - 1372x + 2401 \\ \hline
\end{array}
$$

2.

$$
\begin{array}{l|l}
 & 16 - \frac{32}{m} + \frac{24}{m^2} - \frac{8}{m^3} + \frac{1}{m^4} \;\Big(\; 4 - \frac{4}{m} + \frac{1}{m^2} \\
 & 16 \\ \hline
8 - \frac{4}{m} & -\frac{32}{m} + \frac{24}{m^2} \\
 & -\frac{32}{m} + \frac{16}{m^2} \\ \hline
8 - \frac{8}{m} + \frac{1}{m^2} & \frac{8}{m^2} - \frac{8}{m^3} + \frac{1}{m^4} \\
 & \frac{8}{m^2} - \frac{8}{m^3} + \frac{1}{m^4} \\ \hline
\end{array}
$$

3.

$$
\begin{array}{l|l}
 & a^4 + 8a^3x + 24a^2x^2 + 32ax^3 + 16x^4 \;(\; a^2 + 4ax + 4x^2 \\
 & a^4 \\ \hline
2a^2 + 4ax & 8a^3x + 24a^2x^2 \\
 & 8a^3x + 16a^2x^2 \\ \hline
2a^2 + 8ax + 4x^2 & 8a^2x^2 + 32ax^3 + 16x^4 \\
 & 8a^2x^2 + 32ax^3 + 16x^4 \\ \hline
\end{array}
$$

4.

$$\begin{array}{l|l}
 & 1+4x+2x^2-8x^3-5x^4+8x^5+2x^6-4x^7+x^8 \;(\; 1+2x-x^2-2x^3+x^4 \\
 & 1 \\
\hline
2+2x & 4x+2x^2 \\
 & 4x+4x^2 \\
\hline
2+4x-x^2 & -2x^2-8x^3-5x^4 \\
 & -2x^2-4x^3+x^4 \\
\hline
2+4x-2x^2-2x^3 & -4x^3-6x^4+8x^5+2x^6 \\
 & -4x^3-8x^4+4x^5+4x^6 \\
\hline
2+4x-2x^2-4x^3+x^4 & 2x^4+4x^5-2x^6-4x^7+x^8 \\
 & 2x^4+4x^5-2x^6-4x^7+x^8 \\
\hline
\end{array}$$

5.

$$\begin{array}{l|l}
 & 1+8x+20x^2+8x^3-26x^4-8x^5+20x^6-8x^7+x^8 \;(\; 1+4x+2x^2-4x^3+x^4 \\
 & 1 \\
\hline
2+4x & 8x+20x^2 \\
 & 8x+16x^2 \\
\hline
2+8x+2x^2 & 4x^2+8x^3-26x^4 \\
 & 4x^2+16x^3+4x^4 \\
\hline
2+8x+4x^2-4x^3 & -8x^3-30x^4-8x^5+20x^6 \\
 & -8x^3-32x^4-16x^5+16x^6 \\
\hline
2+8x+4x^2-8x^3+x^4 & 2x^4+8x^5+4x^6-8x^7+x^8 \\
 & 2x^4+8x^5+4x^6-8x^7+x^8 \\
\hline
\end{array}$$

In Examples 6—8 it will be sufficient to find the square root; the remaining step is evident.

6.

$$\begin{array}{l|l}
 & 1+6x+15x^2+20x^3+15x^4+6x^5+x^6 \;(\; 1+3x+3x^2+x^3 \\
 & 1 \\
\hline
2+3x & 6x+15x^2 \\
 & 6x+9x^2 \\
\hline
2+6x+3x^2 & 6x^2+20x^3+15x^4 \\
 & 6x^2+18x^3+9x^4 \\
\hline
2+6x+6x^2+x^3 & 2x^3+6x^4+6x^5+x^6 \\
 & 2x^3+6x^4+6x^5+x^6 \\
\hline
\end{array}$$

7.

$$
\begin{array}{l}
x^6 - 12ax^5 + 60a^2x^4 - 160a^3x^3 + 240a^4x^2 - 192a^5x + 64a^6 \;(\; x^3 - 6ax^2 + 12a^2x - 8a^3 \\
x^6 \\
2x^3 - 6ax^2 \;\big|\; -12ax^5 + 60a^2x^4 \\
\qquad\qquad\; -12ax^5 + 36a^2x^4 \\
2x^3 - 12ax^2 + 12a^2x \;\big|\; 24a^2x^4 - 160a^3x^3 + 240a^4x^2 \\
\qquad\qquad\qquad\qquad\; 24a^2x^4 - 144a^3x^3 + 144a^4x^2 \\
2x^3 - 12ax^2 + 24a^2x - 8a^3 \;\big|\; -16a^3x^3 + 96a^4x^2 - 192a^5x + 64a^6 \\
\qquad\qquad\qquad\qquad\qquad\; -16a^3x^3 + 96a^4x^2 - 192a^5x + 64a^6
\end{array}
$$

8.

$$
\begin{array}{l}
a^6 - 18a^5x + 135a^4x^2 - 540a^3x^3 + 1215a^2x^4 - 1458ax^5 + 729x^6 \;(\; a^3 - 9a^2x + 27ax^2 - 27x^3 \\
a^6 \\
2a^3 - 9a^2x \;\big|\; -18a^5x + 135a^4x^2 \\
\qquad\qquad\; -18a^5x + 81a^4x^2 \\
2a^3 - 18a^2x + 27ax^2 \;\big|\; 54a^4x^2 - 540a^3x^3 + 1215a^2x^4 \\
\qquad\qquad\qquad\qquad\; 54a^4x^2 - 486a^3x^3 + 729a^2x^4 \\
2a^3 - 18a^2x + 54ax^2 - 27x^3 \;\big|\; -54a^3x^3 + 486a^2x^4 - 1458ax^5 + 729a^6 \\
\qquad\qquad\qquad\qquad\qquad\; -54a^3x^3 + 486a^2x^4 - 1458ax^5 + 729a^6
\end{array}
$$

9.

$$
\begin{array}{l}
x^8 - 8x^7y + 28x^6y^2 - 56x^5y^3 + 70x^4y^4 - 56x^3y^5 + 28x^2y^6 - 8xy^7 + y^8 \;(\; x^4 - 4x^3y + 6x^2y^2 - 4xy^3 + y^4 \\
x^8 \\
2x^4 - 4x^3y \;\big|\; -8x^7y + 28x^6y^2 \\
\qquad\qquad\; -8x^7y + 16x^6y^2 \\
2x^4 - 8x^3y + 6x^2y^2 \;\big|\; 12x^6y^2 - 56x^5y^3 + 70x^4y^4 \\
\qquad\qquad\qquad\qquad\; 12x^6y^2 - 48x^5y^3 + 36x^4y^4 \\
2x^4 - 8x^3y + 12x^2y^2 - 4xy^3 \;\big|\; -8x^5y^3 + 34x^4y^4 - 56x^3y^5 + 28x^2y^6 \\
\qquad\qquad\qquad\qquad\qquad\; -8x^5y^3 + 32x^4y^4 - 48x^3y^5 + 16x^2y^6 \\
2x^4 - 8x^3y + 12x^2y^2 - 8xy^3 + y^4 \;\big|\; 2x^4y^4 - 8x^3y^5 + 12x^2y^6 - 8xy^7 + y^8 \\
\qquad\qquad\qquad\qquad\qquad\qquad\; 2x^4y^4 - 8x^3y^5 + 12x^2y^6 - 8xy^7 + y^8
\end{array}
$$

The square root of $x^4 - 4x^3y + 6x^2y^2 - 4xy^3 + y^4$ is $x^2 - 2xy + y^2$, and the square root of this last expression is $x - y$.

10. The eighth root is the square root of the fourth root.

$$x^4+2(p-1)x^3+(p^2-2p-1)x^2-2(p-1)x+1 \;\big|\; x^2+(p-1)x-1$$
$$x^4$$
$$2x^2+(p-1)x \;\big|\; 2(p-1)x^3+(p^2-2p-1)x^2$$
$$ 2(p-1)x^3+(p^2-2p+1)x^2$$
$$2x^2+2(p-1)x-1 \;\big|\; -2x^2-2(p-1)x+1$$
$$ -2x^2-2(p-1)x+1$$

11. $1+x \left(1+\frac{x}{2}-\frac{x^2}{8}+\frac{x^3}{16}\right.$

$$1$$
$$2+\frac{x}{2} \;\Big|\; x$$
$$\phantom{2+\frac{x}{2} \;\Big|\;} x+\frac{x^2}{4}$$
$$2+x-\frac{x^2}{8} \;\Big|\; -\frac{x^2}{4}$$
$$\phantom{2+x-\frac{x^2}{8} \;\Big|\;} -\frac{x^2}{4}-\frac{x^3}{8}+\frac{x^4}{64}$$
$$2+x-\frac{x^2}{4}+\frac{x^3}{16} \;\Big|\; \frac{x^3}{8}-\frac{x^4}{64}$$
$$\phantom{2+x-\frac{x^2}{4}+\frac{x^3}{16} \;\Big|\;} \frac{x^3}{8}+\frac{x^4}{16}-\frac{x^5}{64}+\frac{x^6}{256}$$
$$-\frac{5x^4}{16}+\frac{x^5}{64}-\frac{x^6}{256}$$

12. $1-2x \left(1-x-\frac{x^2}{2}-\frac{x^3}{2}\right.$

$$1$$
$$2-x \;\Big|\; -2x$$
$$ -2x+x^2$$
$$2-2x-\frac{x^2}{2} \;\Big|\; -x^2$$
$$\phantom{2-2x-\frac{x^2}{2} \;\Big|\;} -x^2+x^3+\frac{x^4}{4}$$
$$2-2x-x^2-\frac{x^3}{2} \;\Big|\; -x^3-\frac{x^4}{4}$$
$$\phantom{2-2x-x^2-\frac{x^3}{2} \;\Big|\;} -x^3+x^4+\frac{x^5}{2}+\frac{x^6}{4}$$
$$-\frac{5x^4}{4}-\frac{x^5}{2}-\frac{x^6}{4}$$

13. $4+2x \left(2+\frac{x}{2}-\frac{x^2}{16}+\frac{x^3}{64}\right.$

$$4$$
$$4+\frac{x}{2} \;\Big|\; 2x$$
$$\phantom{4+\frac{x}{2} \;\Big|\;} 2x+\frac{x^2}{4}$$
$$4+x-\frac{x^2}{16} \;\Big|\; -\frac{x^2}{4}$$
$$\phantom{4+x-\frac{x^2}{16} \;\Big|\;} -\frac{x^2}{4}-\frac{x^3}{16}+\frac{x^4}{256}$$
$$4+x-\frac{x^2}{8}+\frac{x^3}{64} \;\Big|\; \frac{x^3}{16}-\frac{x^4}{256}$$
$$\phantom{4+x-\frac{x^2}{8}+\frac{x^3}{64} \;\Big|\;} \frac{x^3}{16}+\frac{x^4}{64}-\frac{x^5}{512}+\frac{x^6}{4096}$$
$$-\frac{5x^4}{64}+\frac{x^5}{512}-\frac{x^6}{4096}$$

14. $1-x-x^2 \left(1-\frac{x}{2}-\frac{5x^2}{8}-\frac{5x^3}{16}\right.$

$$1$$
$$2-\frac{x}{2} \;\Big|\; -x-x^2$$
$$\phantom{2-\frac{x}{2} \;\Big|\;} -x+\frac{x^2}{4}$$
$$2-x-\frac{5x^2}{8} \;\Big|\; -\frac{5x^2}{4}$$
$$\phantom{2-x-\frac{5x^2}{8} \;\Big|\;} -\frac{5x^2}{4}+\frac{5x^3}{8}+\frac{25x^4}{64}$$
$$2-x-\frac{5x^2}{4}-\frac{5x^3}{16} \;\Big|\; -\frac{5x^3}{8}-\frac{25x^4}{64}$$
$$\phantom{2-x-\frac{5x^2}{4}-\frac{5x^3}{16} \;\Big|\;} -\frac{5x^3}{8}+\frac{5x^4}{16}+\frac{25x^5}{64}+\frac{25x^6}{256}$$
$$-\frac{45x^4}{16}-\frac{25x^5}{64}-\frac{25x^6}{256}$$

15.

$$a^2 - x \left(a - \frac{x}{2a} - \frac{x^2}{8a^3} - \frac{x^3}{16a^5} \right.$$

$$\underline{a^2}$$

$$2a - \frac{x}{2a} \quad \Big| \; -x$$

$$\underline{\qquad -x + \frac{x^2}{4a^2}}$$

$$2a - \frac{x}{a} - \frac{x^2}{8a^3} \quad \Big| \; -\frac{x^2}{4a^2}$$

$$\underline{\qquad -\frac{x^2}{4a^2} + \frac{x^3}{8a^4} + \frac{x^4}{64a^6}}$$

$$2a - \frac{x}{a} - \frac{x^2}{4a^3} - \frac{x^3}{16a^5} \quad \Big| \; -\frac{x^3}{8a^4} - \frac{x^4}{64a^6}$$

$$\underline{\qquad -\frac{x^3}{8a^4} + \frac{x^4}{16a^6} + \frac{x^5}{64a^8} + \frac{x^6}{256a^{10}}}$$

$$-\frac{5x^4}{64a^6} - \frac{x^5}{64a^8} - \frac{x^6}{256a^{10}}$$

16.

$$x^2 + a^2 \left(x + \frac{a^2}{2x} - \frac{a^4}{8x^3} + \frac{a^6}{16x^5} \right.$$

$$\underline{x^2}$$

$$2x + \frac{a^2}{2x} \quad \Big| \; a^2$$

$$\underline{\qquad a^2 + \frac{a^4}{4x^2}}$$

$$2x + \frac{a^2}{x} - \frac{a^4}{8x^3} \quad \Big| \; -\frac{a^4}{4x^2}$$

$$\underline{\qquad -\frac{a^4}{4x^2} - \frac{a^6}{8x^4} + \frac{a^8}{64x^6}}$$

$$2x + \frac{a^2}{x} - \frac{a^4}{4x^3} + \frac{a^6}{16x^5} \quad \Big| \; \frac{a^6}{8x^4} - \frac{a^8}{64x^6}$$

$$\underline{\qquad \frac{a^6}{8x^4} + \frac{a^8}{16x^6} - \frac{a^{10}}{64x^8} + \frac{a^{12}}{256x^{10}}}$$

$$-\frac{5a^8}{16x^6} + \frac{a^{10}}{64x^8} - \frac{a^{12}}{256x^{10}}$$

17.

$$a^4 - 3x^2 \left(a^2 - \frac{3x^2}{2a^2} - \frac{9x^4}{8a^6} - \frac{27x^6}{16a^{10}} \right.$$

$$a^4$$

$$2a^2 - \frac{3x^2}{2a^2} \;\Big|\; -3x^2$$

$$-3x^2 + \frac{9x^4}{4a^4}$$

$$2a^2 - \frac{3x^2}{a^2} - \frac{9x^4}{8a^6} \;\Big|\; -\frac{9x^4}{4a^4}$$

$$-\frac{9x^4}{4a^4} + \frac{27x^6}{8a^8} + \frac{81x^8}{64a^{12}}$$

$$2a^2 - \frac{3x^2}{a^2} - \frac{9x^4}{4a^6} - \frac{27x^6}{16a^{10}} \;\Big|\; -\frac{27x^6}{8a^8} - \frac{81x^8}{64a^{12}}$$

$$-\frac{27x^6}{8a^8} + \frac{81x^8}{16a^{12}} + \frac{243x^{10}}{64a^{16}} + \frac{729x^{12}}{256a^{20}}$$

$$-\frac{405x^8}{64a^{12}} - \frac{243x^{10}}{64a^{16}} - \frac{729x^{12}}{256a^{20}}$$

18.

$$9a^2 + 12ax \left(3a + 2x - \frac{2x^2}{3a} + \frac{4x^3}{9a^2} \right.$$

$$9a^2$$

$$6a + 2x \;\Big|\; 12ax$$

$$12ax + 4x^2$$

$$6a + 4x - \frac{2x^2}{3a} \;\Big|\; -4x^2$$

$$-4x^2 - \frac{8x^3}{3a} + \frac{4x^4}{9a^2}$$

$$6a + 4x - \frac{4x^2}{3a} + \frac{4x^3}{9a^2} \;\Big|\; \frac{8x^3}{3a} - \frac{4x^4}{9a^2}$$

$$\frac{8x^3}{3a} + \frac{16x^4}{9a^2} - \frac{16x^5}{27a^3} + \frac{16x^6}{81a^4}$$

$$-\frac{20x^4}{9a^2} + \frac{16x^5}{27a^3} - \frac{16x^6}{81a^4}$$

19.

$$x^3 - a^3 \left(x - \frac{a^3}{3x^2} - \frac{a^6}{9x^5} \right.$$

$$x^3$$

$$-a^3$$

$$3 \times x^2 = 3x^2$$

$$3 \times x \times \left(-\frac{a^3}{3x^2}\right) = \quad - \frac{a^3}{x}$$

$$\left(-\frac{a^3}{3x^2}\right)^2 = \quad + \frac{a^6}{9x^4}$$

$$3x^2 - \frac{a^3}{x} + \frac{a^6}{9x^4} \qquad -a^3 + \frac{a^6}{3x^3} - \frac{a^9}{27x^6}$$

$$-\frac{a^6}{3x^3} + \frac{a^9}{27x^6}$$

$$3 \times \left(x - \frac{a^3}{3x^2}\right)^2 = 3x^2 - \frac{2a^3}{x} + \frac{a^6}{3x^4}$$

$$3 \times \left(x - \frac{a^3}{3x^2}\right) \times \left(-\frac{a^6}{9x^5}\right) = \quad - \frac{a^6}{3x^4} + \frac{a^9}{9x^7}$$

$$\left(-\frac{a^6}{9x^5}\right)^2 = \quad + \frac{a^{12}}{81x^{10}}$$

$$3x^2 - \frac{2a^3}{x} + \frac{a^9}{9x^7} + \frac{a^{12}}{81x^{10}} \qquad -\frac{a^6}{3x^3} + \frac{2a^9}{9x^6} - \frac{a^{15}}{81x^{12}} - \frac{a^{18}}{729x^{15}}$$

$$-\frac{5a^9}{27x^6} + \frac{a^{15}}{81x^{12}} + \frac{a^{18}}{729x^{15}}$$

20.

$$8 + x \left(2 + \frac{x}{12} - \frac{x^2}{288} \right.$$

$$8$$

$$x$$

$$3 \times (2)^2 = 12$$

$$3 \times 2 \times \frac{x}{12} = \quad + \frac{x}{2}$$

$$\left(\frac{x}{12}\right)^2 = \quad + \frac{x^2}{144}$$

$$12 + \frac{x}{2} + \frac{x^2}{144} \qquad x + \frac{x^2}{24} + \frac{x^3}{1728}$$

$$- \frac{x^2}{24} - \frac{x^3}{1728}$$

$$3\left(2 + \frac{x}{12}\right)^2 = 12 + x + \frac{x^2}{48}$$

$$3\left(2 + \frac{x}{12}\right) \times \left(-\frac{x^2}{288}\right) = \quad - \frac{x^2}{48} - \frac{x^3}{1152}$$

$$\left(-\frac{x^2}{288}\right)^2 = \quad + \frac{x^4}{(288)^2}$$

$$12 + x \quad - \frac{x^3}{1152} + \frac{x^4}{(288)^2} \qquad - \frac{x^2}{24} - \frac{x^3}{288} + \frac{x^5}{1152 \times 288} - \frac{x^6}{(288)^3}$$

$$\frac{5x^3}{1728} - \frac{x^5}{1152 \times 288} + \frac{x^6}{(288)^3}$$

21.

$$\begin{aligned}
3 \times \frac{1}{a^2} &= \frac{3}{a^2} \\
3 \times \frac{1}{a} \times 3a^2x &= \quad + 9ax \\
(3a^2x)^2 &= \quad\quad\quad + 9a^4x^2 \\
\hline
& \frac{3}{a^2} + 9ax + 9a^4x^2
\end{aligned}$$

$$\begin{aligned}
3 \times \left(\frac{1}{a} + 3a^2x\right)^2 &= \frac{3}{a^2} + 18ax + 27a^4x^2 \\
3 \times \left(\frac{1}{a} + 3a^2x\right)(-9a^5x^2) &= \quad\quad - 27a^4x^2 - 81a^7x^3 \\
(-9a^5x^2)^2 &= \quad\quad\quad\quad + 81a^{10}x^4 \\
\hline
& \frac{3}{a^2} + 18ax \quad - 81a^7x^3 + 81a^{10}x^4
\end{aligned}$$

$$\begin{array}{l}
\frac{1}{a^3} + 9x \; \left(\frac{1}{a} + 3a^2x - 9a^5x^2 \right. \\
\frac{1}{a^3} \\
\hline
\quad 9x \\
\quad 9x + 27a^3x^2 + 27a^6x^3 \\
\hline
\quad - 27a^3x^2 - 27a^6x^3 \\
\quad - 27a^3x^2 - 162a^6x^3 + 729a^{12}x^5 - 729a^{15}x^6 \\
\hline
\quad\quad 135a^6x^3 - 729a^{12}x^5 + 729a^{15}x^6
\end{array}$$

22.

$$1 - 6x + 21x^2 \;(\; 1 - 2x + 3x^2$$
$$1$$
$$-6x + 21x^2$$

$$\begin{aligned} 3 \times (1)^2 &= 3 \\ 3 \times 1 \times (-2x) &= \quad -\; 6x \\ (-2x)^2 &= \qquad\quad +\; 4x^2 \\ \hline & \;\; 3 - 6x + 4x^2 \end{aligned} \qquad -6x + 12x^2 - 8x^3$$

$$9x^2 + 8x^3$$

$$\begin{aligned} 3 \times (1 - 2x)^2 &= 3 - 12x + 12x^2 \\ 3 \times (1 - 2x) \times 3x^2 &= \qquad\qquad 9x^2 - 18x^3 \\ (3x^2)^2 &= \qquad\qquad\qquad\qquad +\; 9x^4 \\ \hline & \;\; 3 - 12x + 21x^2 - 18x^3 + 9x^4 \end{aligned} \qquad 9x^2 - 36x^3 + 63x^4 - 54x^5 + 27x^6$$

$$44x^3 - 63x^4 + 54x^5 - 27x^6$$

23.

$$27x^6 - 27x^5 - 18x^4 \;(\; 3x^2 - x - 1$$
$$27x^6$$
$$-27x^5 - 18x^4$$

$$\begin{aligned} 3 \times (3x^2)^2 &= 27x^4 \\ 3 \times (3x^2) \times (-x) &= \qquad -\; 9x^3 \\ (-x)^2 &= \qquad\qquad +\; x^2 \\ \hline & \;\; 27x^4 - 9x^3 + x^2 \end{aligned} \qquad -27x^5 + 9x^4 - x^3$$

$$-27x^4 + x^3$$

$$\begin{aligned} 3\,(3x^2 - x)^2 &= 27x^4 - 18x^3 + 3x^2 \\ 3\,(3x^2 - x) \times (-1) &= \qquad\qquad -9x^2 + 3x \\ (-1)^2 &= \qquad\qquad\qquad\qquad +1 \\ \hline & \;\; 27x^4 - 18x^3 - 6x^2 + 3x + 1 \end{aligned} \qquad -27x^4 + 18x^3 + 6x^2 - 3x - 1$$

$$-17x^3 - 6x^2 + 3x + 1$$

24.

$$64 - 48x + 9x^2 \;\Big(\; 4 - x - \frac{x^2}{16}$$
$$64$$
$$-48x + 9x^2$$

$$\begin{aligned} 3 \times (4)^2 &= 48 \\ 3 \times 4\,(-x) &= \quad -12x \\ (-x)^2 &= \qquad\quad +\; x^2 \\ \hline & \;\; 48 - 12x + x^2 \end{aligned} \qquad -48x + 12x^2 - x^3$$

$$-\;3x^2 + x^3$$

$$\begin{aligned} 3\,(4 - x)^2 &= 48 - 24x + 3x^2 \\ 3\,(4 - x) \times \left(-\frac{x^2}{16}\right) &= \qquad\qquad -\frac{3x^2}{4} + \frac{3x^3}{16} \\ \left(-\frac{x^2}{16}\right)^2 &= \qquad\qquad\qquad\qquad + \frac{x^4}{256} \\ \hline & \;\; 48 - 24x + \frac{9x^2}{4} + \frac{3x^3}{16} + \frac{x^4}{256} \end{aligned} \qquad -\;3x^2 + \frac{3x^3}{2} - \frac{9x^4}{64} - \frac{3x^5}{256} - \frac{x^6}{4096}$$

$$-\frac{x^3}{2} + \frac{9x^4}{64} + \frac{3x^5}{256} + \frac{x^6}{4096}$$

EXAMPLES. XXIX. c. PAGE 230.

1. $b(x^3+a^3)+ax(x^2-a^2)+a^3(x+a)=b(x^3+a^3)+ax^3-a^3x+a^3x+a^4$
$$=b(x^3+a^3)+a(x^3+a^3)$$
$$=(a+b)(x^3+a^3)$$
$$=(a+b)(x+a)(x^2-ax+a^2).$$

2. $(ax+by)^2+(ay-bx)^2+c^2x^2+c^2y^2$
$$=a^2x^2+2abxy+b^2y^2+a^2y^2-2abxy+b^2x^2+c^2(x^2+y^2)$$
$$=a^2(x^2+y^2)+b^2(x^2+y^2)+c^2(x^2+y^2)$$
$$=(a^2+b^2+c^2)(x^2+y^2).$$

3. $(x+y)^3+3(x+y)^2z+3(x+y)z^2+z^3=\{(x+y)+z\}^3=\{(x+z)+y\}^3$
$$=(x+z)^3+3(x+z)^2y+3(x+z)y^2+y^3.$$

4. $(a+b+c)(ab+bc+ca)-abc=(a+b)(ab+bc+ca)+c(ab+bc+ca)-abc$
$$=(a+b)(ab+bc+ca)+abc+bc^2+c^2a-abc$$
$$=(a+b)(ab+bc+ca)+c^2(a+b)$$
$$=(a+b)(ab+bc+ca+c^2)$$
$$=(a+b)(b+c)(c+a).$$

5. The left-hand side
$$=a^2+b^2+c^2+2ab+2ac+2bc-ab-ac+a^2-ab-bc+b^2-ac-bc+c^2$$
$$=2a^2+2b^2+2c^2=2(a^2+b^2+c^2).$$

6. $(x-y)^3+3(x-y)^2(x+y)+3(x-y)(x+y)^2+(x+y)^3$
$$=\{(x-y)+(x+y)\}^3=(2x)^3=8x^3.$$

7. The left-hand side $=(y-z)x^2+y^2z-xy^2+xz^2-yz^2+(y-z)(z-x)(x-y)$
$$=(y-z)x^2+yz(y-z)-x(y^2-z^2)-(y-z)(x-z)(x-y)$$
$$=(y-z)(x^2+yz-xy-xz)-(y-z)(x^2+yz-xy-xz)=0.$$

8. $a^3(b-c)+b^3(c-a)+c^3(a-b)=a^3(b-c)+b^3c-ab^3+ac^3-bc^3$
$$=a^3(b-c)+bc(b^2-c^2)-a(b^3-c^3)$$
$$=(b-c)\{a^3+bc(b+c)-a(b^2+bc+c^2)\}$$
$$=(b-c)\{a^3-ac^2-ab^2+b^2c-abc+bc^2\}$$
$$=(b-c)\{a(a^2-c^2)-b^2(a-c)-bc(a-c)\}$$
$$=(b-c)(a-c)\{a(a+c)-b^2-bc\}$$
$$=(b-c)(a-c)(a^2-b^2+ac-bc)$$
$$=(b-c)(a-c)(a-b)(a+b+c)$$
$$=-(b-c)(c-a)(a-b)(a+b+c).$$

9. One of the factors of $x^3+y^3+z^3-3xyz$ is $x+y+z$;

$\therefore$ if $x+y+z=0$, $\qquad x^3+y^3+z^3-3xyz=0$;

that is, $\qquad x^3+y^3+z^3=3xyz$.

10. One of the factors of $(b-c)^3+(c-a)^3+(a-b)^3-3(b-c)(c-a)(a-b)$ is $(b-c)+(c-a)+(a-b)$, which is zero identically;

$$\therefore\ (b-c)^3+(c-a)^3+(a-b)^3-3(b-c)(c-a)(a-b)=0;$$

$$\therefore\ (b-c)^3+(c-a)^3+(a-b)^3=3(b-c)(c-a)(a-b).$$

11. $(s-a)^2+(s-b)^2+(s-c)^2+s^2=s^2-2as+a^2+s^2-2bs+b^2+s^2-2cs+c^2+s^2$

$$=4s^2-2(a+b+c)s+a^2+b^2+c^2$$
$$=4s^2-2.2s.s+a^2+b^2+c^2$$
$$=a^2+b^2+c^2.$$

12. $(s-a)^3+(s-b)^3+(s-c)^3+3abc$

$$=s^3-3as^2+3a^2s-a^3+s^3-3bs^2+3b^2s-b^3+s^3-3cs^2+3c^2s-c^3+3abc$$
$$=3s^3-3(a+b+c)s^2+3(a^2+b^2+c^2)s-(a^3+b^3+c^3-3abc)$$
$$=3s^3-3.2s.s^2+3(a^2+b^2+c^2)s-(a^2+b^2+c^2-bc-ac-ab)(a+b+c)$$
$$=3s^3-6s^3+3(a^2+b^2+c^2)s-(a^2+b^2+c^2-bc-ac-ab)(2s)$$
$$=-3s^3+s\{3a^2+3b^2+3c^2-2a^2-2b^2-2c^2+2bc+2ac+2ab\}$$
$$=-3s^3+s\{a^2+b^2+c^2+2bc+2ac+2ab\}$$
$$=-3s^3+s.(2s)^2$$
$$=s^3.$$

13. $16s(s-a)(s-b)(s-c)=2s(2s-2a)(2s-2b)(2s-2c)$

$$=(a+b+c)(b+c-a)(a+c-b)(a+b-c)$$
$$=\{(b+c)^2-a^2\}\{a^2-(b-c)^2\}$$
$$=a^2\{(b+c)^2+(b-c)^2\}-a^4-(b^2-c^2)^2$$
$$=a^2(2b^2+2c^2)-a^4-b^4+2b^2c^2-c^4$$
$$=2a^2b^2+2a^2c^2+2b^2c^2-a^4-b^4-c^4.$$

14. $2(s-a)(s-b)(s-c)+a(s-b)(s-c)+b(s-c)(s-a)+c(s-a)(s-b)$

$$=(s-b)(s-c)(2s-a)+(s-a)(bs+cs-2bc)$$
$$=\{s^2-(b+c)s+bc\}(b+c)+(b+c)(s^2-as)-2bc(s-a)$$
$$=(b+c)\{2s^2-(a+b+c)s+bc\}-bc(b+c-a)$$
$$=bc(b+c)-bc(b+c-a)$$
$$=abc.$$

15. $(2a-b)^3+(2b-c)^3+(2c-a)^3-3(2a-b)(2b-c)(2c-a)$ is of the form $x^3+y^3+z^3-3xyz$, and therefore has a factor

$$(2a-b)+(2b-c)+(2c-a)=2a-b+2b-c+2c-a=a+b+c=0;$$

$$\therefore\ (2a-b)^3+(2b-c)^3+(2c-a)^3-3(2a-b)(2b-c)(2c-a)=0;$$

$$\therefore\ (2a-b)^3+(2b-c)^3+(2c-a)^3=3(2a-b)(2b-c)(2c-a).$$

16. Since $a+b+c=0$, we may express the denominators in a more convenient form; thus

$$2a^2+bc=a^2+a^2+bc=a^2-(b+c)a+bc=(a-b)(a-c).$$

We have now to shew that

$$\frac{a^2}{(a-b)(a-c)}+\frac{b^2}{(b-c)(b-a)}+\frac{c^2}{(c-a)(c-b)}=1.$$

This is proved in XXIX. d. Ex. 3.

17. $(x+y+z)^3=(x+y)^3+3(x+y)^2z+3(x+y)z^2+z^3$,

$(x+y-z)^3=(x+y)^3-3(x+y)^2z+3(x+y)z^2-z^3$,

$(x-y+z)^3=(x-y)^3+3(x-y)^2z+3(x-y)z^2+z^3$,

$(x-y-z)^3=(x-y)^3-3(x-y)^2z+3(x-y)z^2-z^3$;

$$\begin{aligned}\therefore\ \text{the sum}&=2(x+y)^3+2(x-y)^3+6(x+y)z^2+6(x-y)z^2\\&=2(x^3+3x^2y+3xy^2+y^3+x^3-3x^2y+3xy^2-y^3)+12xz^2\\&=4(x^3+3xy^2)+12xz^2\\&=4x(x^2+3y^2+3z^2).\end{aligned}$$

18. The expression on the left-hand side is of the form $x^3+y^3+z^3-3xyz$, $\therefore$ it has a factor

$$(s-3a)+(s-3b)+(s-3c)=3s-3(a+b+c)=0;$$

$\therefore$ the expression $=0$.

19. $X^3+Y^3+Z^3-3XYZ$ has a factor

$$X+Y+Z=(b+c-2a)+(c+a-2b)+(a+b-2c)=0;$$

$$\therefore\ X^3+Y^3+Z^3-3XYZ=0.$$

20. $a(a^2+bc)+b(b^2+ac)-c(c^2-ab)=a^3+b^3-c^3+3abc=(a+b-c)(a^2+\ldots)$; but $a+b-c=\cdot 7+\cdot 08-\cdot 78=0$; hence the expression $=0$.

21. $(a-b)^2+(b-c)^2+(c-a)^2+2(b-c)(c-a)+2(a-b)(b-c)+2(a-b)(c-a)$

$$=\{(a-b)+(b-c)+(c-a)\}^2=0;$$

$$\begin{aligned}\therefore (a-b)^2+(b-c)^2+(c-a)^2&=-2(b-c)(c-a)-2(a-b)(b-c)-2(a-b)(c-a)\\&=2(c-b)(c-a)+2(b-a)(b-c)+2(a-b)(a-c).\end{aligned}$$

22. $a^2(b^3-c^3)+b^2(c^3-a^3)+c^2(a^3-b^3)$

$$=a^2(b-c)(b^2+bc+c^2)-(b^3c^2-b^2c^3)-a^3(b^2-c^2)$$
$$=(b-c)\{a^2b^2+a^2bc+a^2c^2-b^2c^2-a^3(b+c)\}$$
$$=(b-c)\{-b^2(c^2-a^2)+a^2b(c-a)+a^2c(c-a)\}$$
$$=(b-c)(c-a)\{-b^2(c+a)+a^2b+a^2c\}$$
$$=(b-c)(c-a)\{ab(a-b)+c(a^2-b^2)\}$$
$$=(b-c)(c-a)(a-b)(ab+ac+bc).$$

Again the expression

$$=a^2\{(b-c)^3+3bc(b-c)\}+b^2\{(c-a)^3+3ca(c-a)\}+c^2\{(a-b)^3+3ab(a-b)\}$$
$$=a^2(b-c)^3+b^2(c-a)^3+c^2(a-b)^3+3abc\{a(b-c)+b(c-a)+c(a-b)\}$$
$$=a^2(b-c)^3+b^2(c-a)^3+c^2(a-b)^3.$$

Again the expression

$$=a^2b^3-a^2c^3+b^2c^3-b^2a^3+c^2a^3-c^2b^3$$
$$=-a^3b^2+a^2b^3-a^2c^3+a^3c^2-b^3c^2+b^2c^3$$
$$=-[a^2b^2(a-b)+c^2a^2(c-a)+b^2c^2(b-c)].$$

23. If $(a+b)^2+(b+c)^2+(c+d)^2=4ab+4bc+4cd,$

$$a^2+2ab+b^2-4ab+b^2+2bc+c^2-4bc+c^2+2cd+d^2-4cd=0,$$

or $$(a^2-2ab+b^2)+(b^2-2bc+c^2)+(c^2-2cd+d^2)=0,$$

or $$(a-b)^2+(b-c)^2+(c-d)^2=0\,;$$

$\therefore\ a-b=0,\ b-c=0,$ and $c-d=0$; [Art. 223, Ex. 3]

$$\therefore\ a=b=c=d.$$

24. $$x^2+y^2+z^2-yz-zx-xy=\tfrac{1}{2}\{(y-z)^2+(z-x)^2+(x-y)^2\}$$
$$=\tfrac{1}{2}\{(b-c)^2+(c-a)^2+(a-b)^2\}$$
$$=a^2+b^2+c^2-bc-ca-ab.$$

25. If $a+b+c=0$, then $a^2=b^2+2bc+c^2$, and $b^2+c^2-a^2=-2bc$.

Thus the expression $=-\dfrac{1}{2bc}-\dfrac{1}{2ac}-\dfrac{1}{2ab}=-\dfrac{a+b+c}{2abc}=0.$

26. As in the last example $b^2+c^2-a^2=-2bc$, &c.;

$\therefore$ the expression $=\dfrac{b+c}{bc}(-2bc)+\dfrac{c+a}{ca}(-2ca)+\dfrac{a+b}{ab}(-2ab)$

$$=-2(b+c)-2(c+a)-2(a+b)$$
$$=-4(a+b+c)=0.$$

27. $(x-a)^2+(y-b)^2+(a^2+b^2-1)(x^2+y^2-1)$

$$=x^2-2ax+a^2+y^2-2by+b^2+a^2x^2+a^2y^2-a^2+b^2x^2+b^2y^2-b^2-x^2-y^2+1$$
$$=(a^2x^2+b^2y^2+1-2ax-2by+2abxy)+(b^2x^2+a^2y^2-2abxy)$$
$$=(ax+by-1)^2+(bx-ay)^2;$$

$\therefore$ the equation becomes $(ax+by-1)^2+(bx-ay)^2=0$;

$$\therefore\ ax+by-1=0, \text{ and } bx-ay=0 \textit{ simultaneously.}$$

From these equations we obtain

$$x=\frac{a}{a^2+b^2}, \text{ and } y=\frac{b}{a^2+b^2}.$$

28. If $\qquad 2(x^2+a^2-ax)(y^2+b^2-by)=x^2y^2+a^2b^2,$

$$2(x^2+a^2)(y^2+b^2)-2ax(y^2+b^2)-2by(x^2+a^2)+2abxy$$
$$=(x^2+a^2)(y^2+b^2)-a^2y^2-b^2x^2;$$

$$\therefore\ (x^2+a^2)(y^2+b^2)-2ax(y^2+b^2)-2by(x^2+a^2)+4abxy+(a^2y^2+b^2x^2-2abxy)$$
$$=0;$$

$$\therefore\ (x^2+a^2-2ax)(y^2+b^2-2by)+(ay-bx)^2=0,$$

or $\qquad (x-a)^2(y-b)^2+(ay-bx)^2=0.$

From this equation we see that

$$(x-a)(y-b)=0, \text{ and } ay-bx=0 \textit{ simultaneously.}$$

From the first equation we have $x=a$ *or* $y=b$; and by combining this result with the second equation we see that the only solutions are $x=a, y=b$.

EXAMPLES. XXIX. d. Page 232.

1. The expression $=-\dfrac{a}{(a-b)(c-a)}-\dfrac{b}{(b-c)(a-b)}-\dfrac{c}{(c-a)(b-c)}$

$$=-\frac{a(b-c)+b(c-a)+c(a-b)}{(a-b)(b-c)(c-a)}$$

$=0.$ [See Art. 224, Note.]

2. The expression $=-\dfrac{bc}{(a-b)(c-a)}-\dfrac{ca}{(b-c)(a+b)}-\dfrac{ab}{(c-a)(b-c)}$

$$=-\frac{bc(b-c)+ca(c-a)+ab(a-b)}{(a-b)(b-c)(c-a)}$$

$=+\dfrac{(a-b)(b-c)(c-a)}{(a-b)(b-c)(c-a)}$ [Art. 224, Note]

$=1.$

3. The expression $= -\dfrac{a^2}{(a-b)(c-a)} - \dfrac{b^2}{(b-c)(a-b)} - \dfrac{c^2}{(c-a)(b-c)}$

$$= -\frac{a^2(b-c)+b^2(c-a)+c^2(a-b)}{(a-b)(b-c)(c-a)}$$

$$= +\frac{(a-b)(b-c)(c-a)}{(a-b)(b-c)(c-a)}$$ [Art. 224, Note]

$$=1.$$

4. The expression $= -\dfrac{a^3}{(a-b)(c-a)} - \dfrac{b^3}{(b-c)(a-b)} - \dfrac{c^3}{(c-a)(b-c)}$

$$= -\frac{a^3(b-c)+b^3(c-a)+c^3(a-b)}{(a-b)(b-c)(c-a)}$$

$$= +\frac{(a-b)(b-c)(c-a)(a+b+c)}{(a-b)(b-c)(c-a)}$$ [XXIX. c. Ex. 8]

$$=a+b+c.$$

5. The expression $= \dfrac{a(b+c)(b-c)+b(c+a)(c-a)+c(a+b)(a-b)}{(a-b)(b-c)(c-a)}$

$$= \frac{a(b^2-c^2)+b(c^2-a^2)+c(a^2-b^2)}{(a-b)(b-c)(c-a)}$$

$$=1.$$ [Art. 224, Note.]

6. The expression $= -\dfrac{1}{a(a-b)(c-a)} - \dfrac{1}{b(b-c)(a-b)} - \dfrac{1}{c(c-a)(b-c)}$

$$= -\frac{bc(b-c)+ca(c-a)+ab(a-b)}{abc(a-b)(b-c)(c-a)}$$

$$= +\frac{(a-b)(b-c)(c-a)}{abc(a-b)(b-c)(c-a)}$$ [Art. 224, Note]

$$= \frac{1}{abc}.$$

7. The expression $= -\dfrac{bc}{a(a^2-b^2)(c^2-a^2)} - \dfrac{ca}{b(b^2-c^2)(a^2-b^2)} - \dfrac{ab}{c(c^2-a^2)(b^2-c^2)}$

$$= -\frac{b^2c^2(b^2-c^2)+c^2a^2(c^2-a^2)+a^2b^2(a^2-b^2)}{abc(a^2-b^2)(b^2-c^2)(c^2-a^2)}$$

$$= +\frac{(a^2-b^2)(b^2-c^2)(c^2-a^2)}{abc(a^2-b^2)(b^2-c^2)(c^2-a^2)}$$ [Art. 224, Note]

$$= \frac{1}{abc}.$$

8. The expression $= -\dfrac{(x-b)(x-c)}{(a-b)(c-a)} - \dfrac{(x-c)(x-a)}{(b-c)(a-b)} - \dfrac{(x-a)(x-b)}{(c-a)(b-c)}$.

The L. C. D. $=(a-b)(b-c)(c-a)$.

The numerator

$= -(x-b)(x-c)(b-c) - (x-c)(x-a)(c-a) - (x-a)(x-b)(a-b)$

$= -(x^2 - \overline{b+c}\,.\,x + bc)(b-c) - \ldots\ldots - \ldots\ldots$

$= -x^2(b-c+c-a+a-b) + x(b^2-c^2+c^2-a^2+a^2-b^2) - bc(b-c) - ca(c-a)$
$- ab(a-b)$

$= -bc(b-c) - ca(c-a) - ab(a-b)$

$= +(a-b)(b-c)(c-a)$.

Hence the expression $=1$.

9. The expression

$$= -\frac{bc(a+d)}{(a-b)(c-a)} - \frac{ca(b+d)}{(b-c)(a-b)} - \frac{ab(c+d)}{(c-a)(b-c)}$$

$$= -\frac{(abc+bcd)(b-c)+(abc+cad)(c-a)+(abc+abd)(a-b)}{(a-b)(b-c)(c-a)}$$

$$= -\frac{abc(b-c+c-a+a-b)+d\{bc(b-c)+ca(c-a)+ab(a-b)\}}{(a-b)(b-c)(c-a)}$$

$$= +\frac{d(a-b)(b-c)(c-a)}{(a-b)(b-c)(c-a)} = d.$$

10. The expression

$$= -\frac{1}{(a-b)(c-a)(x-a)} - \frac{1}{(b-c)(a-b)(x-b)} - \frac{1}{(c-a)(b-c)(x-c)}.$$

The L. C. D. $=(a-b)(b-c)(c-a)(x-a)(x-b)(x-c)$.

The numerator

$= -(b-c)(x-b)(x-c) - \ldots\ldots - \ldots\ldots$

$= -(b-c)(x^2 - \overline{b+c}\,.\,x + bc) - \ldots\ldots - \ldots\ldots$

$= -(b-c+c-a+a-b)x^2 + (b^2-c^2+c^2-a^2+a^2-b^2)x - bc(b-c) - ca(c-a)$
$- ab(a-b)$

$= +(a-b)(b-c)(c-a)$.

∴ The expression

$$= \frac{(a-b)(b-c)(c-a)}{(a-b)(b-c)(c-a)(x-a)(x-b)(x-c)} = \frac{1}{(x-a)(x-b)(x-c)}.$$

11. The expression

$$= -\frac{a^2}{(a-b)(c-a)(x+a)} - \frac{b^2}{(b-c)(a-b)(x+b)} - \frac{c^2}{(c-a)(b-c)(x+c)}.$$

The L. C. D. $= (a-b)(b-c)(c-a)(x+a)(x+b)(x+c)$.

The numerator $= -a^2(b-c)(x+b)(x+c) - \ldots\ldots - \ldots\ldots$

$$= -a^2(b-c)\{x^2+(b+c)x+bc\} - \ldots\ldots - \ldots\ldots$$

The coefficient of x^2

$$= -a^2(b-c) - b^2(c-a) - c^2(a-b)$$
$$= +(a-b)(b-c)(c-a);$$

the coefficient of x $= -a^2(b^2-c^2) - b^2(c^2-a^2) - c^2(a^2-b^2)$

$$=0;$$

the terms which do not contain x

$$= -a^2(b-c)bc - b^2(c-a)ca - c^2(a-b)ab$$
$$= -abc\{a(b-c)+b(c-a)+c(a-b)\}$$
$$=0.$$

$\therefore$ the expression $= \dfrac{(a-b)(b-c)(c-a)x^2}{(a-b)(b-c)(c-a)(x+a)(x+b)(x+c)}$

$$= \frac{x^2}{(x+a)(x+b)(x+c)}.$$

12. The expression

$$= -\frac{a^2(a+b)(a+c)}{(a-b)(c-a)} - \frac{b^2(b+c)(b+a)}{(b-c)(a-b)} - \frac{c^2(c+a)(c+b)}{(c-a)(b-c)}.$$

The L. C. D. $= (a-b)(b-c)(c-a)$.

If $a+b+c=s$,

the numerator $= -a^2(a+b)(a+c)(b-c) - \ldots\ldots - \ldots\ldots$

$$= -a^2(s-c)(s-b)(b-c) - \ldots\ldots - \ldots\ldots$$
$$= -a^2(b-c)\{s^2-(b+c)s+bc\} - \ldots\ldots - \ldots\ldots$$

The coefficient of $s^2 = -a^2(b-c) - b^2(c-a) - c^2(a-b) = (a-b)(b-c)(c-a)$;

the coefficient of s $= -a^2(b^2-c^2) - b^2(c^2-a^2) - c^2(a^2-b^2) = 0$;

and the terms which do not contain s

$$= -a^2bc(b-c) - b^2ac(c-a) - c^2ab(a-b)$$
$$= -abc\{a(b-c)+b(c-a)+c(a-b)\}$$
$$=0.$$

$\therefore$ the expression $= \dfrac{(a-b)(b-c)(c-a)s^2}{(a-b)(b-c)(c-a)} = s^2$, or $(a+b+c)^2$.

13. $a^3(b-c)+b^3(c-a)+c^3(a-b)=-(b-c)(c-a)(a-b)(a+b+c)$ [XXIX. c. Ex. 8.]

Also $(b-c)^3+(c-a)^3+(a-b)^3=3(b-c)(c-a)(a-b)$. [XXIX. c. Ex. 10.]

$\therefore$ the fraction $=-\dfrac{(b-c)(c-a)(a-b)(a+b+c)}{3(b-c)(c-a)(a-b)}=-\dfrac{a+b+c}{3}$.

14. $a^2(b-c)+b^2(c-a)+c^2(a-b)=-(a-b)(b-c)(c-a)$;

and $(b-c)^3+(c-a)^3+(a-b)^3=3(b-c)(c-a)(a-b)$.

On dividing numerator and denominator by $(a-b)(b-c)(c-a)$,

the fraction $=\dfrac{-1+2}{3}=\dfrac{1}{3}$.

15. The numerator $=-(b-c)(c-a)(a-b)(a+b+c)$. [XXIX. c. Ex. 8.]

The denominator $=-(a-b)(b-c)(c-a)$.

$\therefore$ the fraction $=a+b+c$.

16. $a^2(b-c)^3+b^2(c-a)^3+c^2(a-b)^3=(a-b)(b-c)(c-a)(bc+ca+ab)$; [XXIX. c. Ex. 22.]

$\therefore$ the fraction $=\dfrac{(a-b)(b-c)(c-a)(bc+ac+ab)}{(a-b)(b-c)(c-a)}=bc+ac+ab$.

17. The numerator

$$=\frac{1}{abc}\{bc(b-c)+ca(c-a)+ab(a-b)\}=-\frac{(a-b)(b-c)(c-a)}{abc};$$

the denominator $=\left(\dfrac{1}{a}-\dfrac{1}{b}\right)\left(\dfrac{1}{b}-\dfrac{1}{c}\right)\left(\dfrac{1}{c}-\dfrac{1}{a}\right)$ being of the same form as the last identity in Art. 224, Note,

$$=\frac{(b-a)(c-b)(a-c)}{ab\,.\,bc\,.\,ac}=-\frac{(a-b)(b-c)(c-a)}{a^2b^2c^2};$$

$\therefore$ the fraction $=\dfrac{(a-b)(b-c)(c-a)}{abc}\div\dfrac{(a-b)(b-c)(c-a)}{a^2b^2c^2}=abc$.

18. The numerator $=\dfrac{a^4(b^2-c^2)+b^4(c^2-a^2)+c^4(a^2-b^2)}{a^2b^2c^2}$

$$=\frac{-(b^2-c^2)(c^2-a^2)(a^2-b^2)}{a^2b^2c^2}.$$

The denominator $=-\left(\dfrac{1}{c}-\dfrac{1}{b}\right)\left(\dfrac{1}{a}-\dfrac{1}{c}\right)\left(\dfrac{1}{b}-\dfrac{1}{a}\right)$

$$=\frac{-(b-c)(c-a)(a-b)}{a^2b^2c^2};$$

$\therefore$ the expression $=(b+c)(c+a)(a+b)$.

EXAMPLES. XXIX. e. PAGE 238.

1.

$$\begin{array}{r|l}
 & x^4+6x^3+13x^2+13x-1\ (\,x^2+3x+2 \\
 & x^4 \\
2x^2+3x & 6x^3+13x^2 \\
 & 6x^3+\ \ 9x^2 \\
2x^2+6x+2 & 4x^2+13x-1 \\
 & 4x^2+12x+4 \\
 & \qquad\qquad x-5
\end{array}$$

Here in order that the expression may be a perfect square the remainder $x-5$ must $=0$;

$$\therefore\ x=5.$$

2.

$$\begin{array}{r|l}
 & x^4+6x^3+11x^2+3x+31\ (\,x^2+3x+1 \\
 & x^4 \\
2x^2+3x & 6x^3+11x^2 \\
 & 6x^3+\ \ 9x^2 \\
2x^2+6x+1 & 2x^2+3x+31 \\
 & 2x^2+6x+\ \ 1 \\
 & \qquad -3x+30
\end{array}$$

Here in order that the expression may be a perfect square the remainder $-3x+30$ must $=0$;

$$\therefore\ x=10.$$

3.

$$\begin{array}{r|l}
 & x^4-2ax^3+(a^2+2b)\,x^2-3abx+2b^2\ (\,x^2-ax+b \\
 & x^4 \\
2x^2-ax & -2ax^3+(a^2+2b)\,x^2 \\
 & -2ax^3+\ a^2x^2 \\
2x^2-2ax+b & 2bx^2-3abx+2b^2 \\
 & 2bx^2-2abx+\ b^2 \\
 & \qquad\ -\ abx+\ b^2
\end{array}$$

Here in order that the expression may be a perfect square the remainder $-abx+b^2$ must $=0$;

$$\therefore\ x=\frac{b}{a}.$$

4.

$$\begin{array}{r|l}
 & 4p^2x^4-4pqx^3+(q^2+2p^2)\,x^2-5pqx+\frac{p^2}{2}\ \left(\,2px^2-qx+\frac{p}{2}\right. \\
 & 4p^2x^4 \\
4px^2-qx & -4pqx^3+(q^2+2p^2)\,x^2 \\
 & -4pqx^3+q^2x^2 \\
4px^2-2qx+\frac{p}{2} & 2p^2x^2-5pqx+\frac{p^2}{2} \\
 & 2p^2x^2-\ \ pqx+\frac{p^2}{4} \\
 & \qquad -4pqx+\frac{p^2}{4}
\end{array}$$

Here in order that the expression may be a perfect square, the remainder $-4pqx+\frac{p^2}{4}$ must $=0$;

$$\therefore\ x=\frac{p^2}{16pq}=\frac{p}{16q}.$$

5.

$$\frac{a^2x^6}{9} - \frac{abx^4}{2} + \frac{2acx^3}{3} + \frac{9b^2x^2}{16} - \frac{5bcx}{2} + 6c^2 \left(\frac{ax^3}{3} - \frac{3bx}{4} \right.$$

$$\frac{a^2x^6}{9}$$

$$\frac{2ax^3}{3} - \frac{3bx}{4} \;\Big|\; -\frac{abx^4}{2} + \frac{2acx^3}{3} + \frac{9b^2x^2}{16}$$

$$-\frac{abx^4}{2} + \frac{9b^2x^2}{16}$$

$$\frac{2ax^3}{3} - \frac{3bx}{2} + c \;\Big|\; \frac{2acx^3}{3} - \frac{5bcx}{2} + 6c^2$$

$$\frac{2acx^3}{3} - \frac{3bcx}{2} + c^2$$

$$-bcx + 5c^2$$

Here in order that the expression may be a perfect square, the remainder $-bcx + 5c^2$ must $=0$;

$$\therefore\ bcx = 5c^2,$$

$$x = \frac{5c}{b}.$$

6.

$$x^4 + 2ax^3 + 3a^2x^2 + cx + d\ (\,x^2 + ax + a^2$$

$$x^4$$

$$2x^2 + ax \;\Big|\; 2ax^3 + 3a^2x^2$$

$$2ax^3 + a^2x^2$$

$$2x^2 + 2ax + a^2 \;\Big|\; 2a^2x^2 + cx + d$$

$$2a^2x^2 + 2a^3x + a^4$$

$$(c - 2a^3)\,x + d - a^4$$

Here in order that the expression may be a perfect square, the remainder $(c-2a^3)x+d-a^4$ must $=0$;

$$\therefore\ x = \frac{d - a^4}{2a^3 - c}.$$

7.

$$x^4 - ax^3 + bx^2 - cx + 1 \;\Big|\; x^2 - \frac{1}{2}ax + \frac{1}{2}\left(b - \frac{a^2}{4}\right)$$

$$x^4$$

$$2x^2 - \frac{1}{2}ax \;\Big|\; -ax^3 + bx^2$$

$$-ax^3 + \frac{a^2x^2}{4}$$

$$2x^2 - ax + \frac{1}{2}\left(b - \frac{a^2}{4}\right) \;\Big|\; \left(b - \frac{a^2}{4}\right)x^2 - cx + 1$$

$$\left(b - \frac{a^2}{4}\right)x^2 - \frac{ax}{2}\left(b - \frac{a^2}{4}\right) + \frac{1}{4}\left(b - \frac{a^2}{4}\right)^2$$

$$\left\{\frac{a}{2}\left(b - \frac{a^2}{4}\right) - c\right\}x + 1 - \frac{1}{4}\left(b - \frac{a^2}{4}\right)^2$$

Here in order that the expression may be a perfect square, the remainder $\left\{\frac{a}{2}\left(b - \frac{a^2}{4}\right) - c\right\}x + 1 - \frac{1}{4}\left(b - \frac{a^2}{4}\right)^2$ must $=0$. For this to be 0 for all values of x, we must have $\frac{a}{2}\left(b - \frac{a^2}{4}\right) - c = 0$ and $1 - \frac{1}{4}\left(b - \frac{a^2}{4}\right)^2 = 0$. From the second of these results we get $b - \frac{a^2}{4} = \pm 2$, or $b = \frac{a^2}{4} \pm 2$; and from the first $a = \pm c$.

8.

$$\begin{array}{rr|l}
 & & 8x^3-36x^2+56x-39\ (\ 2x-3 \\
 & & 8x^3 \\
3\times 4x^2=12x^2 & & -36x^2+56x-39 \\
3\times 2x\times(-3)= & -18x & \\
(-3)^2= & +9 & \\
 & 12x^2-18x+9 & -36x^2+54x-27 \\
 & & 2x-12
\end{array}$$

Here in order that the expression may be a perfect cube the remainder $2x-12$ must $=0$;

$$\therefore\ x=6.$$

9.

$$\begin{array}{rr|l}
 & & \dfrac{x^6}{27}-\dfrac{a^2x^4}{3}+4a^4x^2-28a^6\ \left(\ \dfrac{x^2}{3}-a^2\right. \\
 & & \dfrac{x^6}{27} \\
3\times\dfrac{x^4}{9}=\dfrac{x^4}{3} & & -\dfrac{a^2x^4}{3}+4a^4x^2-28a^6 \\
3\times\dfrac{x^2}{3}\times(-a^2)= & -a^2x^2 & \\
(-a^2)^2= & +a^4 & \\
 & \dfrac{x^4}{3}-a^2x^2+a^4 & -\dfrac{a^2x^4}{3}+a^4x^2-a^6 \\
 & & 3a^4x^2-27a^6
\end{array}$$

Here in order that the expression may be a perfect cube the remainder $3a^4x^2-27a^6$ must $=0$;

$$\therefore\ x=\pm 3a.$$

10.

$$\begin{array}{rr|l}
 & & m^3x^6-9m^2nx^4+39mn^2x^2-51n^3\ (\ mx^2-3n \\
 & & m^3x^6 \\
3\,(mx^2)^2=3m^2x^4 & & -9m^2nx^4+39mn^2x^2-51n^3 \\
3\,(mx^2)\times(-3n)= & -9mnx^2 & \\
(-3n)^2= & +9n^2 & \\
 & 3m^2x^4-9mnx^2+9n^2 & -9m^2nx^4+27mn^2x^2-27n^3 \\
 & & 12mn^2x^2-24n^3
\end{array}$$

Here in order that the expression may be a perfect cube the remainder $12mn^2x^2-24n^3$ must $=0$;

$$\therefore\ x^2=\frac{2n}{m}\ \text{ or }\ x=\pm\sqrt{\frac{2n}{m}}.$$

11. If x^3+3ax^2+bx+c is a perfect cube it must be of the form

$$x^3+3ax^2+3a^2x+a^3,$$

and therefore we must have $b=3a^2,\ c=a^3$.

$$\therefore\ \left(\frac{b}{3}\right)^3=c^2,\ \text{that is, } b^3=27c^2.$$

12.

$$x^6+3ax^5+3bx^4+a(6b-5a^2)x^3+3b(b-a^2)x^2+3cx+d\;(x^2+ax+b-a^2$$
$$x^6$$

$$\begin{array}{ll}
3\,.\,(x^2)^2 = 3x^4 & \\
3\,.\,x^2\,.\,ax = \quad +3ax^3 & \\
\overline{ax}|^2 = \qquad +a^2x^2 & 3ax^5+3bx^4+a(6b-5a^2)x^3 \\
\hline
3x^4+3ax^3+a^2x^2 & 3ax^5+3a^2x^4+a^3x^3
\end{array}$$

$$\begin{array}{ll}
3(x^2+ax)^2 = 3x^4+6ax^3+3a^2x^2 & 3(b-a^2)x^4+6a(b-a^2)x^3+3b(b-a^2)x^2+3cx+d \\
3(x^2+ax)(b-a^2) = \quad +3(b-a^2)x^2+3a(b-a^2)x & \\
(b-a^2)^2 = \qquad +(b-a^2)^2 & \\
\hline
3x^4+6ax^3+3bx^2+3a(b-a^2)x+(b-a^2)^2 & 3(b-a^2)x^4+6a(b-a^2)x^3+3b(b-a^2)x^2+3a(b-a^2)^2x+(b-a^2)^3 \\
\hline
 & 3\{c-a(b-a^2)^2\}x+d-(b-a^2)^3
\end{array}$$

Here in order that the expression may be a perfect cube this remainder must $=0$ for all values of x;

$$\therefore\; c-a(b-a^2)^2=0, \text{ and } d-(b-a^2)^3=0,$$

$$\text{or } c=a(b-a^2)^2 \text{ and } d=(b-a^2)^3.$$

13. Dividing x^3+2x^2 by $x+4$, the quotient is x^2-2x+8 and the remainder is -32, therefore if 32 be added to the expression the remainder is 0.

14. If $x+a$ is a factor of x^2+px+q and of x^2+lx+m, it is a factor of their difference $(l-p)x+m-q$, or of $x+\frac{m-q}{l-p}$.

Hence evidently $$a=\frac{m-q}{l-p}.$$

15. $x^3-6x^2+11x-6$ vanishes when $x=1$; $\therefore$ $x-1$ is a factor, and $x^3-6x^2+11x-6=x^2(x-1)-5x(x-1)+6(x-1)$
$$=(x-1)(x^2-5x+6)=(x-1)(x-2)(x-3).$$

16. $x^3-5x^2-2x+24$ vanishes when $x=3$; $\therefore$ $x-3$ is a factor, and $x^3-5x^2-2x+24=(x-3)(x^2-2x-8)=(x-3)(x-4)(x+2)$.

17. $x^3+9x^2+26x+24$ vanishes when $x=-2$; $\therefore$ $x+2$ is a factor, and $x^3+9x^2+26x+24=(x+2)(x^2+7x+12)=(x+2)(x+3)(x+4)$.

18. $x^3-x^2-41x+105$ vanishes when $x=3$; $\therefore$ $x-3$ is a factor, and $x^3-x^2-41x+105=(x-3)(x^2+2x-35)=(x-3)(x-5)(x+7)$.

19. $x^3-39x+70$ vanishes when $x=2$; $\therefore$ $x-2$ is a factor, and $x^3-39x+70=(x-2)(x^2+2x-35)=(x-2)(x-5)(x+7)$.

20. $x^3-8x^2-31x-22$ vanishes when $x=-1$; $\therefore$ $x+1$ is a factor, and $x^3-8x^2-31x-22=(x+1)(x^2-9x-22)=(x+1)(x+2)(x-11)$.

21. $6x^3+7x^2-x-2$ vanishes when $x=-1$; $\therefore$ $x+1$ is a factor, and $6x^3+7x^2-x-2=(x+1)(6x^2+x-2)=(x+1)(3x+2)(2x-1)$.

22. $6x^3+x^2-19x+6$ vanishes when $x=-2$; $\therefore$ $x+2$ is a factor, and $6x^3+x^2-19x+6=(x+2)(6x^2-11x+3)=(x+2)(3x-1)(2x-3)$.

27.
$$\begin{array}{r|l}
 & x^4+(2a-4)x^3+(a^2-2a+4)x^2+(2a^2-4a)x+a^2\ (\ x^2+(a-2)x+a \\
 & x^4 \\
2x^2+(a-2)x & (2a-4)x^3+(a^2-2a+4)x^2 \\
 & (2a-4)x^3+(a^2-4a+4)x^2 \\
2x^2+2(a-2)x+a & 2ax^2+2a(a-2)x+a^2 \\
 & 2ax^2+2a(a-2)x+a^2
\end{array}$$

28. The expression may be written
$$(a+1)^2x^4+(2a^2+2a)x^3+\ldots+\ldots+(a-3)^2.$$

Hence the square root is of the form $(a+1)x^2+Px+(a-3)$. By squaring this expression and equating the coefficient of x^3 to that in the original expression, we find $P=a$. It will be found that this value of P satisfies the other conditions.

29. If the expression is a perfect square it must be of the form

$$(x\sqrt{3m}+2\sqrt{2})^2;$$

$$\therefore\ 3mx^2+6(m-2)x+8=3mx^2+4\sqrt{2}\,.\,\sqrt{3m}\,.\,x+8.$$

$$\therefore\ 6(m-2)=4\sqrt{6m};$$

whence $$9m^2-36m+36=24m,$$

or $$3m^2-20m+12=0.$$

From this equation we obtain $m=\frac{2}{3}$, or 6.

30. Assume $4x^4+12x^3y+Px^2y^2+6xy^3+y^4=(2x^2+lxy+y^2)^2$.

Expanding the right-hand side and equating the coefficients of like terms, we obtain

$$P=\text{coefficient of } x^2y^2=4+l^2,$$
$$12=\text{coefficient of } x^3y=4l,$$
$$6=\text{coefficient of } xy^3=2l;$$

all these conditions are satisfied if $l=3$, and $P=9+4=13$.

31. Substituting 1 for x, the expression vanishes; $\therefore\ x-1$ is a factor.

32. Substituting ± 2 for x, the expression vanishes; $\therefore\ x+2$ and $x-2$ are factors; and therefore $(x+2)(x-2)$ or x^2-4 is a factor.

33. Substituting 3 and -2 for x, the expression vanishes; $\therefore\ x+2$ and $x-3$ are factors, and $\therefore\ (x+2)(x-3)$ or x^2-x-6 is a factor.

34. The remainder $=5^5-5\,.\,5^2+5=3125-125+5=3005$.

35. The remainder $=(-2a)^3-7(-2a)^2a+8(-2a)a^2+15a^3$

$$=-8a^3-28a^3-16a^3+15a^3=-37a^3.$$

36. Their common factor must be a factor of their difference, that is of dx^3-ax^2, or of $\left(x-\frac{a}{d}\right)dx^2$. As the common factor is clearly not x, it must be $x-\frac{a}{d}$.

Since ax^2-bx+c is divisible by $x-\frac{a}{d}$,

$$a\left(\frac{a}{d}\right)^2-b\left(\frac{a}{d}\right)+c=0;$$

or $$a^3-abd+cd^2=0.$$

37. $5^{2n}-1=25^n-1$ which is always divisible by $25-1$ or 24.

[Art. 229 I.]

38. The expression $=(1-x)(1-x^n)$, but $1-x^n$ is divisible by $1-x$, and therefore the expression is divisible by $(1-x)^2$ or $1-2x+x^2$.

39. Proceeding to find their H. C. F. we see that if the two expressions have a common factor, $3x^2+p$ must be divisible by $x+\dfrac{3r}{2p}$.

$$\therefore 3\left(-\frac{3r}{2p}\right)^2+p=0;$$

or
$$\frac{27r^2}{4p^2}+p=0;$$

or
$$\frac{r^2}{4}+\frac{p^3}{27}=0.$$

40.
$$x^2-(ay+bz)x+abyz=(x-ay)(x-bz).$$

If $x-ay$ is a factor of $x^n+py^n+qz^n$, then $(ay)^n+py^n+qz^n=0$ ......(1);

and if $x-bz$ ...., then $(bz)^n+py^n+qz^n=0$ ......(2).

From (1),
$$\frac{y^n}{z^n}=-\frac{q}{a^n+p}.$$

From (2),
$$\frac{y^n}{z^n}=-\frac{b^n+q}{p}.$$

Equating these two values we obtain

$$a^nb^n+qa^n+pb^n+pq=pq.$$

$$\therefore 1+\frac{q}{b^n}+\frac{p}{a^n}=0.$$

EXAMPLES. XXX. a. PAGE 244.

41. $16^{\frac{3}{4}}=(\sqrt[4]{16})^3=2^3=8.$

42. $4^{-\frac{5}{2}}=\dfrac{1}{(\sqrt{4})^5}=\dfrac{1}{2^5}=\dfrac{1}{32}.$

43. $125^{\frac{2}{3}}=(\sqrt[3]{125})^2=5^2=25.$

44. $8^{-\frac{2}{3}}=\dfrac{1}{(\sqrt[3]{8})^2}=\dfrac{1}{4}.$

45. $36^{-\frac{3}{2}}=\dfrac{1}{(\sqrt{36})^3}=\dfrac{1}{6^3}=\dfrac{1}{216}.$

46. $\dfrac{1}{25^{-2}}=25^2=625.$

47. $243^{\frac{2}{5}}=(\sqrt[5]{243})^2=3^2=9.$

48. $\left(\dfrac{8}{27}\right)^{-\frac{1}{3}}=\left(\dfrac{27}{8}\right)^{\frac{1}{3}}=\dfrac{\sqrt[3]{27}}{\sqrt[3]{8}}=\dfrac{3}{2}$

49. $\left(\frac{81}{16}\right)^{\frac{3}{4}} = \left(\sqrt[4]{\frac{81}{16}}\right)^3 = \left(\frac{3}{2}\right)^3 = \frac{27}{8}.$

50. $\left(\frac{32}{243}\right)^{-\frac{7}{5}} = \left(\frac{243}{32}\right)^{\frac{7}{5}} = \left(\sqrt[5]{\frac{243}{32}}\right)^7 = \left(\frac{3}{2}\right)^7 = \frac{2187}{128}.$

EXAMPLES. XXX. b. Page 247.

1. $(\sqrt{a^2b^3})^6 = (ab^{\frac{3}{2}})^6 = a^6b^9.$

2. $(\sqrt[9]{x^{-4}y^3})^{-3} = (x^{-\frac{4}{9}}y^{\frac{1}{3}})^{-3} = x^{\frac{4}{3}}y^{-1} = \frac{x^{\frac{4}{3}}}{y}.$

3. $(x^a y^{-b})^3 \times (x^3 y^2)^{-a} = x^{3a}y^{-3b} \times x^{-3a}y^{-2a} = x^0 y^{-(2a+3b)} = \frac{1}{y^{2a+3b}}.$

4. $\left(\frac{16x^2}{y^{-2}}\right)^{-\frac{1}{4}} = \frac{(4x)^{-\frac{1}{2}}}{y^{\frac{1}{2}}} = \frac{1}{2x^{\frac{1}{2}}y^{\frac{1}{2}}}.$

5. $\left(\frac{27x^3}{8a^{-3}}\right)^{-\frac{2}{3}} = \left(\frac{8a^{-3}}{27x^3}\right)^{\frac{2}{3}} = \left\{\sqrt[3]{\frac{8}{27a^3x^3}}\right\}^2 = \left(\frac{2}{3ax}\right)^2 = \frac{4}{9a^2x^2}.$

6. $\left(\frac{a^{-\frac{1}{2}}}{4c^2}\right)^{-2} = \frac{a}{(2c)^{-4}} = a\,(2c)^4 = 16ac^4.$

7. $\{\sqrt[4]{(x^{-\frac{2}{3}}y^{\frac{1}{2}})^3}\}^{-\frac{2}{3}} = \{\sqrt[4]{x^{-2}y^{\frac{3}{2}}}\}^{-\frac{2}{3}} = (x^{-2}y^{\frac{3}{2}})^{-\frac{1}{6}} = x^{\frac{1}{3}}y^{-\frac{1}{4}} = \frac{x^{\frac{1}{3}}}{y^{\frac{1}{4}}}.$

8. $\sqrt[4]{x\sqrt[3]{x^{-1}}} = (x\,.\,x^{-\frac{1}{3}})^{\frac{1}{4}} = (x^{\frac{2}{3}})^{\frac{1}{4}} = x^{\frac{1}{6}}.$

9. $(4a^{-2} \div 9x^2)^{-\frac{1}{2}} = \left(\frac{4a^{-2}}{9x^2}\right)^{-\frac{1}{2}} = \left(\frac{4}{9a^2x^2}\right)^{-\frac{1}{2}} = \left(\frac{9a^2x^2}{4}\right)^{\frac{1}{2}} = \frac{3ax}{2}.$

10. $(x \div \sqrt[n]{x})^n = \left(\frac{x}{x^{\frac{1}{n}}}\right)^n = \frac{x^n}{x} = x^{n-1}.$

11. $(x \times \sqrt[n]{x^{-\frac{1}{n}}})^{\frac{n^2}{1-n}} = (x \times x^{-\frac{1}{n^2}})^{\frac{n^2}{1-n}} = (x^{\frac{n^2-1}{n^2}})^{\frac{n^2}{1-n}} = x^{-(n+1)} = \frac{1}{x^{n+1}}.$

12. $(\sqrt[b]{x^b} \div \sqrt[a]{x})^{\frac{1}{1-a}} = \left(\frac{x}{x^{\frac{1}{a}}}\right)^{\frac{1}{1-a}} = (x^{\frac{a-1}{a}})^{\frac{1}{1-a}} = x^{-\frac{1}{a}} = \frac{1}{x^{\frac{1}{a}}}.$

13. $\sqrt{a^{-2}b} \times \sqrt[3]{ab^{-3}} = a^{-1}b^{\frac12} \times a^{\frac13}b^{-1} = a^{-\frac23}b^{-\frac12} = \frac{1}{a^{\frac23}b^{\frac12}}$.

14. $\sqrt[3]{ab^{-1}c^{-2}} \times (a^{-1}b^{-2}c^{-4})^{-\frac16} = a^{\frac13}b^{-\frac13}c^{-\frac23} \times a^{\frac16}b^{\frac13}c^{\frac23} = a^{\frac12}$.

15. $\sqrt[6]{a^{4b}x^6} \times (a^{\frac23}x^{-1})^{-b} = a^{\frac{2b}{3}}x \times a^{-\frac{2b}{3}}x^b = x^{b+1}$.

16. $\sqrt[3]{x^{-1}\sqrt{y^3}} \div \sqrt{y\sqrt[3]{x}} = \frac{\sqrt[3]{x^{-1}y^{\frac32}}}{\sqrt{yx^{\frac13}}} = \frac{x^{-\frac13}y^{\frac12}}{y^{\frac12}x^{\frac16}} = \frac{x^{-\frac13}}{x^{\frac16}} = \frac{1}{x^{\frac13}\cdot x^{\frac16}} = \frac{1}{x^{\frac12}}$.

17. $(a^{-\frac12}\sqrt[3]{x})^{-3} \times \sqrt{x^{-2}\sqrt{a^{-6}}} = (a^{-\frac12}x^{\frac13})^{-3} \times \sqrt{x^{-2}a^{-3}}$

$= a^{\frac32}x^{-1} \times x^{-1}a^{-\frac32} = x^{-2} = \frac{1}{x^2}$.

18. $\sqrt[n]{a^{n+k}b^{2n-k}} \div (a^{\frac1n}b^{-\frac1n})^k = a^{\frac{n+k}{n}}b^{\frac{2n-k}{n}} \div a^{\frac kn}b^{-\frac kn} = a^{\frac{n+k}{n}-\frac kn}b^{\frac{2n-k}{n}+\frac kn} = ab^2$.

19. $\sqrt[3]{(a+b)^5} \times (a+b)^{-\frac23} = (a+b)^{\frac53} \times (a+b)^{-\frac23} = a+b$.

20. $\{(x-y)^{-3}\}^n \div \{(x+y)^n\}^3 = (x-y)^{-3n} \div (x+y)^{3n} = \frac{1}{(x-y)^{3n}} \div (x+y)^{3n}$

$= \frac{1}{(x-y)^{3n}(x+y)^{3n}} = \frac{1}{(x^2-y^2)^{3n}}$.

21. $\left(\frac{a^{-2}b}{a^3b^{-4}}\right)^{-3} \div \left(\frac{ab^{-1}}{a^{-3}b^2}\right)^5 = \left(\frac{b^5}{a^5}\right)^{-3} \div \left(\frac{a^4}{b^3}\right)^5 = \left(\frac{a^5}{b^5}\right)^3 \times \left(\frac{b^3}{a^4}\right)^5$

$= \frac{a^{15}b^{15}}{b^{15}a^{20}} = \frac{1}{a^5}$.

22. $\left\{\frac{\sqrt[3]{a}}{\sqrt[4]{b^{-1}}} \cdot \left(\frac{b^{\frac14}}{a^{\frac13}}\right)^2 \div \frac{a^{-\frac13}}{b^{-\frac12}}\right\}^6 = \left(\frac{a^{\frac13}}{b^{-\frac14}} \cdot \frac{b^{\frac12}}{a^{\frac23}} \cdot \frac{a^{\frac13}}{b^{\frac12}}\right)^6 = \left(\frac{1}{b^{-\frac14}}\right)^6 = b^{\frac32}$.

23. $\left(a^{-\frac12}x^{\frac13}\sqrt{ax^{-\frac13}\sqrt[4]{x^{\frac43}}}\right)^{\frac13} = \left(a^{-\frac12}x^{\frac13}\sqrt{ax^{-\frac13}x^{\frac13}}\right)^{\frac13} = (a^{-\frac12}x^{\frac13}a^{\frac12})^{\frac13} = (x^{\frac13})^{\frac13} = x^{\frac19}$.

24. $\sqrt[4]{(a+b)^6} \times (a^2-b^2)^{-\frac12} = (a+b)^{\frac32} \times (a+b)^{-\frac12}(a-b)^{-\frac12} = \frac{a+b}{(a-b)^{\frac12}}$

25. $\left(\frac{a^{-3}}{b^{-\frac23}c}\right)^{-\frac32} \div \left(\frac{\sqrt{a^{-\frac12}} \cdot \sqrt[6]{b^3}}{a^2c^{-1}}\right)^{-2} = \left(\frac{b^{\frac23}}{a^3c}\right)^{-\frac32} \div \left(\frac{a^{-\frac14}b^{\frac12}}{a^2c^{-1}}\right)^{-2}$

$= \frac{a^{\frac92}c^{\frac32}}{b} \times \frac{a^{-4}c^2}{a^{\frac12}b^{-1}} = \frac{a^{\frac12}c^{\frac72}}{a^{\frac12}} = c^{\frac72}$.

26. $\left(\dfrac{a^{-\frac{2}{3}}x^{\frac{1}{2}}}{x^{-1}a}\right)^2 \div \sqrt[3]{\dfrac{a^{-1}}{x^{-3}}} = \left(\dfrac{x^{\frac{3}{2}}}{a^{\frac{5}{3}}}\right)^2 \div \dfrac{a^{-\frac{1}{3}}}{x^{-1}} = \dfrac{x^3}{a^{\frac{10}{3}}} \div \dfrac{x}{a^{\frac{1}{3}}} = \dfrac{x^2}{a^3}.$

27. $\left(\sqrt[5]{\dfrac{a^{\frac{1}{2}}x^{-2}}{x^{\frac{1}{2}}a^{-2}}} \times \sqrt[3]{\dfrac{a\sqrt{x}}{x^{-1}\sqrt{a}}}\right)^{-4} = \left(\sqrt[5]{\dfrac{a^{\frac{5}{2}}}{x^{\frac{5}{2}}}} \times \sqrt[3]{\dfrac{ax^{\frac{1}{2}}}{x^{-1}a^{\frac{1}{2}}}}\right)^{-4}$

$= \left(\dfrac{a^{\frac{1}{2}}}{x^{\frac{1}{2}}} \times \sqrt[3]{a^{\frac{1}{2}}x^{\frac{3}{2}}}\right)^{-4} = \left(\dfrac{a^{\frac{1}{2}}}{x^{\frac{1}{2}}} \cdot a^{\frac{1}{6}}x^{\frac{1}{2}}\right)^{-4} = (a^{\frac{2}{3}})^{-4} = \dfrac{1}{a^{\frac{8}{3}}}.$

28. $\dfrac{\sqrt[3]{a^3b^3+a^6}}{\sqrt[3]{(b^6-a^3b^3)^{-1}}} = \dfrac{(a^3b^3+a^6)^{\frac{1}{3}}}{(b^6-a^3b^3)^{-\frac{1}{3}}} = \{a^3(a^3+b^3)\}^{\frac{1}{3}}\{b^3(b^3-a^3)\}^{\frac{1}{3}}$

$= a(a^3+b^3)^{\frac{1}{3}} \cdot b(b^3-a^3)^{\frac{1}{3}} = ab(b^6-a^6)^{\frac{1}{3}}.$

29. $(a^{n^2-1})^{\frac{n}{n+1}} + \dfrac{\sqrt[n]{a^{2n}}}{a} = a^{\frac{n(n^2-1)}{n+1}} + \dfrac{a^2}{a} = a^{n(n-1)} + a.$

30. $(x^{\frac{n}{n+1}})^{n^2-1} + \dfrac{\sqrt{x^{2n}}}{x} = x^{\frac{n(n^2-1)}{n+1}} + \dfrac{x^n}{x} = x^{n(n-1)} + x^{n-1}.$

31. $\left\{\dfrac{a^{p-q}}{\sqrt[q]{a^{q^2-pq}}} \times a^{2(p-q)}\right\}^n = \left\{\dfrac{a^{p-q}}{a^{q-p}} \times a^{2(p-q)}\right\}^n = \{a^{2(p-q)} \times a^{2(p-q)}\}^n = a^{4n(p-q)}.$

32. $(x^{\frac{a}{b}}y^{-1})^b \div \left(\dfrac{x^{a^2-b^2}}{y^{ab+b^2}}\right)^{\frac{1}{a+b}} = x^a y^{-b} \div \dfrac{x^{a-b}}{y^b} = \dfrac{x^a}{y^b} \times \dfrac{y^b}{x^{a-b}} = \dfrac{x^a}{x^{a-b}} = x^b.$

33. $\left(\dfrac{x^{-2}y^3}{x^3y^{-2}}\right)^{-\frac{1}{5}} \times \left(\dfrac{y^3x^{-3}}{x^3y^{-3}}\right)^{-1} = \left(\dfrac{y^5}{x^5}\right)^{-\frac{1}{5}} \times \left(\dfrac{y^6}{x^6}\right)^{-1} = \dfrac{y^{-1}}{x^{-1}} \times \dfrac{y^{-6}}{x^{-6}} = \dfrac{x^7}{y^7}.$

34. $\left(\dfrac{y^{-3}}{x^{\frac{2}{7}}z^{-1}}\right)^{-\frac{3}{2}} \times \left(\dfrac{y^{\frac{14}{3}}x^{-1}}{z^{-\frac{21}{4}}}\right)^{\frac{2}{7}} = \dfrac{y^{\frac{9}{2}}}{x^{-\frac{3}{7}}z^{\frac{3}{2}}} \times \dfrac{y^{\frac{4}{3}}x^{-\frac{2}{7}}}{z^{-\frac{3}{2}}} = \dfrac{y^{\frac{9}{2}+\frac{4}{3}}}{x^{-\frac{1}{7}}} = x^{\frac{1}{7}}y^{\frac{35}{6}}.$

35. $\dfrac{2^n \times (2^{n-1})^n}{2^{n+1} \times 2^{n-1}} \times \dfrac{1}{4^{-n}} = \dfrac{2^n \times 2^{n^2-n}}{2^{n+1+n-1}} \times 4^n = \dfrac{2^{n^2}}{2^{2n}} \times (2^2)^n = 2^{n^2}.$

36. $\dfrac{2^{n+1}}{(2^n)^{n-1}} \div \dfrac{4^{n+1}}{(2^{n-1})^{n+1}} = \dfrac{2^{n+1}}{2^{n^2-n}} \times \dfrac{2^{n^2-1}}{(2^2)^{n+1}} = \dfrac{2^{n^2+n}}{2^{n^2+n+2}} = \dfrac{1}{2^2} = \dfrac{1}{4}.$

[*For Examples* 37, 38, *see Appendix, p.* 368.]

EXAMPLES. XXX. c. Page 250.

1.
$$\begin{array}{r}
3x^{\frac{1}{3}} - 5 + 8x^{-\frac{1}{3}} \\
4x^{\frac{1}{3}} + 3x^{-\frac{1}{3}} \\
\hline
12x^{\frac{2}{3}} - 20x^{\frac{1}{3}} + 32 \qquad\qquad \\
9 - 15x^{-\frac{1}{3}} + 24x^{-\frac{2}{3}} \\
\hline
12x^{\frac{2}{3}} - 20x^{\frac{1}{3}} + 41 - 15x^{-\frac{1}{3}} + 24x^{-\frac{2}{3}}
\end{array}$$

2.
$$\begin{array}{r}
3a^{\frac{2}{5}} - 4a^{\frac{1}{5}} - a^{-\frac{1}{5}} \qquad\qquad \\
3a^{\frac{2}{5}} + a^{-\frac{1}{5}} - 6a^{-\frac{3}{5}} \qquad\qquad \\
\hline
9a^{\frac{4}{5}} - 12a^{\frac{3}{5}} - 3 \qquad\qquad\qquad \\
3a^{\frac{2}{5}} - 4 - a^{-\frac{2}{5}} \qquad\quad \\
-18 + 24a^{-\frac{2}{5}} + 6a^{-\frac{4}{5}} \\
\hline
9a^{\frac{4}{5}} - 9a^{\frac{3}{5}} - 25 + 23a^{-\frac{2}{5}} + 6a^{-\frac{4}{5}}
\end{array}$$

3.
$$\begin{array}{l}
c^{x} - 7 + 2c^{-x} \\
2c^{x} + 5 - 3c^{-x} \\
\hline
2c^{2x} - 14c^{x} + 4 \\
\qquad\qquad 5c^{x} - 35 + 10c^{-x} \\
\qquad\qquad\qquad - 3 + 21c^{-x} - 6c^{-2x} \\
\hline
2c^{2x} - 9c^{x} - 34 + 31c^{-x} - 6c^{-2x}
\end{array}$$

4.
$$\begin{array}{l}
2x^{2a} + 5 + 3x^{-2a} \\
4x^{a} - 3x^{-a} \\
\hline
8x^{3a} + 20x^{a} + 12x^{-a} \\
\qquad\quad - 6x^{a} - 15x^{-a} - 9x^{-3a} \\
\hline
8x^{3a} + 14x^{a} - 3x^{-a} - 9x^{-3a}
\end{array}$$

5.
$$\begin{array}{l}
3x^{\frac{1}{3}} + 1 \,)\, 21x + x^{\frac{2}{3}} + x^{\frac{1}{3}} + 1 \,(\, 7x^{\frac{2}{3}} - 2x^{\frac{1}{3}} + 1 \\
\qquad\qquad 21x + 7x^{\frac{2}{3}} \\
\hline
\qquad\qquad\quad - 6x^{\frac{2}{3}} + x^{\frac{1}{3}} \\
\qquad\qquad\quad - 6x^{\frac{2}{3}} - 2x^{\frac{1}{3}} \\
\hline
\qquad\qquad\qquad\qquad 3x^{\frac{1}{3}} + 1 \\
\qquad\qquad\qquad\qquad 3x^{\frac{1}{3}} + 1
\end{array}$$

6.
$$\begin{array}{l}
5a^{\frac{2}{3}} + 4 \,)\, 15a - 3a^{\frac{1}{3}} - 2a^{-\frac{1}{3}} + 8a^{-1} \,(\, 3a^{\frac{1}{3}} - 3a^{-\frac{1}{3}} + 2a^{-1} \\
\qquad\qquad 15a + 12a^{\frac{1}{3}} \\
\hline
\qquad\qquad\quad - 15a^{\frac{1}{3}} - 2a^{-\frac{1}{3}} \\
\qquad\qquad\quad - 15a^{\frac{1}{3}} - 12a^{-\frac{1}{3}} \\
\hline
\qquad\qquad\qquad\qquad 10a^{-\frac{1}{3}} + 8a^{-1} \\
\qquad\qquad\qquad\qquad 10a^{-\frac{1}{3}} + 8a^{-1}
\end{array}$$

7.
$$\begin{array}{l}
2a^{-1} - 1 \,)\, 16a^{-3} + 6a^{-2} + 5a^{-1} - 6 \,(\, 8a^{-2} + 7a^{-1} + 6 \\
\qquad\qquad 16a^{-3} - 8a^{-2} \\
\hline
\qquad\qquad\qquad 14a^{-2} + 5a^{-1} \\
\qquad\qquad\qquad 14a^{-2} - 7a^{-1} \\
\hline
\qquad\qquad\qquad\qquad 12a^{-1} - 6 \\
\qquad\qquad\qquad\qquad 12a^{-1} - 6
\end{array}$$

8.
$$\begin{array}{l}
b^{\frac{1}{6}} - 2b^{-\frac{1}{6}} \,)\, 5b^{\frac{2}{3}} - 6b^{\frac{1}{3}} - 5 - 4b^{-\frac{1}{3}} - 4b^{-\frac{2}{3}} \,(\, 5b^{\frac{1}{2}} + 4b^{\frac{1}{6}} + 3b^{-\frac{1}{6}} + 2b^{-\frac{1}{2}} \\
\qquad\qquad 5b^{\frac{2}{3}} - 10b^{\frac{1}{3}} \\
\hline
\qquad\qquad\qquad 4b^{\frac{1}{3}} - 5 \\
\qquad\qquad\qquad 4b^{\frac{1}{3}} - 8 \\
\hline
\qquad\qquad\qquad\qquad 3 - 4b^{-\frac{1}{3}} \\
\qquad\qquad\qquad\qquad 3 - 6b^{-\frac{1}{3}} \\
\hline
\qquad\qquad\qquad\qquad\quad 2b^{-\frac{1}{3}} - 4b^{-\frac{2}{3}} \\
\qquad\qquad\qquad\qquad\quad 2b^{-\frac{1}{3}} - 4b^{-\frac{2}{3}}
\end{array}$$

9.

$$\begin{array}{rl}
3a^x - 5\ ) & 21a^{3x} - 26a^{2x} - 27a^x + 20\ (\ 7a^{2x} + 3a^x - 4 \\
 & \underline{21a^{3x} - 35a^{2x}} \\
 & \qquad 9a^{2x} - 27a^x \\
 & \qquad \underline{9a^{2x} - 15a^x} \\
 & \qquad\qquad -12a^x + 20 \\
 & \qquad\qquad \underline{-12a^x + 20}
\end{array}$$

10.

$$\begin{array}{rl}
5c^n - 3c^{-n}\ ) & 5c^{3n} - 8c^n + 8c^{-n} - 3c^{-3n}\ (\ c^{2n} - 1 + c^{-2n} \\
 & \underline{5c^{3n} - 3c^n} \\
 & \quad -5c^n + 8c^{-n} \\
 & \quad \underline{-5c^n + 3c^{-n}} \\
 & \qquad\qquad 5c^{-n} - 3c^{-3n} \\
 & \qquad\qquad \underline{5c^{-n} - 3c^{-3n}}
\end{array}$$

11.

$$\begin{array}{r|l}
 & 9x - 12x^{\frac{1}{2}} + 10 - 4x^{-\frac{1}{2}} + x^{-1}\ (\ 3x^{\frac{1}{2}} - 2 + x^{-\frac{1}{2}} \\
 & \underline{9x} \\
6x^{\frac{1}{2}} - 2 & -12x^{\frac{1}{2}} + 10 \\
 & \underline{-12x^{\frac{1}{2}} + 4} \\
6x^{\frac{1}{2}} - 4 + x^{-\frac{1}{2}} & 6 - 4x^{-\frac{1}{2}} + x^{-1} \\
 & \underline{6 - 4x^{-\frac{1}{2}} + x^{-1}}
\end{array}$$

12.

$$\begin{array}{r|l}
 & 25a^{\frac{4}{3}} - 30a + 49a^{\frac{2}{3}} - 24a^{\frac{1}{3}} + 16\ (\ 5a^{\frac{2}{3}} - 3a^{\frac{1}{3}} + 4 \\
 & \underline{25a^{\frac{4}{3}}} \\
10a^{\frac{2}{3}} - 3a^{\frac{1}{3}} & -30a + 49a^{\frac{2}{3}} \\
 & \underline{-30a + 9a^{\frac{2}{3}}} \\
10a^{\frac{2}{3}} - 6a^{\frac{1}{3}} + 4 & 40a^{\frac{2}{3}} - 24a^{\frac{1}{3}} + 16 \\
 & \underline{40a^{\frac{2}{3}} - 24a^{\frac{1}{3}} + 16}
\end{array}$$

13.

$$\begin{array}{r|l}
 & 4x^n - 16x^{\frac{n}{2}} + 28 - 24x^{-\frac{n}{2}} + 9x^{-n}\ (\ 2x^{\frac{n}{2}} - 4 + 3x^{-\frac{n}{2}} \\
 & \underline{4x^n} \\
4x^{\frac{n}{2}} - 4 & -16x^{\frac{n}{2}} + 28 \\
 & \underline{-16x^{\frac{n}{2}} + 16} \\
4x^{\frac{n}{2}} - 8 + 3x^{-\frac{n}{2}} & 12 - 24x^{-\frac{n}{2}} + 9x^{-n} \\
 & \underline{12 - 24x^{-\frac{n}{2}} + 9x^{-n}}
\end{array}$$

14.

$$\begin{array}{r|l}
 & a^{4x} - 6a^{3x} + 5a^{2x} + 12a^x + 4\ (\ a^{2x} - 3a^x - 2 \\
 & \underline{a^{4x}} \\
2a^{2x} - 3a^x & -6a^{3x} + 5a^{2x} \\
 & \underline{-6a^{3x} + 9a^{2x}} \\
2a^{2x} - 6a^x - 2 & -4a^{2x} + 12a^x + 4 \\
 & \underline{-4a^{2x} + 12a^x + 4}
\end{array}$$

15.
$$\begin{array}{l}
a^{\frac{3}{2}} - 2a^{\frac{1}{2}} + 4a^{-\frac{1}{2}} - 8a^{-\frac{3}{2}} \\
a^{\frac{1}{2}} + 4a^{-\frac{1}{2}} + 4a^{-\frac{3}{2}} \\ \hline
a^2 - 2a + 4 - 8a^{-1} \\
\quad 4a - 8 + 16a^{-1} - 32a^{-2} \\
\qquad + 4 - 8a^{-1} + 16a^{-2} - 32a^{-3} \\ \hline
a^2 + 2a \qquad - 16a^{-2} - 32a^{-3} \\ \hline
\end{array}$$

16.
$$\begin{array}{l}
1 - 2x^{\frac{1}{3}} - 2x^{\frac{1}{2}} \\
1 - x^{\frac{1}{6}} \\ \hline
1 - 2x^{\frac{1}{3}} - 2x^{\frac{1}{2}} \\
\quad - x^{\frac{1}{6}} + 2x^{\frac{1}{2}} + 2x^{\frac{2}{3}} \\ \hline
1 - x^{\frac{1}{6}} - 2x^{\frac{1}{3}} + 2x^{\frac{2}{3}} \\ \hline
\end{array}$$

18.
$$\begin{array}{l}
x^{\frac{1}{6}} + 4x^{-\frac{1}{6}} + 4x^{-\frac{1}{2}} \;) \; x^{\frac{2}{3}} + 2x^{\frac{1}{3}} - 16x^{-\frac{2}{3}} - 32x^{-1} \; (\; x^{\frac{1}{2}} - 2x^{\frac{1}{6}} + 4x^{-\frac{1}{6}} - 8x^{-\frac{1}{2}} \\
\qquad x^{\frac{2}{3}} + 4x^{\frac{1}{3}} + 4 \\ \hline
\qquad - 2x^{\frac{1}{3}} - 4 \\
\qquad - 2x^{\frac{1}{3}} - 8 - 8x^{-\frac{1}{3}} \\ \hline
\qquad\quad 4 + 8x^{-\frac{1}{3}} - 16x^{-\frac{2}{3}} \\
\qquad\quad 4 + 16x^{-\frac{1}{3}} + 16x^{-\frac{2}{3}} \\ \hline
\qquad\quad - 8x^{-\frac{1}{3}} - 32x^{-\frac{2}{3}} - 32x^{-1} \\
\qquad\quad - 8x^{-\frac{1}{3}} - 32x^{-\frac{2}{3}} - 32x^{-1} \\ \hline
\end{array}$$

17.
$$\begin{array}{l}
2a^{\frac{5}{3}} - a^{\frac{1}{3}} - 3a^{-1} \\
2a - 3a^{-\frac{1}{3}} - a^{-\frac{5}{3}} \\ \hline
4a^{\frac{8}{3}} - 2a^{\frac{4}{3}} - 6 \\
\quad - 6a^{\frac{4}{3}} + 3 + 9a^{-\frac{4}{3}} \\
\qquad - 2 + a^{-\frac{4}{3}} + 3a^{-\frac{8}{3}} \\ \hline
4a^{\frac{8}{3}} - 8a^{\frac{4}{3}} - 5 + 10a^{-\frac{4}{3}} + 3a^{-\frac{8}{3}} \\ \hline
\end{array}$$

19.
$$\begin{array}{l}
1 - a^{\frac{1}{2}} \;) \; 1 - a^{\frac{1}{2}} - 2a + 2a^2 \; (\; 1 - 2a - 2a^{\frac{3}{2}} \\
\qquad 1 - a^{\frac{1}{2}} \\ \hline
\qquad - 2a + 2a^2 \\
\qquad - 2a + 2a^{\frac{3}{2}} \\ \hline
\qquad\quad - 2a^{\frac{3}{2}} + 2a^2 \\
\qquad\quad - 2a^{\frac{3}{2}} + 2a^2 \\ \hline
\end{array}$$

20.
$$\begin{array}{l}
2x^{\frac{5}{12}} - x^{\frac{1}{12}} - 3x^{-\frac{1}{4}} \;) \; 4x^{\frac{2}{3}} - 8x^{\frac{1}{3}} - 5 + 10x^{-\frac{1}{3}} + 3x^{-\frac{2}{3}} \; (\; 2x^{\frac{1}{4}} - 3x^{-\frac{1}{12}} - x^{-\frac{5}{12}} \\
\qquad 4x^{\frac{2}{3}} - 2x^{\frac{1}{3}} - 6 \\ \hline
\qquad - 6x^{\frac{1}{3}} + 1 + 10x^{-\frac{1}{3}} \\
\qquad - 6x^{\frac{1}{3}} + 3 + 9x^{-\frac{1}{3}} \\ \hline
\qquad\quad - 2 + x^{-\frac{1}{3}} + 3x^{-\frac{2}{3}} \\
\qquad\quad - 2 + x^{-\frac{1}{3}} + 3x^{-\frac{2}{3}} \\ \hline
\end{array}$$

21.
$$\begin{array}{l|l}
 & 9x^{-4} - 18x^{-3}y^{\frac{1}{2}} + 15x^{-2}y - 6x^{-1}y^{\frac{3}{2}} + y^2 \; (\; 3x^{-2} - 3x^{-1}y^{\frac{1}{2}} + y \\
 & 9x^{-4} \\ \hline
6x^{-2} - 3x^{-1}y^{\frac{1}{2}} & - 18x^{-3}y^{\frac{1}{2}} + 15x^{-2}y \\
 & - 18x^{-3}y^{\frac{1}{2}} + 9x^{-2}y \\ \hline
6x^{-2} - 6x^{-1}y^{\frac{1}{2}} + y & 6x^{-2}y - 6x^{-1}y^{\frac{3}{2}} + y^2 \\
 & 6x^{-2}y - 6x^{-1}y^{\frac{3}{2}} + y^2 \\ \hline
\end{array}$$

22.

$$4x^{\frac{3}{2}} - 12x^{\frac{3}{4}}y^{\frac{1}{4}} + 25y^{\frac{1}{2}} - 24x^{-\frac{1}{4}}y^{\frac{3}{4}} + 16x^{-\frac{1}{2}}y \;\big(\; 2x^{\frac{3}{4}} - 3y^{\frac{1}{4}} + 4x^{-\frac{1}{4}}y^{\frac{1}{2}}$$
$$4x^{\frac{3}{2}}$$
$$4x^{\frac{3}{4}} - 3y^{\frac{1}{4}} \;\Big|\; -12x^{\frac{3}{4}}y^{\frac{1}{4}} + 25y^{\frac{1}{2}}$$
$$-12x^{\frac{3}{4}}y^{\frac{1}{4}} + 9y^{\frac{1}{2}}$$
$$4x^{\frac{3}{4}} - 6y^{\frac{1}{4}} + 4x^{-\frac{1}{4}}y^{\frac{1}{2}} \;\Big|\; 16y^{\frac{1}{2}} - 24x^{-\frac{1}{4}}y^{\frac{3}{4}} + 16x^{-\frac{1}{2}}y$$
$$16y^{\frac{1}{2}} - 24x^{-\frac{1}{4}}y^{\frac{3}{4}} + 16x^{-\frac{1}{2}}y$$

23. The expression $= 81\,(x^{\frac{4}{3}}y^{-2} + 1) + 36x^{\frac{1}{3}}y^{-\frac{1}{2}}\,(x^{\frac{2}{3}}y^{-1} - 1) - 158x^{\frac{2}{3}}y^{-1}$; multiply out, and arrange in descending powers of x.

$$81x^{\frac{4}{3}}y^{-2} + 36xy^{-\frac{3}{2}} - 158x^{\frac{2}{3}}y^{-1} - 36x^{\frac{1}{3}}y^{-\frac{1}{2}} + 81 \;\big(\; 9x^{\frac{2}{3}}y^{-1} + 2x^{\frac{1}{3}}y^{-\frac{1}{2}} - 9$$
$$81x^{\frac{4}{3}}y^{-2}$$
$$18x^{\frac{2}{3}}y^{-1} + 2x^{\frac{1}{3}}y^{-\frac{1}{2}} \;\Big|\; 36xy^{-\frac{3}{2}} - 158x^{\frac{2}{3}}y^{-1}$$
$$36xy^{-\frac{3}{2}} + 4x^{\frac{2}{3}}y^{-1}$$
$$18x^{\frac{2}{3}}y^{-1} + 4x^{\frac{1}{3}}y^{-\frac{1}{2}} - 9 \;\Big|\; -162x^{\frac{2}{3}}y^{-1} - 36x^{\frac{1}{3}}y^{-\frac{1}{2}} + 81$$
$$-162x^{\frac{2}{3}}y^{-1} - 36x^{\frac{1}{3}}y^{-\frac{1}{2}} + 81$$

24. The expression $= \dfrac{x^{-2}}{16} + 1 + 9y^{\frac{2}{3}} + \dfrac{x^{-1}}{2}\,(1 - 3y^{\frac{1}{3}}) - 6y^{\frac{1}{3}}$; arrange it according to ascending powers of x.

$$\frac{x^{-2}}{16} + \frac{x^{-1}}{2} - \frac{3x^{-1}y^{\frac{1}{3}}}{2} + 1 - 6y^{\frac{1}{3}} + 9y^{\frac{2}{3}} \;\Big(\; \frac{x^{-1}}{4} + 1 - 3y^{\frac{1}{3}}$$
$$\frac{x^{-2}}{16}$$
$$\frac{x^{-1}}{2} + 1 \;\Big|\; \frac{x^{-1}}{2} - \frac{3x^{-1}y^{\frac{1}{3}}}{2} + 1$$
$$\frac{x^{-1}}{2} \qquad + 1$$
$$\frac{x^{-1}}{2} + 2 - 3y^{\frac{1}{3}} \;\Big|\; -\frac{3x^{-1}y^{\frac{1}{3}}}{2} - 6y^{\frac{1}{3}} + 9y^{\frac{2}{3}}$$
$$-\frac{3x^{-1}y^{\frac{1}{3}}}{2} - 6y^{\frac{1}{3}} + 9y^{\frac{2}{3}}$$

EXAMPLES. XXX. d. PAGE 252.

25. $(x + x^{\frac{1}{2}} - 4)\,(x + x^{\frac{1}{2}} + 4) = (x + x^{\frac{1}{2}})^2 - (4)^2 = x^2 + 2x^{\frac{3}{2}} + x - 16.$

26. $(2x^{\frac{1}{3}} + 4 + 3x^{-\frac{1}{3}})\,(2x^{\frac{1}{3}} + 4 - 3x^{-\frac{1}{3}}) = (2x^{\frac{1}{3}} + 4)^2 - (3x^{-\frac{1}{3}})^2$

$$= 4x^{\frac{2}{3}} + 16x^{\frac{1}{3}} + 16 - 9x^{-\frac{2}{3}}.$$

27. $(2-x^{\frac{1}{3}}+x)(2+x^{\frac{1}{3}}+x)=(2+x)^2-(x^{\frac{1}{3}})^2=4+4x+x^2-x^{\frac{2}{3}}$.

28. $(a^x+7+3a^{-x})(a^x-7-3a^{-x})=(a^x)^2-(7+3a^{-x})^2=a^{2x}-49-42a^{-x}-9a^{-2x}$.

29. $$\frac{a^{\frac{4}{3}}-8a^{\frac{1}{3}}b}{a^{\frac{2}{3}}+2\sqrt[3]{ab}+4b^{\frac{2}{3}}}=\frac{a^{\frac{1}{3}}(a-8b)}{a^{\frac{2}{3}}+2a^{\frac{1}{3}}b^{\frac{1}{3}}+4b^{\frac{2}{3}}}=\frac{a^{\frac{1}{3}}(a^{\frac{1}{3}}-2b^{\frac{1}{3}})(a^{\frac{2}{3}}+2a^{\frac{1}{3}}b^{\frac{1}{3}}+4b^{\frac{2}{3}})}{a^{\frac{2}{3}}+2a^{\frac{1}{3}}b^{\frac{1}{3}}+4b^{\frac{2}{3}}}$$
$$=a^{\frac{1}{3}}(a^{\frac{1}{3}}-2b^{\frac{1}{3}}).$$

30. $$\frac{x-7x^{\frac{1}{2}}}{x-5\sqrt{x}-14}\div\left(1+\frac{2}{\sqrt{x}}\right)^{-1}=\frac{x^{\frac{1}{2}}(x^{\frac{1}{2}}-7)}{x-5x^{\frac{1}{2}}-14}\times\left(1+\frac{2}{x^{\frac{1}{2}}}\right)$$
$$=\frac{x^{\frac{1}{2}}(x^{\frac{1}{2}}-7)(x^{\frac{1}{2}}+2)}{(x^{\frac{1}{2}}-7)(x^{\frac{1}{2}}+2)x^{\frac{1}{2}}}=1.$$

31. $$\frac{x^{\frac{2}{3}}-4\sqrt[3]{x^{-2}}}{\sqrt[3]{x^2}+4+4x^{-\frac{2}{3}}}=\frac{x^{\frac{2}{3}}-4x^{-\frac{2}{3}}}{x^{\frac{2}{3}}+4+4x^{-\frac{2}{3}}}=\frac{(x^{\frac{1}{3}}+2x^{-\frac{1}{3}})(x^{\frac{1}{3}}-2x^{-\frac{1}{3}})}{(x^{\frac{1}{3}}+2x^{-\frac{1}{3}})^2}=\frac{x^{\frac{1}{3}}-2x^{-\frac{1}{3}}}{x^{\frac{1}{3}}+2x^{-\frac{1}{3}}}$$
$$=\frac{x^{\frac{2}{3}}-2}{x^{\frac{2}{3}}+2}.$$

32. $$\frac{a^{\frac{3}{2}}+ab}{ab-b^3}-\frac{\sqrt{a}}{\sqrt{a}-b}=\frac{a(a^{\frac{1}{2}}+b)}{b(a-b^2)}-\frac{a^{\frac{1}{2}}}{a^{\frac{1}{2}}-b}=\frac{a(a^{\frac{1}{2}}+b)}{b(a^{\frac{1}{2}}+b)(a^{\frac{1}{2}}-b)}-\frac{a^{\frac{1}{2}}}{a^{\frac{1}{2}}-b}$$
$$=\frac{a}{b(a^{\frac{1}{2}}-b)}-\frac{a^{\frac{1}{2}}}{a^{\frac{1}{2}}-b}=\frac{a-a^{\frac{1}{2}}b}{b(a^{\frac{1}{2}}-b)}=\frac{a^{\frac{1}{2}}}{b}.$$

EXAMPLES. XXXI. b. PAGE 256.

33. $3\sqrt{45}-\sqrt{20}+7\sqrt{5}=9\sqrt{5}-2\sqrt{5}+7\sqrt{5}=14\sqrt{5}$.

34. $4\sqrt{63}+5\sqrt{7}-8\sqrt{28}=12\sqrt{7}+5\sqrt{7}-16\sqrt{7}=\sqrt{7}$.

35. $\sqrt{44}-5\sqrt{176}+2\sqrt{99}=2\sqrt{11}-20\sqrt{11}+6\sqrt{11}=-12\sqrt{11}$

36. $2\sqrt{363}-5\sqrt{243}+\sqrt{192}=22\sqrt{3}-45\sqrt{3}+8\sqrt{3}=-15\sqrt{3}$.

37. $2\sqrt[3]{189}+3\sqrt[3]{875}-7\sqrt[3]{56}=6\sqrt[3]{7}+15\sqrt[3]{7}-14\sqrt[3]{7}=7\sqrt[3]{7}$.

38. $5\sqrt[3]{81}-7\sqrt[3]{192}+4\sqrt[3]{648}=15\sqrt[3]{3}-28\sqrt[3]{3}+24\sqrt[3]{3}=11\sqrt[3]{3}$.

39. $3\sqrt[4]{162}-7\sqrt[4]{32}+\sqrt[4]{1250}=9\sqrt[4]{2}-14\sqrt[4]{2}+5\sqrt[4]{2}=0$.

40. $5\sqrt[3]{-54}-2\sqrt[3]{-16}+4\sqrt[3]{686}=-15\sqrt[3]{2}+4\sqrt[3]{2}+28\sqrt[3]{2}=17\sqrt[3]{2}$.

41. $4\sqrt{128}+4\sqrt{75}-5\sqrt{162}=32\sqrt{2}+20\sqrt{3}-45\sqrt{2}=20\sqrt{3}-13\sqrt{2}.$

42. $5\sqrt{24}-2\sqrt{54}-\sqrt{6}=10\sqrt{6}-6\sqrt{6}-\sqrt{6}=3\sqrt{6}.$

43. $\sqrt{252}-\sqrt{294}-48\sqrt{\frac{1}{6}}=6\sqrt{7}-7\sqrt{6}-48\frac{\sqrt{6}}{6}=6\sqrt{7}-7\sqrt{6}-8\sqrt{6}$
$=6\sqrt{7}-15\sqrt{6}.$

44. $3\sqrt{147}-\frac{7}{3}\sqrt{\frac{1}{3}}-\sqrt{\frac{1}{27}}=21\sqrt{3}-\frac{7}{3}\cdot\frac{\sqrt{3}}{3}-\frac{\sqrt{3}}{\sqrt{81}}=21\sqrt{3}-\frac{7}{9}\sqrt{3}-\frac{1}{9}\sqrt{3}$
$=\frac{181\sqrt{3}}{9}.$

EXAMPLES. XXXI. c. PAGE 259.

1. $2\sqrt{14}\times\sqrt{21}=2\sqrt{2\cdot 7}\times\sqrt{3\cdot 7}=2\sqrt{2\cdot 3\cdot 7^2}=14\sqrt{6}.$

2. $3\sqrt{8}\times\sqrt{6}=6\sqrt{2}\times\sqrt{6}=6\sqrt{12}=12\sqrt{3}.$

4. $2\sqrt{15}\times 3\sqrt{5}=6\sqrt{3\cdot 5^2}=30\sqrt{3}.$

5. $8\sqrt{12}\times 3\sqrt{24}=16\sqrt{3}\times 6\sqrt{6}=96\sqrt{18}=288\sqrt{2}.$

7. The quotient $=\frac{21\sqrt{384}}{8\sqrt{98}}=\frac{21\times 8\sqrt{6}}{8\times 7\sqrt{2}}=\frac{3\sqrt{6}}{\sqrt{2}}=3\sqrt{3}.$

8. The quotient $=\frac{5\sqrt{27}}{3\sqrt{24}}=\frac{15\sqrt{3}}{6\sqrt{6}}=\frac{5}{2\sqrt{2}}=\frac{5\sqrt{2}}{4}.$

9. The quotient $=\frac{-13\sqrt{125}}{5\sqrt{65}}=-\frac{13\times 5\sqrt{5}}{5\times\sqrt{65}}=-13\sqrt{\frac{5}{65}}=-\frac{13}{\sqrt{13}}=-\sqrt{13}.$

10. $\sqrt[3]{168}\times\sqrt[3]{147}=\sqrt[3]{8\cdot 7\cdot 3\cdot 7^2\cdot 3}=2\cdot 7\sqrt[3]{9}=14\sqrt[3]{9}.$

11. $5\sqrt[3]{128}\times 2\sqrt[3]{432}=20\sqrt[3]{2}\times 12\sqrt[3]{2}=240\sqrt[3]{4}.$

12. The quotient $=\frac{6\sqrt{14}}{2\sqrt{21}}=\frac{3\sqrt{2\cdot 7}}{\sqrt{3\cdot 7}}=\frac{3\sqrt{2}}{\sqrt{3}}=\sqrt{3}\times\sqrt{2}=\sqrt{6}.$

13. $a\sqrt{b^3}\times b^2\sqrt{a}=ab^2\sqrt{b^3a}=ab^3\sqrt{ab}.$

14. $\frac{3\sqrt{11}}{2\sqrt{98}}\div\frac{5}{7\sqrt{22}}=\frac{3\sqrt{11}}{14\sqrt{2}}\times\frac{7\sqrt{22}}{5}=\frac{33\sqrt{2}}{10\sqrt{2}}=\frac{33}{10}.$

15. $\frac{3\sqrt{48}}{5\sqrt{112}}\div\frac{6\sqrt{84}}{\sqrt{392}}=\frac{1}{10}\sqrt{\frac{48\times 392}{112\times 84}}=\frac{\sqrt{2}}{10}.$

16. $\frac{3}{x}\sqrt{\frac{a^2}{x}}\times\frac{4}{3}\sqrt{\frac{x^3}{2a^4}}=\frac{4}{x}\sqrt{\frac{x^2}{2a^2}}=\frac{4}{a\sqrt{2}}=\frac{2\sqrt{2}}{a}.$

17. $\dfrac{3}{a-b}\sqrt{\dfrac{2x}{a-b}}\div\sqrt{\dfrac{18x^3}{(a-b)^5}}=\dfrac{3}{a-b}\sqrt{\dfrac{(a-b)^4}{9x^2}}=\dfrac{3}{a-b}\cdot\dfrac{(a-b)^2}{3x}.$

18. $\dfrac{14}{\sqrt{2}}=\dfrac{14\sqrt{2}}{2}=7\sqrt{2}=7\times1\cdot41421=9\cdot89947.$

19. $\dfrac{25}{\sqrt{5}}=\dfrac{25\sqrt{5}}{5}=5\sqrt{5}=5\times2\cdot23607=11\cdot18035.$

20. $\dfrac{10}{\sqrt{7}}=\dfrac{10\sqrt{7}}{7}=\dfrac{10\times2\cdot64575}{7}=\dfrac{26\cdot4575}{7}=3\cdot77964.$

21. $\dfrac{48}{\sqrt{6}}=\dfrac{48\sqrt{6}}{6}=8\sqrt{6}=8\times2\cdot44949=19\cdot59592.$

22. $\dfrac{60}{\sqrt{5}}=\dfrac{60\sqrt{5}}{5}=12\sqrt{5}=12\times2\cdot23607=26\cdot83284.$

23. $144\div\sqrt{6}=144\sqrt{6}\div6=24\sqrt{6}=24\times2\cdot44949=58\cdot78776.$

24. $\sqrt{2}\div\sqrt{3}=\sqrt{6}\div(\sqrt{3})^2=\sqrt{6}\div3=2\cdot44949\div3=\cdot81649.$

25. $\dfrac{1}{2\sqrt{3}}=\dfrac{\sqrt{3}}{6}=\dfrac{1\cdot73205}{6}=\cdot28867.$

26. $\dfrac{1}{\sqrt{500}}=\dfrac{1}{10\sqrt{5}}=\dfrac{\sqrt{5}}{50}=\dfrac{2\cdot23607}{50}=\cdot04472.$

27. $\dfrac{4}{\sqrt{243}}=\dfrac{4}{9\sqrt{3}}=\dfrac{4\sqrt{3}}{27}=\dfrac{4\times1\cdot73205}{27}=\dfrac{6\cdot9282}{27}=\cdot2566.$

28. $\dfrac{25}{\sqrt{252}}=\dfrac{25}{6\sqrt{7}}=\dfrac{25\sqrt{7}}{42}=\dfrac{25\times2\cdot64575}{42}=\dfrac{66\cdot14375}{42}=1\cdot57485.$

29. $\sqrt{\dfrac{256}{1575}}=\dfrac{16}{15\sqrt{7}}=\dfrac{16\sqrt{7}}{105}=\dfrac{16\times2\cdot64575}{105}=\dfrac{42\cdot332}{105}=\cdot40316.$

EXAMPLES. XXXI. d. PAGE 260.

5. $(2\sqrt{3}+3\sqrt{2})^2=12+12\sqrt{6}+18=30+12\sqrt{6}.$

6. $(\sqrt{7}+5\sqrt{3})(2\sqrt{7}-4\sqrt{3})=14-4\sqrt{21}+10\sqrt{21}-60=6\sqrt{21}-46.$

7. $(3\sqrt{5}-4\sqrt{2})(2\sqrt{5}+3\sqrt{2})=30+9\sqrt{10}-8\sqrt{10}-24=6+\sqrt{10}.$

11. $(\sqrt{a+x}-2\sqrt{a})^2=a+x-4\sqrt{a}\cdot\sqrt{a+x}+4a=5a+x-4\sqrt{a^2+ax}.$

12. $(2\sqrt{a}-\sqrt{1+4a})^2=4a-4\sqrt{a}\cdot\sqrt{1+4a}+1+4a=8a+1-4\sqrt{a+4a^2}.$

13. $(\sqrt{a+x}-\sqrt{a-x})^2=a+x-2\sqrt{a+x}\sqrt{a-x}+a-x=2a-2\sqrt{a^2-x^2}.$

15. $(\sqrt{2}+\sqrt{3}-\sqrt{5})(\sqrt{2}+\sqrt{3}+\sqrt{5})=(\sqrt{2}+\sqrt{3})^2-(\sqrt{5})^2=2+2\sqrt{6}+3-5$
$=2\sqrt{6}.$

16. $(\sqrt{5}+3\sqrt{2}+\sqrt{7})(\sqrt{5}+3\sqrt{2}-\sqrt{7})=(\sqrt{5}+3\sqrt{2})^2-(\sqrt{7})^2$
$=5+6\sqrt{10}+18-7=16+6\sqrt{10}.$

EXAMPLES. XXXI. e. PAGE 262.

7. $(\sqrt{a+x}-\sqrt{a})(\sqrt{a+x}+\sqrt{a})=(\sqrt{a+x})^2-(\sqrt{a})^2=a+x-a=x.$

8. $(\sqrt{2p+3q}-2\sqrt{q})(\sqrt{2p+3q}+2\sqrt{q})=(\sqrt{2p+3q})^2-(2\sqrt{q})^2$
$=2p+3q-4q=2p-q.$

9. $(\sqrt{a+x}+\sqrt{a-x})(\sqrt{a+x}-\sqrt{a-x})=(\sqrt{a+x})^2-(\sqrt{a-x})^2$
$=a+x-(a-x)=2x.$

10. $(5\sqrt{x^2-3y^2}+7a)(5\sqrt{x^2-3y^2}-7a)=(5\sqrt{x^2-3y^2})^2-(7a)^2$
$=25(x^2-3y^2)-49a^2.$

11. $29\div(11+3\sqrt{7})=\dfrac{29(11-3\sqrt{7})}{(11+3\sqrt{7})(11-3\sqrt{7})}=\dfrac{29(11-3\sqrt{7})}{121-63}=\dfrac{11-3\sqrt{7}}{2}.$

12. $17\div(3\sqrt{7}+2\sqrt{3})=\dfrac{17(3\sqrt{7}-2\sqrt{3})}{(3\sqrt{7}+2\sqrt{3})(3\sqrt{7}-2\sqrt{3})}=\dfrac{17(3\sqrt{7}-2\sqrt{3})}{63-12}$
$=\dfrac{3\sqrt{7}-2\sqrt{3}}{3}.$

13. $(3\sqrt{2}-1)\div(3\sqrt{2}+1)=\dfrac{3\sqrt{2}-1}{3\sqrt{2}+1}=\dfrac{(3\sqrt{2}-1)^2}{(3\sqrt{2}+1)(3\sqrt{2}-1)}$
$=\dfrac{18-6\sqrt{2}+1}{18-1}=\dfrac{19-6\sqrt{2}}{17}.$

14. The quotient $=\dfrac{2\sqrt{3}+7\sqrt{2}}{5\sqrt{3}-4\sqrt{2}}=\dfrac{(2\sqrt{3}+7\sqrt{2})(5\sqrt{3}+4\sqrt{2})}{(5\sqrt{3}-4\sqrt{2})(5\sqrt{3}+4\sqrt{2})}$
$=\dfrac{30+8\sqrt{6}+35\sqrt{6}+56}{75-32}=\dfrac{86+43\sqrt{6}}{43}=2+\sqrt{6}.$

15. The quotient $=\dfrac{2x-\sqrt{xy}}{2\sqrt{xy}-y}=\dfrac{\sqrt{x}(2\sqrt{x}-\sqrt{y})}{\sqrt{y}(2\sqrt{x}-\sqrt{y})}=\dfrac{\sqrt{x}}{\sqrt{y}}=\dfrac{\sqrt{xy}}{y}.$

16. The quotient $=\dfrac{(3+\sqrt{5})(\sqrt{5}-2)}{5-\sqrt{5}}=\dfrac{3\sqrt{5}-6+5-2\sqrt{5}}{\sqrt{5}(\sqrt{5}-1)}$
$=\dfrac{\sqrt{5}-1}{\sqrt{5}(\sqrt{5}-1)}=\dfrac{1}{\sqrt{5}}=\dfrac{\sqrt{5}}{5}.$

17. The quotient $=\dfrac{\sqrt{a}}{\sqrt{a}-\sqrt{x}}\times\dfrac{\sqrt{x}}{\sqrt{a}+\sqrt{x}}=\dfrac{\sqrt{a}\cdot\sqrt{x}}{(\sqrt{a})^2-(\sqrt{x})^2}=\dfrac{\sqrt{ax}}{a-x}.$

18. The quotient $=\dfrac{2\sqrt{15}+8}{5+\sqrt{15}}\times\dfrac{5\sqrt{3}-3\sqrt{5}}{8\sqrt{3}-6\sqrt{5}}=\dfrac{2(\sqrt{15}+4)}{\sqrt{5}(\sqrt{5}+\sqrt{3})}\times\dfrac{\sqrt{15}(\sqrt{5}-\sqrt{3})}{2\sqrt{3}(4-\sqrt{15})}$

$$=\frac{(\sqrt{15}+4)(\sqrt{5}-\sqrt{3})^2}{(\sqrt{5}+\sqrt{3})(\sqrt{5}-\sqrt{3})(4-\sqrt{15})}=\frac{(\sqrt{15}+4)(8-2\sqrt{15})}{2(4-\sqrt{15})}$$

$$=\sqrt{15}+4.$$

19. $\dfrac{25\sqrt{3}-4\sqrt{2}}{7\sqrt{3}-5\sqrt{2}}=\dfrac{(25\sqrt{3}-4\sqrt{2})(7\sqrt{3}+5\sqrt{2})}{(7\sqrt{3}-5\sqrt{2})(7\sqrt{3}+5\sqrt{2})}$

$$=\frac{525-28\sqrt{6}+125\sqrt{6}-40}{147-50}=\frac{485+97\sqrt{6}}{97}=5+\sqrt{6}.$$

20. $\dfrac{10\sqrt{6}-2\sqrt{7}}{3\sqrt{6}+2\sqrt{7}}=\dfrac{(10\sqrt{6}-2\sqrt{7})(3\sqrt{6}-2\sqrt{7})}{(3\sqrt{6}+2\sqrt{7})(3\sqrt{6}-2\sqrt{7})}$

$$=\frac{180-6\sqrt{42}-20\sqrt{42}+28}{54-28}=\frac{208-26\sqrt{42}}{26}=8-\sqrt{42}.$$

21. $\dfrac{\sqrt{7}+\sqrt{2}}{9+2\sqrt{14}}=\dfrac{(\sqrt{7}+\sqrt{2})(9-2\sqrt{14})}{(9+2\sqrt{14})(9-2\sqrt{14})}=\dfrac{9\sqrt{7}+9\sqrt{2}-2\sqrt{98}-2\sqrt{28}}{81-56}$

$$=\frac{9\sqrt{7}+9\sqrt{2}-14\sqrt{2}-4\sqrt{7}}{25}=\frac{5\sqrt{7}-5\sqrt{2}}{25}=\frac{\sqrt{7}-\sqrt{2}}{5}.$$

22. $\dfrac{2\sqrt{3}+3\sqrt{2}}{5+2\sqrt{6}}=\dfrac{(2\sqrt{3}+3\sqrt{2})(5-2\sqrt{6})}{(5+2\sqrt{6})(5-2\sqrt{6})}=\dfrac{10\sqrt{3}+15\sqrt{2}-4\sqrt{18}-6\sqrt{12}}{25-24}$

$$=10\sqrt{3}+15\sqrt{2}-12\sqrt{2}-12\sqrt{3}=3\sqrt{2}-2\sqrt{3}.$$

23. $\dfrac{y^2}{x+\sqrt{x^2-y^2}}=\dfrac{y^2(x-\sqrt{x^2-y^2})}{(x+\sqrt{x^2-y^2})(x-\sqrt{x^2-y^2})}=\dfrac{y^2(x-\sqrt{x^2-y^2})}{x^2-(x^2-y^2)}$

$$=\frac{y^2(x-\sqrt{x^2-y^2})}{y^2}=x-\sqrt{x^2-y^2}.$$

24. $\dfrac{x^2}{\sqrt{x^2+a^2}+a}=\dfrac{x^2(\sqrt{x^2+a^2}-a)}{(\sqrt{x^2+a^2}+a)(\sqrt{x^2+a^2}-a)}=\dfrac{x^2(\sqrt{x^2+a^2}-a)}{(x^2+a^2)-a^2}$

$$=\frac{x^2(\sqrt{x^2+a^2}-a)}{x^2}=\sqrt{x^2+a^2}-a.$$

25. $\dfrac{\sqrt{1+x^2}-\sqrt{1-x^2}}{\sqrt{1+x^2}+\sqrt{1-x^2}}=\dfrac{(\sqrt{1+x^2}-\sqrt{1-x^2})^2}{(\sqrt{1+x^2}+\sqrt{1-x^2})(\sqrt{1+x^2}-\sqrt{1-x^2})}$

$$=\frac{(1+x^2)-2\sqrt{(1+x^2)(1-x^2)}+(1-x^2)}{(1+x^2)-(1-x^2)}$$

$$=\frac{2-2\sqrt{1-x^4}}{2x^2}=\frac{1-\sqrt{1-x^4}}{x^2}.$$

26. $$\frac{2\sqrt{a+b}+3\sqrt{a-b}}{2\sqrt{a+b}-\sqrt{a-b}}=\frac{(2\sqrt{a+b}+3\sqrt{a-b})(2\sqrt{a+b}+\sqrt{a-b})}{(2\sqrt{a+b}-\sqrt{a-b})(2\sqrt{a+b}+\sqrt{a-b})}$$
$$=\frac{4(a+b)+6\sqrt{a^2-b^2}+2\sqrt{a^2-b^2}+3(a-b)}{4(a+b)-(a-b)}=\frac{7a+b+8\sqrt{a^2-b^2}}{3a+5b}.$$

27. $$\frac{\sqrt{9+x^2}-3}{\sqrt{9+x^2}+3}=\frac{(\sqrt{9+x^2}-3)^2}{(\sqrt{9+x^2}+3)(\sqrt{9+x^2}-3)}=\frac{9+x^2-6\sqrt{9+x^2}+9}{(9+x^2)-9}$$
$$=\frac{18+x^2-6\sqrt{9+x^2}}{x^2}.$$

28. $$\frac{3+\sqrt{6}}{5\sqrt{3}-2\sqrt{12}-\sqrt{32}+\sqrt{50}}=\frac{\sqrt{3}(\sqrt{3}+\sqrt{2})}{5\sqrt{3}-4\sqrt{3}-4\sqrt{2}+5\sqrt{2}}=\frac{\sqrt{3}(\sqrt{3}+\sqrt{2})}{\sqrt{3}+\sqrt{2}}$$
$$=\sqrt{3}.$$

29. $$\frac{1}{2+\sqrt{3}}=\frac{2-\sqrt{3}}{(2+\sqrt{3})(2-\sqrt{3})}=\frac{2-\sqrt{3}}{4-3}=2-1{\cdot}73205={\cdot}26795.$$

30. $$\frac{3+\sqrt{5}}{\sqrt{5}-2}=\frac{(3+\sqrt{5})(\sqrt{5}+2)}{(\sqrt{5}-2)(\sqrt{5}+2)}=\frac{3\sqrt{5}+5+6+2\sqrt{5}}{5-4}=11+5\sqrt{5}$$
$$=11+11{\cdot}18035=22{\cdot}18035.$$

31. $$\frac{\sqrt{5}+\sqrt{3}}{4+\sqrt{15}}=\frac{(\sqrt{5}+\sqrt{3})(4-\sqrt{15})}{(4+\sqrt{15})(4-\sqrt{15})}=\frac{4\sqrt{5}+4\sqrt{3}-5\sqrt{3}-3\sqrt{5}}{16-15}$$
$$=4\sqrt{5}+4\sqrt{3}-5\sqrt{3}-3\sqrt{5}=\sqrt{5}-\sqrt{3}=2{\cdot}23607-1{\cdot}73205={\cdot}50402.$$

32. $$\frac{\sqrt{5}-2}{9-4\sqrt{5}}=\frac{(\sqrt{5}-2)(9+4\sqrt{5})}{(9-4\sqrt{5})(9+4\sqrt{5})}=\frac{9\sqrt{5}-18+20-8\sqrt{5}}{81-80}=\sqrt{5}+2$$
$$=2{\cdot}23607+2=4{\cdot}23607.$$

33. $$\frac{7\sqrt{5}+15}{\sqrt{5}-1}\times\frac{\sqrt{5}-2}{3+\sqrt{5}}=\frac{35+15\sqrt{5}-14\sqrt{5}-30}{3\sqrt{5}-3+5-\sqrt{5}}$$
$$=\frac{5+\sqrt{5}}{2(\sqrt{5}+1)}$$
$$=\frac{\sqrt{5}}{2}=\frac{2{\cdot}23607}{2}=1{\cdot}118035.$$

34. $$\frac{(2-\sqrt{3})(7-4\sqrt{3})}{3\sqrt{3}-5}=\frac{(26-15\sqrt{3})(3\sqrt{3}+5)}{(3\sqrt{3}-5)(3\sqrt{3}+5)}=\frac{3\sqrt{3}-5}{2}=\frac{5{\cdot}19615-5}{2}$$
$$={\cdot}098075.$$

EXAMPLES. XXXI. f. PAGE 266.

1. Assume $\sqrt{7-2\sqrt{10}}=\sqrt{x}-\sqrt{y}$.

Then $7-2\sqrt{10}=x-2\sqrt{xy}+y$;

$\therefore\ x+y=7$ ....(1),

and $2\sqrt{xy}=2\sqrt{10}$;

whence $x-y=\pm 3$ ....(2).

From (1) and (2) we obtain $x=5$, or 2; $y=2$, or 5.

That is, the required root is $\sqrt{5}-\sqrt{2}$.

2. Assume $\sqrt{13+2\sqrt{30}}=\sqrt{x}+\sqrt{y}$.

Then $13+2\sqrt{30}=x+2\sqrt{xy}+y$;

$\therefore\ x+y=13$ ....(1),

and $2\sqrt{xy}=2\sqrt{30}$;

whence $x-y=\pm 7$ ....(2).

From (1) and (2) we obtain $x=10$, or 3; $y=3$, or 10.

That is, the required root is $\sqrt{10}+\sqrt{3}$.

3. Assume $\sqrt{8-2\sqrt{7}}=\sqrt{x}-\sqrt{y}$.

Then $8-2\sqrt{7}=x-2\sqrt{xy}+y$;

$\therefore\ x+y=8$, and $2\sqrt{xy}=2\sqrt{7}$.

Hence $x-y=\pm 6$, and we obtain $x=7$, or 1; $y=1$, or 7.

That is, the required root is $\sqrt{7}-1$.

4. Assume $\sqrt{5+2\sqrt{6}}=\sqrt{x}+\sqrt{y}$.

Then $5+2\sqrt{6}=x+2\sqrt{xy}+y$;

$\therefore\ x+y=5$, and $2\sqrt{xy}=2\sqrt{6}$.

Hence $x-y=\pm 1$, and we obtain $x=3$, or 2; $y=2$, or 3.

That is, the required root is $\sqrt{3}+\sqrt{2}$.

5. Assume $\sqrt{75+12\sqrt{21}}=\sqrt{x}+\sqrt{y}$.

Then $75+12\sqrt{21}=x+2\sqrt{xy}+y$;

$\therefore\ x+y=75$, and $2\sqrt{xy}=12\sqrt{21}$.

Hence $x-y=\pm 51$, and we obtain $x=63$, or 12; $y=12$, or 63.

That is, the required root is $\sqrt{63}+\sqrt{12}=3\sqrt{7}+2\sqrt{3}$.

6. Assume $\sqrt{18-8\sqrt{5}}=\sqrt{x}-\sqrt{y}$.

Then $18-8\sqrt{5}=x-2\sqrt{xy}+y$;

$\therefore\ x+y=18$, and $2\sqrt{xy}=8\sqrt{5}$.

Hence $x-y=\pm 2$, and we obtain $x=10$, or 8; and $y=8$, or 10.

That is, the required root is $\sqrt{10}-\sqrt{8}=\sqrt{10}-2\sqrt{2}$.

7. Assume $\sqrt{41-24\sqrt{2}}=\sqrt{x}-\sqrt{y}$.

Then $41-24\sqrt{2}=x-2\sqrt{xy}+y$;

$\therefore\ x+y=41$, and $2\sqrt{xy}=24\sqrt{2}$.

Hence $x-y=\pm 23$, and we obtain $x=32$, or 9; and $y=9$, or 32.

That is, the required root is $\sqrt{32}-\sqrt{9}=4\sqrt{2}-3$.

8. Assume $\sqrt{83+12\sqrt{35}}=\sqrt{x}+\sqrt{y}$.

Then $83+12\sqrt{35}=x+2\sqrt{xy}+y$;

$\therefore\ x+y=83$, and $2\sqrt{xy}=12\sqrt{35}$.

Hence $x-y=\pm 43$, and we obtain $x=63$, or 20; and $y=20$, or 63.

That is, the required root is $\sqrt{63}+\sqrt{20}=3\sqrt{7}+2\sqrt{5}$.

9. Assume $\sqrt{47-4\sqrt{33}}=\sqrt{x}-\sqrt{y}$.

Then $47-4\sqrt{33}=x-2\sqrt{xy}+y$;

$\therefore\ x+y=47$, and $2\sqrt{xy}=4\sqrt{33}$.

Hence $x-y=\pm 41$, and we obtain $x=44$, or 3; and $y=3$, or 44.

That is, the required root is $\sqrt{44}-\sqrt{3}=2\sqrt{11}-\sqrt{3}$.

10. Assume $\sqrt{2\tfrac{1}{4}+\sqrt{5}}=\sqrt{x}+\sqrt{y}$.

Then $2\tfrac{1}{4}+\sqrt{5}=x+2\sqrt{xy}+y$;

$\therefore\ x+y=\frac{9}{4}$, and $2\sqrt{xy}=\sqrt{5}$.

Hence $x-y=\pm\frac{1}{4}$, and we obtain $x=\frac{5}{4}$, or 1; and $y=1$, or $\frac{5}{4}$.

That is, the required root is $\sqrt{\frac{5}{4}}+1=\frac{1}{2}\sqrt{5}+1$.

11. Assume $\sqrt{4\tfrac{1}{3}-\frac{4}{3}\sqrt{3}}=\sqrt{x}-\sqrt{y}$.

Then $4\tfrac{1}{3}-\frac{4}{3}\sqrt{3}=x-2\sqrt{xy}+y$;

$\therefore\ x+y=\frac{13}{3}$, and $2\sqrt{xy}=\frac{4}{3}\sqrt{3}$.

Hence $x-y=\pm\frac{11}{3}$, and we obtain $x=4$, or $\frac{1}{3}$; and $y=\frac{1}{3}$, or 4.

That is, the required root is $\sqrt{4}-\sqrt{\frac{1}{3}}=2-\frac{\sqrt{3}}{3}$.

12. Assume $\sqrt{16+5\sqrt{7}}=\sqrt{x}+\sqrt{y}$.

Then $16+5\sqrt{7}=x+2\sqrt{xy}+y$;

$\therefore\ x+y=16$, and $2\sqrt{xy}=5\sqrt{7}$.

Hence $x-y=\pm 9$, and we obtain

$$x=\frac{25}{2},\ \text{or}\ \frac{7}{2};\ \text{and}\ y=\frac{7}{2},\ \text{or}\ \frac{25}{2}.$$

That is, the required root is $\sqrt{\frac{25}{2}}+\sqrt{\frac{7}{2}}=5\sqrt{\frac{1}{2}}+\sqrt{\frac{7}{2}}$.

13. $\sqrt{27}+2\sqrt{6}=\sqrt{3}(\sqrt{9}+2\sqrt{2})=\sqrt{3}(3+2\sqrt{2})$;

$\therefore$ the square root of $\sqrt{27}+2\sqrt{6}=\sqrt[4]{3}\cdot\sqrt{3+2\sqrt{2}}$.

Also $\sqrt{3+2\sqrt{2}}=\sqrt{2}+1$, by inspection.

Thus the required root is $\sqrt[4]{3}(\sqrt{2}+1)$.

14. $\sqrt{32}-\sqrt{24}=\sqrt{2}(\sqrt{16}-\sqrt{12})=\sqrt{2}(4-2\sqrt{3})$;

$\therefore$ the square root of $\sqrt{32}-\sqrt{24}=\sqrt[4]{2}\cdot\sqrt{4-2\sqrt{3}}$.

Also $\sqrt{4-2\sqrt{3}}=\sqrt{3}-1$, by inspection.

Thus the required root is $\sqrt[4]{2}(\sqrt{3}-1)$.

15. $3\sqrt{5}+\sqrt{40}=\sqrt{5}(3+\sqrt{8})$;

$\therefore$ the square root of $3\sqrt{5}+\sqrt{40}=\sqrt[4]{5}\cdot\sqrt{3+\sqrt{8}}$

$=\sqrt[4]{5}\cdot\sqrt{3+2\sqrt{2}}=\sqrt[4]{5}(\sqrt{2}+1)$, by inspection.

16. Assume $\sqrt{17+12\sqrt{2}}=\sqrt{x}+\sqrt{y}$;

then $17+12\sqrt{2}=x+2\sqrt{xy}+y$.

$\therefore\ x+y=17$, and $2\sqrt{xy}=12\sqrt{2}$.

From which we obtain $x=9$, or 8; and $y=8$, or 9.

$\therefore$ the square root of the expression is $\sqrt{9}+\sqrt{8}=3+2\sqrt{2}=(\sqrt{2}+1)^2$;

$\therefore$ the required fourth root is $1+\sqrt{2}$.

17. Assume $\sqrt{56+24\sqrt{5}}=\sqrt{x}+\sqrt{y}$.

Then $56+24\sqrt{5}=x+2\sqrt{xy}+y$;

$\therefore x+y=56$, and $2\sqrt{xy}=24\sqrt{5}$.

From which we obtain $x=36$, or 20; $y=20$, or 36.

$\therefore$ the square root of the expression is $\sqrt{36}+\sqrt{20}=6+2\sqrt{5}=(\sqrt{5}+1)^2$;

$\therefore$ the required fourth root is $1+\sqrt{5}$.

18. Assume $\sqrt{\frac{3}{2}\sqrt{5}+\frac{7}{2}}=\sqrt{x}+\sqrt{y}$.

Then $\frac{3}{2}\sqrt{5}+\frac{7}{2}=x+2\sqrt{xy}+y$.

$\therefore x+y=\frac{7}{2}$, and $2\sqrt{xy}=\frac{3}{2}\sqrt{5}$.

From which we obtain $x=\frac{9}{4}$, or $\frac{5}{4}$; $y=\frac{5}{4}$, or $\frac{9}{4}$.

$\therefore$ the square root of the expression is $\sqrt{\frac{9}{4}}+\sqrt{\frac{5}{4}}=\frac{3}{2}+\frac{\sqrt{5}}{2}$.

But $\frac{3+\sqrt{5}}{2}=\frac{6+2\sqrt{5}}{4}=\frac{(1+\sqrt{5})^2}{2^2}$;

$\therefore$ the square root of $\frac{3}{2}+\frac{\sqrt{5}}{2}=\frac{1+\sqrt{5}}{2}$, which is the required fourth root.

19. Assume $\sqrt{14+8\sqrt{3}}=\sqrt{x}+\sqrt{y}$.

Then $14+8\sqrt{3}=x+2\sqrt{xy}+y$;

$\therefore x+y=14$, and $2\sqrt{xy}=8\sqrt{3}$.

From which we obtain $x=8$, or 6; $y=6$, or 8.

$\therefore$ the square root of the expression is $\sqrt{8}+\sqrt{6}=\sqrt{2}\,(2+\sqrt{3})$, or $\frac{1}{\sqrt{2}}(4+2\sqrt{3})$. Now, by inspection, the square root of $4+2\sqrt{3}$ is $1+\sqrt{3}$;

$\therefore$ the required fourth root is $\frac{1}{\sqrt[4]{2}}(1+\sqrt{3})$.

20. Assume $\sqrt{49-20\sqrt{6}}=\sqrt{x}-\sqrt{y}$.

Then $49-20\sqrt{6}=x-2\sqrt{xy}+y$;

$\therefore x+y=49$, and $2\sqrt{xy}=20\sqrt{6}$.

From which we obtain $x=25$, or 24; $y=24$, or 25.

$\therefore$ the square root of the expression is $\sqrt{25}-\sqrt{24}=5-2\sqrt{6}=(\sqrt{3}-\sqrt{2})^2$;

$\therefore$ the required fourth root is $\sqrt{3}-\sqrt{2}$.

21. Assume $\sqrt{248+32\sqrt{60}}=\sqrt{x}+\sqrt{y}$.

Then $248+32\sqrt{60}=x+2\sqrt{xy}+y$;

$$\therefore\ x+y=248, \text{ and } 2\sqrt{xy}=32\sqrt{60}.$$

From which we obtain $x=128$, or 120; $y=120$, or 128.

$\therefore$ the square root of the expression

$$=\sqrt{128}+\sqrt{120}=\sqrt{2}\,(\sqrt{64}+\sqrt{60})=\sqrt{2}\,(8+2\sqrt{15});$$

and the square root of $(8+2\sqrt{15})=(\sqrt{5}+\sqrt{3})$;

$\therefore$ the fourth root required is $\sqrt[4]{2}\,(\sqrt{5}+\sqrt{3})$.

25. $19+8\sqrt{3}=19+2\sqrt{16\,.\,3}=16+3+2\sqrt{16\,.\,3}$.

29. $15-4\sqrt{14}=15-2\sqrt{4\,.\,14}=8+7-2\sqrt{8\,.\,7}$.

30. $29+6\sqrt{22}=29+2\sqrt{9\,.\,22}=18+11+2\sqrt{18\,.\,11}$.

EXAMPLES. XXXI. g. PAGE 268.

6. Transposing, $2\sqrt{3-7x}=3\sqrt{8x-12}$.

Squaring both sides, $4(3-7x)=9(8x-12)$;

whence $x=\dfrac{120}{100}=\dfrac{6}{5}$.

7. Cubing both sides, $8(5x-35)=125(2x-7)$;

whence $x=\dfrac{595}{210}=\dfrac{17}{6}$.

8. Squaring both sides, $9x^2-11x-5=9x^2-12x+4$;

$$\therefore\ x=9.$$

10. Squaring both sides, $4x^2-7x+1=4x^2-\dfrac{36}{5}x+\dfrac{81}{25}$;

whence $x=\dfrac{56}{5}$.

11. Squaring both sides, $x+25=1+2\sqrt{x}+x$;

whence $\sqrt{x}=12$; and $x=144$.

12. By transposition and squaring,

$$8x+33=9+12\sqrt{2x}+8x;$$
$$-12\sqrt{2x}=-24;$$
$$\therefore \sqrt{2x}=2; \text{ and } x=2.$$

13. By transposition and squaring,

$$x+3=25-10\sqrt{x}+x;$$
$$\therefore 10\sqrt{x}=22;$$
$$\therefore x=\frac{121}{25}.$$

14. By transposition and squaring,

$$100-60\sqrt{x}+9x=25+9x;$$
$$\therefore 60\sqrt{x}=75;$$
$$\therefore 4\sqrt{x}=5; \text{ and } x=\frac{25}{16}.$$

15. Squaring both sides,

$$x-4+6\sqrt{x-4}+9=x+11;$$
$$\therefore 6\sqrt{x-4}=6; \text{ and } x=5.$$

16. Squaring both sides, $9x-8=9x+36-12\sqrt{x+4}+4;$

$$\therefore 12\sqrt{x+4}=48; \text{ and } x=12.$$

17. Squaring both sides,

$$4x+5-2\sqrt{x(4x+5)}+x=x+3;$$
$$\therefore \sqrt{4x^2+5x}=2x+1.$$

Squaring, $4x^2+5x=4x^2+4x+1;$

$$\therefore x=1.$$

18. By transposition and squaring,

$$25x-29=4x-11+6\sqrt{x(4x-11)}+9x,$$
$$3-2x=\sqrt{4x^2-11x}.$$

Squaring, $4x^2-11x=4x^2-12x+9$

whence $x=9.$

19. By transposition and squaring,

$$8x+17=2x+2\sqrt{2x(2x+9)}+2x+9;$$

$$\therefore \sqrt{4x^2+18x}=2x+4.$$

Squaring, $$4x^2+18x=4x^2+16x+16;$$

whence $$x=8.$$

20. Squaring both sides,

$$3x-11+2\sqrt{3x(3x-11)}+3x=12x-23;$$

$$\therefore \sqrt{9x^2-33x}=3x-6.$$

Squaring, $$9x^2-33x=9x^2-36x+36;$$

whence $$x=12.$$

21. Squaring both sides,

$$12x-5+2\sqrt{(12x-5)(3x-1)}+3x-1=27x-2;$$

$$\therefore \sqrt{36x^2-27x+5}=6x+2.$$

Squaring, $$36x^2-27x+5=36x^2+24x+4;$$

whence $$x=\frac{1}{51}.$$

22. By transposition and squaring,

$$x+3+2\sqrt{(x+3)(x+8)}+x+8=4x+21;$$

$$\therefore \sqrt{x^2+11x+24}=x+5.$$

Squaring, $$x^2+11x+24=x^2+10x+25;$$

$$\therefore x=1.$$

23. By transposition and squaring,

$$x+2+2\sqrt{(x+2)(4x+1)}+4x+1=9x+7;$$

$$\therefore \sqrt{4x^2+9x+2}=2x+2.$$

Squaring, $$4x^2+9x+2=4x^2+8x+4;$$

$$\therefore x=2.$$

24. Squaring both sides, $$x+4ab=4a^2+4a\sqrt{x}+x;$$

$$\therefore 4a(b-a)=4a\sqrt{x};$$

$$\therefore x=(b-a)^2.$$

25. Squaring both sides,

$$x+2\sqrt{x(4a+x)}+4a+x=4(b+x);$$

$$\therefore \sqrt{4ax+x^2}=x-2(a-b).$$

Squaring, $$4ax+x^2=x^2-4(a-b)x+4(a-b)^2;$$

$$\therefore ax+(a-b)x=(a-b)^2;$$

$$\therefore x=\frac{(a-b)^2}{2a-b}.$$

[*For Examples* 26–30, *see Appendix, p.* 369.]

EXAMPLES. XXXI. h. Page 269.

1. Multiplying across,

$$24x-162\sqrt{x}+273=24x-145\sqrt{x}+154.$$

$$\therefore -17\sqrt{x}=119; \text{ and } x=49.$$

2. Multiplying across,

$$18x-100\sqrt{x}+138=18x-99\sqrt{x}+136;$$

$$\therefore -\sqrt{x}=-2; \text{ and } x=4.$$

3. Multiplying across,

$$3x-4\sqrt{x}-39=3x-11\sqrt{x}+10;$$

$$\therefore \sqrt{x}=7; \text{ and } x=49.$$

4. By simplifying the left side, we obtain

$$\frac{\sqrt{x}+1}{\sqrt{x}+2}=\frac{\sqrt{x}+9}{\sqrt{x}+7};$$

$$\therefore x+8\sqrt{x}+7=x+11\sqrt{x}+18;$$

$$\therefore 3\sqrt{x}=-11; \text{ and } x=\frac{121}{9}.$$

5. Multiplying across,

$$2x-\sqrt{x}-\frac{8}{3}\sqrt{x}+\frac{4}{3}=2x-4\sqrt{x}+\frac{4}{3}\sqrt{x}-\frac{8}{3};$$

$$\therefore \sqrt{x}=4, \text{ or } x=16.$$

6. By simplifying the left side, we have

$$\frac{\sqrt{x}-2}{\sqrt{x}-1}=\frac{7\sqrt{x}-26}{7\sqrt{x}-21}.$$

Multiplying across, $$7x-35\sqrt{x}+42=7x-33\sqrt{x}+26;$$

$$\therefore \sqrt{x}=8, \text{ or } x=64.$$

7. We have $$\frac{12\sqrt{x}-11}{12\sqrt{x}-14}=\frac{6\sqrt{x}+5}{6\sqrt{x}+2}.$$

Multiplying across, $72x-42\sqrt{x}-22=72x-24\sqrt{x}-70$;

$$\therefore \sqrt{x}=\frac{8}{3}, \text{ or } x=\frac{64}{9}.$$

8. Clearing of fractions,

$$1+x+\sqrt{x(1+x)}=2;$$

$$\therefore \sqrt{x+x^2}=1-x.$$

Squaring both sides, $x+x^2=1-2x+x^2$;

$$\therefore x=\frac{1}{3}.$$

9. Clearing of fractions,

$$\sqrt{x(x-1)}+x=2;$$

$$\sqrt{x^2-x}=2-x.$$

Squaring both sides, $x^2-x=4-4x+x^2$

$$\therefore x=\frac{4}{3}.$$

10. Clearing of fractions,

$$\sqrt{x(x-8)}-(x-8)=2;$$

$$\sqrt{x^2-8x}=x-6.$$

Squaring both sides, $x^2-8x=x^2-12x+36$;

$$\therefore x=9.$$

11. Clearing of fractions,

$$\sqrt{x(x+5)}+x=10;$$

$$\sqrt{x^2+5x}=10-x.$$

Squaring both sides, $x^2+5x=100-20x+x^2$

$$\therefore x=4.$$

12. Clearing of fractions,

$$2\sqrt{x(4x-3)}-(4x-3)=1;$$

$$\therefore \sqrt{4x^2-3x}=2x-1.$$

Squaring both sides, $4x^2-3x=4x^2-4x+1$;

$$\therefore x=1.$$

13. Clearing of fractions,

$$3\sqrt{x(9x-32)}=8+9x-32;$$

$$\therefore \sqrt{9x^2-32x}=3x-8.$$

Squaring both sides, $$9x^2-32x=9x^2-48x+64;$$

$$\therefore x=4.$$

14. Clearing of fractions,

$$(\sqrt{x}-7)(\sqrt{x}+7)=1;$$

$$\therefore x-49=1, \text{ or } x=50.$$

15. $$(\sqrt{x}+11)(\sqrt{x}-11)+110=0;$$

$$x-121+110=0;$$

$$\therefore x=11.$$

16. Clearing of fractions,

$$4x-6\sqrt{x}=12-6\sqrt{x};$$

$$\therefore x=3.$$

17. We have $$3\sqrt{x}-7=\frac{5}{3\sqrt{x}+7};$$

$$\therefore 9x-49=5;$$

$$\therefore x=6.$$

18. $$\frac{x-1}{\sqrt{x}-1}=3+\frac{\sqrt{x}+1}{2},$$

$$\therefore \sqrt{x}+1=3+\frac{\sqrt{x}+1}{2};$$

whence $$\sqrt{x}=5, \text{ or } x=25.$$

19. Here $$-\frac{1}{x-1}+\frac{1}{\sqrt{x}+1}+\frac{1}{\sqrt{x}-1}=0.$$

Clearing of fractions, $$-1+\sqrt{x}-1+\sqrt{x}+1=0;$$

$$\therefore \sqrt{x}=\frac{1}{2}, \text{ or } x=\frac{1}{4}.$$

20. We have

$$2\sqrt{2+x}-2\sqrt{2-x}=\sqrt{2+x}+\sqrt{2-x}:$$

$$\sqrt{2+x}=3\sqrt{2-x}.$$

Squaring both sides, $$2+x=18-9x;$$

$$\therefore x=\frac{8}{5}.$$

21. Clearing of fractions,

$$2x-3=2(x-2)+\sqrt{x-2}-1;$$

$$\therefore \sqrt{x-2}=2.$$

Whence $$x=6.$$

22. We have

$$\frac{2}{(\sqrt{x}-2)(\sqrt{x}+3)}+\frac{3}{\sqrt{x}-2}=\frac{4}{\sqrt{x}+3}.$$

Clearing of fractions, $2+3\sqrt{x}+9=4\sqrt{x}-8$;

$$\therefore \sqrt{x}=19, \text{ or } x=361.$$

EXAMPLES. XXXII. a. PAGE 274.

7. We have $$\frac{x+y}{y-x}=\frac{\frac{x}{y}+1}{1-\frac{x}{y}}=\frac{\frac{5}{7}+1}{1-\frac{5}{7}}=6.$$

8. $$\frac{x-3y}{2x-5y}=\frac{\frac{x}{y}-3}{\frac{2x}{y}-5}=\frac{3\frac{1}{3}-3}{6\frac{2}{3}-5}=\frac{1}{5}.$$

9. $$\frac{2a-3b}{3b-a}=\frac{2-3\cdot\frac{b}{a}}{3\cdot\frac{b}{a}-1}=\frac{2-3\times\frac{2}{5}}{3\times\frac{2}{5}-1}=4.$$

10. $$\frac{3ax-by}{4by-7ax}=\frac{3\frac{ax}{by}-1}{4-\frac{7ax}{by}}=\frac{\frac{45}{28}-1}{4-\frac{105}{28}}=\frac{17}{7}.$$

11. Here $$\frac{7x-4y}{3x+y}=\frac{5}{13}.$$

Multiplying across, $91x-52y=15x+5y$;

$$\therefore 76x=57y;$$

$$\therefore \frac{x}{y}=\frac{57}{76}=\frac{3}{4}.$$

12. Multiplying across,

$$82a^2 - 123b^2 = 2a^2 + 2b^2;$$
$$\therefore\ 80a^2 = 125b^2;$$
$$16a^2 = 25b^2.$$

That is, $$4a = \pm 5b, \text{ or } \frac{a}{b} = \pm\frac{5}{4}.$$

13. Here $$2x : 3y = (2x-m)^2 : (3y-m)^2;$$
$$\therefore\ \frac{2x}{3y} = \frac{(2x-m)^2}{(3y-m)^2} = \frac{4x^2 - 4mx + m^2}{9y^2 - 6my + m^2}.$$

Multiplying across,

$$18xy^2 - 12mxy + 2m^2x = 12x^2y - 12mxy + 3m^2y;$$
$$18xy^2 - 12x^2y = 3m^2y - 2m^2x;$$
$$6xy(3y - 2x) = m^2(3y - 2x);$$
$$\therefore\ 6xy = m^2,$$

unless $3y - 2x$ is zero.

14. Here $$P : Q = \sqrt{P-x} : \sqrt{Q-x};$$
$$\therefore\ \frac{P}{Q} = \frac{\sqrt{P-x}}{\sqrt{Q-x}}.$$

Squaring both sides, $$\frac{P^2}{Q^2} = \frac{P-x}{Q-x};$$
$$\therefore\ P^2Q - P^2x = PQ^2 - Q^2x;$$
$$\therefore\ x = \frac{PQ^2 - P^2Q}{Q^2 - P^2} = \frac{PQ}{P+Q},$$

unless $Q - P$ is zero.

15. Put $$a = bk, \quad c = dk, \quad e = fk;$$

then $$\sqrt[3]{\frac{2a^2c + 3c^3e + 4e^2c}{2b^2d + 3d^3e + 4f^2d}} = \sqrt[3]{\frac{2k^2b^2 . kd + 3k^3d^3e + 4k^2f^2dk}{2b^2d + 3d^3e + 4f^2d}}$$
$$= \sqrt[3]{\frac{k^3(2b^2d + 3d^3e + 4f^2d)}{2b^2d + 3d^3e + 4f^2d}}$$
$$= \sqrt[3]{k^3} = k$$
$$= \frac{a}{b} = \frac{c}{d} = \frac{e}{f}.$$

16. Let the numbers be denoted by $3x$ and $4x$.

Then $$\frac{3x-7}{4x-7}=\frac{2}{3};$$

$$\therefore\ 9x-21=8x-14;$$

$$x=7.$$

$\therefore$ the numbers are 21 and 28.

17. Let x be the number required.

Then $$\frac{27-x}{35-x}=\frac{2}{3};$$

$$\therefore\ 81-3x=70-2x;$$

$$x=11.$$

18. Let x be the number required.

Then $$\frac{37+x}{29+x}=\frac{8}{7};$$

$$\therefore\ 259+7x=232+8x;$$

$$x=27.$$

19. Put each ratio equal to k; then $p=(b-c)k$, $q=(c-a)k$, $r=(a-b)k$;

$$\therefore\ p+q+r=k(b-c+c-a+a-b)=0.$$

[Ex. 20 may be proved in the same way.]

21. Let $$\frac{a}{b}=\frac{c}{d}=\frac{e}{f}=k.$$

Then $$a=bk,\quad c=dk,\quad e=fk.$$

$$\sqrt{\frac{a^6b-2c^5e+3a^4c^3e^2}{b^7-2d^5f+3b^4cd^2e^2}}=\sqrt{\frac{k^6b^7-2k^5d^5kf+3k^4b^4\,.\,k^2d^2\,.\,c^2e^2}{b^7-2d^5f+3b^4d^2ce^2}}$$

$$=\sqrt[6]{\frac{k^6(b^7-2d^5f+3b^4d^2ce^2)}{b^7-2d^5f+3b^4d^2ce^2}}$$

$$=\sqrt{k^6}=k^3=\frac{a}{b}\cdot\frac{c}{d}\cdot\frac{e}{f}.$$

22. When the ratios $\frac{a}{b}$, $\frac{c}{d}$, $\frac{e}{f}$ are equal the example is included in Art. 294. If the ratios are not equal, suppose $\frac{a}{b}$, $\frac{c}{d}$, $\frac{e}{f}$ are in descending order of magnitude and denote the greatest by k; so that

$$\frac{a}{b}=k,\qquad \frac{c}{d}<k,\qquad \frac{e}{f}<k.$$

Then $$a=bk,\qquad c<dk,\qquad e<fk;$$

$$\therefore\ la=lbk,\quad mc<mdk,\quad ne<nfk;$$

$\therefore$ by addition, $\qquad la+mc+ne<k(lb+md+nf)$;

that is, $\qquad \dfrac{la+mc+ne}{lb+md+nf}<k<\dfrac{a}{b}$.

Similarly it can be shewn that the given ratio is greater than $\dfrac{e}{f}$.

23. Each fraction

$$=\frac{(bx-ay)+(cx-az)+a(z+y)}{(cy-az)+(by-ax)+a(x+z)}=\frac{bx-ay+cx-az+az+ay}{cy-az+by-ax+ax+az}$$

$$=\frac{(b+c)x}{(b+c)y}=\frac{x}{y};$$

since $b+c$ is not zero.

24. Multiply each term of the 1st fraction by 1, each of the 2nd fraction by -3, and each of the 3rd fraction by 1; then, by Art. 294,

each fraction $\qquad =\dfrac{(2x-3y)-3(z-y)+(x+3z)}{(3z+y)-3(z-x)+(2y-3x)}=\dfrac{3x}{3y}=\dfrac{x}{y}$.

Now since $\dfrac{z-y}{z-x}=\dfrac{x}{y}$, we have $zy-y^2=xz-x^2$;

$$\therefore\ z(y-x)=y^2-x^2;$$

$$\therefore\ \text{either } z=y+x,$$

or $\qquad y-x=0$, that is $y=x$.

EXAMPLES. XXXII. b. Page 280.

12. Since $a:b=b:c$, we have $ac=b^2$.

$$\therefore\ a^2-ac=a^2-b^2;$$

$$\therefore\ a(a-c)=(a-b)(a+b).$$

That is, $\qquad a:a+b=a-b:a-c.$ [Art. 296.]

13. Since $b^2=ac$, we have

$$(b^2+bc+c^2)(ac-bc+c^2)=(b^2+c^2+bc)(b^2+c^2-bc)$$

$$=b^4+b^2c^2+c^4$$

$$=b^4+ac^3+c^4.$$

14. If $\dfrac{a}{b}=\dfrac{c}{d}$, then $\dfrac{b}{d}=\dfrac{a}{c}$.

$$\therefore\ \frac{ab}{cd}=\frac{a^2}{c^2};$$

$$\therefore\ ab:cd=a^2:c^2;$$

$$\therefore\ ab+cd:ab-cd=a^2+c^2:a^2-c^2.$$ [Art. 300 (5).]

15. Let $\frac{a}{b}=\frac{c}{d}=k$, then $a=bk$, $c=dk$.

Then $$\frac{a^2+ac+c^2}{a^2-ac+c^2}=\frac{b^2k^2+bdk^2+d^2k^2}{b^2k^2-bdk^2+d^2k^2}=\frac{b^2+bd+d^2}{b^2-bd+d^2};$$

$$\therefore\ a^2+ac+c^2 : a^2-ac+c^2=b^2+bd+d^2 : b^2-bd+d^2.$$

16. As in Ex. 15,

$$\frac{\sqrt{3a^2+5c^2}}{\sqrt{3b^2+5d^2}}=\frac{\sqrt{3b^2k^2+5d^2k^2}}{\sqrt{3b^2+5d^2}}=\frac{k\sqrt{3b^2+5d^2}}{\sqrt{3b^2+5d^2}}=k=\frac{a}{b};$$

$$\therefore\ a : b=\sqrt{3a^2+5c^2} : \sqrt{3b^2+5d^2}.$$

17. As in Ex. 15,

$$\frac{\frac{a}{p}+\frac{b}{q}}{a}=\frac{\frac{bk}{p}+\frac{b}{q}}{kb}=\frac{\frac{k}{p}+\frac{1}{q}}{k}=\frac{\frac{kd}{p}+\frac{d}{q}}{kd}=\frac{\frac{c}{p}+\frac{d}{q}}{c}.$$

18. As in Ex. 15,

$$\frac{\;\dfrac{\frac{b}{a}+\frac{a}{b}}{\frac{ab}{a^2+b^2}}\;}{}=\frac{\dfrac{\frac{1}{k}+k}{\frac{k}{k^2+1}}}{}=\frac{\dfrac{\frac{d}{c}+\frac{c}{d}}{\frac{kd^2}{k^2d^2+d^2}}}{}=\frac{\dfrac{\frac{d}{c}+\frac{c}{d}}{\frac{cd}{c^2+d^2}}}{}.$$

19. We have $(3x-1)(9x+10)=(6x-7)(7x-10)$; [Art. 296.]

$$\therefore\ 27x^2+21x-10=42x^2-109x+70;$$

$$\therefore\ 3x^2-26x+16=0;$$

whence $$x=8 \text{ or } \frac{2}{3}.$$

20. We have $14x-168=5y+15$, and $28x-266=25y-65$; [Art. 296.]

or $$14x-5y=183,\ 28x-25y=201;$$

whence $$x=17, \text{ and } y=11.$$

21. $$\frac{x^2-2x+3}{2x-3}=\frac{x^2-3x+5}{3x-5}.$$

Componendo, $$\frac{x^2}{2x-3}=\frac{x^2}{3x-5};$$

dividing by x^2 which gives a solution $x=0$, we also obtain

$$\frac{1}{2x-3}=\frac{1}{3x-5};$$

whence $$x=2.$$

22. $$\frac{2x-1}{x^2+2x-1}=\frac{x+4}{x^2+x+4}.$$

Invertendo, $$\frac{x^2+2x-1}{2x-1}=\frac{x^2+x+4}{x+4};$$

Dividendo, $$\frac{x^2}{2x-1}=\frac{x^2}{x+4};$$

dividing by x^2 which gives a solution $x=0$, we also obtain

$$\frac{1}{2x-1}=\frac{1}{x+4};$$

$$x+4=2x-1;$$

$$\therefore\ x=5.$$

23. Here we have $\dfrac{a+b-3c-3d}{a-b-3c+3d}=\dfrac{2a+2b-c-d}{2a-2b-c+d};$

Componendo and Dividendo

$$\frac{2\,(a-3c)}{2\,(b-3d)}=\frac{2\,(2a-c)}{2\,(2b-d)};$$

Alternando, $$\frac{2a-6c}{2a-c}=\frac{2b-6d}{2b-d};$$

Dividendo, $$\frac{-5c}{2a-c}=\frac{-5d}{2b-d};$$

$$\therefore\ \frac{c}{2a-c}=\frac{d}{2b-d}.$$

Componendo, $$\frac{c}{2a}=\frac{d}{2b};$$

$$\therefore\ a:b=c:d.$$

24. Let $\dfrac{a}{b}=\dfrac{b}{c}=\dfrac{c}{d}=k$, then $\dfrac{a}{b}\times\dfrac{b}{c}\times\dfrac{c}{d}=\dfrac{a}{d}=k^3.$

$$\frac{a^3+b^3+c^3}{b^3+c^3+d^3}=\frac{k^3b^3+k^3c^3+k^3d^3}{b^3+c^3+d^3}=k^3=\frac{a}{d};$$

$$\therefore\ a:d=a^3+b^3+c^3:b^3+c^3+d^3.$$

25. Let $\dfrac{a}{b}=\dfrac{b}{c}=k$, then $\dfrac{a}{b}\times\dfrac{b}{c}=\dfrac{a}{c}=k^2.$

$$\frac{4a^2-9b^2}{4b^2-9c^2}=\frac{4b^2k^2-9c^2k^2}{4b^2-9c^2}=k^2=\frac{a^2}{b^2};$$

$\therefore\ 4a^2-9b^2:4b^2-9c^2=a^2:b^2=$ duplicate ratio of $a:b$.

26. Here $$\frac{a}{b}=\frac{b}{c}=\frac{c}{d}.$$

We have $$\frac{a}{b}=\frac{b}{c}=\frac{a+b}{b+c},$$

and $$\frac{b}{c}=\frac{c}{d}=\frac{b+c}{c+d}.$$

$$\therefore \frac{a+b}{b+c}=\frac{b+c}{c+d};$$

$$\therefore a+b : b+c=b+c : c+d;$$

or $b+c$ is a mean proportional between $a+b$ and $c+d$.

27. We have $$(a+b)(d+a)=(b+c)(c+d); \quad \text{[Art. 296.]}$$

$$\therefore ad+bd+a^2+ab=bc+c^2+bd+cd;$$

$$\therefore d(a-c)+(a+c)(a-c)+b(a-c)=0;$$

$$\therefore a-c=0, \text{ or } d+a+c+b=0.$$

EXAMPLES. XXXII. c. PAGE 285.

1. Let $x=my$, where m is constant. When $y=7$, and $x=18$, we obtain $18=7m$;

$$\therefore m=\frac{18}{7}, \text{ and } x=\frac{18}{7}y;$$

$\therefore$ when $y=21$, $$x=\frac{18}{7}\times 21=54.$$

3. Let $A=mBC$, where m is constant. When $A=6$, $B=3$, and $C=2$, $6=3\times 2\times m$; $\therefore m=1$, and $A=BC$;

$\therefore$ when $B=5$, and $C=7$, $$A=5\times 7=35.$$

5. Let $x=\frac{m}{y}$, where m is constant. When $y=4$, and $x=15$, we obtain

$15=\frac{m}{4}$; $$\therefore m=60, \text{ and } x=\frac{60}{y};$$

$\therefore$ when $x=6$, $$6=\frac{60}{y}; \text{ or } y=10.$$

7. Let $A=\frac{mB}{C}$, where m is constant. When $A=10$, $B=15$, and $C=6$,

$10=\frac{15m}{6}$; $$\therefore m=4, \text{ and } A=\frac{4B}{C},$$

$\therefore$ when $B=8$, and $C=2$, $$A=\frac{4\times 8}{2}=16.$$

9. Let $x=\frac{m}{y}$, and $y=\frac{n}{z}$, where m and n are constants; $\therefore\ xy=m$, and $yz=n$;

$$\therefore\ \frac{x}{z}=\frac{m}{n},\text{ or } x=\frac{m}{n}z;$$

$\therefore\ z \propto x$, since $\frac{n}{m}$ is constant.

10. Let $a=mb$, where m is constant. Then $a^n=m^nb^n$; $\therefore\ a^n \propto b^n$, since m^n is constant.

11. Let $x=mz$, and $y=nz$, where m and n are constants.

Then $$x^2-y^2=(m^2-n^2)z^2;$$

$\therefore\ x^2-y^2 \propto z^2$, since m^2-n^2 is constant.

12. Let $3a+7b=m(3a+13b)$, where m is constant. When $a=5$, and $b=3$,
$$15+21=m(15+39);$$
$$\therefore\ m=\frac{36}{54}=\frac{2}{3};$$
$$\therefore\ 3a+7b=\frac{2}{3}(3a+13b);$$
$$\therefore\ 3a=5b.$$

14. Let $x^3=my^2$, where m is constant. When $x=3$, and $y=5$, $27=25m$;
$$\therefore\ m=\frac{27}{25},\text{ and } x^3=\frac{27}{25}y^2;$$
$$\therefore\ 25x^3=27y^2.$$

15. Let $\sqrt{a}=m\sqrt[3]{b}$, where m is constant. When $a=4$, and $b=8$, $2=2m$;
$$\therefore\ m=1;$$
and $$\sqrt{a}=\sqrt[3]{b},\text{ or } a^3=b^2.$$

16. Let $y=\frac{m}{x^2}$, where m is constant. When $y=8$, and $x=3$, $8=\frac{m}{9}$;
$$\therefore\ m=72;\text{ and } y=\frac{72}{x^2};$$
$\therefore$ when $y=2$, $$2=\frac{72}{x^2};$$
$$\therefore\ x^2=36,\text{ and } x=\pm 6.$$

17. Let $x = m(y+a)$, where a and m are constant.

When $x=15$, and $y=1$, $\quad 15 = m(1+a)$..............................(1).

When $x=35$, and $y=5$, $\quad 35 = m(5+a)$..............................(2).

By division, $$\frac{5+a}{1+a} = \frac{35}{15} = \frac{7}{3};$$

$$\therefore a = 2.$$

Substituting in (1), $$m = 5;$$

$$\therefore x = 5y + 10;$$

$\therefore$ when $y=2$, $$x = 20.$$

18. (1) Let $a+b = m(a-b)$, where m is constant.

Then $$a^2 + 2ab + b^2 = m^2(a^2 - 2ab + b^2);$$

or $$2ab + 2abm^2 = m^2(a^2+b^2) - (a^2+b^2);$$

$$\therefore a^2 + b^2 = \frac{2(1+m^2)}{m^2-1}ab;$$

$\therefore a^2 + b^2 \propto ab$, since $\frac{2(1+m^2)}{m^2-1}$ is constant.

(2) Let $a = mb$, where m is constant.

Then $$a^2 - b^2 = (1-m^2)b^2 = \frac{(1-m^2)}{m}ab.$$

$\therefore a^2 - b^2 \propto ab$, since $\frac{1-m^2}{m}$ is constant.

19. Let $y = lx + mx^2 + nx^3$, where l, m, and n are constant.

When $x=1$, and $y=4$, we have $\quad l + m + n = 4$..............................(1).

,, $x=2$, and $y=8$, ,, $\quad 2l + 4m + 8n = 8$,

or $\quad l + 2m + 4n = 4$..............................(2).

,, $x=3$, and $y=18$, we have $\quad 3l + 9m + 27n = 18$,

or $\quad l + 3m + 9n = 6$..............................(3).

Subtracting (1) from (2), $$m + 3n = 0.$$

,, (2) from (3), $$m + 5n = 2;$$

whence $m = -3$, $n = 1$, and from (1) $l = 6$;

$$\therefore y = 6x - 3x^2 + x^3.$$

20. Let the area $= m\,(\text{radius})^2$, where m is constant. When the area $= 154$ sq. ft., and radius $= 7$ ft., $\quad 154 = 49m$;

$$\therefore m = \frac{22}{7};$$

$$\therefore \text{ the area} = \frac{22}{7} (\text{radius})^2;$$

$$\therefore \text{ when the radius} = 10\tfrac{1}{2} \text{ ft., the area} = \frac{22}{7} \times (10\tfrac{1}{2})^2$$

$$= 346\tfrac{1}{2} \text{ sq. ft.}$$

21. We have $A = mD^2$, where A denotes the area in square inches and D the diameter in inches, and m is constant.

Then in the first circle area $= m\,(2\tfrac{1}{2})^2 = \frac{25}{4} m$ sq. inches.

The sum of the other areas $= m\,(1\tfrac{1}{2})^2 + 4m = \left(\frac{9}{4} + 4\right) m$

$$= \frac{25}{4} m \text{ sq. inches.}$$

22. Let p = the pressure in pounds, a the area in sq. feet, and v the velocity in miles per hour, then $p = mav^2$ where m is constant.

When $p = 1$, $a = 1$, $v = 15$, we have $1 = 225m$;

$$\therefore m = \frac{1}{225}, \text{ and } p = \frac{av^2}{225};$$

$$\therefore \text{ when } p = 16, \text{ and } a = 9, \quad 16 = \frac{9v^2}{225} = \frac{v^2}{25};$$

$$\therefore v^2 = 400; \text{ or } v = 20;$$

that is, the velocity is 20 miles an hour.

23. Let $4v$ and v be the values of the coins, $4d$ and $3d$ their diameters and t and t' their thicknesses.

Let $4v = m\,(4d)^2\,t$ where m is constant;

then $\quad v = m\,(3d)^2\,t'$.

By division, $4 = \frac{16}{9} \times \frac{t}{t'}$; so that $\frac{t}{t'} = \frac{9}{4}$.

24. Let V represent the volume, R the radius of the base, and H the height; then $V = mR^2H$, where m is constant.

When $V = 88$ cu. ft., $R = 2$ ft., and $H = 7$ ft., we have $88 = m \,.\, 4 \,.\, 7$;

$$\therefore m = \frac{22}{7}, \text{ and } V = \frac{22}{7} R^2H.$$

Again when $V = 396$ cu. ft., $R = 9$ ft., we have $396 = \frac{22}{7} \times 81H$;

$$\therefore H = \frac{7 \times 396}{22 \times 81} = 1\tfrac{5}{9} \text{ ft.}$$

[*For Examples* 25, 26, *see Appendix*, *p.* 370.]

EXAMPLES. XXXIII. a. PAGE 289.

For Examples 1—6 see Art. 315, Ex. 1.

11. Here $d = \cdot 7$, and the 11th term $= 2\cdot 7 + 10 \times \cdot 7 = 9\cdot 7$.

13. Here the common difference $= a + d - (a - d) = 2d$;

$$\therefore \text{ the 30th term} = a - d + 29 \times 2d = a + 57d.$$

14. Here $d = 4a - 3b - (2a - b) = 2a - 2b$;

$$\therefore \text{ the 40th term} = 2a - b + 39\,(2a - 2b) = 80a - 79b.$$

For Examples 15—30 see formula (2) of Art. 314, and Art. 315, Ex. 2.

18. Here $d = -\frac{1}{4} - \frac{1}{4} = -\frac{1}{2}$;

$$\therefore \text{ the 21st term} = \frac{1}{4} + 20\left(-\frac{1}{2}\right) = -9\tfrac{3}{4}.$$

The sum of 21 terms $= \frac{21}{2}\left\{2 \times \frac{1}{4} + 20\left(-\frac{1}{2}\right)\right\}$

$$= \frac{21}{2}\left(\frac{1}{2} - 10\right) = -99\tfrac{3}{4}.$$

19. Here $d = 1 - 3\tfrac{1}{2} = -2\tfrac{1}{2}$;

$$\therefore \text{ the 19th term} = 3\tfrac{1}{2} + 18\left(-\frac{5}{2}\right) = -41\tfrac{1}{2}.$$

The sum of 19 terms $= \frac{19}{2}\left\{2 \times 3\tfrac{1}{2} + 18\left(-\frac{5}{2}\right)\right\}$

$$= \frac{19}{2}(7 - 45) = -361.$$

25. Here $d = 1 - (-3) = 4$;

$$\therefore \text{ the sum of 17 terms} = \frac{17}{2}\{2\,(-3) + 16 \times 4\}$$

$$= \frac{17}{2} \times 58 = 493.$$

27. Here $d = 2p$;

$$\therefore \text{ the sum of } p \text{ terms} = \frac{p}{2}\{2p + (p-1)\,2p\}$$

$$= \frac{p}{2} \times 2p^2 = p^3.$$

29. Here $d = 0 - a = -a$.

$$\therefore \text{ the sum of } a \text{ terms} = \frac{a}{2}\{2a + (a-1)(-a)\}$$

$$= \frac{a}{2}(3a - a^2) = \frac{a^2}{2}(3 - a).$$

30. Here $d = -q - (-3q) = 2q$.

$$\therefore \text{ the sum of } p \text{ terms} = \frac{p}{2}\{2(-3q) + (p-1)2q\}$$

$$= \frac{p}{2}(-6q + 2pq - 2q)$$

$$= pq(p-4).$$

For Examples 31—38 see Art. 315, Ex. 3.

35. Here $-133 = \frac{n}{2}(-3-16) = -\frac{19n}{2}$; whence $n = 14$.

$\therefore\ -16 = \text{the } 14^{\text{th}} \text{ term} = -3 + 13d$; and $d = -1$.

37. Here $49a = \frac{n}{2}(a + 13a) = 7an$; whence $n = 7$.

$\therefore\ 13a = \text{the } 7^{\text{th}} \text{ term} = a + 6d$; and $d = 2a$.

38. Here $-320x = \frac{n}{2}(3x - 35x) = -16nx$; whence $n = 20$.

$\therefore\ -35x = \text{the } 20^{\text{th}} \text{ term} = 3x + 19d$; and $d = -2x$.

EXAMPLES. XXXIII. b. Page 293.

For Examples 1—9 see Art. 316.

4. Let a be the first term, and d the common difference.

$$\left.\begin{aligned}\text{Then the } 9^{\text{th}} \text{ term} &= a + 8d = -11\\ \text{and the } 102^{\text{nd}} \text{ term} &= a + 101d = -150\tfrac{1}{2}\end{aligned}\right\}; \text{ whence } d = -1\tfrac{1}{2},\ a = 1.$$

$$\therefore \text{ the series is } 1,\ -\frac{1}{2},\ -2, \ldots\ldots$$

6. Let a be the first term, and d the common difference.

$$\left.\begin{aligned}\text{Then the } 16^{\text{th}} \text{ term} &= a + 15d = 214\\ \text{and the } 51^{\text{st}} \text{ term} &= a + 50d = 739\end{aligned}\right\}; \text{ whence } d = 15, \text{ and } a = -11.$$

$$\therefore \text{ the series is } -11,\ 4,\ 19, \ldots\ldots$$

8. Let a be the first term, and d the common difference.

$$\left.\begin{aligned}\text{Then the } 54^{\text{th}} \text{ term} &= a + 53d = -125\\ \text{and the } 4^{\text{th}} \text{ term} &= a + 3d = 0\end{aligned}\right\};\ \text{whence } d = -2\tfrac{1}{2},\ \text{and } a = 7\tfrac{1}{2}.$$

$$\therefore \text{ the } 42^{\text{nd}} \text{ term} = a + 41d = 7\tfrac{1}{2} \times 41(-2\tfrac{1}{2}) = -95.$$

9. Let a be the first term, and d the common difference.

$$\left.\begin{aligned}\text{Then the } 31^{\text{st}} \text{ term} &= a + 30d = \tfrac{1}{2}\\ \text{and the } 2^{\text{nd}} \text{ term} &= a + d = 7\tfrac{3}{4}\end{aligned}\right\};\ \text{whence } d = -\tfrac{1}{4},\ \text{and } a = 8.$$

$$\therefore \text{ the } 59^{\text{th}} \text{ term} = a + 58d = 8 - \frac{58}{4} = -6\tfrac{1}{2}.$$

For Examples 10—14, See Art. 319.

11. Let d be the common difference; the first term is 93, and including the extremes the number of terms is 19.

$$\therefore 69 = \text{the } 19^{\text{th}} \text{ term} = 93 + 18d;\ \text{whence } d = -1\tfrac{1}{3}:$$

$\therefore$ the means are $91\frac{2}{3},\ 90\frac{1}{3}, \ldots\ldots 70\frac{1}{3}$.

12. Let d be the common difference; the first term is $-7\frac{1}{5}$, and including the extremes the number of terms is 16;

$$\therefore -2\tfrac{1}{5} = 16^{\text{th}} \text{ term} = -7\tfrac{1}{5} + 15d;\ \text{whence } d = \frac{1}{3};$$

$\therefore$ the means are $-6\frac{13}{15},\ -6\frac{8}{15}, \ldots\ldots -2\frac{8}{15}$.

13. Let d be the common difference; the first term is $7{\cdot}2$, and including the extremes the number of terms is 18;

$$\therefore -6{\cdot}4 = \text{the } 18^{\text{th}} \text{ term} = 7{\cdot}2 + 17d;\ \text{whence } d = {\cdot}8;$$

$\therefore$ the means are $6{\cdot}4,\ 5{\cdot}6, \ldots\ldots -5{\cdot}6$.

15. Let n be the number of terms; then since the common difference is -3, we have

$$315 = \frac{n}{2}\{2 \times 42 + (n-1)(-3)\} = 42n - \frac{3n}{2}(n-1);$$

$$\therefore 3n^2 - 87n + 630 = 0, \text{ or } n^2 - 29n + 210 = 0;$$

whence $n = 14$, or 15.

16. Let n be the number of terms; then since the common difference is 1, we have

$$-100 = \frac{n}{2}\{(-2) \times 16 + (n-1)\} = \frac{n}{2}(n-33);$$

$$\therefore n^2 - 33n + 200 = 0;$$

whence $n = 8$, or 25.

17. Let n be the number of terms; then since the common difference is $-\frac{1}{3}$, we have

$$129=\frac{n}{2}\left\{2\times 15\tfrac{2}{3}+(n-1)\left(-\frac{1}{3}\right)\right\}$$
$$=\frac{n}{6}(95-n);$$
$$\therefore\ n^2-95n+734=0;$$

whence $n=9$, or 86.

18. Let n be the number of terms; then since the common difference is $-\frac{5}{4}$, we have

$$162\tfrac{1}{2}=\frac{n}{2}\left\{2\times 20+(n-1)\left(-\frac{5}{4}\right)\right\};$$
$$\therefore\ 5n^2-165n+1300=0,\text{ or } n^2-33n+260=0;$$

whence $n=13$, or 20.

19. Let n be the number of terms; then since the common difference is $1\frac{1}{2}$, we have

$$-42=\frac{n}{2}\{2(-10\tfrac{1}{2})+(n-1)\times 1\tfrac{1}{2}\}=\frac{n}{4}(3n-45);$$
$$\therefore\ 3n^2-45n+168=0,\text{ or } n^2-15n+56=0;$$

whence $n=7$, or 8.

20. Let n be the number of terms; then since the common difference is $\frac{2}{5}$, we have

$$-52\tfrac{4}{5}=\frac{n}{2}\left\{2(-6\tfrac{4}{5})+(n-1)\frac{2}{5}\right\}=\frac{n}{5}(n-35);$$
$$\therefore\ n^2-35n+264=0;$$

whence $n=11$, or 24.

21. Let a be the middle number and d the common difference; then the three numbers are $a-d$, a, and $a+d$.

Hence $a-d+a+a+d=39$, and $a=13$.

Thus the three numbers are $13-d$, 13, and $13+d$;

and $13(13-d)(13+d)=2184$;

$$\therefore\ 169-d^2=168;\text{ and } d=\pm 1.$$

The numbers are 12, 13, 14.

22. Let a be the middle number and d the common difference; then the three numbers are $a-d$, a, and $a+d$.

Hence $a-d+a+a+d=12$, and $a=4$.

Thus the numbers are $4-d$, 4, and $4+d$;

and $(4-d)^2+16+(4+d)^2=66$;

$$\therefore\ 48+2d^2=66;$$
$$\therefore\ d^2=9;\text{ and } d=\pm 3.$$

The numbers are 1, 4, 7.

23. Let a be the middle number and d the common difference; then the five numbers are $a-2d$, $a-d$, a, $a+d$, $a+2d$.

Hence $a-2d+a-d+a+a+d+a+2d=75$, and $a=15$.

Thus the numbers are $15-2d$, $15-d$, 15, $15+d$, $15+2d$;

and
$$(15-2d)(15+2d)=161;$$
$$\therefore\ 225-4d^2=161;$$
$$\therefore\ d^2=16;\ \text{and } d=\pm 4.$$

The numbers are 7, 11, 15, 19, 23.

24. Let a be the middle number and d the common difference; then the five numbers are $a-2d$, $a-d$, a, $a+d$, $a+2d$.

Hence $a-2d+a-d+a+a+d+a+2d=40$, and $a=8$.

Thus the numbers are $8-2d$, $8-d$, 8, $8+d$, $8+2d$;

and
$$(8-2d)^2+(8-d)^2+64+(8+d)^2+(8+2d)^2=410;$$
$$\therefore\ d^2=9;\ \text{and } d=\pm 3.$$

The numbers are 2, 5, 8, 11, 14.

25. Let a be the first term, d the common difference, and n the number of terms.

Then the 12th term $=a+11d=38$, and the 85th term $=a+84d=257$; whence $d=3$, and $a=5$;

and the nth term or $a+(n-1)d=395$;

that is, $5+3(n-1)=395$, or $n=131$.

EXAMPLES. XXXIV. a. PAGE 297.

5. Here
$$r=\frac{1}{32}\div\frac{1}{64}=\frac{64}{32}=2;$$

$\therefore$ the 14th term
$$=\frac{1}{64}\times 2^{13}=\frac{2^{13}}{2^6}=2^7=128;$$

and the 7th term
$$=\frac{1}{64}\times 2^6=\frac{2^6}{2^6}=1.$$

6. Here
$$r=\frac{\cdot 04}{\cdot 008}=5;$$

$\therefore$ the 4th term $=\cdot 008\times 5^3=\cdot 008\times 125=1$;

and the 8th term $=\cdot 008\times 5^7=(\cdot 008\times 125)\times 5^4=625$.

12. Here $r=\frac{1}{x}$, and the last term is the 30th.

$\therefore$ the last term $$=x\times\left(\frac{1}{x}\right)^{29}=\frac{1}{x^{28}}.$$

For Examples 13—16 see Art. 324, Ex.

18. Here $r=-\frac{18}{6}=-3$, and there are 6 terms;

$\therefore$ the last term $$=6\times(-3)^5=-6\times 243=-1458,$$

and the sum $$=\frac{6\{1-(-3)^6\}}{1-(-3)}=\frac{6(1-729)}{4}=-1092.$$

19. Here $r=\frac{32}{64}=\frac{1}{2}$, and there are 10 terms;

$\therefore$ the last term $$=64\times\left(\frac{1}{2}\right)^9=\frac{2^6}{2^9}=\frac{1}{2^3}=\frac{1}{8};$$

and the sum $$=\frac{64\left(1-\frac{1}{2^{10}}\right)}{1-\frac{1}{2}}=2^7\left(1-\frac{1}{2^{10}}\right)$$

$$=2^7-\frac{1}{2^3}=128-\frac{1}{8}=127\tfrac{7}{8}.$$

20. Here $r=\frac{2\cdot7}{8\cdot1}=\frac{1}{3}$, and there are 7 terms;

$\therefore$ the last term $$=8\cdot1\times\left(\frac{1}{3}\right)^6=\frac{8\cdot1}{729}=\frac{\cdot1}{9}=\frac{1}{90};$$

and the sum $$=\frac{81}{10}\times\frac{1-\frac{1}{3^7}}{1-\frac{1}{3}}=\frac{3^5\left(1-\frac{1}{3^7}\right)}{20}=\frac{243-\frac{1}{9}}{20}$$

$$=\frac{242\frac{8}{9}}{20}=\frac{1093}{90}=12\tfrac{13}{90}.$$

21. Here $r=\frac{1}{24}\div\frac{1}{72}=3$, and there are 8 terms;

$\therefore$ the last term $$=\frac{1}{72}\times 3^7=\frac{3^5}{8}=30\tfrac{3}{8};$$

and the sum $$=\frac{1}{72}\times\frac{3^8-1}{3-1}=\frac{6561-1}{144}=\frac{6560}{144}=45\tfrac{5}{9}.$$

22. Here $r = 1\frac{1}{2} \div 4\frac{1}{2} = \frac{1}{3}$, and there are 9 terms;

$\therefore$ the last term $= \frac{9}{2} \times \left(\frac{1}{3}\right)^8 = \frac{1}{2 \times 3^6} = \frac{1}{1458}$;

and the sum $= \dfrac{\frac{9}{2}\left(1 - \frac{1}{3^9}\right)}{1 - \frac{1}{3}} = \frac{27}{4}\left(1 - \frac{1}{19683}\right)$

$$= \frac{27 \times 19682}{4 \times 19683} = \frac{9841}{1458} = 6\tfrac{1093}{1458}.$$

25. Here $r = \frac{1}{2} \div \left(-\frac{2}{5}\right) = -\frac{5}{4}$, and there are 6 terms;

$\therefore$ the sum $= \dfrac{-\frac{2}{5}\left\{1 - \left(-\frac{5}{4}\right)^6\right\}}{1 - \left(-\frac{5}{4}\right)} = \dfrac{\frac{2}{5}\left(\frac{15625}{4096} - 1\right)}{\frac{9}{4}}$

$$= \frac{2 \times 4 \times 11529}{5 \times 9 \times 4096} = \frac{1281}{2560}.$$

28. Here $r = -\frac{1}{6} \div \frac{2}{3} = -\frac{1}{4}$, and there are 8 terms;

$\therefore$ the sum $= \dfrac{\frac{2}{3}\left\{1 - \left(-\frac{1}{4}\right)^8\right\}}{1 - \left(-\frac{1}{4}\right)} = \dfrac{\frac{2}{3}\left(1 - \frac{1}{65536}\right)}{\frac{5}{4}}$

$$= \frac{2 \times 4 \times 65535}{3 \times 5 \times 65536} = \frac{4369}{8192}.$$

31. Here $r = 1 \div \frac{1}{\sqrt{3}} = \sqrt{3}$, and there are 8 terms;

$\therefore$ the sum $= \dfrac{\frac{1}{\sqrt{3}}\{(\sqrt{3})^8 - 1\}}{\sqrt{3} - 1} = \dfrac{3^4 - 1}{\sqrt{3}(\sqrt{3} - 1)}$

$$= \frac{80}{3 - \sqrt{3}} = \frac{80(3 + \sqrt{3})}{6}.$$

32. Here $r = \frac{\sqrt{a^3}}{\sqrt{a}} = a$, and there are a terms;

$\therefore$ the sum $= \dfrac{\sqrt{a}(a^a - 1)}{a - 1}$.

33. Here $r = -2 \div \frac{1}{\sqrt{2}} = -2\sqrt{2}$, and there are 7 terms;

$$\therefore \text{ the sum} = \frac{\frac{1}{\sqrt{2}}\{1-(-2\sqrt{2})^7\}}{1-(-2\sqrt{2})} = \frac{\frac{1}{\sqrt{2}}\{(2\sqrt{2})^7+1\}}{2\sqrt{2}+1}$$

$$= \frac{(1024\sqrt{2}+1)(2\sqrt{2}-1)}{\sqrt{2}(8-1)} = \frac{4095-1022\sqrt{2}}{7\sqrt{2}}$$

$$= \frac{585-146\sqrt{2}}{\sqrt{2}} = \frac{585\sqrt{2}-292}{2}.$$

34. Here $r = \frac{\sqrt{6}}{\sqrt{2}} = \sqrt{3}$, and there are 12 terms;

$$\therefore \text{ the sum} = \frac{\sqrt{2}\{(\sqrt{3})^{12}-1\}}{\sqrt{3}-1} = \frac{\sqrt{2}(3^6-1)}{\sqrt{3}-1}$$

$$= \frac{728\sqrt{2}(\sqrt{3}+1)}{3-1} = 364(\sqrt{6}+\sqrt{2}).$$

EXAMPLES. XXXIV. b. PAGE 300.

9.
$$\cdot\dot{3} = \cdot 3333\ldots = \frac{3}{10} + \frac{3}{10^2} + \frac{3}{10^3} + \ldots$$

$$= \frac{3}{10} \div \left(1 - \frac{1}{10}\right) = \frac{3}{9} = \frac{1}{3}.$$

10.
$$\cdot 1\dot{6} = \cdot 1666\ldots = \frac{1}{10} + \left(\frac{6}{10^2} + \frac{6}{10^3} + \frac{6}{10^4} + \ldots\right)$$

$$= \frac{1}{10} + \frac{\frac{6}{100}}{1-\frac{1}{10}} = \frac{1}{10} + \frac{6}{90} = \frac{1}{10} + \frac{1}{15} = \frac{1}{6}.$$

11.
$$\cdot\dot{2}\dot{4} = \cdot 242424\ldots = \frac{24}{100} + \frac{24}{100^2} + \frac{24}{100^3} + \ldots$$

$$= \frac{24}{100} \div \left(1 - \frac{1}{100}\right) = \frac{24}{99} = \frac{8}{33}.$$

12.
$$\cdot 3\dot{7}\dot{8} = \cdot 3787878\ldots = \frac{3}{10} + \left(\frac{78}{1000} + \frac{78}{100000} + \ldots\right)$$

$$= \frac{3}{10} + \frac{78}{1000}\left(1 + \frac{1}{10^2} + \frac{1}{10^4} + \ldots\right)$$

$$= \frac{3}{10} + \frac{78}{1000} \cdot \frac{1}{1-\frac{1}{100}} = \frac{3}{10} + \frac{78}{1000} \times \frac{100}{99}$$

$$= \frac{3}{10} + \frac{78}{990} = \frac{375}{990} = \frac{25}{66}.$$

13. $\cdot\dot{0}3\dot{7} = \cdot037037037\ldots = \frac{37}{1000} + \frac{37}{1000^2} + \frac{37}{1000^3} + \ldots$

$$= \frac{\frac{37}{1000}}{1 - \frac{1}{1000}} = \frac{37}{999} = \frac{1}{27}.$$

15. Let a be the 1st term, and r the common ratio.

Then the 5th term $= ar^4 = \frac{27}{16}$,

and the 9th term $= ar^8 = \frac{1}{3}$;

by division, $r^4 = \frac{16}{81}$; and $r = \frac{2}{3}$.

Hence $a = \frac{27}{16} \div \frac{16}{81} = \frac{2187}{256}$.

$\therefore$ the series is $\frac{2187}{256}, \frac{729}{128}, \frac{243}{64}, \ldots\ldots$

16. Let a be the 1st term, and r the common ratio.

Then the 7th term $= ar^6 = 625$,

and the 4th term $= ar^3 = -5$;

by division, $r^3 = -125$; and $r = -5$.

Hence $a = \frac{-5}{-125} = \frac{1}{25}$.

$\therefore$ the series is $\frac{1}{25}, -\frac{1}{5}, 1, \ldots\ldots$

18. Let a, ar, ar^2 be the 3 parts, then we have

$$a + ar + ar^2 = 183, \quad a + ar^2 = \frac{41}{20} ar.$$

From the second equation $1 + r^2 = \frac{41}{20} r$;

that is, $r = \frac{5}{4}$, or $\frac{4}{5}$.

Hence by substituting in the first equation

$$a\left(\frac{41}{20} \cdot \frac{5}{4} + \frac{5}{4}\right) = 183; \text{ whence } a = 48.$$

$\therefore$ the three parts are 48, 60, 75.

19. Let $a, ar, ar^2, \ldots\ldots ar^{n-1}$ be any n consecutive terms of a G.P. Then if n is odd the middle term is the $\left(\frac{n+1}{2}\right)^{\text{th}}$, and it is equal to

$$ar^{\frac{n+1}{2}-1} \text{ or } ar^{\frac{n-1}{2}}.$$

Now the product of the n terms $= a^n \cdot r^{1+2+3+\ldots+(n-1)}$

$$= a^n r^{\frac{n(n-1)}{2}} = (ar^{\frac{n-1}{2}})^n$$

= the nth power of the middle term.

20. Let a be the first term and r the common ratio.

Then we have $$a + ar = 1 \quad\ldots\ldots\ldots(1);$$

and $$a = 2(ar + ar^2 + ar^3 + \ldots\ldots)$$

$$= \frac{2ar}{1-r} \quad\ldots\ldots\ldots(2).$$

From (1) and (2) we easily obtain $r = \frac{1}{3}$ and $a = \frac{3}{4}$.

Thus the series is $\frac{3}{4}, \frac{1}{4}, \frac{1}{12}, \ldots$

21. The series $= (y^2 + y^4 + y^6 + \ldots) + (2b + 4b + 6b + \ldots)$

$$= \frac{y^2(y^{2n}-1)}{y^2-1} + \frac{n}{2}\{2 \times 2b + (n-1)2b\}$$

$$= \frac{y^2(y^{2n}-1)}{y^2-1} + n(n+1)b.$$

22. Here $r = \frac{3-2\sqrt{2}}{3+2\sqrt{2}}$, so that $1 - r = \frac{4\sqrt{2}}{3+2\sqrt{2}}$.

$$\therefore \text{ the sum} = \frac{3+2\sqrt{2}}{3-2\sqrt{2}} \div \frac{4\sqrt{2}}{3+2\sqrt{2}} = \frac{17+12\sqrt{2}}{4(3\sqrt{2}-4)}$$

$$= \frac{51\sqrt{2}+72+68+48\sqrt{2}}{4 \times 2} = \frac{140+99\sqrt{2}}{8}.$$

23. Here $$r = \frac{\sqrt{2}}{3} \div \frac{\sqrt{3}}{\sqrt{2}} = \frac{2}{3\sqrt{3}}.$$

$\therefore$ the sum to infinity $$= \frac{\frac{\sqrt{3}}{\sqrt{2}}}{1 - \frac{2}{3\sqrt{3}}} = \frac{9}{\sqrt{2}(3\sqrt{3}-2)} = \frac{9\sqrt{2}(3\sqrt{3}+2)}{2 \times 23}$$

$$= \frac{9(3\sqrt{6}+2\sqrt{2})}{46}.$$

24. The series $=(2n+4n+6n+\dots)-\left(\frac{1}{2}-\frac{1}{6}+\frac{1}{18}-\dots\right)$, and each series consists of n terms.

$$\therefore \text{ the sum } =\frac{2n}{2}\{2\times 2n+(2n-1)2n\}-\frac{1}{2}\cdot\frac{1-\left(-\frac{1}{3}\right)^{2n}}{1-\left(-\frac{1}{3}\right)}$$

$$=n(2n+1)2n-\frac{3}{8}\left\{1-\left(\frac{1}{3}\right)^{2n}\right\}$$

$$=2n^2(2n+1)-\frac{3}{8}\left(1-\frac{1}{3^{2n}}\right).$$

EXAMPLES. XXXV. PAGE 304.

1. The corresponding A. P. is $\frac{1}{4}, \frac{1}{2}, \frac{3}{4}, \dots$ and the common difference $=\frac{1}{4}$. Thus the 6th term $=\frac{1}{4}+5\times\frac{1}{4}=\frac{3}{2}$.

$\therefore$ the 6th term of the H. P. $=\frac{2}{3}$.

4. The corresponding A. P. is $\frac{1}{3}, \frac{2}{3}, 1, \dots$ and the common difference $=\frac{1}{3}$. Thus the nth term $=\frac{1}{3}+(n-1)\frac{1}{3}=\frac{n}{3}$.

$\therefore$ the nth term of the H. P. $=\frac{3}{n}$.

5. Let a be the first term and d the common difference of the corresponding A. P.,

then we have $\quad a+14d=25,\ a+22d=41;$

whence $\quad d=2$, and $a=-3$.

$\therefore$ the A. P. is $\quad -3, -1, 1, \dots;$

and the H. P. is $\quad -\frac{1}{3}, -1, 1, \dots.$

Examples 6 and 7 may be solved similarly.

In Examples 8—13 we have only to apply the formula of Art. 332.

14. Here we have to insert two Arith. means between $\frac{1}{4}$ and $\frac{1}{12}$

Let d be the common difference, then as in Art. 319, Ex., we have

$$\frac{1}{12}=\frac{1}{4}+3d; \text{ whence } d=-\frac{1}{18}.$$

Thus the Arith. means are $\frac{7}{36}$, $\frac{5}{36}$.

$\therefore$ the Harm. means are $5\frac{1}{7}$, $7\frac{1}{5}$.

17. From Art. 333 it is evident that we have to prove that

$$\frac{A+G}{2} : \frac{2AG}{A+G}=\frac{B+G}{2} : \frac{2BG}{B+G};$$

or,
$$A+G : B+G=\frac{AG}{A+G} : \frac{BG}{B+G}.$$

Now since $G=\sqrt{AB}$, we have

$$\frac{A+G}{B+G}=\frac{A+\sqrt{AB}}{B+\sqrt{AB}}=\frac{\sqrt{A}\,(\sqrt{A}+\sqrt{B})}{\sqrt{B}\,(\sqrt{B}+\sqrt{A})}$$

$$=\frac{A\,(\sqrt{AB}+B)}{B\,(\sqrt{AB}+A)}=\frac{A\,(B+G)}{B\,(A+G)},$$

which proves the proposition.

18. Let a, ar, ar^2 be the 3 consecutive terms of the G. P.; then we have to shew that $a+ar$, $2ar$, ar^2+ar are in H. P., or that $2ar$ is the Harm. mean between $a\,(1+r)$ and $ar\,(1+r)$.

The Harm. mean $=\dfrac{2a^2r\,(1+r)^2}{(a+ar)\,(1+r)}$ [Art. 333.]

$=2ar.$

19. This is a G. P. in which $a=1$, and $r=\frac{7}{4}$.

$\therefore$ the sum
$$=\frac{\left(\frac{7}{4}\right)^6-1}{\frac{7}{4}-1}=\frac{\frac{117649}{4096}-1}{\frac{3}{4}}$$

$$=\frac{4\times 113553}{3\times 4096}=\frac{37851}{1024}=36\tfrac{987}{1024}.$$

20. This is an A. P. in which $a=1$, $d=\frac{3}{4}$.

$\therefore$ the sum
$$=\frac{6}{2}\left\{2+5\times\frac{3}{4}\right\}=3\,(2+3\tfrac{3}{4})=17\tfrac{1}{4}.$$

21. This is an A. P. in which the first term is $2a+x$, and the common difference is $a-x$.

$$\therefore \text{ the sum } =\frac{p}{2}\{2(2a+x)+(p-1)(a-x)\}$$

$$=\frac{p}{2}\{4a+(p-1)a+2x-(p-1)x\}$$

$$=\frac{p}{2}\{(p+3)a-(p-3)x\}.$$

22. This is a G. P. in which $a=\frac{9}{5}$, and $r=-\frac{2}{3}$.

$$\therefore \text{ the sum } =\frac{\frac{9}{5}\left\{1-\left(-\frac{2}{3}\right)^8\right\}}{1-\left(-\frac{2}{3}\right)}=\frac{\frac{9}{5}\left(1-\frac{256}{6561}\right)}{\frac{5}{3}}$$

$$=\frac{27\times 6305}{25\times 6561}=\frac{1261}{1215}=1\tfrac{46}{1215}.$$

23. This is an A. P. in which $a=1\frac{4}{5}$, and $d=-\frac{3}{5}$.

$$\therefore \text{ the sum } =\frac{12}{2}\left\{2\times 1\tfrac{4}{5}+11\left(-\frac{3}{5}\right)\right\}=6\left(3\tfrac{3}{5}-6\tfrac{3}{5}\right)=-18.$$

24. By supposition, $x-a:y-a=y-a:z-a$.

Componendo and Dividendo,

$$y-x:x+y-2a=z-y:y+z-2a;$$

$$\therefore\ y-x:y-z=x+y-2a:-y-z+2a;$$

$$\therefore\ y-x:y-z=(2y-2a)-(y-x):(y-z)-(2y-2a);$$

hence $y-x$, $2y-2a$, $y-z$ are in H.P. [Art. 329. Def.]

25. From the data we have $cf=ad$,

also $$a-b=b-c=c-d;$$

$$\therefore\ 2(a-b)=b-d, \text{ and } 2(c-d)=a-c;$$

whence $$b=\frac{2a+d}{3}, \text{ and } c=\frac{2d+a}{3}.$$

Again, $$\frac{1}{a}-\frac{1}{g}=\frac{1}{g}-\frac{1}{h}=\frac{1}{h}-\frac{1}{d};$$

whence $$h=\frac{3ad}{2a+d}, \text{ and } g=\frac{3ad}{2d+a}.$$

$$\therefore\ bh=\frac{2a+d}{3}\times\frac{3ad}{2a+d}=ad=cg \text{ similarly.}$$

26. If a^2, b^2, c^2 are in A. P., by adding $bc+ca+ab$ to each,

$$a^2+bc+ca+ab,\ b^2+bc+ca+ab,\ c^2+bc+ca+ab \text{ are in A. P.,}$$

or $$(a+b)(a+c),\ (b+c)(b+a),\ (c+b)(c+a) \text{ are in A. P.} \qquad \text{[Art. 334.]}$$

Divide each expression by $(b+c)(c+a)(a+b)$;

then $$\frac{1}{b+c},\ \frac{1}{c+a},\ \frac{1}{a+b} \text{ are in A. P.;} \qquad \text{[Art. 334.]}$$

$$\therefore\ b+c,\ c+a,\ a+b \text{ are in H. P.}$$

27. We have $$b=\frac{a+c}{2}, \text{ and } \beta=\frac{2\alpha\gamma}{\alpha+\gamma}.$$

$$\therefore\ \frac{a+c}{b\beta}=\frac{(a+c)(\alpha+\gamma)}{(a+c)\,\alpha\gamma}=\frac{\alpha+\gamma}{\alpha\gamma}.$$

28. We have $2a=b+c$, and $b^2=ac$;

$\therefore$ the Harm. mean between a and $b=\dfrac{2ab}{a+b}=\dfrac{(b+c)\,b}{a+b}$

$$=\frac{b^2+bc}{a+b}=\frac{ac+bc}{a+b}$$

$$=c.$$

29. We have $$\frac{2}{b}=\frac{2}{a+b}+\frac{2}{b+c};$$

$$\therefore\ (a+b)(b+c)=b\,(a+2b+c);$$

that is, $$ab+b^2+ac+bc=ab+2b^2+bc,$$

or $$b^2=ac.$$

$\therefore\ a, b, c$ are in G. P.

30. Here $$b^2=ac,\ c^2=bd,\ d^2=ce;$$

$$\therefore\ c\,(a+2c+e)=ac+2c^2+ce=b^2+2bd+d^2$$
$$=(b+d)^2.$$

31. We have $\dfrac{b}{a}=\dfrac{c}{b}=\dfrac{d}{c}=\ldots=r$, suppose;

then in the series $$\frac{1}{a^2-b^2},\ \frac{1}{b^2-c^2},\ \frac{1}{c^2-d^2},\ldots\ldots$$

the ratio of each term to the preceding is of the form

$$\frac{a^2-b^2}{b^2-c^2}=\frac{\dfrac{a^2}{b^2}-1}{1-\dfrac{c^2}{b^2}}=\frac{\dfrac{1}{r^2}-1}{1-r^2}=\frac{1}{r^2}.$$

$\therefore$ the series is a G. P. whose common ratio is $\frac{1}{r^2}$;

and the sum of n terms $=\frac{1}{a^2-b^2}\cdot\frac{1-\left(\frac{1}{r^2}\right)^n}{1-\frac{1}{r^2}}$

$$=\frac{1}{a^2-b^2}\cdot\frac{1-\left(\frac{a^2}{b^2}\right)^n}{1-\frac{a^2}{b^2}}$$

$$=\frac{b^2}{b^{2n}}\cdot\frac{a^{2n}-b^{2n}}{(a^2-b^2)^2}=\frac{a^{2n}-b^{2n}}{(a^2-b^2)^2\,.\,b^{2n-2}}.$$

32. We have $b:d=b-c:c-d$; [Art. 329. Def.]

$\therefore\ b:d=a-b:c-d$; since $b-c=a-b$.

$\therefore\ a-b:b=c-d:d.$

Componendo, $a:b=c:d$,

or $ad=bc$.

Now $\frac{c^2}{d}=\frac{ac}{b}=\frac{ac}{\frac{a+c}{2}}=\frac{2ac}{a+c}$;

$\therefore\ a, \frac{c^2}{d}, c$ are in H. P.

Again, $c=\frac{ad}{b}$, since $ad=bc$;

and b, c, d are in H. P.;

$\therefore\ b, \frac{ad}{b}, d$ are in H. P.

33. Here $a=\frac{m+n}{2}$, $g^2=mn$, $k^2=\frac{m^2+n^2}{2}$,

$$\therefore\ \frac{g^2+k^2}{2}=\frac{mn+\frac{m^2+n^2}{2}}{2}=\frac{m^2+2mn+n^2}{4}=a^2.$$

$\therefore\ a^2$ is the Arith. mean between g^2 and k^2.

34. We have $b^2=ac$, and $c^2=bd$,

also $b^2c^2=abcd$, or $bc=ad$;

$\therefore\ (b-c)^2=b^2+c^2-2bc=ac+bd-2ad.$

35. We have $$b^2=ac,\ c^2=bd,\ ad=bc;$$

$$\therefore\ a^2-ac=a^2-b^2=(a-b)(a+b).$$

Hence $$\frac{(a+d)(a-b)^2}{a(a-d)(a-c)}=\frac{(a+d)(a-b)}{(a-d)(a+b)}$$

$$=\frac{a^2+ad-ab-bd}{a^2-ad+ab-bd}=\frac{a^2+bc-ab-c^2}{a^2-bc+ab-c^2}$$

$$=\frac{(a-b+c)(a-c)}{(a+b+c)(a-c)}=\frac{a-b+c}{a+b+c}.$$

36. Since $\frac{1}{a}$, $\frac{1}{b}$, $\frac{1}{c}$ are in A. P., on multiplying each expression by abc, bc, ca, ab are in A. P. [Art. 334.]

Subtract each expression from $bc+ca+ab$;

then $$ab+ac,\ bc+ab,\ ac+bc \text{ are in A. P.},$$

or $$\frac{ab+ac}{a+b+c},\ \frac{bc+ab}{a+b+c},\ \frac{ac+bc}{a+b+c} \text{ are in A. P.}$$

$$\therefore\ \frac{a+b+c}{ab+ac},\ \frac{a+b+c}{bc+ab},\ \frac{a+b+c}{ac+bc} \text{ are in H. P.},$$

that is, $$\frac{1}{a}+\frac{1}{b+c},\ \frac{1}{b}+\frac{1}{a+c},\ \frac{1}{c}+\frac{1}{a+b} \text{ are in H. P.}$$

[*For Miscellaneous Examples V., see Appendix, p.* 371.]

EXAMPLES. XXXVI. PAGE 317.

For Examples 1—6 see Art. 338, and for Examples 7—12 see Art. 340.

13. The condition for equal roots gives

$$4(1+k)^2=4k^2, \text{ whence } k=-\frac{1}{2}.$$

14. Here $$(2m+3n)^2-4\,.\,3m\,.\,2n=(2m-3n)^2,$$
and therefore the roots are rational. [Art. 338, (4).]

15. Let α and β be the roots; then

$$\alpha+\beta=\frac{4}{3},\ \alpha\beta=-\frac{1}{3}.$$

$$\therefore\ \alpha-\beta=\sqrt{(\alpha+\beta)^2-4\alpha\beta}=\sqrt{\frac{16}{9}+\frac{4}{3}}=\frac{2\sqrt{7}}{3}.$$

Also $$\alpha^2+\beta^2=(\alpha+\beta)^2-2\alpha\beta=\frac{16}{9}+\frac{2}{3}=\frac{22}{9}.$$

16. The equation is $ax^2-(b-c)x-a=0$; and

$$(b-c)^2-4a(-a)=(b-c)^2+4a^2,$$

which is always positive. Therefore the roots are real.

17. The sum of the roots is 6, and their product is 4. Therefore the equation is $x^2-6x+4=0$.

20. Here the sum of the roots is 4, and their product is $\frac{9}{4}$. Therefore the equation is $x^2-4x+\frac{9}{4}=0$.

21. The equation is
$$\left(x-\frac{a+b}{a-b}\right)\left(x-\frac{a-b}{a+b}\right)=0,$$

or
$$\{(a-b)x-(a+b)\}\{x(a+b)-(a-b)\}=0,$$

that is,
$$(a^2-b^2)x^2-2(a^2+b^2)x+a^2-b^2=0.$$

22. The equation is
$$\left(x-\frac{a}{2b}\right)\left(x-\frac{b}{2a}\right)=0,$$

or
$$(2bx-a)(2ax-b)=0,$$

that is,
$$4a^2b^2x^2-2(a^2+b^2)x+ab=0.$$

23. $\alpha^2+\beta^2=(\alpha+\beta)^2-2\alpha\beta=\frac{q^2}{p^2}-\frac{2r}{p}=\frac{q^2-2pr}{p^2}.$

24. $(\alpha-\beta)^2=(\alpha+\beta)^2-4\alpha\beta=\frac{q^2}{p^2}-\frac{4r}{p}=\frac{q^2-4pr}{p^2}.$

25. $\alpha^2\beta+\alpha\beta^2=\alpha\beta(\alpha+\beta)=-\frac{q}{p}\cdot\frac{r}{p}=-\frac{qr}{p^2}.$

26. $\alpha^4+\beta^4=(\alpha^2+\beta^2)^2-2\alpha^2\beta^2=\{(\alpha+\beta)^2-2\alpha\beta\}^2-2\alpha^2\beta^2$

$$=\left(\frac{q^2-2pr}{p^2}\right)^2-2\frac{r^2}{p^2}=\frac{q^4-4pq^2r+2p^2r^2}{p^4}.$$

27. $\alpha^5\beta^2+\alpha^2\beta^5=\alpha^2\beta^2(\alpha^3+\beta^3)=\frac{r^2}{p^2}(\alpha+\beta)(\alpha^2-\alpha\beta+\beta^2)$

$$=-\frac{r^2}{p^2}\cdot\frac{q}{p}\{(\alpha+\beta)^2-3\alpha\beta\}=-\frac{qr^2}{p^3}\left(\frac{q^2}{p^2}-\frac{3r}{p}\right)$$

$$=-\frac{qr^2(q^2-3pr)}{p^5}.$$

28. $$\frac{\alpha^2}{\beta}+\frac{\beta^2}{\alpha}=\frac{\alpha^3+\beta^3}{\alpha\beta}=\frac{(\alpha+\beta)(\alpha^2-\alpha\beta+\beta^2)}{\alpha\beta}$$

$$=-\frac{\frac{q}{p}}{\frac{r}{p}}\{(\alpha+\beta)^2-3\alpha\beta\}=-\frac{q}{r}\cdot\frac{q^2-3pr}{p^2}=\frac{q(3pr-q^2)}{p^2r}.$$

29. Here $\alpha+\beta=p$, $\alpha\beta=q$; and $\alpha^3+\beta^3=P$, $\alpha^3\beta^3=Q$.

$\therefore\ P=\alpha^3+\beta^3=(\alpha+\beta)(\alpha^2-\alpha\beta+\beta^2)=p\{(\alpha+\beta)^2-3\alpha\beta\}=p(p^2-3q).$

$Q=\alpha^3\beta^3=q^3.$

30. Here $\alpha+\beta=a$, and $\alpha\beta=b$. In the required equation the coefficient of x with sign changed

$$=\frac{\alpha}{\beta^2}+\frac{\beta}{\alpha^2}=\frac{\alpha^3+\beta^3}{\alpha^2\beta^2}=\frac{(\alpha+\beta)\{(\alpha+\beta)^2-3\alpha\beta\}}{\alpha^2\beta^2}=\frac{a(a^2-3b)}{b^2};$$

and the absolute term $$=\frac{\alpha}{\beta^2}\times\frac{\beta}{\alpha^2}=\frac{1}{\alpha\beta}=\frac{1}{b}.$$

$\therefore$ the equation is $$x^2-\frac{a(a^2-3b)}{b^2}x+\frac{1}{b}=0,$$

or $$b^2x^2-a(a^2-3b)x+b=0.$$

31. Let α and 2α be the roots of the equation;

then $$\alpha+2\alpha=-\frac{b}{a},\ 2\alpha^2=\frac{c}{a};$$

$$\therefore\ \alpha=-\frac{b}{3a},\ \alpha^2=\frac{c}{2a}.$$

$\therefore$ the condition is $$\frac{c}{2a}=\left(\frac{b}{3a}\right)^2=\frac{b^2}{9a^2};$$

or $$9ac=2b^2.$$

32. Let α and β be the roots of the equation,

then $$\alpha+\beta=a,\ \alpha\beta=\frac{a^2}{2}.$$

In the required equation, the coefficient of x with sign changed

$$=\alpha^3+\beta^3=(\alpha+\beta)\{(\alpha+\beta)^2-3\alpha\beta\}=a\left(a^2+\frac{3a^2}{2}\right)=\frac{5a^3}{2};$$

and the absolute term $$=\alpha^3\beta^3=\left(-\frac{a^2}{2}\right)^3=-\frac{a^6}{8}.$$

$\therefore$ the required equation is $$x^2-\frac{5a^3}{2}x-\frac{a^6}{8}=0,$$

or $$8x^2-20a^3x-a^6=0.$$

33. The condition for real roots is that $(a+b+c)^2-4(a+b)\frac{c}{2}$ should be positive. [Art. 338, (1).]

The expression $=(a+b)^2+2(a+b)c+c^2-2(a+b)c=(a+b)^2+c^2$, which is always positive.

34. The condition for rational roots is that

$$4(a+b)^2-4(a+b+c)(a+b-c)$$

should be a perfect square. [Art. 338, (4).]

The expression $=4(a+b)^2-4\{(a+b)^2-c^2\}=4c^2$.

35. Let α, β be the roots of $x^2-px+q=0$; then the new equation will be

$$x^2-\left(\frac{\alpha+\beta}{2}+\frac{2\alpha\beta}{\alpha+\beta}\right)x+\left(\frac{\alpha+\beta}{2}\times\frac{2\alpha\beta}{\alpha+\beta}\right)=0$$

But

$$\alpha+\beta=p, \text{ and } \alpha\beta=q;$$

$$\therefore \frac{\alpha+\beta}{2}+\frac{2\alpha\beta}{\alpha+\beta}=\frac{p}{2}+\frac{2q}{p}=\frac{p^2+4q}{2p},$$

and

$$\frac{\alpha+\beta}{2}\times\frac{2\alpha\beta}{\alpha+\beta}=\alpha\beta=q.$$

∴ the equation becomes

$$x^2-\frac{p^2+4q}{2p}x+q=0,$$

or

$$2px^2-(p^2+4q)x+2pq=0.$$

36. Let $l\alpha$ and $m\alpha$ be the roots of the equation;

then

$$l\alpha+m\alpha=-\frac{q}{p}, \text{ and } lm\alpha^2=\frac{r}{p};$$

$$\therefore \alpha=-\frac{q}{p(l+m)}, \quad \alpha^2=\frac{r}{lmp};$$

$$\therefore \frac{q^2}{p^2(l+m)^2}=\frac{r}{lmp}$$

$$\therefore pr(l+m)^2-lmq^2=0;$$

or

$$(l^2+m^2)pr+(2pr-q^2)lm=0.$$

37. Let

$$\frac{x^2-15}{2x-8}=y.$$

Multiplying up and transposing, we have

$$x^2-2yx+8y-15=0.$$

If x has real roots, $4y^2-4(8y-15)$ must be positive, that is, $y^2-8y+15$ or $(y-5)(y-3)$ must be positive;

∴ y cannot lie between 3 and 5.

38. Let
$$\frac{3x^2+2}{x^2-2x-1}=y.$$

Multiplying up and transposing, we have
$$(3-y)\,x^2+2yx+2+y=0.$$

If x has real roots, $4y^2-4\,(3-y)\,(2+y)$ must be positive, that is,
$$y^2-(6+y-y^2) \text{ or } (y-2)\,(2y+3)$$
must be positive;

$\therefore$ y cannot lie between 2 and $-\dfrac{3}{2}$.

EXAMPLES. XXXVII. a. PAGE 323.

1.
$$^5P_4=5\,.\,4\,.\,3\,.\,2=120;\ ^7P_6=7\,.\,6\,.\,5\,.\,4\,.\,3\,.\,2=5040;$$
$$^8C_5={}^8C_3=\frac{8\,.\,7\,.\,6}{|\underline{3}}=56;\ ^{25}C_{23}={}^{25}C_2=\frac{25\,.\,24}{1\,.\,2}=300.$$

2. All the letters are different; hence

in (1) the number $={}^7P_5=7\,.\,6\,.\,5\,.\,4\,.\,3=2520$;

and in (2) the number $={}^7P_7=|\underline{7}=5040$.

3.
$$^nC_3=\frac{n\,(n-1)\,(n-2)}{|\underline{3}};\ ^{n-1}C_4=\frac{(n-1)\,(n-2)\,(n-3)\,(n-4)}{|\underline{4}}.$$
$$\therefore\ \frac{n\,(n-1)\,(n-2)\,.\,|\underline{4}}{|\underline{3}\,.\,(n-1)\,(n-2)\,(n-3)\,(n-4)}=\frac{8}{5};$$
$$\frac{4n}{(n-3)\,(n-4)}=\frac{8}{5};$$
$$\therefore\ 2n^2-19n+24=0,\ \text{or } (2n-3)\,(n-8)=0.$$
$$\therefore\ n=8,\ \text{or } \frac{3}{2}.$$

Hence $n=8$, the other value being inadmissible.

4. The number of different coins is 9; therefore the number of selections that can be made is $^9C_4=\dfrac{9\,.\,8\,.\,7\,.\,6}{|\underline{4}}=126$.

5. Here the digit 3 must always stand first, if the number is to lie between 3000 and 4000. We have therefore to find the number of arrangements that can be made with the 3 digits 9, 4, and 6. This $=|\underline{3}=6$.

6. The number of ways in which the 3 consonants can be arranged in the odd places $=|\underline{3}=6$. Similarly the number of ways in which the 3 vowels can be arranged in the even places $=6$. $\therefore$ the number of arrangements
$$=6\times 6=36.$$

7. Here $${}^nP_4 = 14 \times {}^{n-2}P_3.$$
that is, $$n(n-1)(n-2)(n-3) = 14(n-2)(n-3)(n-4);$$
or $$n(n-1) = 14(n-4);$$
$$\therefore\ n^2 - 15n + 56 = 0, \text{ or } (n-8)(n-7) = 0;$$
$$\therefore\ n = 7 \text{ or } 8.$$

8. The three masters can be chosen in ${}^5C_3$ ways and the six boys in ${}^{10}C_6$ ways.

$$\therefore \text{ the number of committee } = \frac{5 \cdot 4}{1 \cdot 2} \times \frac{10 \cdot 9 \cdot 8 \cdot 7}{1 \cdot 2 \cdot 3 \cdot 4}$$
$$= 10 \times 210 = 2100.$$

9. Since ${}^{20}C_r = {}^{20}C_{r-10}$ we must have $r + r - 10 = 20$, and $r = 15$.

[Art. 319.]

Hence $${}^rC_{12} = {}^{15}C_{12} = {}^{15}C_3 = \frac{15 \cdot 14 \cdot 13}{1 \cdot 2 \cdot 3} = 455,$$
$${}^{18}C_r = {}^{18}C_{15} = {}^{18}C_3 = \frac{18 \cdot 17 \cdot 16}{1 \cdot 2 \cdot 3} = 816.$$

10. Since a and e occur in every word, we have only to choose 3 other letters, and then arrange the 5 letters in all possible ways. The 3 letters can be chosen in ${}^{24}C_3$ ways, and now each group of 5 letters admits of $\underline{|5}$ arrangements.

$$\text{Thus the required number } = \frac{24 \cdot 23 \cdot 22}{1 \cdot 2 \cdot 3} \times \underline{|5}$$
$$= 2024 \times 120 = 242880.$$

11. Three consonants can be chosen in ${}^{21}C_3$ ways . ... (1);

two vowels .... ${}^5C_2$,, ... (2).

Each selection in (1) can be associated with each selection in (2) giving ${}^{21}C_3 \times {}^5C_2$ combined groups, each of which admits of $\underline{|5}$ arrangements.

$$\therefore \text{ the required number } = \frac{21 \cdot 20 \cdot 19}{1 \cdot 2 \cdot 3} \times \frac{5 \cdot 4}{1 \cdot 2} \times \underline{|5}$$
$$= 1330 \times 10 \times 120 = 1596000.$$

12. Divide the 7 persons who can sit on either side into two groups containing 3 and 4 respectively. This can be done in ${}^7C_3$ ways.

Now each side admits of $\underline{|5}$ arrangements;

$$\therefore \text{ the required number } = {}^7C_3 \times \underline{|5} \times \underline{|5}$$
$$= \frac{7 \cdot 6 \cdot 5}{1 \cdot 2 \cdot 3} \times 120 \times 120$$
$$= 504000.$$

EXAMPLES. XXXVII. b. Page 328.

1. (1) We have 12 letters, of which 3 are i, 2 are r, 2 are e, 2 are s.

∴ the number of permutations

$$=\frac{\lfloor 12}{\lfloor 3\,\lfloor 2\,\lfloor 2\,\lfloor 2}=12\,.\,11\,.\,10\,.\,9\,.\,7\,.\,5\,.\,4\,.\,3\,.\,2=9979200.$$

(2) We have 10 letters of which 2 are e, 3 are n, 2 are o.

∴ the number of permutations $=\frac{\lfloor 10}{\lfloor 2\,\lfloor 3\,\lfloor 2}=10\,.\,9\,.\,8\,.\,7\,.\,5\,.\,3\,.\,2=151200.$

(3) We have 12 letters of which 6 are t, 2 are l, 2 are e.

∴ the number of permutations $=\frac{\lfloor 12}{\lfloor 6\,\lfloor 2\,\lfloor 2}=3\,.\,11\,.\,10\,.\,9\,.\,8\,.\,7=166320.$

2. (1) We have 7 digits, of which two are 2 and three are 3.

∴ the number of permutations $=\frac{\lfloor 7}{\lfloor 2\,\lfloor 3}=7\,.\,6\,.\,5\,.\,2=420.$

(2) We have 7 digits of which two are 2 and three are 3.

∴ the number of permutations $=\frac{\lfloor 7}{\lfloor 2\,\lfloor 3}=420;$

but of these 0 will stand first in $\frac{\lfloor 6}{\lfloor 2\,\lfloor 3}$ or 60 cases;

∴ the number of permutations that can be formed $=420-60=360.$

3. The vowels can be arranged in the odd places in $\frac{\lfloor 3}{\lfloor 2}$ or 3 ways,

the consonants........................ even places in $\frac{\lfloor 3}{\lfloor 2}$ or 3 ways.

∴ 3×3 or 9 words can be arranged by placing the vowels in the odd places and the consonants in the even places. Similarly 9 words can be arranged by placing the vowels in the even and the consonants in the odd places.

∴ the total number of words that can be formed $=2\times 9=18.$

4. Each arm can be placed in 4 different positions. Therefore two arms can be placed in 4^2 positions, 3 arms in 4^3, 4 arms in 4^4 and 5 arms in 4^5 positions, each of which will form a signal, with the exception of the case where all are in their position of rest.

∴ the number of signals that can be formed $=4^5-1=1024-1=1023.$

5. Any one of the n things can be given in m ways, and any one of the remaining things can be given in m ways. Thus two things can be given in m^2 ways, three in m^3 ways, and so on. Thus the n things can be given away in m^n ways.

6. We have 14 letters, namely 5a's, 3b's, 6c's.

$\therefore$ the number of arrangements $=\dfrac{14!}{5!\,3!\,6!}=14 \cdot 13 \cdot 11 \cdot 3 \cdot 4 \cdot 7=168168.$

7. There are 12 books, 4 of which are exactly alike of one kind, 4 exactly alike of another kind, and 4 exactly alike of a third kind.

$$\therefore \text{ the number of arrangements}=\frac{12!}{4!\,4!\,4!}=11 \cdot 10 \cdot 9 \cdot 7 \cdot 5$$
$$=34650.$$

8. (1) Here we are concerned only with the *relative* positions of the 6 persons. We may therefore suppose one fixed and arrange the others in the remaining 5 seats; this can be done in $5!$ ways.

(2) Let one lady be seated in any place; then the others can sit in alternate places in $3!$ ways. With any arrangement of the ladies the gentlemen can now be placed in $4!$ ways.

$\therefore$ the total number of arrangements $=3! \times 4!=144.$

9. The number of words that can be formed is the number of permutations of 6 letters 4 at a time when each letter may be repeated as many as four times in any word.

$\therefore$ the number of words $=6^4=1296.$ [Art. 353.]

10. The consonants being all different can be arranged in their places in $3!$ ways. Of the 5 vowels, 2 are alike of one kind, and 2 alike of another kind; therefore they can be arranged in their places in $\dfrac{5!}{2!\,2!}$ ways.

$\therefore$ the number of arrangements $=3! \times \dfrac{5!}{2!\,2!}=6 \times 30=180.$

11. There are 6 men who can row on either side. These can be divided into two groups of 3, one for each side, in ${}^6C_3$ ways. Each side now admits of $4!$ arrangements.

$\therefore$ the number of arrangements $=\dfrac{6 \cdot 5 \cdot 4}{1 \cdot 2 \cdot 3} \times 4! \times 4!=20 \times 24 \times 24=11520.$

12.
$$ {}^nC_r+{}^nC_{r-1}=\frac{n!}{r!\,(n-r)!}+\frac{n!}{(r-1)!\,(n-r+1)!}$$
$$=\frac{n!}{r!\,(n-r+1)!}\{n-r+1+r\}=\frac{(n+1)\,n!}{r!\,(n-r+1)!}$$
$$=\frac{(n+1)!}{r!\,(n+1-r)!}={}^{n+1}C_r.$$

EXAMPLES. XXXVIII. a. Page 333.

1. $(x+2)^4 = x^4 + {}^4C_1 x^3 \cdot 2 + {}^4C_2 x^2 \cdot 2^2 + {}^4C_3 x \cdot 2^3 + 2^4$
$= x^4 + 4 \cdot 2x^3 + 6 \cdot 4 \cdot x^2 + 4 \cdot 8 \cdot x + 16$
$= x^4 + 8x^3 + 24x^2 + 32x + 16.$

2. $(x+3)^5 = x^5 + {}^5C_1 x^4 \cdot 3 + {}^5C_2 x^3 \cdot 3^2 + {}^5C_2 x^2 \cdot 3^3 + {}^5C_1 x \cdot 3^4 + 3^5$
$= x^5 + 5 \cdot 3 \cdot x^4 + 10 \cdot 9 \cdot x^3 + 10 \cdot 27 \cdot x^2 + 5 \cdot 81 \cdot x + 243$
$= x^5 + 15x^4 + 90x^3 + 270x^2 + 405x + 243.$

3. $(a+x)^7 = a^7 + {}^7C_1 a^6 x + {}^7C_2 a^5 x^2 + {}^7C_3 a^4 x^3 + {}^7C_3 a^3 x^4 + {}^7C_2 a^2 x^5 + {}^7C_1 a x^6 + x^7$
$= a^7 + 7a^6 x + 21a^5 x^2 + 35a^4 x^3 + \ldots$

4. $(a-x)^5 = a^5 - {}^5C_1 a^4 x + {}^5C_2 a^3 x^2 - {}^5C_2 a^2 x^3 + {}^5C_1 a x^4 - x^5$
$= a^5 - 5a^4 x + 10a^3 x^2 - \ldots$

5. $(1-2y)^5 = 1^5 - {}^5C_1 . 1^4 . 2y + {}^5C_2 . 1^3 . (2y)^2 - {}^5C_2 . 1^2 . (2y)^3 + {}^5C_1 . 1 . (2y)^4 - (2y)^5$
$= 1 - 5 \cdot 2y + 10 \cdot 4y^2 - 10 \cdot 8y^3 + 5 \cdot 16y^4 - 32y^5$
$= 1 - 10y + 40y^2 - 80y^3 + 80y^4 - 32y^5.$

6. $\left(2x + \frac{y}{2}\right)^4 = (2x)^4 + {}^4C_1 (2x)^3 \cdot \frac{y}{2} + {}^4C_2 (2x)^2 \cdot \left(\frac{y}{2}\right)^2 + {}^4C_1 (2x) \cdot \left(\frac{y}{2}\right)^3 + \left(\frac{y}{2}\right)^4$

$$= 16x^4 + 4 \cdot 8x^3 \cdot \frac{y}{2} + 6 \cdot 4x^2 \cdot \frac{y^2}{4} + 4 \cdot 2x \cdot \frac{y^3}{8} + \frac{y^4}{16}$$

$$= 16x^4 + 16x^3 y + 6x^2 y^2 + xy^3 + \frac{y^4}{16}.$$

7. $\left(2 - \frac{x}{2}\right)^6 = 2^6 - {}^6C_1 \cdot 2^5 \cdot \frac{x}{2} + {}^6C_2 \cdot 2^4 \cdot \left(\frac{x}{2}\right)^2 - {}^6C_3 \cdot 2^3 \cdot \left(\frac{x}{2}\right)^3 + {}^6C_2 \cdot 2^2 \cdot \left(\frac{x}{2}\right)^4$

$$- {}^6C_1 \cdot 2 \cdot \left(\frac{x}{2}\right)^5 + \left(\frac{x}{2}\right)^6$$

$$= 64 - 6 \cdot 32 \cdot \frac{x}{2} + 15 \cdot 16 \cdot \frac{x^2}{4} - 20 \cdot 8 \cdot \frac{x^3}{8} + 15 \cdot 4 \cdot \frac{x^4}{16} - 6 \cdot 2 \cdot \frac{x^5}{32} + \frac{x^6}{64}$$

$$= 64 - 96x + 60x^2 - 20x^3 + \frac{15x^4}{4} - \frac{3x^5}{8} + \frac{x^6}{64}.$$

8. $\left(a - \frac{3}{b}\right)^7 = a^7 - {}^7C_1 a^6 \cdot \frac{3}{b} + {}^7C_2 a^5 \cdot \left(\frac{3}{b}\right)^2 - {}^7C_3 a^4 \cdot \left(\frac{3}{b}\right)^3 + {}^7C_3 a^3 \cdot \left(\frac{3}{b}\right)^4$

$$- {}^7C_2 a^2 \cdot \left(\frac{3}{b}\right)^5 + {}^7C_1 a \cdot \left(\frac{3}{b}\right)^6 - \left(\frac{3}{b}\right)^7$$

$$= a^7 - 7 \cdot a^6 \cdot \frac{3}{b} + 21 \cdot a^5 \cdot \frac{9}{b^2} - 35 \cdot a^4 \cdot \frac{27}{b^3} + 35 \cdot a^3 \cdot \frac{81}{b^4} - 21 \cdot a^2 \cdot \frac{243}{b^5}$$

$$+ 7 \cdot a \cdot \frac{729}{b^6} - \frac{2187}{b^7}$$

$$= a^7 - \frac{21a^6}{b} + \frac{189a^5}{b^2} - \frac{945a^4}{b^3} + \frac{2835a^3}{b^4} - \frac{5103a^2}{b^5} + \frac{5103a}{b^6} - \frac{2187}{b^7}.$$

9. $\left(ax+\frac{y}{a}\right)^9=(ax)^9+{}^9C_1(ax)^8.\frac{y}{a}+{}^9C_2(ax)^7.\left(\frac{y}{a}\right)^2+{}^9C_3(ax)^6.\left(\frac{y}{a}\right)^3$

$$+{}^9C_4(ax)^5.\left(\frac{y}{a}\right)^4+{}^9C_4(ax)^4.\left(\frac{y}{a}\right)^5+{}^9C_3(ax)^3.\left(\frac{y}{a}\right)^6$$

$$+{}^9C_2(ax)^2.\left(\frac{y}{a}\right)^7+{}^9C_1\,ax.\left(\frac{y}{a}\right)^8+\left(\frac{y}{a}\right)^9$$

$$=a^9x^9+9a^7x^8y+36a^5x^7y^2+84a^3x^6y^3+126ax^5y^4$$

$$+\frac{126x^4y^5}{a}+\frac{84x^3y^6}{a^3}+\frac{36x^2y^7}{a^5}+\frac{9xy^8}{a^7}+\frac{y^9}{a^9}.$$

10. The 4th term $={}^{12}C_3\,x^3=220x^3$.

11. The 6th term $={}^8C_5\,2^3\,(-y)^5=-56\,.\,8\,.\,y^5=-448y^5$.

12. The 5th term $={}^7C_4\,a^3\,.\,(-5b)^4={}^7C_3\,a^3\,.\,(5b)^4=35a^3\,.\,625b^4=21875a^3b^4$.

13. The 15th term $={}^{17}C_{14}\,(2x)^3\,.\,(-1)^{14}={}^{17}C_3\,.\,8x^3=680\,.\,8x^3=5440x^3$.

14. The 7th term $={}^{10}C_6\left(-\frac{1}{x}\right)^6={}^{10}C_4\,.\,\frac{1}{x^6}=\frac{210}{x^6}$.

15. The 6th term $={}^9C_5\,.\,(3x)^4\left(\frac{a}{2}\right)^5=\frac{126\,.\,81}{32}\,.\,x^4\,.\,a^5=\frac{5103}{16}x^4a^5$.

16. The middle or 4th term $={}^6C_3\left(\frac{2a}{3}\right)^3.\left(-\frac{3}{2a}\right)^3=-20\,.\,\frac{8a^3}{27}\,.\,\frac{27}{8a^3}=-20$.

17. The 23rd term $={}^{25}C_{22}\,(x^2)^3\,.\left(\frac{b}{x}\right)^{22}={}^{25}C_3\,.\,x^6\,.\,\frac{b^{22}}{x^{22}}=\frac{2300b^{22}}{x^{16}}$.

18. The 10th term $={}^{17}C_9\,.\,(x^2)^8\,(-x)^9=-{}^{17}C_8\,x^{16}\,.\,x^9=-24310x^{25}$.

19. Since in the two expansions the 2nd and 4th terms are equal but of opposite sign, we have the required value

$$=2\{x^4+{}^4C_2\,x^2\,.\,(\sqrt{3})^2+(\sqrt{3})^4\}$$
$$=2\{x^4+6x^2\,.\,3+9\}$$
$$=2x^4+36x^2+18.$$

20. The expression $=(y+1)^5-(y-1)^5$, where $y=\sqrt{1-x^2}$.

In the two expansions the 1st, 3rd, and 5th terms destroy each other;

hence the expression $=2\,\{{}^5C_1y^4+{}^5C_3y^2+1\}$

$$=2\,\{5\,(1-x^2)^2+10\,(1-x^2)+1\}$$
$$=2\,\{5-10x^2+5x^4+10-10x^2+1\}$$
$$=32\quad 40x^2+10x^4.$$

21. $(x^2+2x)^{10}=x^{10}(x+2)^{10}$; and, since x^{10} multiplies every term in the expansion, we have to find the coefficient of x^2 in the expansion of $(x+2)^{10}$;

$$\therefore \text{ the required coefficient} = {}^{10}C_2\, 2^8 = 45 \times 256 = 11520.$$

22. $\left(x^2-\dfrac{a}{2x}\right)^{14}=x^{28}\left(1-\dfrac{a}{2x^3}\right)^{14}$;

and the coefficient of x in $\left(x^2-\dfrac{a}{2x}\right)^{14}$ = coefficient of x^{-27} in $\left(1-\dfrac{a}{2x^3}\right)^{14}$

$$= \text{coefficient of } \left(\frac{1}{x^3}\right)^9 \text{ in } \left(1-\frac{a}{2x^3}\right)^{14};$$

$$\therefore \text{ the required coefficient} = {}^{14}C_9\left(-\frac{a}{2}\right)^9 = -{}^{14}C_5\,\frac{a^9}{512} = -\frac{2002a^9}{512} = -\frac{1001a^9}{256}.$$

23. $\left(2x^2-\dfrac{1}{x}\right)^{12}=x^{24}\left(2-\dfrac{1}{x^3}\right)^{12}$;

and the term independent of x in $\left(2x^2-\dfrac{1}{x}\right)^{12}$ = coefficient of x^{-24} in $\left(2-\dfrac{1}{x^3}\right)^{12}$

$$= \text{coefficient of } \left(\frac{1}{x^3}\right)^8 \text{ in } \left(2-\frac{1}{x^3}\right)^{12};$$

$$\therefore \text{ the required coefficient} = {}^{12}C_8 \,.\, 2^4 \,.\, (-1)^8 = {}^{12}C_4 \,.\, 16 = 495 \times 16 = 7920.$$

24. $\left(\dfrac{x^2}{3}-\dfrac{2}{x^3}\right)^{15}=x^{30}\left(\dfrac{1}{3}-\dfrac{2}{x^5}\right)^{15}$;

and the coefficient of x^{-20} in $\left(\dfrac{x^2}{3}-\dfrac{2}{x^3}\right)^{15}$ = coefficient of x^{-50} in $\left(\dfrac{1}{3}-\dfrac{2}{x^5}\right)^{15}$

$$= \text{coefficient of } \left(\frac{1}{x^5}\right)^{10} \text{ in } \left(\frac{1}{3}-\frac{2}{x^5}\right)^{15};$$

$\therefore$ the required coefficient

$$= {}^{15}C_{10}\left(\frac{1}{3}\right)^5(-2)^{10} = {}^{15}C_5 \,.\, \frac{1}{3^5} \,.\, 2^{10} = \frac{3003 \times 1024}{243} = \frac{1025024}{81}.$$

EXAMPLES. XXXVIII. b. PAGE 338.

1. Here $T_{r+1}=\dfrac{17-r+1}{r}\times\dfrac{y}{x}\times T_r=\dfrac{18-r}{r}\times\dfrac{3}{4}\times T_r$;

$$\therefore\ T_{r+1}>T_r, \text{ so long as } \frac{18-r}{r}\times\frac{3}{4}>1;$$

that is, $\quad 54-3r>4r$, or $54>7r$.

The greatest value of r consistent with this is 7; hence the greatest term is the 8th.

2. Here $T_{r+1} = \frac{28-r+1}{r} \times \left(\frac{-y}{x}\right) \times T_r = \frac{29-r}{r} \times \frac{4}{9} \times T_r$, numerically;

$$\therefore\ T_{r+1} > T_r, \text{ so long as } \frac{29-r}{r} \times \frac{4}{9} > 1;$$

that is, $$116 - 4r > 9r, \text{ or } 116 > 13r.$$

The greatest value of r consistent with this is 8; hence the greatest term is the 9th.

3. Here $$T_{r+1} = \frac{4-r+1}{r} \times x \times T_r = \frac{5-r}{r} \times \frac{2}{3} \times T_r;$$

$$\therefore\ T_{r+1} > T_r, \text{ so long as } \frac{5-r}{r} \times \frac{2}{3} > 1;$$

that is, $$10 - 2r > 3r, \text{ or } 10 > 5r.$$

The greatest value of r consistent with this is $r=1$, but when $r=2$, $T_{r+1} = T_r$; thus the 2nd and 3rd terms are equal and greater than any other.

4. Here $T_{r+1} = \frac{15-r+1}{r} \times \frac{4b}{a} \times T_r = \frac{16-r}{r} \times \frac{2}{3} \times T_r$, *numerically;*

$$\therefore\ T_{r+1} > T_r, \text{ so long as } \frac{16-r}{r} \times \frac{2}{3} > 1:$$

that is, $$32 - 2r > 3r, \text{ or } 32 > 5r.$$

The greatest value of r consistent with this is 6; hence the greatest term is the 7th.

5. Here $T_{r+1} = \frac{30-r+1}{r} \times \frac{2y}{7x} \times T_r = \frac{31-r}{r} \times \frac{1}{2} \times T_r = \frac{31-r}{2r} \times T_r;$

$$\therefore\ T_{r+1} > T_r, \text{ so long as } \frac{31-r}{2r} > 1;$$

that is, $$31 - r > 2r, \text{ or } 31 > 3r.$$

The greatest value of r consistent with this is 10; hence the greatest term is the 11th.

6. Here $T_{r+1} = \frac{n-r+1}{r} \times \frac{3}{2x} \times T_r = \frac{15-r+1}{r} \times \frac{3}{5} \times T_r = \frac{16-r}{r} \times \frac{3}{5} \times T_r;$

$$\therefore\ T_{r+1} > T_r, \text{ so long as } \frac{16-r}{r} \times \frac{3}{5} > 1;$$

that is, $$48 - 3r > 5r, \text{ or } 48 > 8r.$$

The greatest value of r consistent with this is $r=5$, but when $r=6$, $T_{r+1} = T_r$; thus the 6th and 7th terms are equal and greater than any other.

7. Here ${}^{25}C_{2r} = {}^{25}C_{r+4}$. Therefore either $2r = r+4$, which makes the two terms the same, or $2r + r + 4 = 25$ which gives $r = 7$.

8. Here ${}^{n}C_{15} = {}^{n}C_{25}$. Therefore $15 + 25 = n$.

9. Here ${}^{3n}C_{r+2}={}^{3n}C_{2r-4}$. Therefore either $r+2=2r-4$, which makes the two terms the same, or $r+2+2r-4=3n$; whence $3r=3n+2$.

10. $$\left(x^2+\frac{1}{x}\right)^{2m}=x^{4m}\left(1+\frac{1}{x^3}\right)^{2m}.$$

$\therefore$ coefficient of x^m in $\left(x^2+\frac{1}{x}\right)^{2m}$ = coefficient of x^{-3m} in $\left(1+\frac{1}{x^3}\right)^{2m}$

$$=\ldots\ldots\left(\frac{1}{x^3}\right)^m\ldots\ldots;$$

$\therefore$ coefficient required $={}^{2m}C_m=\dfrac{\lfloor 2m}{\lfloor m\,\lfloor m}$.

11. There are $2n+1$ terms in the expansion; therefore the $(n+1)^{\text{th}}$ is the middle term.

The $n+1^{\text{th}}$ term $={}^{2n}C_n x^n=\dfrac{\lfloor 2n}{\lfloor n\,\lfloor n}x^n$.

12. Putting $x=y=1$, as in Art. 365, we have the sum of the coefficients $=2^{16}=65536$.

13. Putting $x=y=1$, the sum of the coefficients $=4^9=262144$.

14. The rth term from the beginning

$$={}^nC_{r-1}\,a^{n-r+1}(2x)^{r-1}=\frac{n(n-1)\ldots(n-r+2)}{\lfloor r-1}\,a^{n-r+1}(2x)^{r-1}.$$

The rth term from the end has the same coefficient, but the indices of a and $2x$ are interchanged.

15. $$(a^2+2a+1)^3=\{(a+1)^2\}^3=(a+1)^6$$
$$=a^6+6a^5+15a^4+20a^3+15a^2+6a+1.$$

Regarding $4x-2$, or $2(2x-1)$ as one term,

$$(x^2-4x+2)^3=(x^2)^3-3(x^2)^2.\,2(2x-1)+3x^2.\,4(2x-1)^2-8(2x-1)^3$$
$$=x^6-6x^4(2x-1)+12x^2(4x^2-4x+1)-8(8x^3-12x^2+6x-1)$$
$$=x^6-12x^5+6x^4+48x^4-48x^3+12x^2-64x^3+96x^2-48x+8$$
$$=x^6-12x^5+54x^4-112x^3+108x^2-48x+8.$$

16. $$(1+x)^{\frac{1}{3}}=1+\frac{1}{3}x+\frac{\frac{1}{3}\left(\frac{1}{3}-1\right)}{\lfloor 2}x^2+\frac{\frac{1}{3}\left(\frac{1}{3}-1\right)\left(\frac{1}{3}-2\right)}{\lfloor 3}x^3+\ldots$$
$$=1+\frac{1}{3}x+\frac{(-2)}{2\,.\,9}x^2+\frac{(-2)(-5)}{6\,.\,27}x^3+\ldots$$
$$=1+\frac{1}{3}x-\frac{1}{9}x^2+\frac{5}{81}x^3-\ldots$$

17. $$(1+x)^{\frac{3}{4}}=1+\frac{3}{4}x+\frac{\frac{3}{4}\left(\frac{3}{4}-1\right)}{\underline{|2}}x^2+\frac{\frac{3}{4}\left(\frac{3}{4}-1\right)\left(\frac{3}{4}-2\right)}{\underline{|3}}x^3+\dots$$

$$=1+\frac{3}{4}x+\frac{3(-1)}{2\cdot 4^2}x^2+\frac{3(-1)(-5)}{6\cdot 4^3}x^3+\dots$$

$$=1+\frac{3}{4}x-\frac{3}{32}x^2+\frac{5}{128}x^3\dots$$

18. $$(1+x)^{\frac{2}{5}}=1+\frac{2}{5}x+\frac{\frac{2}{5}\left(\frac{2}{5}-1\right)}{\underline{|2}}x^2+\frac{\frac{2}{5}\left(\frac{2}{5}-1\right)\left(\frac{2}{5}-2\right)}{\underline{|3}}x^3+\dots$$

$$=1+\frac{2}{5}x+\frac{2(-3)}{5^2\cdot\underline{|2}}x^2+\frac{2(-3)(-8)}{5^3\cdot 2\cdot 3}x^3+\dots$$

$$=1+\frac{2}{5}x-\frac{3}{25}x^2+\frac{8}{125}x^3-\dots$$

19. $$(1+3x)^{-2}=1-2(3x)+\frac{(-2)(-3)}{\underline{|2}}(3x)^2+\frac{(-2)(-3)(-4)}{\underline{|3}}(3x)^3+\dots$$

$$=1-6x+\frac{2\cdot 3}{1\cdot 2}\cdot 9x^2-\frac{2\cdot 3\cdot 4}{1\cdot 2\cdot 3}\cdot 27x^3+\dots$$

$$=1-6x+27x^2-108x^3+\dots$$

20. $$(1-x^2)^{-3}=1-3(-x^2)+\frac{(-3)(-4)}{\underline{|2}}(-x^2)^2$$

$$+\frac{(-3)(-4)(-5)}{\underline{|3}}(-x^2)^3+\dots$$

$$=1+3x^2+\frac{3\cdot 4}{1\cdot 2}x^4+\frac{3\cdot 4\cdot 5}{1\cdot 2\cdot 3}x^6+\dots$$

$$=1+3x^2+6x^4+10x^6+\dots$$

21. $$(1+3x)^{-4}=1-4(3x)+\frac{(-4)(-5)}{\underline{|2}}(3x)^2+\frac{(-4)(-5)(-6)}{\underline{|3}}(3x)^3+\dots$$

$$=1-12x+\frac{4\cdot 5}{2}\cdot 9x^2-\frac{4\cdot 5\cdot 6}{2\cdot 3}\cdot 27x^3+$$

$$=1-12x+90x^2-540x^3+\dots$$

22.
$$(2+x)^{-3}=2^{-3}\left(1+\frac{x}{2}\right)^{-3}$$
$$=\frac{1}{8}\left\{1-\frac{3x}{2}+\frac{(-3)(-4)}{\underline{|2}}\left(\frac{x}{2}\right)^2 + \frac{(-3)(-4)(-5)}{\underline{|3}}\left(\frac{x}{2}\right)^3+\ldots\right\}$$
$$=\frac{1}{8}\left\{1-\frac{3x}{2}+\frac{3\,.\,4}{1\,.\,2}\cdot\frac{x^2}{4}-\frac{3\,.\,4\,.\,5}{1\,.\,2\,.\,3}\cdot\frac{x^3}{8}+\ldots\right\}$$
$$=\frac{1}{8}\left\{1-\frac{3x}{2}+\frac{3x^2}{2}-\frac{5x^3}{4}+\ldots\right\}$$
$$=\frac{1}{8}-\frac{3x}{16}+\frac{3x^2}{16}-\frac{5x^3}{32}+\ldots$$

23.
$$(1+2x)^{-\frac{1}{2}}=1-\frac{1}{2}(2x)+\frac{\left(-\frac{1}{2}\right)\left(-\frac{1}{2}-1\right)}{\underline{|2}}(2x)^2 + \frac{\left(-\frac{1}{2}\right)\left(-\frac{1}{2}-1\right)\left(-\frac{1}{2}-2\right)}{\underline{|3}}(2x)^3+\ldots$$
$$=1-x+\frac{1\,.\,3}{2^2\,\underline{|2}}\,2^2\,.\,x^2-\frac{1\,.\,3\,.\,5}{2^3\,\underline{|3}}\,2^3\,.\,x^3+\ldots$$
$$=1-x+\frac{3}{2}x^2-\frac{5}{2}x^3+\ldots$$

24.
$$(a-2x)^{-\frac{3}{2}}=a^{-\frac{3}{2}}\left(1-\frac{2x}{a}\right)^{-\frac{3}{2}}$$
$$=a^{-\frac{3}{2}}\left\{1-\frac{3}{2}\left(-\frac{2x}{a}\right)+\frac{-\frac{3}{2}\left(-\frac{3}{2}-1\right)}{\underline{|2}}\left(-\frac{2x}{a}\right)^2 + \frac{-\frac{3}{2}\left(-\frac{3}{2}-1\right)\left(-\frac{3}{2}-2\right)}{\underline{|3}}\left(-\frac{2x}{a}\right)^3+\ldots\right.$$
$$=a^{-\frac{3}{2}}\left\{1+\frac{3x}{a}+\frac{3\,.\,5}{2^2\,\underline{|2}}\cdot\frac{2^2x^2}{a^2}+\frac{3\,.\,5\,.\,7}{2^3\,\underline{|3}}\cdot\frac{2^3x^3}{a^3}+\ldots\right\}$$
$$=\frac{1}{a^{\frac{3}{2}}}\left\{1+\frac{3x}{a}+\frac{15x^2}{2a^2}+\frac{35x^3}{2a^3}+\ldots\right\}.$$

25. The 5th term $=\dfrac{-\frac{3}{2}\left(-\frac{3}{2}-1\right)\left(-\frac{3}{2}-2\right)\left(-\frac{3}{2}-3\right)}{\lfloor 4}x^4$

$$=\frac{3\,.\,5\,.\,7\,.\,9}{2^4\,.\,2\,.\,3\,.\,4}x^4=\frac{315}{128}x^4.$$

The 10th term $=\dfrac{-\frac{3}{2}\left(-\frac{3}{2}-1\right)\ldots\left(-\frac{3}{2}-10+2\right)}{\lfloor 9}x^9$

$$=-\frac{3\,.\,5\,.\,7\,.\,9\,.\,11\ldots 19}{2^9\lfloor 9}x^9=-\frac{11\,.\,13\,.\,15\,.\,17\,.\,19}{512\,.\,2\,.\,4\,.\,6\,.\,8}x^9$$

$$=-\frac{230945}{65536}x^9.$$

26. The 3rd term $=\dfrac{\frac{11}{2}\left(\frac{11}{2}-1\right)}{\lfloor 2}(2x)^2=\dfrac{11\,.\,9}{8}4x^2=\dfrac{99}{2}x^2.$

The 11th term $=\dfrac{\frac{11}{2}\left(\frac{11}{2}-1\right)\left(\frac{11}{2}-2\right)\ldots\left(\frac{11}{2}-11+2\right)}{\lfloor 10}(2x)^{10}$

$$=\frac{11\,.\,9\,.\,7\,.\,5\,.\,3\,.\,1\,.\,(-1)(-3)(-5)(-7)}{2^{10}\lfloor 10}2^{10}\,.\,x^{10}$$

$$=\frac{11\,.\,3\,.\,5\,.\,7}{10\,.\,8\,.\,6\,.\,4\,.\,2}x^{10}=\frac{77}{256}x^{10}.$$

27. The 4th term $=\dfrac{(-2)(-3)(-4)}{\lfloor 3}x^3=-\dfrac{2\,.\,3\,.\,4}{1\,.\,2\,.\,3}x^3=-4x^3.$

The $(r+1)^{\text{th}}$ term $=\dfrac{(-2)(-3)\ldots(-2-r+1)}{\lfloor r}x^r$

$$=(-1)^r\frac{2\,.\,3\,.\,4\ldots(r+1)}{1\,.\,2\,.\,3\ldots r}x^r=(-1)^r(r+1)x^r.$$

28. The 7th term $=\dfrac{\frac{1}{2}\left(\frac{1}{2}-1\right)\left(\frac{1}{2}-2\right)\ldots\left(\frac{1}{2}-7+2\right)}{\lfloor 6}(-x)^6$

$$=\frac{1(-1)(-3)(-5)(-7)(-9)}{2^6\,.\,\lfloor 6}x^6$$

$$=-\frac{7\,.\,9}{2^6\,.\,2\,.\,4\,.\,6}x^6=-\frac{21}{1024}x^6.$$

The $(r+1)^{\text{th}}$ term $=\dfrac{\frac{1}{2}\left(\frac{1}{2}-1\right)\left(\frac{1}{2}-2\right)\ldots\left(\frac{1}{2}-r+1\right)}{\lfloor r}(-x)^r$

$$=\frac{1(-1)(-3)\ldots(3-2r)}{2^r\lfloor r}(-x)^r$$

$$=(-1)^{r-1}\frac{1\,.\,3\,.\,5\,.\,7\ldots(2r-3)}{2^r\lfloor r}(-x)^r$$

$$=(-1)^{2r-1}\frac{1\,.\,3\,.\,5\,.\,7\ldots(2r-3)}{2^r\lfloor r}x^r$$

$$=-\frac{1\,.\,3\,.\,5\ldots(2r-3)}{2^r\lfloor r}x^r.$$

29. The $(r+1)^{\text{th}}$ term of $a^{-1}\left(1-\dfrac{bx}{a}\right)^{-1}$

$$=a^{-1}\frac{(-1)(-2)(-3)\ldots(-1-r+1)}{\lfloor r}\left(-\frac{bx}{a}\right)^r$$

$$=a^{-1}(-1)^r\frac{1\,.\,2\,.\,3\ldots r}{\lfloor r}(-1)^r\left(\frac{bx}{a}\right)^r$$

$$=(-1)^{2r}\frac{1}{a}\left(\frac{bx}{a}\right)^r=\frac{b^rx^r}{a^{r+1}}.$$

The $(r+1)^{\text{th}}$ term of $(1-nx)^{\frac{1}{n}}$

$$=\frac{\frac{1}{n}\left(\frac{1}{n}-1\right)\left(\frac{1}{n}-2\right)\ldots\left(\frac{1}{n}-r+1\right)}{\lfloor r}(-nx)^r$$

$$=\frac{(-1)^{r-1}(n-1)(2n-1)\ldots\{(r-1)n-1\}}{n^r\lfloor r}(-1)^rn^r\,.\,x^r$$

$$=-\frac{(n-1)(2n-1)\ldots\{(r-1)n-1\}}{\lfloor r}x^r.$$

30. $\sqrt[3]{122}=(5^3-3)^{\frac{1}{3}}=5\left(1-\dfrac{3}{5^3}\right)^{\frac{1}{3}}$

$$=5\left\{1-\frac{1}{3}\cdot\frac{3}{5^3}-\frac{1}{9}\cdot\frac{3^2}{5^6}-\frac{5}{81}\cdot\frac{3^3}{5^9}-\ldots\right\}$$

$$=5-\frac{1}{5^2}-\frac{1}{5^5}-\frac{1}{3}\cdot\frac{1}{5^7}-\ldots$$

$$=5-\frac{2^2}{10^2}-\frac{2^5}{10^5}-\frac{1}{3}\cdot\frac{2^7}{10^7}-\ldots$$

$=5-{\cdot}04-{\cdot}00032-\ldots$

$=4{\cdot}9596$ to four places of decimals.

31 $$\sqrt[4]{620}=(5^4-5)^{\frac{1}{4}}=5\left(1-\frac{1}{5^3}\right)^{\frac{1}{4}}$$
$$=5\left\{1-\frac{1}{4}\cdot\frac{1}{5^3}-\frac{3}{32}\cdot\frac{1}{5^6}-\dots\right\}$$
$$=5-\frac{1}{4}\cdot\frac{1}{5^2}-\frac{3}{32}\cdot\frac{1}{5^5}-\dots$$
$$=5-\frac{1}{4}\cdot\frac{2^2}{10^2}-\frac{3}{32}\cdot\frac{2^5}{10^5}-\dots$$
$$=5-{\cdot}01-{\cdot}00003-$$
$$=4{\cdot}9899 \text{ to four places of decimals.}$$

32. $$\sqrt[5]{31}=(2^5-1)^{\frac{1}{5}}=2\left(1-\frac{1}{2^5}\right)^{\frac{1}{5}}$$
$$=2\left\{1-\frac{1}{5}\cdot\frac{1}{2^5}-\frac{2}{25}\cdot\frac{1}{2^{10}}-\dots\right\}$$
$$=2-\frac{1}{5}\cdot\frac{5^4}{10^4}-\frac{1}{25}\cdot\frac{5^8}{10^8}-\dots$$
$$=2-{\cdot}0125-{\cdot}00015625-$$
$$=1{\cdot}9873 \text{ to four places of decimals.}$$

33. $$1\div\sqrt{99}=99^{-\frac{1}{2}}=(10^2-1)^{-\frac{1}{2}}=\frac{1}{10}\left(1-\frac{1}{10^2}\right)^{-\frac{1}{2}}$$
$$=\frac{1}{10}\left\{1+\frac{1}{2}\cdot\frac{1}{10^2}+\frac{3}{8}\cdot\frac{1}{10^4}+\dots\right\}$$
$$=\frac{1}{10}+\frac{1}{2}\cdot\frac{1}{10^3}+\frac{3}{8}\cdot\frac{1}{10^5}+\dots$$
$$={\cdot}1+\frac{{\cdot}001}{2}+\frac{{\cdot}00003}{8}+\dots$$
$$={\cdot}1+{\cdot}0005+{\cdot}00000375+\dots$$
$$={\cdot}1005 \text{ to four places of decimals.}$$

EXAMPLES. XXXIX. PAGE 347

1. (1) Let $x=\log\sqrt{32}$ to base $\sqrt[3]{2}$; then $(\sqrt[3]{2})^x=\sqrt{32}$,
$$\therefore\ 2^{\frac{x}{3}}=\sqrt{32}=2^{\frac{5}{2}};$$
$$\therefore\ \frac{x}{3}=\frac{5}{2},\ \text{or}\ x=\frac{15}{2}.$$

(2) Let $x=\log \cdot 03125$ to base $\sqrt[3]{2}$; then $(\sqrt[3]{2})^x = \cdot 03125 = \frac{1}{32}$,

$$\therefore \quad 2^{\frac{x}{3}} = \frac{1}{32} = 2^{-5}; \text{ whence } x = -15.$$

(3) Let $x=\log 100$ to base $\cdot 01$; then $(\cdot 01)^x = 100$,

$$\therefore \left(\frac{1}{100}\right)^x = 100, \text{ whence } x = -1.$$

(4) Let $x=\log \cdot 00001$ to base $\cdot 01$; then $(\cdot 01)^x = \cdot 00001$,

$$\therefore \left(\frac{1}{100}\right)^x = \frac{1}{100000}; \text{ or } \left(\frac{1}{10}\right)^{2x} = \left(\frac{1}{10}\right)^5;$$

$$\text{whence } x = \frac{5}{2}.$$

2. (2) Let $\log_5 \cdot 0016 = x$; then $5^x = \cdot 0016 = \frac{1}{625} = 5^{-4}$;

$$\therefore \quad x = -4.$$

4. Denote the expression by $\log X$; then

$$X = a^{\frac{1}{3}} b^{\frac{1}{3}} c^{\frac{2}{3}} \div (a^{-\frac{1}{3}} b^{\frac{1}{3}} c^{\frac{2}{3}}) = a^{\frac{1}{2}};$$

$$\therefore \log X = \frac{1}{2} \log a.$$

5. The expression within brackets is equal to

$$(x^5 y^{-5})^{-3} \div (x^{-3} y^4)^5$$
$$= (x^{-15} y^{15}) \div (x^{-15} y^{20})$$
$$= y^{-5}.$$

Hence the logarithm required $= -5 \log y$.

9. $\log 729 = \log 3^6 = 6 \log 3 = 6 \times \cdot 4771213 = 2 \cdot 8627278.$

10. $\log 8400 = \log (2^2 \times 3 \times 7 \times 100) = 2 \log 2 + \log 3 + \log 7 - 2$

$$= 2 \times \cdot 3010300 + \cdot 4771213 + \cdot 8450980 - 2$$
$$= 3 \cdot 9242793.$$

11. $\log 256 = \log 2^8 = 8 \log 2 = 8 \times \cdot 3010300 = 2 \cdot 4082400.$

$$\therefore \log \cdot 256 = \bar{1} \cdot 4082400.$$

12. $\log 5832 = \log (8 \times 729) = \log (2^3 . 3^6) = 3 \log 2 + 6 \log 3$

$$= 3 \times \cdot 3010300 + 6 \times \cdot 4771213$$
$$= 3 \cdot 7658178.$$
$$\therefore \log 5 \cdot 832 = \cdot 7658178.$$

13. $\log \sqrt[3]{392} = \frac{1}{3} \log (8 \times 49) = \frac{1}{3} \log 2^3 . 7^2 = \frac{1}{3} (3 \log 2 + 2 \log 7)$

$$= \frac{1}{3} (3 \times \cdot 3010300 + 2 \cdot 8450980) = \cdot 8644286.$$

14. $\log \cdot 304\dot{8} = \log \frac{2744}{9000} = \log \frac{2^3 \times 7^3}{3^2 \times 1000} = 3 \log 2 + 3 \log 7 - 2 \log 3 - \log 1000$

$$= 3 \times \cdot 3010300 + 3 \times \cdot 8450980 - 2 \times \cdot 4771213 - 3$$

$$= 2 \cdot 4841414 - 3 = \bar{1} \cdot 4841414.$$

15. $$\log \frac{11}{15} + \log \frac{490}{297} - 2 \log \frac{7}{9} = \log \frac{11}{15} + \log \frac{490}{297} - \log \left(\frac{7}{9}\right)^2$$

$$= \log \left\{\frac{11}{15} \times \frac{490}{297} \times \frac{81}{49}\right\} = \log 2.$$

16. The expression $= \log \left\{\frac{225}{224} \times \frac{512}{81} \div \left(\frac{20}{189}\right)^2\right\}$

$$= \log \left\{\frac{225}{224} \times \frac{512}{81} \times \frac{189 \times 189}{20 \times 20}\right\}$$

$= \log (81 \times 7)$, on reduction,

$$= 4 \log 3 + \log 7 = 2 \cdot 7535832.$$

17. $\log \frac{(10 \cdot 8)^{\frac{1}{2}} \times (\cdot 24)^{\frac{5}{3}}}{90^{-2}} = \frac{1}{2} \log 10 \cdot 8 + \frac{5}{3} \log \cdot 24 + 2 \log 90$

$$= \frac{1}{2} \log \frac{2^2 \, . \, 3^3}{10} + \frac{5}{3} \log \frac{2^3 \, . \, 3}{100} + 2 \log (3^2 \, . \, 10)$$

$$= \frac{1}{2} (2 \log 2 + 3 \log 3 - 1) + \frac{5}{3} (3 \log 2 + \log 3 - 2) + 2 (2 \log 3 + 1)$$

$$= 6 \log 2 + \frac{43}{6} \log 3 - \frac{11}{6}$$

$$= 1 \cdot 8061800 + 3 \cdot 4193693 - 1 \cdot 8\dot{3}$$

$$= 3 \cdot 3922160.$$

18. $\log \frac{\sqrt[3]{126} \, . \, \sqrt{108}}{\sqrt[6]{1008} \, . \, \sqrt[3]{162}} = \frac{1}{3} \log 126 + \frac{1}{2} \log 108 - \frac{1}{6} \log 1008 - \frac{1}{3} \log 162$

$$= \frac{1}{3} \log (2 \, . \, 3^2 \, . \, 7) + \frac{1}{2} \log (2^2 \, . \, 3^3) - \frac{1}{6} \log (2^4 \, . \, 3^2 \, . \, 7) - \frac{1}{3} \log (2 \, . \, 3^4)$$

$$= \frac{1}{3} (\log 2 + 2 \log 3 + \log 7) + \frac{1}{2} (2 \log 2 + 3 \log 3) - \frac{1}{6} (4 \log 2 + 2 \log 3 + \log 7) - \frac{1}{3} (\log 2 + 4 \log 3)$$

$$= \frac{1}{3} \log 2 + \frac{1}{2} \log 3 + \frac{1}{6} \log 7$$

$$= \cdot 4797536.$$

19. $$\log \sqrt[5]{\frac{588 \times 768}{686 \times 972}} = \log \sqrt[5]{\frac{128}{7 \,.\, 27}} = \frac{1}{5} \log \frac{2^7}{7 \,.\, 27}$$
$$= \frac{1}{5}(7 \log 2 - \log 7 - 3 \log 3)$$
$$= \frac{1}{5}(7 \times \cdot 3010300 - \cdot 8450980 - 3 \times \cdot 4771213)$$
$$= \frac{1}{5}(4 \cdot 8307481 - 5)$$
$$= \bar{1} \cdot 9661496.$$

20. $$\log 42^{42} = 42 \log (2 \,.\, 3 \,.\, 7) = 42 (\log 2 + \log 3 + \log 7)$$
$$= 42 (\cdot 3010300 + \cdot 4771213 + \cdot 8450980)$$
$$= 42 \times 1 \cdot 6232493$$
$$= 68 \cdot 1764706.$$

Hence the number of digits is 69. [Art. 381.]

21. $$\log \left(\frac{81}{80}\right)^{1000} = 1000 \log \frac{3^4}{2^3 \,.\, 10}$$
$$= 1000 \{4 \log 3 - 3 \log 2 - 1\}$$
$$= 1000 \{4 \times \cdot 4771213 - 3 \times \cdot 3010300 - 1\}$$
$$= 1000 \times \cdot 0053952$$
$$= 5 \cdot 3952.$$

$\therefore$ the number of integral digits in $\left(\frac{81}{80}\right)^{1000}$ is 6.

22. $$\log \left(\frac{2}{3}\right)^{1000} = 1000 (\log 2 - \log 3) = 1000 (\cdot 3010300 - \cdot 4771213)$$
$$= -176 \cdot 0913 = \overline{177} \cdot 9087.$$

$\therefore$ the number of ciphers is 176. [Art. 382.]

23. $$\log 1008 = \log (2^4 \,.\, 3^2 \,.\, 7) = 4 \log 2 + 2 \log 3 + \log 7$$
$$= 4 \times \cdot 3010300 + 2 \times \cdot 4771213 + \cdot 8450980$$
$$= 3 \cdot 0034606;$$

$$\therefore \log \sqrt[5]{\cdot 01008} = \frac{1}{5} \times \bar{2} \cdot 0034606 = \frac{1}{5}(\bar{5} + 3 \cdot 0034606) = \bar{1} \cdot 6006921.$$
$$\therefore \sqrt[5]{\cdot 01008} = \cdot 398742.$$

24. $$\log 792 = \log (2^3 \,.\, 3^2 \,.\, 11) = 3 \log 2 + 2 \log 3 + \log 11$$
$$= 2 \cdot 8987253.$$

$$\therefore \log \sqrt[7]{\cdot 00792} = \frac{1}{7} \times \bar{3} \cdot 8987253 = \frac{1}{7}(\bar{7} + 4 \cdot 8987253) = \bar{1} \cdot 6998179.$$
$$\therefore \sqrt[7]{\cdot 00792} = \cdot 500977.$$

25. The expression $=\log\left(\frac{75}{49}\right)^2+\log\frac{135}{32}-\log\left(\frac{45}{28}\right)^3$

$$=\log\left\{\frac{75\times 75}{49\times 49}\times\frac{135}{32}\times\frac{28\times 28\times 28}{45\times 45\times 45}\right\}$$

$$=\log\frac{50}{21},\text{ on reduction,}$$

$$=1+\log 5-\log 3-\log 7$$

$$=2-\log 2-\log 3-\log 7$$

$$=\cdot 3767507.$$

26. Taking logarithms of both sides, we have

$$(x+2)\log 3=\log 405=\log\frac{3^4 . 10}{2}$$

$$=4\log 3+1-\log 2;$$

$$\therefore\ x\log 3=2\log 3+1-\log 2;$$

$$\therefore\ x=\frac{2\log 3+1-\log 2}{\log 3}=\frac{1\cdot 653}{\cdot 477}=3\cdot 46\ldots$$

27. Taking logarithms of both sides, we have

$$5-3x=(7-2x)\log 2;$$

$$\therefore\ 3x-2x\log 2=5-7\log 2;$$

$$\therefore\ x=\frac{5-7\log 2}{3-2\log 2}=\frac{2\cdot 893}{2\cdot 398}=1\cdot 206\ldots$$

28. Taking logarithms of both sides, we have

$$(x-3)\log 5=\log 8=3\log 2;$$

$$\therefore\ x=\frac{3(\log 2+\log 5)}{\log 5}=\frac{3\log 10}{1-\log 2}$$

$$=\frac{3}{1-\cdot 301}=4\cdot 29.$$

29. Taking logarithms of both sides, we have

$$(3x-4)\log 12+(7-2x)\log 18=\log 1458=\log 2 . 3^6;$$

$$\therefore\ (3x-4)(2\log 2+\log 3)+(7-2x)(\log 2+2\log 3)=\log 2+6\log 3;$$

whence $$x=\frac{2\log 2-4\log 3}{4\log 2-\log 3}=\frac{-1\cdot 306}{\cdot 727}=-1\cdot 8\text{ nearly.}$$

[*For Examples XXXIX. b., see Appendix, p.* 376.]

EXAMPLES. XL. PAGE 353.

3.
```
     31044
      4302
    ------
    112143
   143242
  224341
 ---------
 244332343
```

4.
```
    9t83
     3t7
   -----
   6392t
  90858
 27t29
 -------
 36641tt
```

5.
```
6541 ) 31664435 ( 3245
       26253
        24114
        16412
         44023
         36124
          45665
          45665
```

6.
```
     3024
     3024
    -----
    22161
   11103
  14132
 --------
 14320241
```

7.
```
9 | 75013
9 | 8334...7
9 | 926...0
9 | 102...8
9 | 11...3
    1...2
```

Thus the required number $=123807$.

```
4 | 5210
4 | 1302...2
4 | 325...2
4 | 81...1
4 | 20 ..1
4 | 5...0
    1...1
```

Thus the required number $=1101122$.

8.
```
T | 987504
T | 82292.. 0
T | 6857...8
T | 571...5
T | 47...7
    3...e
```

Thus the required number $=3e7580$.

9.
```
t | 76543
t | 6211...9
t | 500...9
t | 40...0
    3...2
```

Thus the required number $=32099$.

10.
```
7 | 54321
7 | 4534...3
7 | 412...2
7 | 33...5
    3...0
```

Thus the required number $=30523$.

11.
```
2 | te
2 | 55...1
2 | 28...1
2 | 14 ..0
2 | 8...0
2 | 4...0
2 | 2.. 0
    1...0
```

Thus the required number $=10000011$.

12. (1)

```
2|1001
 2|500...1
  2|250...0
  2|125...0
   2|62...1
   2|31...0
   2|15...1
    2|7...1
    2|3...1
      1...1
```

$\therefore\ 1001 = 2^9 + 2^8 + 2^7 + 2^6 + 2^5 + 2^3 + 1.$

(2)

```
e|100000
 e|9090...t
  e|826...4
   e|75...1
      6...9
```

$\therefore\ 100000 = 6 . 11^4 + 9 . 11^3 + 11^2 + 4 . 11 + 10.$

13.

```
     532
    2106
    3261
      53
  e|6315
   e|405...6
    e|24...3
        1...7
```

```
   2021121
   1221212
  e|22202
    e|210...5
         1...t
```

```
    1736
     1t5
    8368
   15725
   1736
  328108
```

14.

```
t|53774
 t|4314...4
  t|341...2
   t|26...5
       2...2
```

```
t|32875
 t|2887...7
  t|262...5
   t|23...8
       2...1
```

53774 in the scale of 8 = 22524 in the denary scale,

and 32875 in the scale of 9 = 21857 ....;

∴ the difference = 667.

15. Treating the integral and fractional parts separately, we have

```
8|131
 8|16...3
     2...0
```

```
·890625
      8
 7·125
     8
 1·
```

∴ the required number = 203·71.

16. Treating the integral and fractional parts separately, we have

$$\begin{array}{r|l} 9 & 1001 \\ \hline 9 & 10 \ldots 1 \\ & 0 \ldots 3 \end{array} \qquad \begin{array}{r} \cdot 12211 \\ 9 \\ \hline 5 \cdot 211 \\ 9 \\ \hline 7 \cdot 1 \\ 9 \\ \hline 3 \cdot 0 \end{array}$$

$\therefore$ the required number $= 31 \cdot 573$.

17.

$$\begin{array}{r} \cdot 2037 \\ 4 \\ \hline 1 \cdot 0174 \\ 4 \\ \hline \cdot 076 \\ 4 \\ \hline \cdot 37 \\ 4 \\ \hline 1 \cdot 74 \\ 4 \\ \hline 3 \cdot 6 \\ 4 \\ \hline 3 \cdot 0 \end{array}$$

$\therefore$ the required fraction $= \cdot 100133$.

18. (1)

$$\frac{27}{32} \times 6 = \frac{81}{16} = 5 + \frac{1}{16};$$

$$\frac{1}{16} \times 6 = \frac{3}{8};$$

$$\frac{3}{8} \times 6 = \frac{9}{4} = 2 + \frac{1}{4};$$

$$\frac{1}{4} \times 6 = \frac{3}{2} = 1 + \frac{1}{2};$$

$$\frac{1}{2} \times 6 = 3$$

$$\therefore \text{ the required fraction} = \frac{5}{6} + \frac{0}{6^2} + \frac{2}{6^3} + \frac{1}{6^4} + \frac{3}{6^5}$$

$$= \cdot 50213.$$

(2)

$$\frac{500}{729} \times 6 = \frac{1000}{243} = 4 + \frac{28}{243};$$

$$\frac{28}{243} \times 6 = \frac{56}{81};$$

$$\frac{56}{81}\times 6=\frac{112}{27}=4+\frac{4}{27};$$

$$\frac{4}{27}\times 6=\frac{8}{9};$$

$$\frac{8}{9}\times 6=\frac{16}{3}=5+\frac{1}{3};$$

$$\frac{1}{3}\times 6=2.$$

$$\therefore \text{ the required fraction}=\frac{4}{6}+\frac{0}{6^2}+\frac{4}{6^3}+\frac{0}{6^4}+\frac{5}{6^5}+\frac{2}{6^6}$$

$$=\cdot 404052.$$

19. The G.C.M. of 587 and 749 = 91;

also $587\div 91=7$,

and $749\div 91=9$,

$$\therefore \text{ the required fraction}=\frac{7\times 91}{9\times 91}=\frac{7}{9}.$$

$$\begin{array}{r|r|r|l} 3 & 587 & 749 & 1 \\ & 4t6 & 587 & \\ \hline & 91 & 172 & 2 \\ & & 172 & \\ \hline \end{array}$$

20. Let r be the radix of the scale required; then $4r^2=100$, or $r=5$.

21. Let r be the radix of the scale required; then

$$6r^2+4r+7=(2r+5)^2=4r^2+20r+25,$$

$$\therefore\ 2r^2-16r-18=0, \text{ or } (r-9)(r+1)=0;$$

$$\therefore\ r=9, \text{ the value } -1 \text{ being inadmissible.}$$

22. Let r be the radix of the scale required; then

$$565-432=708-565,$$

or

$$5r^2+6r+5-(4r^2+3r+2)=7r^2+8-(5r^2+6r+5);$$

$$\therefore\ r^2-9r=0;$$

$$\therefore\ r=9.$$

23. Let r be the radix; then

$$(2\cdot 6)^2=34\times\cdot 22,$$

or

$$\left(2+\frac{6}{r}\right)^2=(3r+4)\left(\frac{2}{r}+\frac{2}{r^2}\right);$$

$$4+\frac{24}{r}+\frac{36}{r^2}=6+\frac{14}{r}+\frac{8}{r^2};$$

$$r^2-5r-14=0, \text{ or } (r-7)(r+2)=0;$$

$$\therefore\ r=7, \text{ the value } -2 \text{ being inadmissible.}$$

24. (1)

$4\dot{4}3\dot{0}0\dot{1}$ (444
31
134 | 1330
    | 1201
1434 | 12401
     | 12401

(2)

$\dot{2}4\dot{3}4\dot{5}2\dot{4}$ (1425
1
24 | 143
   | 132
312 | 1145
    | 624
3145 | 22124
     | 22124

(3)

$t9\dot{8}5\dot{6}7\dot{9}$ (3333
9
63 | 198
   | 179
663 | 1t56
    | 1879
6663 | 18879
     | 18879

EXAMPLES. XLI. PAGE 359.

1. (1) By writing -2 for x in the formula for e^x, we have

$$e^{-2}=1-2+\frac{(-2)^2}{\underline{|2}}+\frac{(-2)^3}{\underline{|3}}+\frac{(-2)^4}{\underline{|4}}+\frac{(-2)^5}{\underline{|5}}+\dots$$

$$=1-2+2-\frac{2^3}{\underline{|3}}+\frac{2^4}{\underline{|4}}-\frac{2^5}{\underline{|5}}+\dots$$

(2) $\dfrac{e^2-1}{2e}=\dfrac{1}{2}\left(e-\dfrac{1}{e}\right)=\dfrac{1}{2}(e-e^{-1})$.

The expansion now follows from the formulae of Ex. 1, Art. 397.

2.
$$\log\sqrt{1+x}=\log(1+x)^{\frac{1}{2}}=\frac{1}{2}\log(1+x)$$

$$=\frac{1}{2}\left(x-\frac{x^2}{2}+\frac{x^3}{3}-\frac{x^4}{4}+\dots\right).$$

3. By writing 1 for x in the formula for $\log(1+x)$, we have

$$\log_e 2=\left(1-\frac{1}{2}\right)+\left(\frac{1}{3}-\frac{1}{4}\right)+\left(\frac{1}{5}-\frac{1}{6}\right)+\dots$$

4. $\log_{10}\left(\dfrac{1}{1-x}\right)=\log_{10}(1-x)^{-1}=-\log_{10}(1-x)=-\dfrac{\log_e(1-x)}{\log_e 10}$;

and the expansion is obtained by Art. 398.

5. $\log\dfrac{1+x}{1-3x}=\log(1+x)-\log(1-3x)$

$$=\left(x-\frac{x^2}{2}+\frac{x^3}{3}-\frac{x^4}{4}+\dots\right)+\left(3x+\frac{9x^2}{2}+\frac{27x^3}{3}+\frac{81x^4}{4}+\dots\right)$$

$$=4x+4x^2+\frac{28}{3}x^3+20x^4+\dots.$$

6. $\log\sqrt{x^2-1}=\log(x^2-1)^{\frac{1}{2}}=\frac{1}{2}\log(x^2-1)=\frac{1}{2}\log x^2\left(1-\frac{1}{x^2}\right)$

$$=\frac{1}{2}\log x^2+\frac{1}{2}\log\left(1-\frac{1}{x^2}\right)$$

$$=\log x-\frac{1}{2}\left(\frac{1}{x^2}+\frac{1}{2x^4}+\frac{1}{3x^6}+\dots\right).$$

7. The expression on the left side of the identity

$$=\log\left(\sqrt{\frac{1+x}{1-x}}\div\sqrt{\frac{1-x}{1+x}}\right)=\log\frac{1+x}{1-x}.$$ [Now see Art. 400.]

8. From the given condition, we have $a=\log(1+b)$;

$$\therefore\ 1+b=e^a=1+a+\frac{a^2}{\underline{|2}}+\frac{a^3}{\underline{|3}}+\frac{a^4}{\underline{|4}}+\dots$$

9. $$\sqrt{e}=e^{\frac{1}{2}}=1+\frac{1}{2}+\frac{1}{2^2\,\underline{|2}}+\frac{1}{2^3\,\underline{|3}}+\frac{1}{2^4\,\underline{|4}}+\dots$$

1·5
·125
·02083 = preceding line divided by 2 × 3 or 6;
·00260 = ... 2 × 4 or 8;
·00026 = ... 2 × 5 or 10;
·00002 = ... 2 × 6 or 12;
1·64871

10. $\log(1+x-2x^2)=\log\{(1+2x)(1-x)\}=\log(1+2x)+\log(1-x)$

$$=\left(2x-\frac{4x^2}{2}+\frac{8x^3}{3}-\frac{16x^4}{4}+\dots\right)$$

$$-\left(x+\frac{x^2}{2}+\frac{x^3}{3}+\frac{x^4}{4}+\dots\right)$$

$$=x-\frac{5x^2}{2}+\frac{7x^3}{3}-\frac{17x^4}{4}+\dots.$$

The general term is $\dfrac{(-1)^{r-1}(2x)^r}{r}-\dfrac{x^r}{r}$.

11. By writing -1 for x in the formula for e^x, we have

$$e^{-1}=(1-1)+\left(\frac{1}{\underline{|2}}-\frac{1}{\underline{|3}}\right)+\left(\frac{1}{\underline{|4}}-\frac{1}{\underline{|5}}\right)+\left(\frac{1}{\underline{|6}}-\frac{1}{\underline{|7}}\right)$$

$$=\frac{3-1}{\underline{|3}}+\frac{5-1}{\underline{|5}}+\frac{7-1}{\underline{|7}}+\dots.$$

12. By writing 3 in the form $\dfrac{1+\frac{1}{2}}{1-\frac{1}{2}}$, we have

$$\log_e 3 = \log_e\left(1+\frac{1}{2}\right) - \log_e\left(1-\frac{1}{2}\right)$$

$$= \left(\frac{1}{2} - \frac{1}{2 \cdot 2^2} + \frac{1}{3 \cdot 2^3} - \frac{1}{4 \cdot 2^4} + \dots\right) + \left(\frac{1}{2} + \frac{1}{2 \cdot 2^2} + \frac{1}{3 \cdot 2^3} + \frac{1}{4 \cdot 2^4} + \dots\right)$$

$$= 2\left(\frac{1}{2} + \frac{1}{3 \cdot 2^3} + \frac{1}{5 \cdot 2^5} + \frac{1}{7 \cdot 2^7} + \dots\right)$$

13. $\log(1-3x+2x^2)^{-1} = \log\{(1-2x)(1-x)\}^{-1} = -\log(1-2x) - \log(1-x)$

$$= \left(2x + \frac{4x^2}{2} + \frac{8x^3}{3} + \frac{16x^4}{4} + \dots\right) + \left(x + \frac{x^2}{2} + \frac{x^3}{3} + \frac{x^4}{4} + \dots\right)$$

$$= 3x + \frac{5x^2}{2} + 3x^3 + \frac{17x^4}{4} + \dots.$$

The general term is $\dfrac{(2x)^r}{r} + \dfrac{x^r}{r}$.

14. $\log(1-x+x^2) = \log\dfrac{1+x^3}{1+x} = \log(1+x^3) - \log(1+x)$

$$= \left(x^3 - \frac{x^6}{2} - \dots\right) - \left(x - \frac{x^2}{2} + \frac{x^3}{3} - \frac{x^4}{4} + \frac{x^5}{5} - \frac{x^6}{6} + \dots\right)$$

$$= -x + \frac{x^2}{2} + \frac{2x^3}{3} + \frac{x^4}{4} - \frac{x^5}{5} - \frac{x^6}{3} - \dots.$$

15. The left side of the identity

$$= -\log\left(1-\frac{1}{x}\right) = -\log\frac{x-1}{x} = \log\frac{x}{x-1} = \log\left(1+\frac{1}{x-1}\right),$$

which on expansion gives the right side.

EXAMPLES. XLII. a. PAGE 362.

1. We have $(x^2+x)^2 + (x^2+x) - 42 = 0;$

that is, $(x^2+x-6)(x^2+x+7) = 0.$

Thus $x^2+x-6=0$, or $x^2+x+7=0;$

whence $x = 2, \ -3;$ or $\dfrac{-1 \pm \sqrt{-27}}{2}.$

2. Write y for $\frac{x^2-1}{x}$; thus

$$\frac{1}{y}+y=\frac{13}{6}, \text{ or } 6y^2-13y+6=0;$$

that is, $(3y-2)(2y-3)=0$; whence $y=\frac{2}{3}$, or $\frac{3}{2}$.

Thus $$\frac{x^2-1}{x}=\frac{3}{2}, \text{ or } \frac{2}{3};$$

$$\therefore\ 2x^2-3x-2=0, \text{ or } 3x^2-2x-3=0;$$

whence $$x=2, \frac{1}{2}; \text{ or } x=\frac{1\pm\sqrt{10}}{3}.$$

3. By transposition, $\left(x+\frac{1}{x}\right)^2-4\left(x+\frac{1}{x}\right)-5=0$;

that is, $$\left(x+\frac{1}{x}-5\right)\left(x+\frac{1}{x}+1\right)=0.$$

Thus $$x^2-5x+1=0, \text{ or } x^2+x+1=0;$$

whence $$x=\frac{5\pm\sqrt{21}}{2}, \text{ or } \frac{-1\pm\sqrt{-3}}{2}.$$

4. We have $$(x^3+8)(8x^3+1)=0;$$

whence $$x^3=-8, \text{ or } -\frac{1}{8}; \text{ and } x=-2, \text{ or } -\frac{1}{2}.$$

5. By transposition, we have

$$\frac{5}{x+4}-\frac{3x+14}{3x+8}-\frac{x+8}{x+12}=\left(1+\frac{6}{3x+8}\right)-\left(1-\frac{4}{x+12}\right);$$

that is, $$\frac{5}{x+4}=\frac{6}{3x+8}+\frac{4}{x+12}=\frac{18x+104}{(3x+8)(x+12)}.$$

$$\therefore\ 15x^2+220x+480=18x^2+176x+416;$$

whence $$3x^2-44x-64=0, \text{ or } (x-16)(3x+4)=0.$$

6. Since $4^x=2^{2x}$, we have $2^{2x}-9\cdot2^x+8=0$.
Thus $(2^x-8)(2^x-1)=0$; whence $2^x=8=2^3$, or $2^x=1=2^0$.

7. The given equation may be written

$$\left(-3+\frac{9}{5-x}\right)+\left(\frac{1}{2}+\frac{6}{10-4x}\right)=3\tfrac{1}{2};$$

whence by transposition and division by 3, we have

$$\frac{3}{5-x}+\frac{1}{5-2x}=2, \text{ or } \frac{20-7x}{25-15x+2x^2}=2,$$

that is, $$4x^2-23x+30=0, \text{ or } (x-2)(4x-15)=0.$$

8. Multiply both sides by $x^{\frac{1}{2}}$ or $\sqrt{x}$ and transpose; thus

$$3x - 8\sqrt{x} - 3 = 0; \text{ or } (3\sqrt{x}+1)(\sqrt{x}-3)=0.$$

9. On transposition, we have

$$\left(x - \frac{6}{x}\right)^2 + 4\left(x - \frac{6}{x}\right) - 5 = 0;$$

that is, $$\left(x - \frac{6}{x} + 5\right)\left(x - \frac{6}{x} - 1\right) = 0.$$

Thus $$x^2 + 5x - 6 = 0, \text{ or } x^2 - x - 6 = 0;$$

whence $$x = 1, \; -6; \text{ or } 3, \; -2.$$

10. By transposition, $27x^{\frac{3}{2}} - 26x^{\frac{3}{4}} - 1 = 0$;

that is, $$(27x^{\frac{3}{4}} + 1)(x^{\frac{4}{3}} - 1) = 0;$$

whence $$x^{\frac{3}{4}} = -\frac{1}{27}, \text{ or } 1; \text{ and } x = \left(-\frac{1}{27}\right)^{\frac{4}{3}} = \frac{1}{81}, \text{ or } x = 1^{\frac{4}{3}} = 1.$$

11. By transposition and squaring, we have

$$49(x-8) + 12 - 28\sqrt{3(x-8)} = 21x + 12;$$

from which by transposition and division by 28,

$$x - 14 = \sqrt{3(x-8)}.$$

Thus $$x^2 - 28x + 196 = 3x - 24, \text{ or } x^2 - 31x + 220 = 0;$$

whence $$x = 20, \text{ or } 11.$$

12. Since $4^{2x+1} = 4 \cdot 4^{2x}$, we have $4 \cdot 4^{2x} - 65 \cdot 4^x + 16 = 0$;

whence $$(4^x - 16)(4 \cdot 4^x - 1) = 0, \text{ and } 4^x = 16 = 4^2, \text{ or } 4^x = \frac{1}{4} = 4^{-1}.$$

13. By squaring, $x^2 + 4x + 4 = 4 + x\sqrt{8-x}$.

On division by x (which corresponds to a root $x = 0$), we have

$$x + 4 = \sqrt{8-x}, \text{ or } x^2 + 8x + 16 = 8 - x;$$

whence $$x^2 + 9x + 8 = 0, \text{ or } (x+8)(x+1).$$

14. Multiply each side by $3^{\frac{x}{2}}$ and transpose; thus

$$3^x - 2 \cdot 3^{\frac{x}{2}} + 1 = 0, \text{ or } (3^{\frac{x}{2}} - 1)^2 = 0;$$

whence $$3^{\frac{x}{2}} = 1 = 3^0, \text{ and } x = 0.$$

15. On transposition, $2x^2 - 7x + 6 + 2\sqrt{2x^2 - 7x + 6} = 0.$

By putting $2x^2 - 7x + 6 = y^2$, we have

$$y^2 + 2y = 0; \text{ whence } y = 0, \text{ or } -2.$$

Thus $$2x^2 - 7x + 6 = 0, \text{ or } 4;$$

$$\therefore\ 2x^2 - 7x + 6 = 0, \text{ or } 2x^2 - 7x + 2 = 0;$$

whence $$(2x - 3)(x - 2) = 0, \text{ or } x = \frac{7 \pm \sqrt{33}}{4}.$$

The latter pair of roots (corresponding to the negative value of y) satisfies the equation

$$2x^2 - 2x - 2\sqrt{2x^2 - 7x + 6} = 5x - 6.$$

16. On transposition, $x^2 - 2x - 11 + 6\sqrt{x^2 - 2x + 5} = 0.$

By putting $x^2 - 2x + 5 = y^2$, we have $(y^2 - 5) - 11 + 6y = 0$;

that is, $y^2 + 6y - 16 = 0$; whence $y = 2$, or -8.

Thus $$x^2 - 2x + 5 = 4, \text{ or } 64;$$

$$\therefore\ x^2 - 2x + 1 = 0, \text{ or } x^2 - 2x - 59 = 0;$$

whence $$(x - 1)^2 = 0, \text{ or } x = 1 \pm \sqrt{60}.$$

The latter pair of roots (corresponding to the negative value of y) satisfies the equation

$$x^2 - 6\sqrt{x^2 - 2x + 5} = 11 + 2x.$$

17. On transposition, $x^2 - 6x - 1 + 2\sqrt{x^2 - 6x + 2} = 0.$

By putting $x^2 - 6x + 2 = y^2$, we have $(y^2 - 2) - 1 + 2y = 0$;

that is, $$y^2 + 2y - 3 = 0;$$

whence $$y = 1, \text{ or } -3.$$

Thus $$x^2 - 6x + 2 = 1, \text{ or } 9.$$

$$\therefore\ x^2 - 6x + 1 = 0, \text{ or } x^2 - 6x - 7 = 0;$$

whence $$x = 3 \pm 2\sqrt{2}, \text{ or } x = 7, \text{ or } -1.$$

The latter pair of roots (corresponding to the negative value of y) satisfies the equation

$$-2\sqrt{x^2 - 6x + 2} + 4x + 1 = x^2 - 2x.$$

18. On transposition, $3(4x^2+2x)-119-\sqrt{4x^2+2x+7}=0$.

By puttting $4x^2+2x+7=y^2$, we have $3(y^2-7)-119-y=0$;

that is, $$3y^2-y-140=0,$$

whence $$(y-7)(3y+20)=0.$$

Thus $$4x^2+2x+7=49, \text{ or } \frac{400}{9}.$$

$$\therefore\ 2x^2+x-21=0, \text{ or } 36x^2+18x-337=0;$$

whence $$(2x+7)(x-3)=0, \text{ or } x=\frac{-3\pm\sqrt{1357}}{12}.$$

The latter pair of roots (corresponding to the negative value of y) satisfies the equation

$$-\sqrt{4x^2+2x+7}=12x^2+6x-119.$$

19. On transposition, $3(x^2-3x)+11-4\sqrt{x^2-3x+5}=0$.

By putting $x^2-3x+5=y^2$, we have $3(y^2-5)+11-4y=0$;

that is, $$3y^2-4y-4=0, \text{ or } (y-2)(3y+2)=0.$$

Thus $$x^2-3x+5=4, \text{ or } \frac{4}{9}.$$

$$\therefore\ x^2-3x+1=0, \text{ or } 9x^2-27x+41=0;$$

whence $$x=\frac{3\pm\sqrt{5}}{2}, \text{ or } \frac{9\pm\sqrt{-83}}{2}.$$

The latter pair of roots (corresponding to the negative value of y) satisfies the equation

$$3x(3-x)=11+4\sqrt{x^2-3x+5}.$$

20. Here $$2x^2-3x-14+6\sqrt{2x^2-3x+2}=0.$$

By putting $2x^2-3x+2=y^2$, we have $(y^2-2)-14+6y=0$;

that is, $$y^2+6y-16=0;$$

whence $$y=2, \text{ or } -8.$$

Thus $$2x^2-3x+2=4, \text{ or } 64.$$

$$\therefore\ 2x^2-3x-2=0, \text{ or } 2x^2-3x-62=0;$$

whence $$(2x+1)(x-2)=0, \text{ or } x=\frac{3\pm\sqrt{505}}{4}.$$

The latter pair of roots (corresponding to the negative value of y) satisfies the equation

$$x^2-x-3\sqrt{2x^2-3x+2}=\frac{x}{2}+7.$$

21. By putting $\sqrt{\frac{2-x}{3x}}=y$, we have $y-\frac{1}{y}=\frac{3}{2}$;

that is, $$2y^2-3y-2=0; \text{ or } (y-2)(2y+1)=0.$$

Thus $$\frac{2-x}{3x}=4, \text{ or } \frac{1}{4};$$

whence $$x=\frac{2}{13}, \text{ or } \frac{8}{7}.$$

The latter root (corresponding to the negative value of y) satisfies the equation $$-\sqrt{\frac{2-x}{3x}}+\sqrt{\frac{3x}{2-x}}=\frac{3}{2}.$$

22. By putting $\sqrt{\frac{x}{a}}=y$, we have $\frac{1}{y}-y=\frac{a^2-1}{a}$;

that is, $$ay^2+a^2y-y-a=0, \text{ or } (y+a)(ay-1)=0.$$

Thus $$\frac{x}{a}=a^2, \text{ or } \frac{1}{a^2};$$

whence $$x=a^3, \text{ or } \frac{1}{a}.$$

The former root satisfies the equation $$\sqrt{\frac{a}{x}}+\sqrt{\frac{x}{a}}=\frac{a^2-1}{a}$$

23. By inspection, $x=1$ satisfies the equation. Also the product of the two roots is $\frac{c-a}{a-b}$, which is therefore the value of the second root.

24. By inspection, $x=1$ is a root. Also the product of the roots is $\frac{c(a-b)}{a(b-c)}$, which is therefore the value of the other root.

25. By squaring, we obtain $$a-x+b-x+2\sqrt{(a-x)(b-x)}=a+b-2x;$$

that is, $$2\sqrt{(a-x)(b-x)}=0;$$

whence $$x=a, \text{ or } b.$$

26. By transposition, $$\frac{1}{a-x}-\frac{1}{a-c}=\frac{1}{b-c}-\frac{1}{b-x};$$

that is, $$\frac{x-c}{(a-x)(a-c)}=-\frac{x-c}{(b-x)(b-c)};$$

$$\therefore\ x-c=0, \text{ or } (a-x)(a-c)=-(b-x)(b-c);$$

whence $$x=c, \text{ or } (a-c+b-c)\,x=a(a-c)+b(b-c).$$

27. By transposition,

$$\sqrt{x-p}-\frac{q}{\sqrt{x-p}}=\frac{p}{\sqrt{x-q}}-\sqrt{x-q};$$

that is,
$$\frac{x-p-q}{\sqrt{x-p}}=-\frac{x-p-q}{\sqrt{x-q}};$$

$$\therefore\ x-p-q=0,\ \text{or}\ \sqrt{x-p}=-\sqrt{x-q}.$$

The latter equation cannot hold unless $p=q$. Thus $x=p+q$.

28. On multiplying throughout by $\sqrt{x-3}$, we have

$$(x-3)\sqrt{x-2}+5\sqrt{x-2}=\sqrt{x-3}\sqrt{x^2+6x+8};$$

that is,
$$(x+2)\sqrt{x-2}=\sqrt{x-3}\sqrt{(x+2)(x+4)}.$$

Divide by $\sqrt{x+2}$, thus removing the root $x=-2$; then

$$\sqrt{x+2}\sqrt{x-2}=\sqrt{x-3}\sqrt{x+4}.$$

By squaring,
$$x^2-4=x^2+x-12;\ \text{whence}\ x=8.$$

29. We have *identically*, $(x^2+4x-4)-(x^2+4x-10)=6$..............(1).

But from the question, $\sqrt{x^2+4x-4}-\sqrt{x^2+4x-10}=6.$

By division, $\sqrt{x^2+4x-4}+\sqrt{x^2+4x-10}=1$..............(2).

By adding (1) and (2), $2\sqrt{x^2+4x-4}=7;$

that is,
$$4x^2+16x-65=0;$$

whence
$$(2x-5)(2x+13)=0.$$

30. If $u-v=c$, then $(u-v)^3=u^3-v^3-3uv(u-v)=u^3-v^3-3uvc.$

By cubing each side of the given equation, we have therefore

$$(x-a)-(x-b)-3\sqrt[3]{x-a}\sqrt[3]{x-b}\sqrt[3]{b-a}=b-a;$$

that is,
$$-3\sqrt[3]{x-a}\sqrt[3]{x-b}\sqrt[3]{b-a}=0;$$

whence
$$x=a,\ \text{or}\ b.$$

EXAMPLES. XLII. b. PAGE 365.

1. Dividing the second equation by the first, we have $3x+2y=19$. By combination with the first equation, we have $3x=15$, and $2y=4$.

2. Multiply the second equation by 3 and add to the first;

thus
$$(x+y)^3=343;\ \text{whence}\ x+y=7.$$

But $xy(x+y)=84$; so that $xy=12$.

From $x+y=7$ and $xy=12$, we obtain $x=4$, or 3; $y=3$, or 4.

3. On multiplying the second equation by 3 and subtracting from the first, we obtain

$$(x-y)^3=125; \text{ whence } x-y=5.$$

But $xy(x-y)=70$; so that $xy=14$.

The solution is now obtained from $x-y=5$ and $xy=14$.

4. Since $x^2+xy+y^2=(x+y)^2-xy=(x+y+\sqrt{xy})(x+y-\sqrt{xy})$, we have on dividing the first equation by the second $x+y-\sqrt{xy}=6$.

By combination with the second equation, we get

$$x+y=10, \text{ and } \sqrt{xy}=4, \text{ or } xy=16.$$

5. Proceeding as in Example 4, we have $x+y+\sqrt{xy}=21$.

By combination with the second equation, we find

$$x+y=15, \text{ and } \sqrt{xy}=6, \text{ or } xy=36.$$

6. The first equation may be written $\left(\frac{3}{x}+\frac{2}{y}\right)\left(\frac{1}{x}-\frac{1}{y}\right)=\frac{2}{9}$;

therefore $$\frac{1}{x}-\frac{1}{y}=\frac{2}{9}\div\frac{4}{3}=\frac{1}{6}.$$

Solving from this equation and $\frac{3}{x}+\frac{2}{y}=\frac{4}{3}$, we find $x=3$, $y=6$.

7. We have $$\left(\frac{2}{x}+\frac{1}{y}\right)\left(\frac{1}{x}-\frac{2}{y}\right)=17.$$

But $$\frac{1}{x}-\frac{2}{y}=1;$$

therefore $$\frac{2}{x}+\frac{1}{y}=17.$$

From these equations $x=\frac{1}{7}$, and $y=\frac{1}{3}$.

8. We have $$xy(x+y)=20, \text{ and } \frac{x+y}{xy}=\frac{5}{4}.$$

By multiplication, $(x+y)^2=25$; whence $x+y=\pm 5$.

If $x+y=+5$, then $xy=+4$.

If $x+y=-5$, then $xy=-4$.

9. Let $\frac{2x+y}{x-3y}=t$; then from the second equation,

$$t-\frac{1}{t}=\frac{8}{3}, \text{ or } 3t^2-8t-3=0; \text{ that is, } (t-3)(3t+1)=0.$$

Thus $$\frac{2x+y}{x-3y}=3;\ \text{whence } x=10y;$$

or $$\frac{2x+y}{x-3y}=-\frac{1}{3};\ \text{whence } x=0.$$

If $x=10y$, from $x^2-7xy+4y^2=34$, we obtain $y=\pm 1$, $x=\pm 10$.

If $x=0$, we have $y=\pm\frac{1}{2}\sqrt{34}$.

10. By subtraction, $x^2-2xy+y^2)+(x-y)=20$;

that is, $$(x-y)^2+(x-y)-20=0;$$

whence $$(x-y+5)(x-y-4)=0.$$

If $x-y=4$, the equation $x(x-y)+x=35$ becomes $5x=35$.

If $x-y=-5$, the same equation becomes $-4x=35$.

11. Multiply the second equation by 4 and subtract from the first; thus
$$\{(x+y)^2-4xy\}-9(x-y)=-14;$$

that is, $$(x-y)^2-9(x-y)+14=0;$$

whence $$x-y=7, \text{ or } x-y=2.$$

Put $x-y=7$ in the first equation; then $(x+y)^2=9$, or $x+y=\pm 3$.

Put $x-y=2$ in the same equation; then $(x+y)^2=24$, or $x+y=\pm 2\sqrt{6}$.

12. From the second equation we have
$$(x^2-y^2)-(x-y)=0; \text{ or } (x-y)(x+y-1)=0;$$

whence $$x-y=0, \text{ or } x+y=1.$$

The first equation gives $(x-y)^2+2(x+y)=3$.

If $x-y=0$; then $x+y=\frac{3}{2}$.

If $x+y=1$; then $(x-y)^2=1$, or $x-y=\pm 1$.

13. Multiply the second equation by 3, and add to the first; thus
$$x^3+3x^2+3x+1=27(y^3+3y^2+3y+1);$$

that is, $$(x+1)^3=27(y+1)^3, \text{ or } x+1=3(y+1).$$

Substitute $x=3y+2$ in the equation $x(x+1)=9(y^3+1)$; then
$$3(3y+2)(y+1)=9(y^3+1);$$

that is, $$y+1=0, \text{ or } 3y+2=3(y^2-y+1).$$

14. (1) By proceeding as in Example 1, Art. 402, we have

$$x^2+y^2=25-2xy, \text{ and } x^3+y^3=125-15xy=5(25-3xy).$$

Thus $$(25-2xy)(25-3xy)=91;$$

that is, $$6x^2y^2-125xy+534=0, \text{ or } (xy-6)(6xy-89)=0.$$

The rational roots are obtained from $x+y=5$ and $xy=6$.

(2) As in Example 1, Art. 402, it is easy to see that

$$x^2+y^2=(x-y)^2+2xy=2(2+xy);$$

and $$x^3-y^3=(x-y)^3+3xy(x-y)=2(4+3xy).$$

Thus $$(2+xy)(4+3xy)=65;$$

that is, $$3x^2y^2+10xy-57=0; \text{ that is, } (xy-3)(3xy+19)=0.$$

The rational roots are obtained from $x-y=2$ and $xy=3$.

15. Let $\sqrt{\frac{x}{y}}=t$; then $t+\frac{1}{t}=\frac{17}{4}$;

that is, $$4t^2-17t+4=0; \text{ or } (t-4)(4t-1)=0.$$

Thus $\sqrt{\frac{x}{y}}=4$, or $\frac{1}{4}$; that is, $x=16y$, or $x=\frac{y}{16}$.

If $x=16y$, we have from the second equation

$$16\sqrt{y}+\frac{\sqrt{y}}{4}=16\tfrac{1}{4}; \text{ whence } \sqrt{y}=1.$$

Similarly, if $y=16x$, we obtain $\sqrt{x}=1$.

16. Let $\sqrt{\frac{x+y}{x-y}}=t$; then $t+\frac{1}{t}=\frac{34}{15}$;

that is, $$15t^2-34t+15=0, \text{ or } (3t-5)(5t-3)=0.$$

Thus $\sqrt{\frac{x+y}{x-y}}=\frac{5}{3}$, or $\frac{3}{5}$; that is, $\frac{x+y}{x-y}=\frac{25}{9}$, or $\frac{9}{25}$.

If $\frac{x+y}{x-y}=\frac{25}{9}$, we have from the second equation

$$\frac{25}{9}\sqrt{x-y}+\frac{3}{5}\sqrt{x-y}=\frac{152}{15};$$

whence $\sqrt{x-y}=3$, or $x-y=9$. Therefore $x+y=25$.

Similarly, if $\frac{x-y}{x+y}=\frac{25}{9}$, we obtain $x+y=9$. Therefore $x-y=25$.

17. By multiplication,

$$x^4y^4z^4=(24)^3\times 3\times 2\times 4=(24)^4;$$

whence $xyz=\pm 24$. Divide each of the given equations by this last equation.

18. By multiplication,

$$x^3y^3z^3u^3 = (30)^2 \times 4 \times (4)^2 \times 5 \times 6 = (30)^3 \times (4)^3;$$

whence $xyzu = 30 \times 4 = 120$. Divide this equation by each of the given equations.

19. By solving the given equations as simple simultaneous equations in yz, zx, xy, we obtain

$$yz = 4, \quad zx = 9, \quad xy = 16; \text{ whence } xyz = \pm 24.$$

20. Solving these as simple simultaneous equations in yz, zx, xy, we obtain

$$yz = 77, \quad zx = 55, \quad xy = 35; \text{ whence } xyz = \pm 11 \times 7 \times 5 = \pm 385.$$

21. By multiplying all the equations together, we have

$$(x+a)^2 (y-b)^2 (z-c)^2 = 36; \text{ whence } (x+a)(y-b)(z+c) = \pm 6.$$

Divide this equation by each of the given equations;

thus $$x+a = \pm 2, \quad y-b = \pm 1, \quad z+c = \pm 3.$$

22. By multiplying all the equations together, we have

$$(y+z)^2 (z+x)^2 (x+y)^2 = (21)^2 \times 3 \times 2 \times 54;$$

whence $$(y+z)(z+x)(x+y) = \pm 21 \times 6 \times 3.$$

Divide this equation by each of the given equations;

thus $$y+z = \pm 6, \quad z+x = \pm 9; \quad x+y = \pm 7.$$

23. These equations may be written

$$x(x-y-z) = 14, \quad y(x-y-z) = 6, \quad z(x-y-z) = 4.$$

Multiply them by 1, -1, -1 respectively and add; thus

$$(x-y-z)(x-y-z) = 4; \text{ whence } x-y-z = \pm 2.$$

24. Solving these equations as simple simultaneous equations in yz, zx, xy, we obtain

$$yz = -8, \quad zx = 6, \quad xy = -12; \text{ whence } xyz = \pm 24.$$

25. By adding unity to each side of the given equations, we have

$$(x+1)(y+1) = 12, \quad (y+1)(z+1) = 4, \quad (z+1)(x+1) = 3;$$

whence $$(x+1)(y+1)(z+1) = \pm 12.$$

$$\therefore x+1 = \pm 3, \quad y+1 = \pm 4, \quad z+1 = \pm 1.$$

26. From the first equation, $(x+y+z)(x+y-z) = 65$; but $x+y-z = 5$;

therefore $$x+y+z = 13.$$

From the second equation, $(x+y+z)(x-y-z) = 13$;

therefore $$x-y-z = 1.$$

Thus $$x+y+z = 13, \quad x-y-z = 1, \quad x+y-z = 5.$$

27. From the given equations, we have

$$\frac{1}{xy}+\frac{1}{yz}=9,\ \frac{1}{yz}+\frac{1}{zx}=5,\ \frac{1}{zx}+\frac{1}{xy}=8.$$

From these we obtain $\frac{1}{yz}=3$, $\frac{1}{zx}=2$, $\frac{1}{xy}=6$, and $\frac{1}{xyz}=\pm 6$.

Divide each of the first three equations of the preceding line by the last. It is obvious that $x=0$, $y=0$, $z=0$ is another solution.

28. From the last equation $2xy-2z^2=0$.

Add to the first equation; thus

$$(x+y)^2-z^2=84,\ \text{or}\ (x+y+z)(x+y-z)=84.$$

But $x+y+z=14$; therefore $x+y-z=6$; whence $x+y=10$, $z=4$. Also $xy=z^2=16$.

29. Multiply the last equation by 2, and subtract from the second; thus

$$(x-z)^2-y^2=-9,\ \text{or}\ (x-z+y)(x-z-y)=-9.$$

But $x+y-z=1$, whence $x-y-z=-9$; whence $x-z=-4$, $y=5$. Also $xz=12$.

30. Multiply the last equation by 2, and subtract from the second; thus

$$x^2-(y+z)^2=9,\ \text{or}\ (x+y+z)(x-y-z)=9.$$

But $y+z-x=9$; therefore $y+z+x=-1$; whence $y+z=4$, $x=-5$. Also $yz=3$.

31. From the last equation, $2yz-2x^2=0$.

Add to the first equation; thus

$$(y+z)^2-x^2=133,\ \text{or}\ (y+z+x)(y+z-x)=133.$$

But $y+z-x=7$; therefore $y+z+x=19$; whence $y+z=13$, $x=6$. Also $yz=x^2=36$.

32. Since $3^x=(3^2)^{y-1}=3^{2y-2}$, therefore $x=2y-2$;

and since $(2^4)^{3-x}=(2^3)^{y-2}$,

therefore $4(3-x)=3(y-2)$.

From these equations, we find

$$x=\frac{30}{11},\ \text{and}\ y=\frac{26}{11}.$$

33. From $2^{y-1}=(2^4)^{x-1}$, we have $y-1=4(x-1)$.

From $3^{\frac{1}{x}}=(3^2)^{\frac{1}{z}}$, we have $\frac{1}{x}=\frac{2}{z}$, or $z=2x$.

From $\sqrt[x]{2^{y-3}}=\sqrt[2x]{(2^3)^{z-2}}$, we have

$$\frac{y-3}{x}=\frac{3(z-2)}{2x}, \text{ or } 2(y-3)=3(z-2).$$

From this last equation, we obtain $y-3=3(x-1)$.

Thus $3(y-1)=4(y-3)$; that is, $y=9$.

34. These equations may be written

$$(x+y-z)(x-y+z)=a^2,$$
$$(-x+y+z)(x+y-z)=b^2,$$
$$(x-y+z)(-x+y+z)=c^2.$$

By multiplying all the equations together and taking the square root, we have $$(-x+y+z)(x-y+z)(x+y-z)=\pm abc.$$

Divide this by each of the given equations; thus

$$-x+y+z=\pm\frac{bc}{a}, \quad x-y+z=\pm\frac{ca}{b}, \quad x+y-z=\pm\frac{ab}{c}.$$

EXAMPLES. XLIII. Page 370.

1. Here $$M=1000\left(\frac{104}{100}\right)^{300};$$

$$\therefore \log M=\log 1000+300(\log 104-\log 100)$$
$$=3+300(\cdot 0170333)$$
$$=8\cdot 10999;$$

whence $$M=128855000.$$

2. Let n be the number of years, and P the sum; then

$$100P=P\left(\frac{1055}{1000}\right)^n;$$

$$\therefore \log 100=n(\log 1055-\log 1000);$$

that is, $$2=n(\cdot 023);$$

whence $$23n=2000, \text{ and } n=87 \text{ nearly.}$$

3. Let V denote the present value; then

$$6000=V(1\cdot 08)^{20};$$

$$\therefore \log 6000=\log V+20(2\log 2+3\log 3-2);$$

$$\therefore \log V=3+\cdot 77815-20(\cdot 03342)=3\cdot 10975;$$

whence $$V=1287\cdot 5.$$

4. Let £R denote the amount of £1 in a year; then

$$20000 = 1200R^{15}.$$

$$\therefore\ 15 \log R = \log 100 - \log 6 = 2 - \cdot77815 = 1\cdot22185;$$

whence $\log R = \cdot08145$, and $R = 1\cdot2063$.

Hence $r = \cdot2063$; that is, the interest on £1 is £·2063, and therefore the interest on £100 is £20·63.

5. We have $M = 100\,\dfrac{(1\cdot04)^{15} - 1}{\cdot04} = 2500\{(1\cdot04)^{15} - 1\}$.

Now $\log(1\cdot04)^{15} = 15 \times \cdot01703 = \cdot25545 = \log 1\cdot80075$;

whence $(1\cdot04)^{15} = 1\cdot80075$.

Thus $M = 2500 \times \cdot80075 = 2001\cdot875$.

6. Here £7000 is the present value of a perpetual annuity of £280; hence as in Art. 408, we have $7000 = \dfrac{280}{r}$; whence $7000r = 280$, and $100r = 4$; so that the rate is 4 per cent.

7. The rate of interest is $\dfrac{100}{40} = 2\frac{1}{2}$. [Art. 408.]

Hence if M be the amount; then

$$M = 300\,\frac{(1\cdot025)^{10} - 1}{\cdot025} = 12000\{(1\cdot025)^{10} - 1\}.$$

Now $\log(1\cdot025)^{10} = 10\,(\cdot01072) = \cdot1072 = \log 1\cdot280$,

$$\therefore\ M = 12000\,(\cdot280) = 3360.$$

8. If V be the present value; then

$$V = 900\,\frac{1 - (1\cdot045)^{-20}}{\cdot045} = 20000\,\{1 - (1\cdot045)^{-20}\}.$$

Now $\log(1\cdot045)^{-20} = -20\,(\cdot01912) = -\cdot3824 = \bar{1}\cdot6176 = \log \cdot41458$;

whence $(1\cdot045)^{-20} = \cdot41458$.

$$\therefore\ V = 20000\,(1 - \cdot41458) = 11708\cdot4.$$

9. Let £A be the value of each instalment; then £20000 is the present value of an annuity of £A payable for 20 years.

Hence $20000 = A\,\dfrac{1 - (1\cdot05)^{-20}}{\cdot05}$. [Art. 408.]

$$\therefore\ A\{1 - (1\cdot05)^{-20}\} = 1000.$$

Now $\log(1\cdot05)^{-20} = -20\,(\cdot0212) = -\cdot424 = \bar{1}\cdot576 = \log \cdot3767$;

whence $(1\cdot05)^{-20} = \cdot3767$.

$$\therefore\ A\,(1 - \cdot3767) = 1000;$$

that is, $A = \dfrac{1000}{\cdot6233} = 1604$ nearly.

10. Let n be the number of years; then £100000 is the present value of an annuity of £7000 payable for n years.

$$\therefore\ 100000 = 7000\,\frac{1-(1\cdot035)^{-n}}{\cdot035};$$

that is, $$7000\,\{1-(1\cdot035)^{-n}\} = 3500;$$

$$\therefore\ 1-(1\cdot035)^{-n} = \frac{1}{2};$$

whence $$(1\cdot035)^{-n} = \frac{1}{2},$$

and $$(1\cdot035)^{n} = 2.$$

$$\therefore\ n\log 1\cdot035 = \log 2.$$

But $$\log 1\cdot035 = \log 207 - \log 200 = 2\log 3 + \log 23 - 2 - \log 2 = \cdot015.$$

$$\therefore\ n = \frac{\cdot301}{\cdot015} = 20 \text{ nearly.}$$

[*For Examples in Chap. XLIV., see Appendix, p.* 379.]

MISCELLANEOUS EXAMPLES. VI. Page 371.

1. The expression $= b - \{b - a \quad b - [b - (b \quad a + b)] + 2a\}$

$$= b - \{ \quad a - [b \quad b + a - b] + 2a\}$$
$$= b - \{- a \quad a + b + 2a\}$$
$$= b - b = 0.$$

3.
$$\begin{array}{l}
\frac{1}{2}x + \frac{2}{3}y \\
\quad x - \frac{1}{3}y \\
\hline
\frac{1}{2}x^2 + \frac{2}{3}xy \\
\qquad\quad \frac{1}{6}xy - \frac{2}{9}y^2 \\
\hline
\frac{1}{2}x^2 + \frac{1}{2}xy \quad \frac{2}{9}y^2
\end{array}$$

4. The value $= \sqrt[3]{(2\times 6) + (3\times 4) + 3} = \sqrt[3]{27} = 3.$

6. We have $$\frac{x+3}{x-1} - 1 = 1 - \frac{x-4}{x-6};$$

that is, $$\frac{4}{x-1} = \frac{-2}{x-6}, \text{ or } 4x - 24 = -2x + 2.$$

7. The difference of the two expressions $=a^2-2a=a\,(a-2)$. Obviously a is not a common factor; but $a-2$ divides a^3-2a-4, and therefore also divides a^3-a^2-4.

8. The expression $=\dfrac{2a\,(a-b)+2b\,(a+b)-(a^2+b^2)}{a^2-b^2}$

$$=\frac{2a^2-2ab+2ab+2b^2-a^2-b^2}{a^2-b^2}=\frac{a^2+b^2}{a^2-b^2}.$$

9. $$12x+5y=260 \ldots\ldots\ldots\ldots (1);$$
$$8x-3y=\;72 \ldots\ldots\ldots\ldots (2).$$

Multiply (2) by 3, and (1) by 2; then by subtraction $19y=304$, or $y=16$.

$$\therefore\; 8x=72+48, \text{ or } x=15.$$

10. Let x be the digit in the tens' place and y the digit in the units' place, so that the number is $10x+y$.

Then $$10x+y+18=10y+x,$$
or $$x-y=-2 \ldots\ldots\ldots\ldots (1).$$
Again, $$10x+y+10y+x=44,$$
or $$x+y=4. \ldots\ldots\ldots\ldots (2).$$

From (1) and (2) we obtain $x=1$, $y=3$; that is, the number is 13.

11. The value $=\dfrac{(1)^2(-2)^2+(-2)^2(3)+(-4)\{1-(-2)\}}{10-(3-2)^2}$

$$=\frac{4+12-12}{10-1}=\frac{4}{9}.$$

12. The sum $$=\frac{1}{12}x^2+\frac{9}{20}y^2+\frac{2}{3}z^2$$
$$-x^2+\quad y^2-\quad z^2$$
$\therefore$ the difference $$=\frac{13}{12}x^2-\frac{11}{20}y^2+\frac{5}{3}z^2.$$

14. $$\frac{x^2+xy}{x^2+y^2}\times\frac{x^4-y^4}{xy+y^2}\times\frac{y}{x}=\frac{x\,(x+y)}{x^2+y^2}\times\frac{(x^2+y^2)\,(x^2-y^2)}{y\,(x+y)}\times\frac{y}{x}=x^2-y^2.$$

15. Clearing of fractions, we have

$$9\,(2x-7)-10\,(x-8)=4x+1+60;$$
$$18x-63-10x+80=4x+1+60;$$
whence $$x=11.$$

16.

$$\begin{array}{lll|ll}
x & x^4+x^3+2x-4 & & x^3+3x^2-4 & 1 \\
 & x^4+3x^3-4x & & x^3-3x+2 & \\
 & -2)\ -2x^3+6x-4 & & 3)\ 3x^2+3x-6 & \\
x & x^3-3x+2 & & x^2+x-2 & \\
 & x^3+x^2-2x & & & \\
-1 & -x^2-x+2 & & & \\
 & -x^2-x+2 & & &
\end{array}$$

$$\therefore \text{H. C. F.} = x^2+x-2=(x+2)(x-1).$$

Again, $x^4+x^3+2x-4=(x^2+x-2)(x^2+2)=(x-1)(x+2)(x^2+2)$,

$$x^3+3x^2-4=(x^2+x-2)(x+2)=(x-1)(x+2)^2.$$

$$\therefore \text{L. C. M.}=(x-1)(x^2+2)(x+2)^2.$$

17.

$$\begin{array}{r|l}
 & 4a^4-12a^3+21a^2-18a+9\ (\ 2a^2-3a+3 \\
 & 4a^4 \\
4a^2-3a & -12a^3+21a^2 \\
 & -12a^3+9a^2 \\
4a^2-6a+3 & 12a^2-18a+9 \\
 & 12a^2-18a+9
\end{array}$$

18. By clearing of fractions and transposing, we have

$$-3x+6y=3a+2b,$$

and

$$6x-3y=2a+3b.$$

By addition, $$3x+3y=5a+5b.$$

Combine this with each of the given equations; thus

$$9x=7a+8b, \text{ and } 9y=8a+7b.$$

19. $$\left(\frac{a}{x+a}-\frac{x}{x-a}\right)\div\frac{x^2+a^2}{x^2+ax}=\frac{a(x-a)-x(x+a)}{(x+a)(x-a)}\times\frac{x^2+ax}{x^2+a^2}$$

$$=\frac{-(a^2+x^2)}{(x+a)(x-a)}\times\frac{x(x+a)}{x^2+a^2}$$

$$=-\frac{x}{x-a}=\frac{x}{a-x}.$$

20. Let x be the numerator and y the denominator, so that the fraction is $\frac{x}{y}$;

then we have $$\frac{x+1}{y+1}=\frac{3}{2};\text{ and }\frac{x-1}{y-1}=2;$$

that is, $$2x-3y=1, \text{ and } x-2y=-1;$$

whence $$x=5 \text{ and } y=3.$$

22.

$$x-\frac{1}{4}y\Big)\,x^2-xy+\frac{3}{16}y^2\Big(\,x-\frac{3}{4}y$$

$$\frac{x^2-\frac{1}{4}xy}{\qquad -\frac{3}{4}xy+\frac{3}{16}y^2}$$

$$\frac{-\frac{3}{4}xy+\frac{3}{16}y^2}{}$$

23.

$$18\left\{\frac{2x}{9}-\frac{1}{6}\left(\frac{2y}{3}+z\right)\right\}=18\left\{\frac{2x}{9}-\frac{y}{9}-\frac{z}{6}\right\}=4x-2y-3z;$$

$$24\left(\frac{3x}{8}-\frac{2y-3z}{12}\right)=9x-2(2y-3z)=9x-4y+6z;$$

$$30\left\{\frac{7z}{15}-\frac{4}{5}(2x-y)\right\}=30\left\{\frac{7z}{15}-\frac{8x}{5}+\frac{4y}{5}\right\}=-48x+24y+14z;$$

$$\therefore \text{ the sum}=-35x+18y+17z.$$

24. (2) $729x^6-y^6=(27x^3+y^3)(27x^3-y^3)$

$$=(3x+y)(9x^2-3xy+y^2)(3x-y)(9x^2+3xy+y^2).$$

25. Clearing of fractions, $187(2x-1)+55(5x+3)=2805-85(4x-118)$;

$$374x-187+275x+165=2805-340x+10030;$$

whence $989x=12857$; and $x=13$.

26. The expression $=(-3b)(-b)-b\{-c(-b)-b^2c\}$

$$=3b^2-b^2c+b^3c=3-\frac{1}{2}-\frac{1}{2}=2.$$

27. $7x^3-10x^2-7x+10=x^2(7x-10)-(7x-10)=(7x-10)(x^2-1)$.

$$2x^3-x^2-2x+1=x^2(2x-1)-(2x-1)=(2x-1)(x^2-1).$$

$$\therefore \text{ H. C. F.}=x^2-1.$$

28. The expression $=\dfrac{(x-3y)(x-4y)}{(x+2y)(x+3y)}\times\dfrac{(x+2y)(x-y)}{(x-y)(x-4y)}$

$$=\frac{x-3y}{x+3y}.$$

29. Multiply the first equation by 3 and subtract the second, then

$$5abx=10b;$$

whence $x=\dfrac{2}{a}$; and $y=3b$.

30. Let x minutes past 7 be the required time; then the minute-hand has moved x minute divisions, and the hour hand has moved $\frac{x}{12}$ minute divisions from its place at 7 o'clock, and is therefore $\frac{x}{12}+35$ minute divisions from 12 o'clock.

$$\therefore \text{ either } \left(\frac{x}{12}+35\right)-x=15, \text{ or } x-\left(\frac{x}{12}+35\right)=15;$$

that is, $$x=\frac{240}{11}=21\tfrac{9}{11}, \text{ or } x=\frac{600}{11}=54\tfrac{6}{11}.$$

31. The value $=\sqrt{(-4)^2-4(-2)+1^2}-\sqrt{3^3+(-2)^3+1-4}$

$$=\sqrt{16+8+1}-\sqrt{27-8+1-4}$$
$$=\sqrt{25}-\sqrt{16}=5-4=1.$$

32.
$$\begin{array}{l} \frac{1}{4}x^2-\frac{1}{2}xy+y^2 \\ \frac{1}{2}x+y \\ \hline \frac{1}{8}x^3-\frac{1}{4}x^2y+\frac{1}{2}xy^2 \\ \qquad +\frac{1}{4}x^2y-\frac{1}{2}xy^2+y^3 \\ \hline \frac{1}{8}x^3 \qquad\qquad\qquad +y^3 \end{array}$$

$$\left(\frac{1}{8}x^3+y^3\right)\times(x^3-8y^3)$$
$$=\frac{x^3+8y^3}{8}\times(x^3-8y^3)$$
$$=\frac{x^6-64y^6}{8}=\frac{x^6}{8}-8y^6.$$

33. The expression

$$=a^4-\{4a^3-6a^2+4a-1\}-[-2-\{a^4-(-4a^3-6a^2+4a)\}-8a+1]$$
$$=a^4-4a^3+6a^2-4a+1-[-2-\{a^4+4a^3+6a^2-4a\}-8a+1]$$
$$=a^4-4a^3+6a^2-4a+1+2+a^4+4a^3+6a^2-4a+8a-1$$
$$=2a^4+12a^2+2.$$

34. The remainder $=5\,.\,4^4-7\,.\,4^3+3\,.\,4^2-4+8$ [Art. 230.]

$$=1280-448+48-4+8=884.$$

35.
$$\frac{x^2+y^2}{x^2-xy}\times\frac{xy-y^2}{x^4-y^4}\times\frac{x}{y}=\frac{x^2+y^2}{x(x-y)}\times\frac{y(x-y)}{(x^2+y^2)(x^2-y^2)}\times\frac{x}{y}$$
$$=\frac{1}{x^2-y^2}.$$

36. Clearing of fractions, and transposing, we obtain

$$x+3y=65, \text{ and } 8x+y=129.$$

Multiply the second equation by 3 and subtract it from the first, then

$$-23x=-322; \text{ whence } x=14;$$

and $$y=129-112=17.$$

37.

$$\begin{array}{r|l}
 & 4x^6-12x^4+28x^3+9x^2-42x+49\ (\,2x^3-3x+7 \\
 & 4x^6 \\
\hline
4x^3-3x & -12x^4+28x^3+9x^2 \\
 & -12x^4 \qquad\quad +9x^2 \\
\hline
4x^3-6x+7 & 28x^3 \qquad -42x+49 \\
 & 28x^3 \qquad -42x+49 \\
\hline
\end{array}$$

38.
$$\cdot006x-\cdot491+\cdot723x=-\cdot005.$$

$$\therefore\ (\cdot006+\cdot723)\,x=\cdot491-\cdot005,$$

$$\cdot729x=\cdot486;$$

$$x=\frac{\cdot486}{\cdot729}=\frac{2}{3}.$$

39.
$$x^3+y^3=(x+y)(x^2-xy+y^2);$$

$$3x^2+2xy-y^2=(3x-y)(x+y);$$

$$x^3-x^2y+xy^2=x(x^2-xy+y^2).$$

$$\therefore\ \text{L. C. M.}=x(x+y)(x^2-xy+y^2)(3x-y)=x(x^3+y^3)(3x-y).$$

40. Let x be the number of crowns, and y the number of half-guineas.

Now x crowns $+\,y$ half-guineas $=\left(5x+\dfrac{21}{2}y\right)$ shillings;

therefore $$5x+\frac{21}{2}y=25\times21=525,$$

or $$10x+21y=1050.$$

Again, $2y=3x+17$; or $-3x+2y=17$.

From these equations we obtain $x=21$, $y=40$.

41. $(a+b+c)^2-(a-b+c)^2=(a+b+c+a-b+c)\{a+b+c-(a-b+c)\}$

$$=2(a+c)\times2b=4ab+4bc;$$

and $(a+b-c)^2-(-a+b+c)^2=(a+b-c-a+b+c)\{a+b-c-(-a+b+c)\}$

$$=2b\times2(a-c)=4ab-4bc;$$

$$\therefore\ \text{the sum}=8ab.$$

42.
$$
\begin{array}{l}
a^2-ab+2b^2\,)\,a^4-3a^3b+2a^2b^2-\;b^4\,(\,a^2-2ab-2b^2\\
a^4-a^3b+2a^2b^2\\
-2a^3b-\;b^4\\
-2a^3b+2a^2b^2-4ab^3\\
-2a^2b^2+4ab^3-b^4\\
-2a^2b^2+2ab^3-4b^4\\
2ab^3+3b^4
\end{array}
$$

43. By erasing every term which contains a,

the expression $= -2bcd\sqrt[3]{-c^3bd+3} = 12\sqrt[3]{24+3} = 36.$

44. The factors of $4x^2+3x-10$ are $x+2$ and $4x-5$.

Evidently $x+2$ is not a factor of $4x^3+7x^2-3x-15$, but by trial we see $4x-5$ divides it without remainder.

45. The expression $= \dfrac{\dfrac{a^2}{a-b}}{\dfrac{a^4-a^2b^2}{a^2+b^2}} \times \left(\dfrac{1}{a}+\dfrac{1}{b}\right)$

$$= \frac{a^2}{a-b} \times \frac{a^2+b^2}{a^2(a^2-b^2)} \times \frac{a+b}{ab} = \frac{a^2+b^2}{ab(a-b)^2}.$$

46.
$$
\begin{array}{rrl}
 & & 8x^3-2x^2y+\dfrac{xy^2}{6}-\dfrac{y^3}{216}\left(2x-\dfrac{y}{6}\right.\\
 & & 8x^3\\
3\times(2x)^2=12x^2 & & -2x^2y+\dfrac{xy^2}{6}-\dfrac{y^3}{216}\\
3\times 2x\times\left(-\dfrac{y}{6}\right)= & -xy & \\
\left(\dfrac{y}{6}\right)^2= & +\dfrac{y^2}{36} & \\
 & 12x^2-xy+\dfrac{y^2}{36} & -2x^2y+\dfrac{xy^2}{6}-\dfrac{y^3}{216}
\end{array}
$$

47. We have $\dfrac{8}{x}+\dfrac{9}{y}=43$, and $\dfrac{9}{x}+\dfrac{8}{y}=42$.

By addition and subtraction respectively we obtain

$$\frac{1}{x}+\frac{1}{y}=5,\text{ and }\frac{1}{x}-\frac{1}{y}=-1;\text{ whence }\frac{1}{x}=2,\text{ and }\frac{1}{y}=3.$$

The given equations are obviously satisfied by $x=0$ and $y=0$.

48. The expression $= \dfrac{3x-15-2(x-4)}{(x-4)(x-5)} - \dfrac{x-7}{(x-2)(x-3)}$

$$= \frac{x-7}{(x-4)(x-5)} - \frac{x-7}{(x-2)(x-3)}$$

$$= (x-7)\left\{\frac{1}{(x-4)(x-5)} - \frac{1}{(x-2)(x-3)}\right\}$$

$$= (x-7) \times \frac{x^2-5x+6-(x^2-9x+20)}{(x-2)(x-3)(x-4)(x-5)}$$

$$= \frac{2(x-7)(2x-7)}{(x-2)(x-3)(x-4)(x-5)}.$$

49. The difference between the expressions is

$$x(2x^2+51x+72) = x(2x+3)(x+24).$$

If there is a common factor it can only be $2x+3$. By division, we get

$$8x^3+38x^2+59x+30 = (2x+3)(4x+5)(x+2);$$

$$6x^3-13x^2-13x+30 = (2x+3)(3x-5)(x-2).$$

$$\therefore \text{L. C. M.} = (2x+3)(4x+5)(3x-5)(x+2)(x-2).$$

50. Suppose that to commence with he had x shillings. In the first shop he spent $\dfrac{x}{2}$ shillings; in the second shop $\dfrac{x}{6}$ shillings; and in the third he spent $\dfrac{1}{5} \times \dfrac{x}{3}$ shillings.

$$\therefore \frac{x}{2} + \frac{x}{6} + \frac{x}{15} + 1 = x; \text{ whence } x = 3\tfrac{3}{4}.$$

51.

$$
\begin{array}{l}
x^2-5x+4\,)\,x^7-10x^6+8x^5-7x^3+3x-11\,(\,x^5-5x^4-21x^3-85x^2-348x-1400 \\
\qquad x^7-5x^6+4x^5 \\
\qquad\quad -5x^6+4x^5 \\
\qquad\quad -5x^6+25x^5-20x^4 \\
\qquad\qquad -21x^5+20x^4-7x^3 \\
\qquad\qquad -21x^5+105x^4-84x^3 \\
\qquad\qquad\quad -85x^4+77x^3 \\
\qquad\qquad\quad -85x^4+425x^3-340x^2 \\
\qquad\qquad\qquad -348x^3+340x^2+3x \\
\qquad\qquad\qquad -348x^3+1740x^2-1392x \\
\qquad\qquad\qquad\quad -1400x^2+1395x-11 \\
\qquad\qquad\qquad\quad -1400x^2+7000x-5600 \\
\qquad\qquad\qquad\qquad -5605x+5589
\end{array}
$$

52. The expression $= \left\{2a-3\left(b-\dfrac{4c}{3}\right)\right\}\{(2a-b)+4(b-c)\}$

$$= (2a-3b+4c)(2a+3b-4c)$$

$$= (2a)^2-(3b-4c)^2.$$

53. $$(a-\sqrt{b})(\sqrt{a}+b)\sqrt{a-b}=\left(\frac{25}{16}-1\right)\left(\sqrt{\frac{25}{16}}+1\right)\sqrt{\frac{25}{16}-1}$$
$$=\frac{9}{16}\cdot\frac{9}{4}\cdot\frac{3}{4}=\frac{243}{256}.$$
$$\frac{3c^4}{\sqrt{a-c^2}}=\frac{3\left(\frac{3}{4}\right)^4}{\sqrt{\frac{25}{16}-\frac{9}{16}}}=\frac{\frac{3^5}{4^4}}{\sqrt{1}}=\frac{243}{256}.$$

54. $$x^2-7x+12=(x-3)(x-4);$$
$$3x^2-6x-9=3(x^2-2x-3)=3(x-3)(x+1);$$
$$2x^2-6x-8=2(x^2-3x-4)=2(x-4)(x+1);$$
$$\therefore \text{ L. C. M.}=6(x-3)(x-4)(x+1).$$

55. The sum $=(a^2x^2+2abxy+b^2y^2)+(b^2x^2-2abxy+a^2y^2)$
$$+(a^2y^2+2abxy+b^2x^2)+(b^2y^2-2abxy+a^2x^2)$$
$$=2a^2x^2+2b^2x^2+2b^2y^2+2a^2x^2=2(a^2+b^2)(x^2+y^2).$$

56. By clearing of fractions, we have
$$2x+3y=12,\quad 3x-5z=4,\quad 7y+2z=16.$$
By eliminating x from the first two equations, we find $9y+10z=28$. Multiply the third equation by 5 and subtract this last equation from the product; thus $26y=52$; whence $y=2$.

57. The expression $=\dfrac{a^2-ab+b^2}{(a-b)(a^2+b^2)}-\dfrac{1}{a-b}-\dfrac{a-b}{2(a^2+b^2)}+\dfrac{1}{2(a-b)}$
$$=\frac{2a^2-2ab+2b^2-2a^2-2b^2-a^2+2ab-b^2+a^2+b^2}{2(a-b)(a^2+b^2)}=0.$$

58. Multiply by 30 to clear of fractions; thus we have
$$30x-90x+6x+15=10x+335+50+10x;$$
whence $$-74x=370;\ \text{and}\ x=-5.$$

59. The sum $=\dfrac{2}{x^2+xy+y^2}-\dfrac{4x}{x^3-y^3}+\dfrac{x^2}{y^2(x-y)^2}-\dfrac{x^2}{y(x^3-y^3)}$
$$=\frac{2y^2(x-y)^2-4xy^2(x-y)+x^2(x^2+xy+y^2)-x^2y(x-y)}{y^2(x^2+xy+y^2)(x-y)^2}$$
$$=\frac{2x^2y^2-4xy^3+2y^4-4x^2y^2+4xy^3+x^4+x^3y+x^2y^2-x^3y+x^2y^2}{y^2(x^2+xy+y^2)(x-y)^2}$$
$$=\frac{x^4+2y^4}{y^2(x^2+xy+y^2)(x-y)^2}.$$

60. Suppose that he worked for x days and thus obtained $\frac{10}{3}x$ shillings. He was absent $30-x$ days, and therefore forfeited $\frac{3}{2}(30-x)$ shillings.

Hence
$$\frac{10}{3}x-\frac{3}{2}(30-x)=71;$$

from which we find
$$x=24.$$

61.
$$\begin{array}{l}
\frac{x^2}{2}-x+3\Big)\ \frac{3x^5}{4}-4x^4+\frac{77x^3}{8}-\frac{43x^2}{4}-\frac{33x}{4}+27\ \Big(\frac{3x^3}{2}-5x^2+\frac{x}{4}+9 \\
\qquad\qquad\quad \frac{3x^5}{4}-\frac{3x^4}{2}+\frac{9x^3}{2} \\
\qquad\qquad\qquad -\frac{5x^4}{2}+\frac{41x^3}{8}-\frac{43x^2}{4} \\
\qquad\qquad\qquad -\frac{5x^4}{2}+5x^3-15x^2 \\
\qquad\qquad\qquad\qquad \frac{x^3}{8}+\frac{17x^2}{4}-\frac{33x}{4} \\
\qquad\qquad\qquad\qquad \frac{x^3}{8}-\frac{x^2}{4}+\frac{3x}{4} \\
\qquad\qquad\qquad\qquad\qquad \frac{9x^2}{2}-9x+27 \\
\qquad\qquad\qquad\qquad\qquad \frac{9x^2}{2}-9x+27
\end{array}$$

62. The expression $=\frac{4y}{5}(y-x)-7(3x-4y)+\frac{7}{2}\left\{3x-5x+\frac{20y}{7}\right\}$

$$=\frac{8}{5}\left(\frac{5}{2}\right)-7\left(-\frac{19}{2}\right)+\frac{7}{2}\left\{1+\frac{40}{7}\right\}$$

$$=4+\frac{133}{2}+\frac{47}{2}=94.$$

63.
$$\frac{10x-11}{3(x+1)(x-1)}-\frac{10x-1}{3(x^2+x+1)}+\frac{x^2-2x+5}{(x-1)(x^2+x+1)(x+1)}$$

$$=\frac{(10x-11)(x^2+x+1)-(10x-1)(x^2-1)+3(x^2-2x+5)}{3(x+1)(x-1)(x^2+x+1)}$$

$$=\frac{10x^3-x^2-x-11-10x^3+x^2+10x-1+3x^2-6x+15}{3(x^2-1)(x^2+x+1)}$$

$$=\frac{3(x^2+x+1)}{3(x^2-1)(x^2+x+1)}=\frac{1}{x^2-1}.$$

64.

$$\frac{a^3c^3}{b^3}x^6 - \frac{3a^2c}{b}x^5 + \frac{3ab}{c}x^4 - \frac{b^3}{c^3}x^3 \left(\frac{ac}{b}x^2 - \frac{b}{c}x \right.$$

$$\frac{a^3c^3}{b^3}x^6$$

$$- \frac{3a^2c}{b}x^5 + \frac{3ab}{c}x^4 - \frac{b^3}{c^3}x^3$$

$$3\left(\frac{ac}{b}x^2\right)^2 = \frac{3a^2c^2}{b^2}x^4$$

$$3\left(\frac{ac}{b}x^2\right)\left(-\frac{b}{c}x\right) = \qquad -3ax^3$$

$$\left(-\frac{b}{c}x\right)^2 = \qquad +\frac{b^2}{c^2}x^2$$

$$\frac{3a^2c^2}{b^2}x^4 - 3ax^3 + \frac{b^2}{c^2}x^2 \quad \Big| \quad -3\frac{a^2c}{b}x^5 + \frac{3ab}{c}x^4 - \frac{b^3}{c^3}x^3$$

65. The equation may be written

$$\left(4 - \frac{1}{x-4}\right) + \left(5 + \frac{2}{2x-3}\right) = \left(4 - \frac{2}{2x-7}\right) + \left(5 + \frac{1}{x-1}\right);$$

that is, $$\frac{2}{2x-3} - \frac{1}{x-4} = \frac{1}{x-1} - \frac{2}{2x-7};$$

$$\therefore \frac{-5}{2x^2-11x+12} = \frac{-5}{2x^2-9x+7}$$

whence $$2x^2-11x+12 = 2x^2-9x+7; \text{ and } x = \frac{5}{2}.$$

67. By clearing of fractions, the equations become

$$x+y+6z=63, \;\; 6x-y-z=130, \;\; 3x+y-z=76.$$

Add the second equation to the first; thus $7x+5z=193$.

Subtract the third equation from the first; thus $-2x+7z=-13$.

From these equations, we obtain $x=24, \; z=5$.

68. The expression $= \dfrac{7(x+2y)}{2x-7y} - \dfrac{3x^2+63xy+70y^2}{(2x-7y)(x+5y)}$

$$= \frac{7x^2+49xy+70y^2-3x^2-63xy-70y^2}{(2x-7y)(x+5y)}$$

$$= \frac{2x(2x-7y)}{(2x-7y)(x+5y)} = \frac{2x}{x+5y}.$$

69. The expression $=(3b-2c-2a)^3\{3b-2(a+c)\}=(3b-2c-2a)^4$.

Thus the square root is $(3b-2c-2a)^2$.

70. Let x be the number of children, and y the sum of their ages; then the united ages of the parents amount to $6y$.

Two years ago the aggregate age of the children is $y-2x$, and that of the parents is $6y-4$.

Six years hence the sum of the ages of the children is $y+6x$, and that of the parents is $6y+12$.

Hence $\qquad 6y-4=10(y-2x)$, and $6y+12=3(y+6x)$;

that is, $\qquad 5x-y=1$, and $6x-y=4$;

whence $x=3$, and $y=14$.

72. The first expression

$$=a^2+ab-ax-bx-ab+b^2+ax-bx=a^2+b^2-2bx.$$

$\therefore$ the difference $=a^2+b^2-2bx-(a^2+2ab+b^2-2bx)=-2ab$.

73. The value $=\sqrt[3]{6.5.4.3+(4+3)^3+(3+5)^3+(5+4)^3-(5+4+3)^3}$

$$=\sqrt[3]{360+343+512+729-1728}=\sqrt[3]{216}=6.$$

75. By putting $y=\dfrac{p}{q}x$, we have $(p+q)x-\dfrac{p}{q}(q-p)x=r$,

that is, $\qquad x\{q(p+q)-p(q-p)\}=r$; or $x(p^2+q^2)=r$.

76. The expression $=\dfrac{2x+y}{4x^2+2xy+y^2}-\dfrac{2x^2-y^2}{8x^3-y^3}$

$$=\frac{(2x+y)(2x-y)-2x^2+y^2}{8x^3-y^3}=\frac{2x^2}{8x^3-y^3}.$$

77. By simplification of the left side, we have

$$\frac{2x-13}{2(x+7)}=\frac{2x-15}{2x-6}, \text{ or } \frac{2x-13}{x+7}=\frac{2x-15}{x-3};$$

that is, $2-\dfrac{27}{x+7}=2-\dfrac{9}{x-3}$; whence $\dfrac{3}{x+7}=\dfrac{1}{x-3}$, and $x=8$.

78. $\dfrac{x^4-x^2-2x+2}{2x^3-x-1}=\dfrac{(x-1)(x^3+x^2-2)}{(x-1)(2x^2+2x+1)}=\dfrac{x^3+x^2-2}{2x^2+2x+1}.$

79. The sum $=\dfrac{1}{2(x-1)^2}+\dfrac{1}{2(x+1)^2}-\dfrac{1}{x^2-1}$

$$=\frac{(x+1)^2+(x-1)^2-2(x^2-1)}{2(x-1)^2(x+1)^2}$$

$$=\frac{x^2+2x+1+x^2-2x+1-2x^2+2}{2(x^2-1)^2}=\frac{4}{2(x^2-1)^2}=\frac{2}{(x^2-1)^2}.$$

80. Let x be the digit in the hundreds' place, y the digit in the tens' place; then the number is $100x+10y$.

On the first supposition, $100y+10x=100x+10y-180$, or $x-y=2$.

On the second supposition, $50x+y=100x+10y-336$, or $50x+9y-336$.

From these equations we find $x=6$, $y=4$.

81.
$$1-x-\frac{14}{15}x^2\Big)\,1-5x+\frac{152}{15}x^3-\frac{106}{225}x^4-\frac{28}{9}x^5\,\Big(\,1-4x-\frac{46}{15}x^2+\frac{10}{3}x^3$$
$$1-x-\frac{14}{15}x^2$$
$$-4x+\frac{14}{15}x^2+\frac{152}{15}x^3$$
$$-4x+4x^2+\frac{56}{15}x^3$$
$$-\frac{46}{15}x^2+\frac{96}{15}x^3-\frac{106}{225}x^4$$
$$-\frac{46}{15}x^2+\frac{46}{15}x^3+\frac{644}{225}x^4$$
$$\frac{10}{3}x^3-\frac{10}{3}x^4-\frac{28}{9}x^5$$
$$\frac{10}{3}x^3-\frac{10}{3}x^4-\frac{28}{9}x^5$$

82. The expression $=\dfrac{(p^2+q^2)-(p-q)(p+q)}{2p+q-p+q-p}$
$$=\frac{p^2+q^2-p^2+q^2}{2q}=\frac{2q^2}{2q}=q=\frac{1}{2}.$$

83.
$$\frac{3x^3}{2}-5x^2+\frac{x}{4}+9$$
$$\frac{x^2}{2}-x+3$$
$$\frac{3x^5}{4}-\frac{5x^4}{2}+\frac{x^3}{8}+\frac{9x^2}{2}$$
$$-\frac{3x^4}{2}+5x^3-\frac{x^2}{4}-9x$$
$$+\frac{9x^3}{2}-15x^2+\frac{3x}{4}+27$$
$$\frac{3x^5}{4}-4x^4+\frac{77x^3}{8}-\frac{43x^2}{4}-\frac{33x}{4}+27$$

84.

$$(a^2b - 2ab^2)^2 = a^2b^2\,(a - 2b)^2;$$

$$2a^2 - 3ab - 2b^2 = (2a + b)\,(a - 2b);$$

$$2\,(2a^2 + ab)^2 = 2a^2\,(2a + b)^2.$$

$$\therefore \text{ L. C. M.} = 2a^2b^2\,(2a + b)^2\,(a - 2b)^2.$$

85. The equation may be written

$$2 + \frac{1}{x+1} = 1 + \frac{1}{4x+4} + 1 + \frac{2}{3x+1};$$

$$\therefore \frac{1}{x+1} - \frac{1}{4\,(x+1)} = \frac{2}{3x+1};$$

that is, $$\frac{3}{4\,(x+1)} = \frac{2}{3x+1}; \text{ whence } x = 5.$$

86.

$$\begin{array}{r|rrr|rrrr|l}
 & 5x^3 - & 14x^2 + & 16 & 3x^3 - & 2x^2 + 16x - & & 48 & \\
 & 2 & & & 5 & & & & \\ \hline
5x & 10x^3 - & 28x^2 + & 32 & 15x^3 - & 10x^2 + 80x - & & 240 & 3 \\
 & 10x^3 + & 25x^2 - & 90x & 15x^3 - & 42x^2 & + & 48 & \\ \hline
 & - & 53x^2 + & 90x + 32 & 16\,) & 32x^2 + 80x - & & 288 & \\
 & - & 2 & & & 2x^2 + 5x - & & 18 & 2x \\
53 & & 106x^2 - & 180x - 64 & & 2x^2 - 4x & & & \\
 & & 106x^2 + & 265x - 954 & & 9x - & & 18 & 9 \\
 & & -445\,) & -445x + 890 & & 9x - & & 18 & \\
 & & & x - 2 & & & & &
\end{array}$$

$$\therefore \text{ H. C. F.} = x - 2.$$

And $$\frac{5x^3 - 14x^2 + 16}{3x^3 - 2x^2 + 16x - 48} = \frac{(x-2)\,(5x^2 - 4x - 8)}{(x-2)\,(3x^2 + 4x + 24)} = \frac{5x^2 - 4x - 8}{3x^2 + 4x + 24}.$$

87.

$$\begin{array}{r|l}
 & 4a^4 + 12a^3 + 9a^2 + 12a + 18 + \dfrac{9}{a^2}\Big(\,2a^2 + 3a + \dfrac{3}{a} \\
 & 4a^4 \\ \hline
4a^2 + 3a & 12a^3 + 9a^2 \\
 & 12a^3 + 9a^2 \\ \hline
4a^2 + 6a + \dfrac{3}{a} & 12a + 18 + \dfrac{9}{a^2} \\
 & 12a + 18 + \dfrac{9}{a^2} \\ \hline
\end{array}$$

88. Multiply the first equation by 6; thus $\dfrac{3x}{a} + \dfrac{2y}{b} = 6a + 6b$.

From this on adding the second equation, we have $\dfrac{6x}{a} = 12b$; whence $x = 2ab$; and therefore $y = 3ab$.

89. The product $=\left(3x+4y+\dfrac{22xy}{2x-3y}\right)\left(10x-3y-\dfrac{44xy}{x+4y}\right)$

$$=\frac{(6x^2+21xy-12y^2)(10x^2-7xy-12y^2)}{(2x-3y)(x+4y)}$$

$$=\frac{3(x+4y)(2x-y)(5x+4y)(2x-3y)}{(2x-3y)(x+4y)}$$

$$=3(2x-y)(5x+4y).$$

90. Let x and y represent the number of shillings and half-crowns respectively; then x shillings plus y half-crowns are equivalent to 100 shillings; that is, $$x+\frac{5}{2}y=100.$$

Also $$3(x-17)=y-6, \text{ or } 3x-y=45.$$

From these equations, we have $x=25$, $y=30$.

91. The value $=5a-5b-2\{3a-a-b\}+7\{a-2b-5a+2b\}$

$$=5a-5b-2(2a-b)+7(-4a)$$

$$=5a-5b-4a+2b-28a=-27a-3b=\left(27\times\frac{b}{9}\right)-3b=0.$$

92.

$$\begin{array}{r|l}
3x-2 & 3x^4-5x^3+7x^2-11x-13\ \left(x^3-x^2+\frac{5}{3}x-\frac{23}{9}\right. \\
 & \underline{3x^4-2x^3} \\
 & \quad -3x^3+7x^2 \\
 & \quad \underline{-3x^3+2x^2} \\
 & \qquad\quad 5x^2-11x \\
 & \qquad\quad \underline{5x^2-\frac{10}{3}x} \\
 & \qquad\qquad -\frac{23}{3}x-13 \\
 & \qquad\qquad \underline{-\frac{23}{3}x+\frac{46}{9}} \\
 & \qquad\qquad\qquad -\frac{163}{9}
\end{array}$$

93. $15(p^3+q^3)=15(p+q)(p^2-pq+q^2)$, and $6(p^2-q^2)=6(p+q)(p-q)$;

∴ L. C. M. of the four given expressions

$$=60(p+q)(p^2-pq+q^2)(p-q)(p^2+pq+q^2)$$

$$=60(p^3-q^3)(p^3+q^3)=60(p^6-q^6).$$

94. (2) $x^4 - x^2 + 2x - 1 = (x^2)^2 - (x-1)^2 = (x^2 + x - 1)(x^2 - x + 1)$.

95. By multiplying across, $x^2 + 2ax + bx + a^2 + ab = x^2 + 3ax + bx + 3ab$;

$$\therefore\ ax = a^2 - 2ab;\ \text{whence}\ x = a - 2b.$$

96. (1) $$\frac{35a^2b^2c^2 - 49b^3c^3}{65a^5bc - 91a^3b^2c^2} = \frac{7b^2c^2(5a^2 - 7bc)}{13a^3bc(5a^2 - 7bc)} = \frac{7bc}{13a^3}.$$

(2) $$\frac{y^4 - 7y^3 + 8y^2 - 12y}{2y^2 - 2y - 60} = \frac{y(y^3 - 7y^2 + 8y - 12)}{2(y-6)(y+5)}$$

$$= \frac{y(y^2 - y + 2)(y-6)}{2(y-6)(y+5)} = \frac{y(y^2 - y + 2)}{2(y+5)}.$$

97. The equations are equivalent to

$$7x - 9y + 4z = 16,\ x + y + 3z = 0,\ 2x - 3y + 4z = 5.$$

Multiply the third equation by 3 and subtract from the first; thus

$$x - 8z = 1.$$

Multiply the second equation by 3 and add to the third; thus

$$5x + 13z = 5.$$

From these equations we obtain $x = 1$, and $z = 0$.

98. The expression

$$= \frac{y^3 - y^2 - 2y}{y - 1} \times \frac{y+1}{y^3 + y^2 - 2y} \times \left(\frac{y^2 - 6y + 5}{y^2 - 5y - 6} \times \frac{y+2}{y-2}\right)$$

$$= \frac{y(y+1)(y-2)}{y-1} \times \frac{y+1}{y(y+2)(y-1)} \times \frac{(y-5)(y-1)}{(y-6)(y+1)} \times \frac{y+2}{y-2}$$

$$= \frac{(y+1)(y-5)}{(y-1)(y-6)}.$$

99.

$$\begin{array}{r|l}
 & 4a^2 - 12ab + 4ac + 9b^2 - 6bc + c^2\ (\ 2a - 3b + c \\
 & 4a^2 \\
\hline
4a - 3b & -12ab + 4ac + 9b^2 \\
 & -12ab + 9b^2 \\
\hline
4a - 6b + c & 4ac - 6bc + c^2 \\
 & 4ac - 6bc + c^2 \\
\hline
\end{array}$$

$\therefore$ the square root $$= \frac{2a - 3b + c}{\sqrt{4a^2 - 12ac + 9c^2}} = \frac{2a - 3b + c}{2a - 3c}.$$

100. Suppose that the trains meet at x minutes past three. The express takes 180 minutes over the journey, and the ordinary train 270 minutes.

$\therefore$ at x minutes after 3 the express train will have gone $\frac{x}{180}$ of the journey;

....ordinary train .... $\frac{90+x}{270}$ ....

$$\therefore \frac{x}{180}+\frac{90+x}{270}=1; \text{ whence } x=72.$$

Thus they meet at 72 minutes past 3, or at 12 minutes past 4.

101. (1) By converting the decimals into fractions, we have

$$\frac{6}{9}x+\frac{3}{4}x-\frac{15}{90}=x-\frac{525}{900}x+5;$$

that is,
$$\frac{2}{3}x+\frac{3}{4}x-\frac{1}{6}=x-\frac{7}{12}x+5;$$

whence
$$8x+9x-2=12x-7x+60; \text{ and } 6x=31.$$

(2) The equation may be written

$$\frac{37}{(x-2)(x-3)}+\frac{4}{x-2}=-\frac{7}{x-3};$$

whence
$$37+4(x-3)=-7(x-2); \text{ and } 11x=-11.$$

102. (1) The sum $=\dfrac{(a+x)(a^2-ax+x^2)+(a-x)(a^2+ax+x^2)+2x^3}{(a^2+ax+x^2)(a^2-ax+x^2)}$

$$=\frac{a^3+x^3+a^3-x^3+2x^3}{(a^2+ax+x^2)(a^2-ax+x^2)}$$

$$=\frac{2(a^3+x^3)}{(a^2+ax+x^2)(a^2-ax+x^2)}$$

$$=\frac{2(a+x)}{a^2+ax+x^2}.$$

(2) The divisor $=1+\dfrac{x(1+x+x^2)}{(1-x)(1+x+x^2)+x}=1+\dfrac{x(1+x+x^2)}{1-x^3+x}$

$$=\frac{1+2x+x^2}{1+x-x^3}.$$

$\therefore$ the quotient $=(1+x)^2\times\dfrac{1+x-x^3}{1+2x+x^2}=1+x-x^3.$

103.

$$a^6-6a^4+15a^2-20+\frac{15}{a^2}-\frac{6}{a^4}+\frac{1}{a^6}\left(a^3-3a+\frac{3}{a}-\frac{1}{a^3}\right.$$

$$\begin{array}{r|l}
 & a^6 \\ \hline
2a^3-3a & -6a^4+15a^2 \\
 & -6a^4+\ \ 9a^2 \\ \hline
2a^3-6a+\frac{3}{a} & 6a^2-20+\frac{15}{a^2} \\
 & 6a^2-18+\frac{9}{a^2} \\ \hline
2a^3-6a+\frac{6}{a}-\frac{1}{a^3} & -\ 2+\frac{6}{a^2}-\frac{6}{a^4}+\frac{1}{a^6} \\
 & -\ 2+\frac{6}{a^2}-\frac{6}{a^4}+\frac{1}{a^6} \\ \hline
\end{array}$$

$$a^3-3a+\frac{3}{a}-\frac{1}{a^3}\left(a-\frac{1}{a}\right.$$

$$\begin{array}{lr|l}
 & & a^3 \\ \hline
3a^2 & =3a^2 & -3a+\frac{3}{a}-\frac{1}{a^3} \\
3a\times\left(-\frac{1}{a}\right)= & -3 & \\
\left(-\frac{1}{a}\right)^2= & +\frac{1}{a^2} & \\ \hline
 & 3a^2-3+\frac{1}{a^2} & -3a+\frac{3}{a}-\frac{1}{a^3} \\ \hline
\end{array}$$

104.

$$\begin{array}{l}
1+3x\,)\,1-2x\,(\,1-5x+15x^2-45x^3 \\
\qquad\quad 1+3x \\ \hline
\qquad\quad -5x \\
\qquad\quad -5x-15x^2 \\ \hline
\qquad\qquad\quad 15x^2 \\
\qquad\qquad\quad 15x^2+45x^3 \\ \hline
\qquad\qquad\qquad\quad -45x^3 \\
\qquad\qquad\qquad\quad -45x^3-135x^4 \\ \hline
\qquad\qquad\qquad\qquad\qquad 135x^4
\end{array}$$

105. Suppose that the horse cost £x; then the carriage cost £$(75-x)$

The gain on the horse is £$\frac{x}{20}$; that on the carriage is £$\frac{1}{5}(75-x)$; and the whole gain is £$\left(\frac{16}{100}\times 75\right)$, or £12.

Hence $\frac{x}{20}+\frac{1}{5}(75-x)=12$; and therefore $x=20$.

106. The divisor

$$=\frac{(4a^2+7ab+5b^2)^2-b^2(9a+11b)^2}{8(a+2b)^2}$$

$$=\frac{\{(4a^2+7ab+5b^2)+b(9a+11b)\}\{(4a^2+7ab+5b^2)-b(9a+11b)\}}{8(a+2b)^2}$$

$$=\frac{(4a^2+16ab+16b^2)(4a^2-2ab-6b^2)}{8(a+2b)^2}$$

$$=\frac{8(a+2b)^2(2a^2-ab-3b^2)}{8(a+2b)^2}=2a^2-ab-3b^2.$$

107. (1) By transposition, $5x^2-17x+14=0$, or $(5x-7)(x-2)=0$.

(2) On clearing of fractions, we have

$$1+6(x^2-3x+2)=3(x-1)+2(x-2);$$

that is, $$6x^2-23x+20=0, \text{ or } (3x-4)(2x-5)=0.$$

108. Here $x-a=b+\dfrac{(a-b)^2}{4(a+b)}=\dfrac{4ab+4b^2+a^2-2ab+b^2}{4(a+b)}=\dfrac{(a+b)^2+4b^2}{4(a+b)}$;

and $$y-b=\frac{a+b}{4}+\frac{ab}{a+b}-b=\frac{(a+b)^2+4ab-(4ab+4b^2)}{4(a+b)}=\frac{(a+b)^2-4b^2}{4(a+b)}.$$

Now $$(x-a)^2-(y-b)^2=\{(x-a)+(y-b)\}\{(x-a)-(y-b)\}$$

$$=\frac{2(a+b)^2}{4(a+b)}\times\frac{8b^2}{4(a+b)}=b^2.$$

109.

$$\begin{array}{r|l}
 & 49x^4-\frac{14x^3}{5}+\frac{1051x^2}{25}-\frac{6x}{5}+9\ \big(\ 7x^2-\frac{x}{5}+3 \\
 & 49x^4 \\ \hline
14x^2-\frac{x}{5} & -\frac{14x^3}{5}+\frac{1051x^2}{25} \\
 & -\frac{14x^3}{5}+\frac{x^2}{25} \\ \hline
14x^2-\frac{2x}{5}+3 & 42x^2-\frac{6x}{5}+9 \\
 & 42x^2-\frac{6x}{5}+9 \\ \hline
\end{array}$$

110. On clearing of fractions, we have

$$x(a+x)(a^2-ax+x^2)+x(a-x)(a^2+ax+x^2)=3a;$$

that is, $$x(a^3+x^3)+x(a^3-x^3)=3a, \text{ or } 2a^3x=3a.$$

111. The difference $=\dfrac{x+4}{(x-4)(x+3)}-\dfrac{x+3}{(x+4)(x-3)}$

$$=\frac{(x^3+5x^2-8x-48)-(x^3+2x^2-15x-36)}{(x-4)(x+3)(x+4)(x-3)}$$

$$=\frac{3x^2+7x-12}{(x-4)(x+3)(x+4)(x-3)}.$$

The quotient $=\dfrac{3x^2+7x-12}{(x-4)(x+3)(x+4)(x-3)}\div\dfrac{3x^2+7x-12}{(x+4)(x+3)}=\dfrac{1}{(x-4)(x-3)}$.

112. $2x^2+(6a-10b)x-30ab=2\{x^2+(3a-5b)x-15ab\}$

$$=2(x-5b)(x+3a),$$

and $3x^2-(9a+15b)x+45ab=3\{x^2-(3a+5b)x+15ab\}$

$$=3(x-3a)(x-5b);$$

$\therefore$ H. C. F. $=x-5b$; and L. C. M. $=6(x+3a)(x-3a)(x-5b)$.

113. (1) $2cx^2-4cdx-abx+2abd=0$;

that is, $2cx(x-2d)-ab(x-2d)=0$, or $(2cx-ab)(x-2d)=0$.

(2) On clearing of fractions, we have

$12x(x-3)-53(x^2-9)=24x^2-6(8x-1)(x+3)$;

$12x^2-36x-53x^2+477=24x^2-48x^2-138x+18$;

that is, $17x^2-102x-459=0$, or $x^2-6x-27=0$.

114. The value $=\dfrac{1^2+2^3+3^4}{2^1+3^2+4^3+(1+2)(2+3)}+3(1^1+2^2+3^3)\left(\dfrac{1}{1}+\dfrac{1}{2}+\dfrac{1}{3}\right)$

$$=\frac{1+8+81}{2+9+64+15}+3(1+4+27)\left(1+\frac{1}{2}+\frac{1}{3}\right)$$

$$=\frac{90}{90}+(3\times32\times1\tfrac{5}{6})=1+176=177.$$

115. Let x miles be the distance of the whole journey. Now I rode $\dfrac{x}{3}$ miles at 10 miles an hour taking $\dfrac{x}{30}$ hours, $\dfrac{x}{3}$ miles at 9 miles an hour taking $\dfrac{x}{27}$ hours, and $\dfrac{x}{3}$ miles at 8 miles an hour taking $\dfrac{x}{24}$ hours.

On the second supposition I should have ridden $\dfrac{x}{2}$ miles at 10 miles an hour taking $\dfrac{x}{20}$ hours, and $\dfrac{x}{2}$ miles at 8 miles an hour taking $\dfrac{x}{16}$ hours.

The difference in time between the journeys is $\dfrac{1}{2}$ minute or $\dfrac{1}{120}$ hour.

$$\therefore\ \frac{x}{20}+\frac{x}{16}-\frac{x}{30}-\frac{x}{27}-\frac{x}{24}=\frac{1}{120};$$

that is, $108x+135x-72x-80x-90x=18$; whence $x=18$.

116. $(3x+2y)^3-(2x+3y)^3$

$$=\{(3x+2y)-(2x+3y)\}\{(3x+2y)^2+(3x+2y)(2x+3y)+(2x+3y)^2\}$$
$$=(x-y)\{9x^2+12xy+4y^2+6x^2+13xy+6y^2+4x^2+12xy+9y^2\}$$
$$=(x-y)(19x^2+37xy+19y^2).$$

$\therefore$ the other factor is $\quad 19x^2+37xy+19y^2$.

117. On the supposition that $a+b=1$,

$$(a^2-b^2)^2=(a-b)^2=a^2+b^2-2ab=(a^2-ab+b^2)(a+b)-ab=a^3+b^3-ab.$$

118. (1) $x^3+y^3+3xy(x+y)=x^3+3x^2y+3xy^2+y^3=(x+y)^3$;

(2) $m^3-n^3-m(m^2-n^2)+n(m-n)^2$

$$=m^3-n^3-m^3+mn^2+m^2n-2mn^2+n^3=m^2n-mn^2=mn(m-n).$$

119. (1) By division, $x-y=4$. Hence $x^3-y^3=28$, and $x-y=4$. [See Art. 205, Example 1.]

(2) By squaring the second equation, $x^2-6xy+9y^2=1$.

On subtraction from the first equation, $2y^2=8$, or $y=\pm 2$.

120. The expression

$$=(a-b)^4-2(a^2+b^2)(a-b)^2+(a^2+b^2)^2+a^4-2a^2b^2+b^4$$
$$=\{(a-b)^2-(a^2+b^2)\}^2+a^4-2a^2b^2+b^4$$
$$=(a^2-2ab+b^2-a^2-b^2)^2+a^4-2a^2b^2+b^4$$
$$=4a^2b^2+a^4-2a^2b^2+b^4$$
$$=a^4+2a^2b^2+b^4=(a^2+b^2)^2.$$

121. (1) $$\frac{1}{a^2-\dfrac{a^3-1}{a+\dfrac{1}{a+1}}}=\frac{1}{a^2-\dfrac{(a^3-1)(a+1)}{a^2+a+1}}=\frac{1}{a^2-(a-1)(a+1)}$$

$$=\frac{1}{a^2-a^2+1}=1.$$

(2) $$\frac{\left(1+\dfrac{1}{x}\right)\left(1-\dfrac{1}{x}\right)^2}{x-\dfrac{1}{x}}=\frac{\dfrac{x+1}{x}\times\dfrac{(x-1)^2}{x^2}}{\dfrac{x^2-1}{x}}=\frac{(x+1)(x-1)^2}{x^2(x^2-1)}=\frac{x-1}{x^2}.$$

122. $a^2b+b^2c-abc-ab^2=b(a^2-ac-ab+bc)=b(a-b)(a-c)$;

and $\quad ax^2+ab-a^2-bx^2=(a-b)x^2-(a-b)a=(a-b)(x^2-a)$.

$\therefore$ the H. C. F. $=a-b$.

123. Let x = the total number of voters; of these $\frac{2x}{3}$ are Conservatives.

Now $$\frac{2x}{3} - 25 - 60 = \frac{x}{3} + 60;$$

whence $$\frac{x}{3} = 145, \text{ and } x = 435.$$

124. (1) Clearing of fractions, $x^2 + a^2 - b^2 = 2ax$;

$$\therefore\ x^2 - 2ax + a^2 = b^2;$$

$$\therefore\ x - a = \pm b.$$

(2) $$\frac{3}{x} + \frac{2}{y} = 2, \text{ and } \frac{6}{y} - \frac{3}{x} = 2.$$

By addition $$\frac{8}{y} = 4, \text{ and } y = 2.$$

125. The quotient $$= \frac{y^2 + 2yz + z^2 - x^2}{2yz} \times \frac{2xy}{z^2 - (x^2 - 2xy + y^2)}$$

$$= \frac{(y+z+x)(y+z-x)}{z} \times \frac{x}{(z+x-y)(z-x+y)}$$

$$= \frac{x(x+y+z)}{z(x-y+z)}.$$

The fraction $$= \frac{(x^3 + 3x^2 + 3x + 1) - (x^3 - 3x^2 + 3x - 1)}{\{(x+1)^2 + (x-1)^2\}\{(x+1)^2 - (x-1)^2\}}$$

$$= \frac{2(3x^2+1)}{2(x^2+1)(4x)} = \frac{3x^2+1}{4x(x^2+1)}.$$

126.
$$\begin{array}{l|l|l}
x^2 - 3x + a + 2 & x^4 + (a-1)x^3 - (2a+1)x^2 + (a^2+4a-5)x + 3a + 6 & x^2 + (a+2)x + 3 \\
 & \underline{x^4 - 3x^3 + (a+2)x^2} & \\
 & \quad (a+2)x^3 - (3a+3)x^2 + (a^2+4a-5)x & \\
 & \quad \underline{(a+2)x^3 - (3a+6)x^2 + (a^2+4a+4)x} & \\
 & \qquad\qquad 3x^2 - 9x + 3a + 6 & \\
 & \qquad\qquad \underline{3x^2 - 9x + 3a + 6} &
\end{array}$$

127. (1) $x^2 + 5xy - 24y^2 + x - 3y = (x - 3y)(x + 8y) + x - 3y = (x - 3y)(x + 8y + 1)$.

(2) $x^3 - \frac{4}{x} = x\left(x^2 - \frac{4}{x^2}\right) = x\left(x + \frac{2}{x}\right)\left(x - \frac{2}{x}\right)$.

128. $p^2 - 3q \left(p - \dfrac{3q}{2p} - \dfrac{9q^2}{8p^3} \right.$.

$$\begin{array}{r|l} & p^2 \\ \hline 2p - \dfrac{3q}{2p} & -3q \\ & -3q + \dfrac{9q^2}{4p^2} \\ \hline & \quad -\dfrac{9q^2}{4p^2} \end{array}$$

The third term of the square root

$$= -\frac{9q^2}{4p^2} \div 2p = -\frac{9q^2}{8p^3}.$$

129. (1) By simplifying each side separately, we have

$$\frac{(x^2-12x+35)-(x^2-12x+36)}{(x-6)(x-7)} = \frac{(x^2-4x+3)-(x^2-4x+4)}{(x-2)(x-3)};$$

$$\therefore \frac{-1}{x^2-13x+42} = \frac{-1}{x^2-5x+6};$$

whence $\qquad x^2-13x+42 = x^2-5x+6$; and $8x=36$.

(2) Here $ax = by$, and $ax+1 = ay+bx$.

Substitute for y, then $\qquad ax+1 = \dfrac{a^2}{b}x + bx$;

whence $$x = \frac{b}{a^2-ab+b^2}.$$

130. $3x^2+(4a-2b)x+a(a-2b) = \{3x+(a-2b)\}\{x+a\}$.
On trial we find that $x+a$ divides the second expression.

131. (1) $$\frac{(x^a)^3}{x^{b+c}} \times \frac{(x^b)^3}{x^{c+a}} \times \frac{(x^c)^3}{x^{a+b}} = \frac{x^{3a} \times x^{3b} \times x^{3c}}{x^{b+c} \times x^{c+a} \times x^{a+b}} = \frac{x^{3(a+b+c)}}{x^{2(a+b+c)}} = x^{a+b+c}.$$

(2) The expression $= x^{\frac{1}{2}} y^{\frac{1}{3}} \times (y^{\frac{1}{2}} x^{-\frac{1}{3}}) \div (x^{-\frac{1}{4}} y^{-\frac{1}{4}})$

$$= x^{\frac{1}{2}-\frac{1}{3}+\frac{1}{4}} y^{\frac{1}{3}+\frac{1}{2}+\frac{1}{4}} = x^{\frac{5}{12}} y^{\frac{13}{12}}.$$

132. Suppose the charge for each person was x shillings. The contractor's expenses were $21x+1$ shillings, and he expected to receive $24x$ shillings.

$$\therefore 24x = \frac{112\frac{1}{2}}{100}(21x+1) = \frac{9}{8}(21x+1);$$

whence $\qquad 192x = 189x+9$, and $x=3$.

133. We have $y = \dfrac{a-2}{2}$, and $x = \dfrac{y}{y+1} = \dfrac{a-2}{a}$; so that $\dfrac{x}{y} = \dfrac{2}{a}$.

$$\therefore x(y+2) + \frac{x}{y} + \frac{y}{x} = \left(\frac{a-2}{a}\right)\left(\frac{a+2}{2}\right) + \frac{2}{a} + \frac{a}{2}$$

$$= \frac{a^2-4+4+a^2}{2a} = a.$$

134.

$$x^3-12x^2+54x-112+\frac{108}{x}-\frac{48}{x^2}+\frac{8}{x^3}\left(x-4+\frac{2}{x}\right.$$

$$x^3$$

$$-12x^2+54x-112$$

$$3(x)^2=3x^2$$
$$3x(-4)=-12x$$
$$(-4)^2=+16$$
$$3x^2-12x+16 \quad\Big|\quad -12x^2+48x-64$$

$$6x-48+\frac{108}{x}-\frac{48}{x^2}+\frac{8}{x^3}$$

$$3(x-4)^2=3x^2-24x+48$$
$$3(x-4)\times\frac{2}{x}=6-\frac{24}{x}$$
$$\left(\frac{2}{x}\right)^2=+\frac{4}{x^2}$$
$$3x^2-24x+54-\frac{24}{x}+\frac{4}{x^2} \quad\Big|\quad 6x-48+\frac{108}{x}-\frac{48}{x^2}+\frac{8}{x^3}$$

135. $x^3+2ax^2+a^2x+2a^3=x^2(x+2a)+a^2(x+2a)=(x^2+a^2)(x+2a)$;

and $x^3-2ax^2+a^2x-2a^3=x^2(x-2a)+a^2(x-2a)=(x^2+a^2)(x-2a)$.

$\therefore$ H. C. F. $=x^2+a^2$; and L. C. M. $=(x^2+a^2)(x+2a)(x-2a)$.

136. (1) The expression

$$=7(4x-3y)-6(3x-4y)-8(3x-2y)+7(2x-3y)$$
$$=28x-21y-18x+24y-24x+16y+14x-21y=-2y.$$

(2) The expression $=\dfrac{(a+4b)(a-3b)+(a-4b)(a+3b)+a^2-3b^2}{a^2-9b^2}$

$$=\frac{a^2+ab-12b^2+a^2-ab-12b^2+a^2-3b^2}{a^2-9b^2}$$

$$=\frac{3a^2-27b^2}{a^2-9b^2}=3.$$

137. The expression $=4a^2x^3+72a^3b^2-32a^5-9b^2x^3$

$$=(4a^2-9b^2)(x^3-8a^3)$$
$$=(2a+3b)(2a-3b)(x-2a)(x^2+2ax+4a^2).$$

138. (1) On squaring both sides, $75x+1-10\sqrt{3x}=75x-29$;

$\therefore$ $10\sqrt{3x}=30$; and $\sqrt{3x}=3$.

(2) We have $\dfrac{x+y}{xy}=\dfrac{1}{70}$, or $\dfrac{1}{x}+\dfrac{1}{y}=\dfrac{1}{70}$.

Similarly $$\frac{1}{x}+\frac{1}{z}=\frac{1}{84}, \text{ and } \frac{1}{y}+\frac{1}{z}=\frac{1}{140}.$$

Add the first two equations together and subtract the last;

thus $$\frac{2}{x}=\frac{1}{70}+\frac{1}{84}-\frac{1}{140}=\frac{6+5-3}{420}=\frac{8}{420}; \text{ whence } x=105.$$

139. The difference $=\left(\dfrac{x}{x-a}-\dfrac{a}{x-a}\right)+\left(\dfrac{x}{x-b}-\dfrac{b}{x-b}\right)+\left(\dfrac{x}{x-c}-\dfrac{c}{x-c}\right)$

$$=\frac{x-a}{x-a}+\frac{x-b}{x-b}+\frac{x-c}{x-c}=1+1+1=3.$$

140. The product $=(x^{\frac{3}{2}})^2-(2y^{\frac{3}{2}}+3z^{\frac{3}{2}})^2$

$$=x^3-(4y^3+12y^{\frac{3}{2}}z^{\frac{3}{2}}+9z^3).$$

141. Let x hours be the time required; then the distance I have walked is $\frac{17}{4}x$ miles; but my friend has walked $3\left(x+\frac{3}{2}\right)$ miles;

$$\therefore \frac{17}{4}x=3\left(x+\frac{3}{2}\right); \text{ that is, } 17x=12x+18.$$

142. (1) The expression $=\{(\sqrt[3]{8})^2+(\sqrt{4})^3\}\dfrac{1}{(\sqrt[4]{16})^3}=\dfrac{\{2^2+2^3\}}{2^3}=\dfrac{1}{2}+1.$

(2) The expression $=\dfrac{\{(3^2)^n\times 3^2\times 3^n\}-(3^3)^n}{3^{3n}\times 3^2}=\dfrac{3^{3n+2}-3^{3n}}{3^{3n+2}}$

$$=1-\frac{1}{3^2}.$$

143.

$$
\begin{array}{r|l}
 & xy^{-1}-2x^{\frac{1}{2}}y^{-\frac{1}{2}}+3-2x^{-\frac{1}{2}}y^{\frac{1}{2}}+x^{-1}y\ (\ x^{\frac{1}{2}}y^{-\frac{1}{2}}-1+x^{-\frac{1}{2}}y^{\frac{1}{2}} \\
 & xy^{-1} \\
\hline
2x^{\frac{1}{2}}y^{-\frac{1}{2}}-1 & -2x^{\frac{1}{2}}y^{-\frac{1}{2}}+3 \\
 & -2x^{\frac{1}{2}}y^{-\frac{1}{2}}+1 \\
\hline
2x^{\frac{1}{2}}y^{-\frac{1}{2}}-2+x^{-\frac{1}{2}}y^{\frac{1}{2}} & 2-2x^{-\frac{1}{2}}y^{\frac{1}{2}}+x^{-1}y \\
 & 2-2x^{-\frac{1}{2}}y^{\frac{1}{2}}+x^{-1}y \\
\hline
\end{array}
$$

144. (1) The expression

$$=\frac{x^2+x-x+1}{x^2-1}\times\frac{(x-1)(x^2+x+1)}{(x^2+1)(x^4-x^2+1)}\times\frac{(x^2-1)^2+x^2}{(x^2+x+1)(x^2-x+1)}$$

$$=\frac{x^2+1}{x+1}\cdot\frac{x^4-x^2+1}{(x^2+1)(x^4-x^2+1)(x^2-x+1)}$$

$$=\frac{1}{(x+1)(x^2-x+1)}=\frac{1}{x^3+1}.$$

(2) $$\frac{a^4-y^4}{a^2-2ay+y^2} \div \frac{a^2+ay}{a-y} = \frac{(a^2+y^2)(a+y)(a-y)}{(a-y)^2} \times \frac{a-y}{a(a+y)}$$
$$= \frac{a^2+y^2}{a};$$

$$\frac{a^5-a^2y^3}{a^3+y^3} \div \frac{a^4-2a^3y+a^2y^2}{a^2-ay+y^2} = \frac{a^3(a+y)(a-y)}{(a+y)(a^2-ay+y^2)} \times \frac{a^2-ay+y^2}{a^2(a-y)^2} = \frac{a}{a-y};$$

the product $$= \frac{a^2+y^2}{a} \times \frac{a}{a-y} = \frac{a^2+y^2}{a-y}.$$

145. (1) $\sqrt{8}+\sqrt{50}-\sqrt{18}+\sqrt{48}=2\sqrt{2}+5\sqrt{2}-3\sqrt{2}+4\sqrt{3}=4\sqrt{2}+4\sqrt{3}$.

(2) Assume $\sqrt{35+14\sqrt{6}}=\sqrt{x}+\sqrt{y}$; then

$x+y=35$, and $2\sqrt{xy}=14\sqrt{6}$.

$\therefore (x-y)^2=(35)^2-(14\sqrt{6})^2=49$; whence $x-y=7$.

Thus $x=21$, and $y=14$.

146. (1) The equation may be written

$$\left(1+\frac{a-b}{x-a}\right)-\left(1-\frac{a-b}{x-b}\right)=\frac{2(a-b)}{x-(a+b)}.$$

On simplifying and dividing by $a-b$, we have

$$\frac{1}{x-a}+\frac{1}{x-b}=\frac{2}{x-(a+b)};$$

that is, $$\frac{2x-(a+b)}{(x-a)(x-b)}=\frac{2}{x-(a+b)};$$

$$\therefore 2x^2-3(a+b)x+(a+b)^2=2x^2-2(a+b)x+2ab;$$

whence $$(a+b)x=a^2+b^2.$$

(2) By squaring the first equation, $4x^2+12xy+9y^2=\frac{9}{4}$;

hence by subtracting the second equation $3xy=-\frac{35}{4}$, and $24xy=-70$.

$$\therefore 4x^2-12xy+9y^2=\frac{9}{4}+70=\frac{289}{4}.$$

Thus $2x-3y=\pm\frac{17}{2}$, while $2x+3y=\frac{3}{2}$.

147. The first fraction $=(a+b)^2+c(a+b)+c^2$;

the second fraction $=(b+c)^2+a(b+c)+a^2$;

the third fraction $=(c+a)^2+b(c+a)+b^2$.

The sum $=3a^2+3b^2+3c^2+4ab+4ac+4bc$

$=a^2+b^2+c^2+2(a+b+c)^2$.

148. $a^{\frac{1}{2}}+2a^{\frac{1}{4}}x^{\frac{1}{4}}-x^{\frac{1}{2}}\,)\; a-4a^{\frac{1}{2}}x^{\frac{1}{2}}+4a^{\frac{1}{4}}x^{\frac{3}{4}}-x\;(\; a^{\frac{1}{2}}-2a^{\frac{1}{4}}x^{\frac{1}{4}}+x^{\frac{1}{2}}$

$$\begin{array}{r}
\underline{a+2a^{\frac{3}{4}}x^{\frac{1}{4}}-\;a^{\frac{1}{2}}x^{\frac{1}{2}}\qquad\qquad\qquad} \\
-2a^{\frac{3}{4}}x^{\frac{1}{4}}-3a^{\frac{1}{2}}x^{\frac{1}{2}}+4a^{\frac{1}{4}}x^{\frac{3}{4}}\qquad \\
\underline{-2a^{\frac{3}{4}}x^{\frac{1}{4}}-4a^{\frac{1}{2}}x^{\frac{1}{2}}+2a^{\frac{1}{4}}x^{\frac{3}{4}}\qquad} \\
a^{\frac{1}{2}}x^{\frac{1}{2}}+2a^{\frac{1}{4}}x^{\frac{3}{4}}-x \\
\underline{a^{\frac{1}{2}}x^{\frac{1}{2}}+2a^{\frac{1}{4}}x^{\frac{3}{4}}-x}
\end{array}$$

149. The expression $=(a-1)^2\{(a-1)^2-2(a^2+1)\}+2(a^4+1)$

$$\begin{aligned}
&=(a-1)^2(a^2-2a+1-2a^2-2)+2(a^4+1)\\
&=-(a-1)^2(a^2+2a+1)+2(a^4+1)\\
&=-(a^2-1)^2+2(a^4+1)\\
&=-a^4+2a^2-1+2a^4+2=a^4+2a^2+1=(a^2+1)^2.
\end{aligned}$$

$\therefore$ the square root $=a^2+1$.

150. Suppose that a score of pears costs x pence; then one penny will buy $\frac{20}{x}$ pears, and therefore a sovereign will buy $\frac{4800}{x}$ pears. Had the price per score been $x-2$ pence, $\frac{4800}{x-2}$ pears could have been bought for a sovereign.

$$\therefore\ \frac{4800}{x-2}-\frac{4800}{x}=120;\ \text{that is, } \frac{40}{x-2}-\frac{40}{x}=1;$$

$$\therefore\ 80=x(x-2),\ \text{or}\ (x-10)(x+8)=0.$$

Thus $x=10$, and therefore 20 pears cost 10 pence, and therefore a gross cost 72 pence.

151. Let x be the digit in the tens' place, y the digit in the units' place; then the number is $10x+y$, the number reversed is $10y+x$.

Thus $$10x+y=6(x+y),\ \text{or}\ 4x=5y.$$

$$\therefore\ 10y+x=5y+5y+x=5y+4x+x=5(x+y).$$

152. The expression $=\frac{1}{(a-b)(b-c)}+\frac{1}{(b-c)(c-a)}+\frac{1}{(c-a)(a-b)}$

$$=\frac{c-a+a-b+b-c}{(a-b)(b-c)(c-a)}=0.$$

153. The first expression $=\dfrac{(3+5x)(4+7x)-(12+41x+36x^2)}{4+7x}$

$$=\frac{12+41x+35x^2-12-41x-36x^2}{4+7x}=\frac{-x^2}{4+7x}.$$

The second expression $=\dfrac{(5-2x)(3-4x)+26x-8x^2-14}{3-4x}$

$$=\frac{15-26x+8x^2+26x-8x^2-14}{3-4x}=\frac{1}{3-4x}.$$

$\therefore$ the product $\quad = -\dfrac{x^2}{(4+7x)(3-4x)}=\dfrac{x^2}{(7x+4)(4x-3)}.$

154. If $\quad 1=x-\dfrac{1}{x};$

then $\quad x^2+\dfrac{1}{x^2}=\left(x-\dfrac{1}{x}\right)^2+2=1+2.$

Also $\quad x^3-\dfrac{1}{x^3}=\left(x-\dfrac{1}{x}\right)^3+3\left(x-\dfrac{1}{x}\right)=1+3.$

155. (1) By clearing of fractions, we have

$$9x(x+4)+759=11(x+5)(x+4);$$

$$\therefore\ 9x^2+36x+759=11x^2+99x+220;$$

that is, $\quad 2x^2+63x-539=0,$ or $(2x+77)(x-7)=0.$

(2) Put $y=mx$, and substitute in both equations; thus

$$x^2(2-3m^2)=23 \text{ and } x^2(2m-3m^2)=3.$$

By division, $\quad \dfrac{2-3m^2}{2m-3m^2}=\dfrac{23}{3},$

$$\therefore\ 30m^2-23m+3=0, \text{ or } (5m-3)(6m-1)=0.$$

If $\quad m=\dfrac{3}{5}$, then $x^2\left(2-\dfrac{27}{25}\right)=23;$

whence $\quad x^2=25.$

If $\quad m=\dfrac{1}{6}$, then $x^2\left(2-\dfrac{3}{36}\right)=23;$

whence $\quad x^2=12.$

156. (1) The expression

$$=\frac{8}{5}(2\sqrt{5})-3\sqrt{5}-\frac{1}{5}\sqrt{5}=\left(\frac{16}{5}-3-\frac{1}{5}\right)\sqrt{5}=0.$$

(2) The expression

$$=\left(x^{\frac{1}{2}}y^{\frac{1}{3}}\right)\times\left(y^{\frac{1}{4}}x^{-\frac{1}{6}}\right)\div\left(y^{-\frac{1}{4}}x^{-\frac{1}{4}}\right)$$

$$=x^{\frac{1}{2}-\frac{1}{6}+\frac{1}{4}}y^{\frac{1}{3}+\frac{1}{4}+\frac{1}{4}}=x^{\frac{7}{12}}y^{\frac{5}{6}}.$$

157. $(p^2-1)x^2+(3p-1)x-p(p-1)=\{(p+1)x-(p-1)\}\{(p-1)x+p\}$;

and $\quad p(p+1)x^2-(p^2-2p-1)x-(p-1)=\{(p+1)x-(p-1)\}\{px+1\}$.

$\therefore$ the H.C.F. $\quad =(p+1)x-(p-1)$.

158. The expression $=\dfrac{axy+a}{xy-1}\times\dfrac{x^2y^2+1}{bx^2y^2-b}\times\dfrac{3(xy-1)^2}{5(x^4y^4-1)}$

$$=\frac{a(xy+1)}{xy-1}\times\frac{x^2y^2+1}{b(xy+1)(xy-1)}\times\frac{3(xy-1)^2}{5(x^2y^2+1)(x^2y^2-1)}$$

$$=\frac{3a}{5b(x^2y^2-1)}.$$

159. (1) The expression

$$=1-2\cdot2^{2n}+(2^2)^{2n}=1-2\cdot2^{2n}+2^{4n}=(1-2^{2n})^2.$$

(2) The expression

$$=(3^2)^n-2\cdot3^n\cdot2^n+(2^2)^n=3^{2n}-2\cdot3^n\cdot2^n+2^{2n}=(3^n-2^n)^2.$$

160. Let x minutes before 6 be the time indicated by the clock; then it has to gain x minutes between 6 A.M. and 7.15 P.M.; that is, in $13\frac{1}{4}$ hrs. But it gains 4 minutes per day;

$$\therefore\ x:4=13\tfrac{1}{4}:24;\ \text{whence } 24x=53.$$

161. If $x=2+\sqrt{2}$; then $\dfrac{1}{x}=\dfrac{1}{2+\sqrt{2}}=\dfrac{2-\sqrt{2}}{2}$.

$$\therefore\ x^2+\frac{4}{x^2}=(2+\sqrt{2})^2+(2-\sqrt{2})^2=2(4+2)=12.$$

162. (1) Multiplying across,

$$x+a\sqrt{x}=x+ab-(a+b)\sqrt{x};$$

that is, $\quad (2a+b)\sqrt{x}=ab.$

(2) Here $\quad \sqrt{1+x}+\sqrt{1-x}=3\sqrt{1+x}-3\sqrt{1-x}$;

$$\therefore\ 2\sqrt{1+x}=4\sqrt{1-x};$$

whence $\quad 1+x=4(1-x)$, or $5x=3$.

163. The expression $= -\frac{a^2}{(a-b)(c-a)} - \frac{b^2}{(b-c)(a-b)} - \frac{c^2}{(c-a)(b-c)}$

$$= -\frac{a^2(b-c)+b^2(c-a)+c^2(a-b)}{(a-b)(b-c)(c-a)}$$

$$= \frac{(a-b)(b-c)(c-a)}{(a-b)(b-c)(c-a)} = 1. \quad \text{[Art. 224, Note.]}$$

164. (1) The product $= \sqrt[6]{\frac{5^3}{5^6}} \times \sqrt[6]{\frac{2^2}{2^6}} \times \sqrt[6]{80} \times \sqrt[6]{25}$

$$= \sqrt[6]{\left(\frac{1}{5^3} \times \frac{1}{2^4} \times 80 \times 25\right)} = \sqrt[6]{1} = 1.$$

(2) The expression $= \frac{4(2-\sqrt{5})}{\sqrt{5}+1} \times \frac{5+\sqrt{7}}{3\sqrt{5}-7}$

$$= \frac{4(2-\sqrt{5})(\sqrt{5}-1)}{5-1} \times \frac{5+\sqrt{7}}{3\sqrt{5}-7}$$

$$= \frac{4(3\sqrt{5}-7)}{4} \times \frac{5+\sqrt{7}}{3\sqrt{5}-7} = 5+\sqrt{7}.$$

165. The expression $= 9y^2(x^6-64) - 4x^2(x^6-64) = (9y^2-4x^2)(x^6-64)$

$$= (3y+2x)(3y-2x)(x^3+8)(x^3-8)$$

$$= (3y+2x)(3y-2x)(x+2)(x^2-2x+4)(x-2)(x^2+2x+4).$$

166. (1) The fraction $= \frac{(x+a)^2-a^2}{(x+a)^3(x-a)} \times \frac{(x-a)(x+a)(x+a)^2}{x(x+2a)}$

$$= \frac{x(x+2a)}{(x+a)^3(x-a)} \times \frac{(x-a)(x+a)^3}{x(x+2a)} = 1.$$

(2) The fraction $= \frac{6x^2y^2}{m+n} \times \frac{7(r+s)}{3(m-n)x} \times \frac{4(r-s)}{21xy^2} \times \frac{4(m^2-n^2)}{r^2-s^2} = \frac{32}{3}.$

167. (1) The expression $= \left(a^{\frac{p+q}{p}}\right)^{\frac{p}{p+q}} \div \sqrt[p]{a^{2p}.a^{-p}} = a \div a = 1.$

(2) $14-\sqrt{132} = 14-2\sqrt{33} = 11+3-2\sqrt{11\times 3} = (\sqrt{11}-\sqrt{3})^2.$

168. $20x^4+x^2-1 = (5x^2-1)(4x^2+1)$;

$25x^4+5x^3-x-1 = (25x^4-1)+x(5x^2-1) = (5x^2-1)(5x^2+1+x)$;

$25x^4-10x^2+1 = (5x^2-1)^2$;

$\therefore$ H.C.F. $= 5x^2-1$; and L.C.M. $= (5x^2-1)^2(4x^2+1)(5x^2+x+1).$

169. (1) By transposition, $\sqrt{2ax+x^2}=b-a-x$;

$$\therefore\ 2ax+x^2=b^2+a^2+x^2-2ab-2bx+2ax;$$

that is, $$2bx=b^2+a^2-2ab.$$

(2) On simplification, $\frac{8x+77}{8}+\frac{7\times 8}{8x+77}=8$.

Let $$\frac{8x+77}{8}=y;\ \text{then we have}\ y+\frac{7}{y}=8;$$

that is, $$y^2-8y+7=0,\ \text{or}\ (y-7)(y-1)=0.$$

Thus $$\frac{8x+77}{8}=7,\ \text{or}\ 1;\ \text{and}\ 8x=-21,\ \text{or}\ -69.$$

170. Suppose that the price is x shillings per dozen; then 1 shilling will buy $\frac{12}{x}$ photographs, and therefore a guinea will buy $\frac{252}{x}$ photographs. Had the price been $x+3$ shillings per dozen, then $\frac{252}{x+3}$ photographs could have been bought for a guinea.

$$\therefore\ \frac{252}{x}-\frac{252}{x+3}=7;\ \text{that is,}\ \frac{36}{x}-\frac{36}{x+3}=1;$$

$$\therefore\ 108=x(x+3),\ \text{or}\ (x+12)(x-9)=0.$$

171. If $\left(a+\frac{1}{a}\right)^2=3$, then $a^2+\frac{1}{a^2}+2=3$, or $a^2-1+\frac{1}{a^2}=0$.

But $$a^3+\frac{1}{a^3}=\left(a+\frac{1}{a}\right)\left(a^2-1+\frac{1}{a^2}\right)=0.$$

172. The fraction

$$-\frac{x+2a}{x-2b}+\frac{x-2a}{x+2b}+\frac{4ab}{x^2-4b^2}=\frac{-(x+2a)(x+2b)+(x-2a)(x-2b)+4ab}{x^2-4b^2}$$

$$=\frac{-x^2-2(a+b)x-4ab+x^2-2(a+b)x+4ab+4ab}{x^2-4b^2}$$

$$=\frac{4\{ab-(a+b)x\}}{x^2-4b^2}=0,\ \text{since}\ (a+b)x=ab.$$

173. (1) The fraction

$$=\frac{yz+zx+xy}{xyz}\div\frac{(x+y+z)^2-(x^2+y^2+z^2-xy-yz-zx)}{(x+y+z)(x^2+y^2+z^2-xy-yz-zx)}+1$$

$$=\frac{yz+zx+xy}{xyz}\times\frac{(x+y+z)(x^2+y^2+z^2-xy-yx-zx)}{3(xy+yz+zx)}+1$$

$$=\frac{x^3+y^3+z^3-3xyz}{3xyz}+1=\frac{x^3+y^3+z^3}{3xyz}.$$

173. (2) By simplifying each factor separately, we have

$$\left(1-\frac{56}{x+4}+\frac{42}{x+3}\right)=\frac{x^2+7x+12-56x-168+42x+168}{(x+4)(x+3)}=\frac{x^2-7x+12}{x^2+7x+12};$$

$$\left(1+\frac{56}{x-4}-\frac{42}{x-3}\right)=\frac{x^2-7x+12+56x-168-42x+168}{(x-4)(x-3)}=\frac{x^2+7x+12}{x^2-7x+12};$$

$\therefore$ the product $$=\frac{x^2-7x+12}{x^2+7x+12}\times\frac{x^2+7x+12}{x^2-7x+12}=1.$$

174. The expression $=(3^3)^{\frac{2}{3}}+(2^4)^{\frac{3}{4}}-2\,(2^3)^{\frac{2}{3}}+2^{\frac{1}{5}}\,(2^2)^{\frac{2}{5}}$

$$=3^2+2^3-2\,(2^2)+2^1=9+8-8+2=11.$$

175. (1) The fraction $=\dfrac{n}{1-x^n}+\dfrac{nx^n}{x^n-1}=\dfrac{n}{1-x^n}-\dfrac{nx^n}{1-x^n}$

$$=\frac{n-nx^n}{1-x^n}=n.$$

(2) Let $$\sqrt{97-56\sqrt{3}}=(\sqrt{x}-\sqrt{y})^2;$$

then $$x+y=97, \text{ and } 2\sqrt{xy}=56\sqrt{3}.$$

$$\therefore\ (x-y)^2=(97)^2-(56\sqrt{3})^2=9409-9408=1.$$

Thus $x-y=1$; and therefore $x=49$, $y=48$.

Hence the square root is $\sqrt{49}-\sqrt{48}$, or $7-4\sqrt{3}$;

and the fourth root is $\sqrt{7-4\sqrt{3}}$; that is, $2-\sqrt{3}$.

176. (1) By transposition, $\dfrac{x-5a}{x-4a}-\dfrac{x+6a}{x-4a}=\dfrac{x+5a}{x-3a}-\dfrac{x-4a}{x-3a}$;

that is, $$\frac{-11a}{x-4a}=\frac{9a}{x-3a};$$

whence $$-11\,(x-3a)=9\,(x-4a), \text{ and } 20x=69a.$$

(2) We have $3\,(x^2+y^2)+xy=8\frac{1}{4}$, and $8\,(x^2+y^2)-3xy=17\frac{3}{4}$.

Multiply the first equation by 3 and add to the second; thus

$$17\,(x^2+y^2)=42\tfrac{1}{2}, \text{ or } x^2+y^2=\frac{5}{2}.$$

From the second equation, we find $xy=\dfrac{3}{4}$.

$$\therefore\ (x+y)^2=4, \text{ and } (x-y)^2=1.$$

178. (1) The expression $=ba^{-\frac{1}{2}}\times a^{\frac{1}{3}}c^{\frac{1}{3}}\times c^{\frac{3}{4}}b^{-\frac{1}{2}}\times b^{-\frac{1}{2}}a^{\frac{1}{6}}$

$$=a^{-\frac{1}{2}+\frac{1}{3}+\frac{1}{6}}\,b^{1-\frac{1}{2}-\frac{1}{2}}\,c^{\frac{1}{3}+\frac{3}{4}}$$

$$=a^0b^0c^{\frac{13}{12}}=c^{\frac{13}{12}}.$$

(2) The expression $=\left(\dfrac{3^{2n+\frac{1}{2}}\times 3^{\frac{1}{2}}\,.\,3^{\frac{n}{2}}}{3\,.\,3^{-\frac{n}{2}}}\right)^{\frac{1}{n}}=\left(\dfrac{3^{2n+\frac{1}{2}+\frac{1}{2}+\frac{n}{2}}}{3^{1-\frac{n}{2}}}\right)^{\frac{1}{n}}$

$$=\left(3^{\frac{5n}{2}+1-1+\frac{n}{2}}\right)^{\frac{1}{n}}=\left(3^{3n}\right)^{\frac{1}{n}}=3^{3}=27.$$

179. Let x miles be the rate of the stream. They row up at the rate of $8-x$ miles an hour and down at $8+x$ miles an hour.

$$\therefore \frac{8}{8-x}+\frac{8}{8+x}=\text{the number of hours taken}=2\tfrac{2}{3};$$

that is,
$$\frac{1}{8-x}+\frac{1}{8+x}=\frac{1}{3};$$

$$\therefore 24+3x+24-3x=64-x^2, \text{ or } x^2=16.$$

180.
$$\begin{aligned}a^2-bc&=(x^2-yz)^2-(y^2-zx)(z^2-xy)\\&=x^4-2x^2yz+y^2z^2-y^2z^2+xy^3+xz^3-x^2yz\\&=x^4-3x^2yz+xy^3+xz^3\\&=x(x^3+y^3+z^3-3xyz)\\&=x\,\{x(x^2-yz)+y(y^2-zx)+z(z^2-xy)\}\\&=x(ax+by+cz).\end{aligned}$$

181. Let x be the quantity required;
then $a-x$, $b-x$, $c-x$ are in continued proportion.

$$\therefore (b-x)^2=(a-x)(c-x);$$

$$\therefore b^2-2bx+x^2=ac-ax-cx+x^2;\text{ and } x(a+c-2b)=ac-b^2.$$

182. (1) The fraction $=\left(x+y-\dfrac{x+y}{x^2+2xy+y^2-xy}\right)\times\dfrac{(x-y)(x^2+xy+y^2)}{(x+y)(x-y)}$

$$\begin{aligned}&=\left(x+y-\frac{x+y}{x^2+xy+y^2}\right)\times\frac{x^2+xy+y^2}{x+y}\\&=\left(1-\frac{1}{x^2+xy+y^2}\right)\times(x^2+xy+y^2)\\&=x^2+xy+y^2-1.\end{aligned}$$

(2) The fraction

$$\begin{aligned}&=\frac{2(7x-4)}{(3x-2)(2x-1)}+\frac{x-10}{(3x-2)(2x+1)}-\frac{2(4x-1)}{(2x+1)(2x-1)}\\&=\frac{2(2x+1)(7x-4)+(x-10)(2x-1)-2(4x-1)(3x-2)}{(3x-2)(2x-1)(2x+1)}\\&=\frac{28x^2-2x-8+2x^2-21x+10-24x^2+22x-4}{(3x-2)(2x-1)(2x+1)}\\&=\frac{6x^2-x-2}{(3x-2)(2x-1)(2x+1)}=\frac{1}{2x-1}.\end{aligned}$$

183. From the subjoined work, we see that the square root of the given expression is

$$\left(27-54x^2+36x^4-8x^6\right.$$

$$\begin{array}{r|l}
 & 729-2916x^2+4860x^4-4320x^6+2160x^8-576x^{10}+64x^{12} \\
 & 729 \\
54-54x^2 & -2916x^2+4860x^4 \\
 & -2916x^2+2916x^4 \\
54-108x^2+36x^4 & 1944x^4-4320x^6+2160x^8 \\
 & 1944x^4-3888x^6+1296x^8 \\
54-108x^2+72x^4-8x^6 & -\ 432x^6+\ 864x^8-576x^{10}+64x^{12} \\
 & -\ 432x^6+\ 864x^8-576x^{10}+64x^{12}
\end{array}$$

By inspection the cube root of $27-54x^2+36x^4-8x^6$ is $3-2x^2$.

184. (1) The fraction $=\dfrac{x-\sqrt{x^2-1}+x+\sqrt{x^2-1}}{(x+\sqrt{x^2-1})(x-\sqrt{x^2-1})}=\dfrac{2x}{x^2-(x^2-1)}=2x.$

(2) $\sqrt[4]{16}+\sqrt[3]{81}-\sqrt[3]{-512}+\sqrt[3]{192}-7\sqrt[6]{9}$

$$=2+3\sqrt[3]{3}-(-8)+4\sqrt[3]{3}-7\sqrt[3]{3}=10.$$

185. (1) The left side $=\dfrac{5}{6-\dfrac{5(6-x)}{31-6x}}=\dfrac{5(31-6x)}{186-36x-30+5x}.$

$$\therefore\ \frac{5(31-6x)}{156-31x}=x;$$

$$155-30x=156x-31x^2;$$

that is, $\quad x^2-6x+5=0$; and $(x-5)(x-1)=0$.

(2) The first equation is equivalent to $(xy-16)(xy-12)=0$.

From $x+y=8$ and $xy=12$, we have $x-y=\pm 4$.

From $x+y=8$ and $xy=16$, we have $x-y=0$.

186. The fraction

$$=\frac{b-c}{(a+b-c)(a-b+c)}+\frac{c-a}{(b+c-a)(b-c+a)}+\frac{a-b}{(c+a-b)(c-a+b)}$$

$$=\frac{(b-c)(b+c-a)+(c-a)(c+a-b)+(a-b)(a+b-c)}{(b+c-a)(c+a-b)(a+b-c)}$$

$$=\frac{b^2-c^2+c^2-a^2+a^2-b^2-a(b-c)-b(c-a)-c(a-b)}{(b+c-a)(c+a-b)(a+b-c)}$$

$$=0.$$

187. (1) Put $x-15\frac{3}{4}=y$; then $y+\frac{5}{y}=6$;

that is, $y^2-6y+5=0$, or $(y-5)(y-1)=0$

Thus $x-15\frac{3}{4}=5$, or 1.

(2) We have $x+\frac{1}{y}=2$, and $\frac{1}{x}-y=\frac{4}{3}$;

hence $\frac{1}{y}=2-x$, so that $y=\frac{1}{2-x}$.

$$\therefore \frac{1}{x}-\frac{1}{2-x}=\frac{4}{3}; \text{ that is, } \frac{2-2x}{x(2-x)}=\frac{4}{3}.$$

$$\therefore 2x^2-7x+3=0, \text{ or } (2x-3)(x-1)=0.$$

188.
$$\left(\frac{x}{a}-\frac{y}{b}\right)\left(\frac{x}{b}-\frac{y}{a}\right)=\frac{x^2+y^2}{ab}-xy\left(\frac{1}{a^2}+\frac{1}{b^2}\right)$$
$$=\frac{x^2-xy+y^2}{ab}-xy\left(\frac{1}{a^2}+\frac{1}{b^2}-\frac{1}{ab}\right)$$
$$=\frac{a^3+b^3}{ab}-\frac{ab(a+b)(a^2-ab+b^2)}{a^2b^2}=0.$$

189. $(2a+1)(a-2)x^2+(a^2+7a+2)x-a(a+2)$
$$=\{(2a+1)x-a\}\{(a-2)x+(a+2)\};$$
$$(2a+1)^2x^2-2a(2a+1)x+a^2=\{(2a+1)x-a\}^2.$$
$$\therefore \text{ H.C.F. is } (2a+1)x-a.$$

190. The product
$$=\left\{\sqrt{2(2x-1)}+\left(\sqrt{2x}-\frac{1}{\sqrt{2x}}\right)\right\}\left\{\sqrt{2(2x-1)}-\left(\sqrt{2x}-\frac{1}{\sqrt{2x}}\right)\right\}$$
$$=2(2x-1)-\left(\sqrt{2x}-\frac{1}{\sqrt{2x}}\right)^2=4x-2-\left(2x-2+\frac{1}{2x}\right)=2x-\frac{1}{2x}.$$

191. $a^2b-a^2c-ab^2+ac^2+b^2c-bc^2=a^2(b-c)-a(b^2-c^2)+bc(b-c)$
$$=(b-c)\{a^2-a(b+c)+bc\}$$
$$=(b-c)(a-b)(a-c).$$

In like manner $a^4b^2-a^4c^2-a^2b^4+a^2c^4+b^4c^2-b^2c^4$
$$=(b^2-c^2)(a^2-b^2)(a^2-c^2).$$

$$\therefore \text{ the quotient}=\frac{(a^2-b^2)(b^2-c^2)(c^2-a^2)}{(a-b)(b-c)(c-a)}=(a+b)(b+c)(c+a).$$

192. (1) The fraction

$$=\frac{21(x-1)(x^2+x+1)-(x+1)(x^2+x+1)-2(x^2-1)(10x-1)}{6(x+1)(x-1)(x^2+x+1)}$$

$$=\frac{21x^3-21-x^3-2x^2-2x-1-20x^3+20x+2x^2-2}{6(x+1)(x^3-1)}$$

$$=\frac{18x-24}{6(x+1)(x^3-1)}=\frac{3x-4}{(x+1)(x^3-1)}.$$

(2) The expression

$$=\frac{x+a-(x-a)}{\sqrt{x-a}\sqrt{x+a}}\times\frac{\sqrt{x-a}\sqrt{x^2+ax+a^2}}{\sqrt{x^2+ax+a^2}}=\frac{2a}{\sqrt{x+a}}.$$

193. Let $p=x-\dfrac{1}{x}$, and $q=x^2-\dfrac{1}{x^2}$.

$$\therefore\ p^2(p^2+4)=\left(x-\frac{1}{x}\right)^2\left\{\left(x-\frac{1}{x}\right)^2+4\right\}=\left(x-\frac{1}{x}\right)^2\left(x^2-2+\frac{1}{x^2}+4\right)$$

$$=\left(x-\frac{1}{x}\right)^2\left(x+\frac{1}{x}\right)^2=\left(x^2-\frac{1}{x^2}\right)^2=q^2.$$

194. As in Art. 188, Example 2, it is easy to shew that the hands are coincident at $16\frac{4}{11}$ minutes past three, and at $27\frac{3}{11}$ minutes past five.

195. Since $2n+1$ is odd, $7^{2n+1}+1$ is divisible by $7+1$; that is, by 8. [See Art. 229, II.]

196. The fraction $=\dfrac{\dfrac{(pq+1)^p}{q^p}\times\dfrac{(pq-1)^q}{q^q}}{\dfrac{(pq+1)^p}{p^p}\times\dfrac{(pq-1)^q}{p^q}}=\dfrac{p^p\cdot p^q}{q^p\cdot q^q}=\left(\dfrac{p}{q}\right)^{p+q}.$

197. (1) The sum $=\dfrac{(7+3\sqrt{5})^2+(7-3\sqrt{5})^2}{(7+3\sqrt{5})(7-3\sqrt{5})}=\dfrac{2(49+45)}{49-45}=\dfrac{94}{2}=47.$

(2) Here $\sqrt{1+x}=\sqrt{\dfrac{b^2+2b+1}{b^2+1}}=\dfrac{b+1}{\sqrt{b^2+1}}.$

Similarly $$\sqrt{1-x}=\frac{b-1}{\sqrt{b^2+1}}.$$

By removing the denominator $\sqrt{b^2+1}$, we have

$$\frac{\sqrt{1+x}+\sqrt{1-x}}{\sqrt{1+x}-\sqrt{1-x}}=\frac{(b+1)+(b-1)}{(b+1)-(b-1)}=b.$$

198. $4(ab+cd)^2-(a^2+b^2-c^2-d^2)^2$

$$=(2ab+2cd+a^2+b^2-c^2-d^2)(2ab+2cd-a^2-b^2+c^2+d^2)$$
$$=\{(a+b)^2-(c-d)^2\}\{(c+d)^2-(a-b)^2\}$$
$$=(a+b+c-d)(a+b-c+d)(c+d+a-b)(c+d-a+b)$$
$$=2(s-d)\,.\,2(s-c)\,.\,2(s-b)\,.\,2(s-a)$$
$$=16(s-a)(s-b)(s-c)(s-d).$$

199. Let x be the number of articles bought; then the cost price of each is $\frac{240}{x}$ pence; and the sale price is $\frac{252}{x-2}$ pence.

$$\therefore\ \frac{252}{x-2}-\frac{240}{x}=2;\ \text{that is, }\frac{126}{x-2}-\frac{120}{x}=1.$$
$$\therefore\ 6x+240=x^2-2x;\ \text{whence }(x-20)(x+12)=0.$$

200. The expression $=(81x^4+y^4-18x^2y^2)$

$$+(81x^4+y^4+18x^2y^2)-2(9x^2+y^2)(3x-y)^2+(3x-y)^4$$
$$=(81x^4+y^4-18x^2y^2)+\{(9x^2+y^2)-(3x-y)^2\}^2$$
$$=(81x^4+y^4-18x^2y^2)+(6xy)^2$$
$$=81x^4+18x^2y^2+y^4=(9x^2+y^2)^2.$$

Thus the square root is $9x^2+y^2$.

201. Let $$\frac{x}{a}=\frac{y}{b}=\frac{z}{c}=k;$$

then $$(bc+ca+ab)^2(x^2+y^2+z^2)=\left(\frac{bz}{k}+\frac{cx}{k}+\frac{ay}{k}\right)^2(a^2k^2+b^2k^2+c^2k^2)$$
$$=(bz+cx+ay)^2(a^2+b^2+c^2).$$

202. Let n be the required number of years. Then 1700 is the sum of n terms of an A. P. whose first term is 20 and common difference 10.

$$\therefore\ 1700=\frac{n}{2}\{40+10(n-1)\};$$

that is, $$340=n(n+3),\ \text{or }(n-17)(n+20)=0.$$

203. The sum of the roots is 5, therefore 1 is the other root.
Again $q=$ the product of the roots $=4$.

204. Suppose that he begins to walk at x miles an hour. To walk the first mile he takes $\frac{1}{x}$ hours, and to walk the last 6 miles he takes $\frac{6}{x+1}$ hours.

If he had not altered his rate he would have taken $\frac{7}{x}$ hours.

$$\therefore\ \frac{7}{x}-\left(\frac{1}{x}+\frac{6}{x+1}\right)=\frac{1}{2};\ \text{that is, }\frac{6}{x(x+1)}=\frac{1}{2};$$
$$\therefore\ x^2+x-12=0,\ \text{or }(x+4)(x-3)=0;$$

whence $x=3$, and the time of walking is $\frac{1}{3}+\frac{6}{4}$ or $\frac{11}{6}$ hours.

205. Here $\dfrac{x}{y}+\dfrac{x}{z}+\dfrac{x}{w}=\dfrac{-a+b+c}{a+b+c}+\dfrac{a-b+c}{a+b+c}+\dfrac{a+b-c}{a+b+c}=1.$

206. Let a be the first term, and r the common ratio;

then $$a+ar=\frac{8}{3}, \text{ and } \frac{a}{1-r}=\frac{25}{6}.$$

Thus $a=\dfrac{25}{6}(1-r)$, and therefore $\dfrac{25}{6}(1+r)(1-r)=\dfrac{8}{3}$;

whence $1-r^2=\dfrac{16}{25}$, and $r=\pm\dfrac{3}{5}$; so that $a=\dfrac{5}{3}$, or $\dfrac{20}{3}$.

207. The expression $=\dfrac{(y+x)^m(x-y)^n}{x^n y^m}\div\dfrac{(x+y)^n(y-x)^m}{x^n y^m}$

$$=\frac{(x+y)^{m-n}}{(-1)^m(x-y)^{m-n}}=(-1)^m\left(\frac{x+y}{x-y}\right)^{m-n}$$

208. The number of ways $={}^{10}P_5=10\,.\,9\,.\,8\,.\,7\,.\,6=30240.$

209. The prices in pence of each successive foot form the A. P. 27, 28, 29,...... to 400 terms.

$\therefore$ cost of the last foot is $27+399$, or 426 pence.

The total cost is $\dfrac{400}{2}(27+426)$, or 90600 pence.

210. Here $$-p=\alpha+\beta, \text{ and } q=\alpha\beta.$$

$$\therefore p+q=-(\alpha+\beta-\alpha\beta), \text{ and } pq=-\alpha\beta(\alpha+\beta);$$

that is, p and q are the roots of $x^2+(\alpha+\beta-\alpha\beta)x-\alpha\beta(\alpha+\beta)=0$.

211.

$$\begin{array}{r} tete \\ ete \\ \hline t0e01 \\ 91e12 \\ t0e01 \\ \hline tte0121 \\ \hline \end{array}$$

212. $\dfrac{x+z}{y}=\dfrac{z}{x}=\dfrac{x}{z-y}=\dfrac{\text{sum of numerators}}{\text{sum of denominators}}=\dfrac{2x+2z}{x+z}=2,$

unless $x+z=0$, in which case $y=0$.

Thus $z=2x$, and therefore $3x=2y$.

213. If a, b, c are in H. P., $\dfrac{2}{b}=\dfrac{1}{a}+\dfrac{1}{c}$.

$$\therefore \left(\frac{3}{a}+\frac{3}{b}-\frac{2}{c}\right)\left(\frac{3}{c}+\frac{3}{b}-\frac{2}{a}\right)+\frac{9}{b^2}=\frac{1}{2}\left(\frac{9}{a}-\frac{1}{c}\right)\cdot\frac{1}{2}\left(\frac{9}{c}-\frac{1}{a}\right)+\frac{9}{4}\left(\frac{1}{a}+\frac{1}{c}\right)^2$$

$$=\frac{1}{4}\left(\frac{82}{ac}-\frac{9}{a^2}-\frac{9}{c^2}\right)+\frac{9}{4}\left(\frac{1}{a^2}+\frac{2}{ac}+\frac{1}{c^2}\right)=\frac{100}{4ac}=\frac{25}{ac}.$$

214. (1) Here, out of 12 letters, 2 are c, 2 are n, 2 are s, and 3 are e; therefore the number of permutations $=\dfrac{\lfloor 12}{\lfloor 2\,\lfloor 2\,\lfloor 2\,\lfloor 3}=9979200.$

(2) Here, out of 9 letters, 4 are a, 2 are n;

$\therefore$ the number of permutations $=\dfrac{\lfloor 9}{\lfloor 4\,\lfloor 2}=7560.$

215. $(2a-3x)^5$

$$=(2a)^5-5(2a)^4(3x)+10(2a)^3(3x)^2-10(2a)^2(3x)^3+5(2a)(3x)^4-(3x)^5$$
$$=32a^5-5\,.\,16a^4\,.\,3x+10\,.\,8a^3\,.\,9x^2-10\,.\,4a^2\,.\,27x^3+5\,.\,2a\,.\,81x^4-243x^5$$
$$=32a^5-240a^4x+720a^3x^2-1080a^2x^3+810ax^4-243x^5.$$

Denote the rth and $(r+1)^{\text{th}}$ terms by T_r and T_{r+1} respectively; then

$$T_{r+1}=\frac{7-r+1}{r}\,.\,\frac{3}{5}\,.\,T_r;$$

hence $$T_{r+1}>T_r,\text{ so long as }\frac{8-r}{r}\,.\,\frac{3}{5}>1;$$

that is, $$24-3r>5r,\text{ or }24>8r.$$

The greatest value of r consistent with this is 3; hence the 3rd and 4th terms are equal, and are the greatest terms in the expansion.

216. The first expression $=\dfrac{2(1+2x)}{2+\sqrt{4+8x}}=\dfrac{2+\sqrt{3}}{2+\sqrt{4+2\sqrt{3}}}$

$$=\frac{2+\sqrt{3}}{2+(\sqrt{3}+1)}=\frac{2+\sqrt{3}}{3+\sqrt{3}}.$$

Similarly, the second expression $=\dfrac{2-\sqrt{3}}{3-\sqrt{3}}.$

$\therefore$ the whole expression $=\dfrac{2+\sqrt{3}}{3+\sqrt{3}}+\dfrac{2-\sqrt{3}}{3-\sqrt{3}}$

$$=\frac{(2+\sqrt{3})(3-\sqrt{3})+(3+\sqrt{3})(2-\sqrt{3})}{9-3}$$

$$=\frac{3+\sqrt{3}+3-\sqrt{3}}{6}=1.$$

217. The expression $=\dfrac{bc-x^2}{(a-b)(c-a)}+\dfrac{ca-x^2}{(b-c)(a-b)}+\dfrac{ab-x^2}{(c-a)(b-c)}.$

The L. C. D. is $(a-b)(b-c)(c-a)$;

and the numerator $=bc(b-c)+ca(c-a)+ab(a-b)-x^2(b-c+c-a+a-b)$

$$=-(b-c)(c-a)(a-b)$$

$\therefore$ the expression $=-1.$

218. (1) By putting $x^2-5x+2=y$, the equation becomes $y^2=y+20$; from which we get $y=5$ or -4.

Thus $$x^2-5x+2=5, \text{ or } -4;$$

that is, $$x^2-5x-3=0, \text{ or } x^2-5x+6=0;$$

whence $$x=\frac{5\pm\sqrt{37}}{2}, \text{ or } 3,\ 2.$$

(2) $$\left(x^2+\frac{1}{x^2}\right)^2+4\left(x^2+\frac{1}{x^2}\right)-12=0;$$

$$\therefore\ \left\{\left(x^2+\frac{1}{x^2}\right)+6\right\}\left\{\left(x^2+\frac{1}{x^2}\right)-2\right\}=0.$$

$$\therefore\ x^4+6x^2+1=0, \text{ or } x^4-2x^2+1=0,$$

whence $$x^2=-3\pm2\sqrt{2}, \text{ or } x^2=1.$$

219. The expression $(y-z)^3+(x-y)^3-(x-z)^3+3(x-y)(x-z)(y-z)$ has a factor $(y-z)+(x-y)-(x-z)$, and is therefore $=0$.

$$\therefore\ (y-z)^3+(x-y)^3+3(x-y)(x-z)(y-z)=(x-z)^3.$$

220. 4 consonants can be selected out of 16 in $\dfrac{\underline{|16}}{\underline{|4}\,\underline{|12}}$ ways, and 2 vowels can be selected out of 5 in $\dfrac{\underline{|5}}{\underline{|2}\,\underline{|3}}$ ways.

$\therefore$ 4 consonants and 2 vowels can be selected in $\dfrac{\underline{|16}\,\underline{|5}}{\underline{|4}\,\underline{|12}\,\underline{|2}\,\underline{|3}}$ ways, and each selection of 6 letters can be arranged in $\underline{|6}$ words.

$$\therefore \text{ the number of words}=\frac{\underline{|16}\,\underline{|5}\,\underline{|6}}{\underline{|4}\,\underline{|12}\,\underline{|2}\,\underline{|3}}=13104000.$$

221. By supposition, $c-a$, $b-a$, $d-a$ are in H. P.

$\therefore\ c-a:d-a=\{(b-a)-(c-a)\}:\{(d-a)-(b-a)\}$ [Art. 329.]

$\therefore\ a-c:a-d=b-c:d-b;$

$\therefore\ a-c:b-c=a-d:d-b;$

$\therefore\ a-c:b-c=\{(a-c)-(d-c)\}:\{(d-c)-(b-c)\};$

that is, $a-c$, $d-c$, $b-c$ are in H.P.

222. The red balls can be selected in $\dfrac{\underline{|4}}{\underline{|2}\,\underline{|2}}$ or 6 ways.

The black balls...................... $\dfrac{\underline{|6}}{\underline{|3}\,\underline{|3}}$ or 20 ways.

The white ball can be selected in 2 ways.

The blue balls .... $\dfrac{\lfloor 5}{\lfloor 2 \ \lfloor 3}$ or 10 ways.

$\therefore$ the number of different selections that can be made $=6 \,.\, 20 \,.\, 2 \,.\, 10=2400$.

Again in each selection of 8 balls there are 2 red, 3 black and 2 blue, and therefore the number of ways in which each selection can be arranged

$$=\frac{\lfloor 8}{\lfloor 2 \lfloor 3 \lfloor 2}=1680.$$

$\therefore$ the total number of arrangements $=1680 \times 2400=4032000$.

223. Let d be the common difference and n the number of terms;
then $\qquad 11=\text{the last term}=1+(n-1)\,d.$

$$36=\text{the sum}=\frac{n}{2}\,(\text{first term}+\text{last term})=6n.$$

$$\therefore\ n=6, \text{ and } d=2.$$

224. $$\left(2-\frac{3a}{4}\right)^5=2^5-5\,(2)^4\left(\frac{3a}{4}\right)+10\,(2)^3\left(\frac{3a}{4}\right)^2$$
$$-10\,(2)^2\left(\frac{3a}{4}\right)^3+5\,(2)\left(\frac{3a}{4}\right)^4-\left(\frac{3a}{4}\right)^5$$
$$=32-5\,.\,16\,.\,\frac{3a}{4}+10\,.\,8\,.\,\frac{9a^2}{16}$$
$$-10\,.\,4\,.\,\frac{27a^3}{64}+5\,.\,2\,.\,\frac{81a^4}{256}-\frac{243a^5}{1024}$$
$$=32-60a+45a^2-\frac{135}{8}\,a^3+\frac{405}{128}\,a^4-\frac{243}{1024}\,a^5.$$

$$\left(1-\frac{2}{3}x\right)^{\frac{3}{2}}=1-\frac{3}{2}\left(\frac{2x}{3}\right)+\frac{\frac{3}{2}\left(\frac{3}{2}-1\right)}{1\,.\,2}\left(\frac{2x}{3}\right)^2$$
$$-\frac{\frac{3}{2}\left(\frac{3}{2}-1\right)\left(\frac{3}{2}-2\right)}{1\,.\,2\,.\,3}\left(\frac{2x}{3}\right)^3+\frac{\frac{3}{2}\left(\frac{3}{2}-1\right)\left(\frac{3}{2}-2\right)\left(\frac{3}{2}-3\right)}{1\,.\,2\,.\,3\,.\,4}\left(\frac{2x}{3}\right)^4-\dots$$
$$=1-x+\frac{3\,.\,1}{1\,.\,2\,.\,2^2}\cdot\frac{2^2x^2}{9}-\frac{3\,.\,1\,.\,(-1)}{1\,.\,2\,.\,3\,.\,2^3}\cdot\frac{2^3x^3}{27}$$
$$+\frac{3\,.\,1\,.\,(-1)\,(-3)}{1\,.\,2\,.\,3\,.\,4\,.\,2^4}\cdot\frac{2^4x^4}{81}-\dots$$
$$=1-x+\frac{1}{6}\,x^2+\frac{1}{54}\,x^3+\frac{1}{216}\,x^4\dots$$

225. Let r be the radix of the scale required; then

$$r^3+5r^2+3r+4=418;$$

$$\therefore\ r^3+5r^2+3r-414=0.$$

The only factors of 414 below ten are 2, 3, 6 and 9; but 2 and 3 are evidently values too small for r, while 9 is too great. By trial it is found that $r=6$ satisfies the equation. Thus the scale is the senary.

226. The expression

$$=\frac{2\sqrt{2}\,.\sqrt{5}}{9\sqrt{3}}\times\frac{15\sqrt{3}\,.\sqrt{7}}{4\sqrt{3}\sqrt{5}}\times\frac{28\sqrt{3}}{5\sqrt{2}\sqrt{7}}=\frac{2\times15\times28}{9\times4\times5}=\frac{14}{3}.$$

Secondly, $\dfrac{1}{3\sqrt{5}-6}=\dfrac{3\sqrt{5}+6}{45-36}=\dfrac{\sqrt{5}+2}{3}=\dfrac{4\cdot236}{3}=1\cdot412.$

227. $\sqrt[3]{128}=(5^3+3)^{\frac{1}{3}}=5\left(1+\dfrac{3}{5^3}\right)^{\frac{1}{3}}$

$$=5\left(1+\frac{1}{3}\cdot\frac{3}{5^3}-\frac{1}{9}\cdot\frac{3^2}{5^6}+\frac{5}{81}\cdot\frac{3^3}{5^9}-\dots\right)$$

$$=5+\frac{1}{5^2}-\frac{1}{5^5}+\frac{1}{3}\cdot\frac{1}{5^7}$$

$$=5+\frac{2^2}{10^2}-\frac{2^5}{10^5}+\frac{1}{3}\cdot\frac{2^7}{10^7}$$

$$=5+\cdot04-\cdot00032+\cdot000004=5\cdot039684.$$

228. In making a selection of the Greek books, each book can either be taken or left out, with the one exception that all may not be left out; thus the selection may be made in 2^4-1 ways. Similarly the selection from the Latin books may be made in 2^3-1 ways, and from the English in 2^2-1 ways.

$\therefore$ the number of selections $=(2^4-1)(2^3-1)(2^2-1)=15\,.\,7\,.\,3=315.$

229. (1) $\dfrac{\sqrt{45x^3}-\sqrt{80x^3}+\sqrt{5a^2x}}{a-x}=\dfrac{3x\sqrt{5x}-4x\sqrt{5x}+a\sqrt{5x}}{a-x}=\sqrt{5x}.$

(2) Multiply each term of dividend and divisor by $x^{\frac{1}{2}}$; thus

$$\text{the expression}=\left(\frac{x+1}{x^2-x+1}-\frac{x-1}{x^2+x+1}\right)\div\left(\frac{x+2}{x^3-1}-\frac{x-2}{x^3+1}\right)$$

$$=\left\{\frac{(x^3+2x^2+2x+1)-(x^3-2x^2+2x-1)}{(x^2-x+1)(x^2+x+1)}\right\}$$

$$\div\left\{\frac{(x^4+2x^3+x+2)-(x^4-2x^3-x+2)}{(x^3-1)(x^3+1)}\right\}$$

$$=\frac{2(2x^2+1)}{(x^2-x+1)(x^2+x+1)}\times\frac{(x^3-1)(x^3+1)}{2x(2x^2+1)}=\frac{(x-1)(x+1)}{x}.$$

230. The sum of the roots $=10$, and their product $=(5+\sqrt{6})(5-\sqrt{6})=19$;

$$\therefore \text{ the quadratic is } x^2-10x+19=0.$$

Secondly, let n and $n+1$ be the roots of the equation $x^2-px+q=0$;

then $$2n+1=p, \text{ and } n^2+n=q.$$

$$\therefore 4q=4n^2+4n=p^2-1.$$

231.

```
      t0̇84̇04̇ ( tet
      84
18e   1884
      1721
19tt   16304
       16304
```

232. $128=2^7=(2^4)^{\frac{7}{4}}=(16)^{\frac{7}{4}}$; hence $\log_{16}128=\frac{7}{4}$.

$\sqrt{128}=\sqrt{2^7}=2^{\frac{7}{2}}=(2^2)^{\frac{7}{4}}=4^{\frac{7}{4}}$; hence $\log_4\sqrt{128}=\frac{7}{4}$

$\frac{1}{4}=\left(\frac{1}{2}\right)^2=2^{-2}$; hence $\log_2\frac{1}{4}=-2$.

$$\log 1728=\log(2^6\times 3^3)=6\log 2+3\log 3$$
$$=6\times \cdot 30103+3\times \cdot 4771213=3\cdot 2375439.$$
$$\therefore \log \cdot 00001728=\bar{5}\cdot 2375439.$$

233. Suppose A to overtake B in x days after A started; A has gone

$$1+2+3\ldots+x \text{ or } \frac{x(x+1)}{2} \text{ miles;}$$

whilst B has gone 12 miles a day for $x-5$ days.

$$\therefore \frac{x(x+1)}{2}=12(x-5);$$

$$\therefore x^2-23x+120=0; \text{ whence } x=8, \text{ or } 15.$$

Thus B overtakes A in 8 days; but after 12 days A begins to travel faster than B, and therefore gradually gains on him. He overtakes him again after 15 days.

234. (1) We have $2^x=2^{3(y+1)}$; that is, $x=3y+3$:

and $3^{2y}=3^{x-9}$; that is, $2y=x-9$.

From these equations, we obtain $x=21$, $y=6$.

(2) From $z^x=y^{2x}$, we have $z=y^2$.

From $2^z=2\times 2^{2x}=2^{2x+1}$, we have $2x+1=z=y^2$.

Substitute for x and z in terms of y in $x+y+z=16$.

Thus $$\frac{y^2-1}{2}+y+y^2=16, \text{ or } 3y^2+2y-33=0;$$

$$\therefore\ (3y+11)(y-3)=0; \text{ that is, } y=3, \text{ or } -\frac{11}{3}.$$

Hence $$z=9, \text{ or } \frac{121}{9}; \text{ and } x=4, \text{ or } \frac{56}{9}.$$

235. Let a be the first term and d the common difference.

The sum of the first 5 terms $=\frac{5}{2}(2a+4d)$;

.... 10 terms $=5(2a+9d)$.

$$\therefore\ \frac{5(2a+9d)}{\frac{5}{2}(2a+4d)}=\frac{13}{4};$$

$$\therefore\ 8(2a+9d)=13(2a+4d); \text{ whence } a=2d.$$

236. Denote the rth and $(r+1)^{\text{th}}$ terms by T_r and T_{r+1} respectively; then

$$T_{r+1}=\frac{\frac{4}{3}+r-1}{r}x\times T_r=\frac{1+3r}{3r}\times\frac{12}{13}\times T_r, \textit{ numerically};$$

hence $T_{r+1}>T_r$, so long as $\frac{1+3r}{3r}\cdot\frac{12}{13}>1$;

that is, $$12+36r>39r, \text{ or } 12>3r.$$

The greatest value of r consistent with this inequality is 4; hence the 4th and 5th terms are equal and are the greatest terms in the expansion.

237. (1) The number of ways in which the places can be occupied is

$${}^6P_3=6\times5\times4=120.$$

(2) The number of selections of the two men is ${}^5C_2=10$.

Any selection of two men and the lady can take their places in $\lfloor 3$ or 6 ways.

Thus the number of ways is 10×6 or 60.

238. $$\log 5=\log 10-\log 2=1-\cdot301030=\cdot698970;$$

$$\therefore\ \log\cdot005=\bar{3}\cdot698970.$$

$$\log 63=\log 9+\log 7=2\log 3+\log 7=\cdot954242+\cdot845098=1\cdot799340.$$

$$\therefore\ \log 6\cdot3=\cdot799340.$$

$$\log\left(\frac{49}{216}\right)^{\frac{1}{3}}=\frac{1}{3}\log 7^2-\frac{1}{3}\log(2^3\times3^3)=\frac{2}{3}\log 7-\log 2-\log 3$$

$$=\cdot563399-\cdot301030-\cdot477121=\bar{1}\cdot785248.$$

Again, $$18^{8-4x}=(54\sqrt{2})^{3x-2}=(18\sqrt{18})^{3x-2}=18^{\frac{3(3x-2)}{2}};$$

$$\therefore\ 8-4x=\frac{3(3x-2)}{2};\ \text{whence } x=\frac{22}{17}.$$

239. Let $P=my^{\frac{1}{2}}z^{\frac{1}{2}}$, and $Q=ny^{\frac{1}{3}}z^{\frac{1}{3}}$, where m and n are constants; then

$$x=P+Q=my^{\frac{1}{2}}z^{\frac{1}{2}}+ny^{\frac{1}{3}}z^{\frac{1}{3}}.$$

When $y=z=64$, and $x=12$, we have $12=64m+16n$;

and when $y=16$, $z=4$, and $x=2$, we have $2=8m+4n$;

from which we obtain $$m=\frac{1}{8},\ n=\frac{1}{4}.$$

$\therefore$ the equation is $$x=\frac{1}{8}(yz)^{\frac{1}{2}}+\frac{1}{4}(yz)^{\frac{1}{3}}.$$

240. The expression $=\log\left(\frac{133}{65}\times\frac{13}{7}\times\frac{13}{7}\times\frac{90}{143}\times\frac{77}{171}\right)=\log 2.$

241. We have $${}^{n}P_4 : {}^{2n}C_3=22:3.$$

$$\therefore\ 3n(n-1)(n-2)(n-3)=\frac{22(2n)(2n-1)(2n-2)}{6};$$

that is, $$9(n-2)(n-3)=44(2n-1);$$

$$9n^2-133n+98=0,\ \text{or } (n-14)(9n-7)=0.$$

242. Since $$\frac{a+c}{ac}=\frac{4b-(a+c)}{4b^2-2(a+c)b+ac};$$

$$\therefore\ 2(a+c)b^2-(a^2+4ac+c^2)b+ac(a+c)=0;$$

$$\therefore\ \{2b-(a+c)\}\{(a+c)b-ac\}=0.$$

Thus $2b=a+c$, in which case $2b$ is the A. M. between $2a$ and $2c$; or $(a+c)b=ac$; whence $2b=\frac{2ac}{a+c}$;

that is, $2b$ is the H. M. between a and c.

243. ${}^{n}C_{r+1}+{}^{n}C_{r-1}+2\times{}^{n}C_r$

$$=\frac{\lfloor n}{\lfloor r+1\,\lfloor n-r-1}+\frac{\lfloor n}{\lfloor r-1\,\lfloor n-r+1}+\frac{2\lfloor n}{\lfloor r\,\lfloor n-r}$$

$$=\frac{\lfloor n}{\lfloor r+1\,\lfloor n-r+1}\{(n-r)(n-r+1)+r(r+1)+2(r+1)(n-r+1)\}.$$

Thus $${}^nC_{r+1} + {}^nC_{r-1} + 2 \times {}^nC_r = \frac{\underline{|n}}{\underline{|r+1}\ \underline{|n-r+1}}(n^2+3n+2)$$

$$= \frac{\underline{|n}}{\underline{|r+1}\ \underline{|n-r+1}}(n+1)(n+2)$$

$$= \frac{\underline{|n+2}}{\underline{|r+1}\ \underline{|n-r+1}} = {}^{n+2}C_{r+1}.$$

244. (1) Since 54 lies between 3^3 and 3^4, the characteristic is 3.

(2) $$\log \sqrt[3]{\cdot 0125} = \frac{1}{3} \log \frac{1}{80} = -\frac{1}{3} \log 8 - \frac{1}{3} \log 10 = -\log 2 - \frac{1}{3}$$

$$= -\cdot 30103 - \cdot\dot{3} = -\cdot 63436 = \bar{1}\cdot 36564.$$

(3) $\log 3^{45} = 45 \log 3 = 45 \times \cdot 4771213 = 21\cdot 4704585$;

$\therefore$ the number of digits is 22.

245. The $(r+1)^{\text{th}}$ term of $(2ax^2)^{\frac{5}{3}}\left(1-\frac{x}{2a}\right)^{\frac{5}{3}}$

$$= (2ax^2)^{\frac{5}{3}}(-1)^r \frac{\frac{5}{3}\left(\frac{5}{3}-1\right)\left(\frac{5}{3}-2\right)\dots\left(\frac{5}{3}-r+1\right)}{\underline{|r}}\left(\frac{x}{2a}\right)^r$$

$$= (-1)^r (2ax^2)^{\frac{5}{3}} \frac{5 \,.\, 2 \,.\, (-1) \,.\, (-4) \dots (8-3r)}{3^r\ \underline{|r}}\left(\frac{x}{2a}\right)^r$$

$$= (-1)^r (-1)^{r-2} \frac{5 \,.\, 2 \,.\, 1 \,.\, 4 \dots (3r-8)}{3^r \,.\, \underline{|r}} x^{r+\frac{10}{3}} (2a)^{\frac{5}{3}-r}.$$

246. The number of ways in which the Government speeches may be arranged is $\underline{|5}$, and number of ways in which the Opposition speeches may be arranged is $\underline{|4}$.

$$\therefore \text{ number of ways} = \underline{|5} \times \underline{|4} = 120 \times 24 = 2880.$$

247. The product of the roots is $2ab$.

The second root $$= \frac{2ab\,(a+b-\sqrt{a^2+b^2})}{(a+b)^2-(a^2+b^2)} = a+b-\sqrt{a^2+b^2};$$

so that the sum of the roots is $2(a+b)$.

Hence the quadratic equation is $x^2 - 2(a+b)x + 2ab = 0$.

248. The time for the nth mile varies inversely as the speed, but the speed varies inversely as $n-1$; the time for the nth mile varies directly as $n-1$.

Let the number of hours taken for the nth mile be $k(n-1)$, where k is some constant.

When $n=2$, the number of hours is 2; hence $2=k$.

Hence the number of hours taken for the nth mile is $2(n-1)$.

249. (1) Substituting a for x, the left-hand side of the equation becomes $a^2(b-c)+a^2(c-a)+a^2(a-b)$, which is identically 0. Hence a is a root of the equation.

Again the product of the roots $=\dfrac{a^2(a-b)}{b-c}$; so that the other root is $\dfrac{a(a-b)}{b-c}$.

(2) We have $(x^2-px+p^2)\{q(x+p)+p^2\}=qx^3+p^2q^2+p^4$;

that is, $$q(x^3+p^3)+p^2(x^2-px+p^2)=qx^3+p^2q^2+p^4;$$

whence $$x^2-px+q(p-q)=0, \text{ or } (x-q)\{x-(p-q)\}=0$$

250. The series $$=1+\frac{3}{2}\cdot\frac{1}{2}+\frac{\frac{3}{2}\cdot\frac{5}{2}}{1.2}\left(\frac{1}{2}\right)^2+\frac{\frac{3}{2}\cdot\frac{5}{2}\cdot\frac{1}{2}}{1.2.3}\left(\frac{1}{2}\right)^3+\dots$$

$$=\left(1-\frac{1}{2}\right)^{-\frac{3}{2}}=\left(\frac{1}{2}\right)^{-\frac{3}{2}}=2^{\frac{3}{2}}=\sqrt{8}.$$

APPENDIX.

Containing Solutions to the New Examples introduced into the Seventh and Eighth Editions of the Elementary Algebra.

EXAMPLES. I. e. PAGE 8_A.

An arrangement similar to that of Ex. 3, p. 7, is recommended here.

1.

x	0	3	6	8	10
x^2	0	9	36	64	100
$9x$	0	27	54	72	90
$x^2-9x+20$	20	2	2	12	30

2.

x	0	1	2	3	4
x^2	0	1	4	9	16
$\frac{x^2}{4}$	0	·25	1	2·25	4
$2x$	0	2	4	6	8
$3+2x+\frac{x^2}{4}$	3	5·25	8	11·25	15

4.

x	2	6	8	10
x^2	4	36	64	100
x^3	8	216	512	1000

$\frac{x^3}{100}$	·08	2·16	5·12	10
$\frac{x^2}{10}$	·4	3·6	6·4	10
$2x$	4	12	16	20

Hence the values of $\frac{x^3}{100}+\frac{x^2}{10}+2x$ are 4·48, 16·76, 27·52, and 40.

7.

x	1	4	8
x^2	1	16	64
x^3	1	64	512

x^3	1	64	512
$13x^2$	13	208	832
$44x$	44	176	352

$\therefore\ x^3-13x^2+44x=32$ in each case.

12. When $x=\frac{1}{3}$, $6x^3-11x^2+3x=\frac{6}{27}-\frac{11}{9}+1=0.$

When $x=\frac{3}{2}$, $6x^3-11x^2+3x=6\,.\,\frac{27}{8}-11\,.\,\frac{9}{4}+\frac{9}{2}$

$$=\frac{81}{4}-\frac{99}{4}+\frac{18}{4}=0.$$

When $x=\frac{1}{10}$, $6x^3-11x^2+3x=\cdot 006-11\times\cdot 01+3\times\cdot 1$

$$=\cdot 006+\cdot 3-\cdot 11=\cdot 306-\cdot 11=\cdot 196.$$

MISCELLANEOUS EXAMPLES. I. Page 23.

3. (1) $x^2+\frac{3}{2}y^3-xyz^3=3^2+\frac{3}{2}\,.\,8-0=9+12=21.$

(2) $\frac{1}{4}x^3y^4+\frac{5z^2}{6}=\frac{1}{4}\,.\,3^3\,.\,2^4+0=27\,.\,4=108.$

8. See Art. 22. I.

12. $5x^2-2x^3z+3y^4=5\,.\,1^2-2\,.\,1^3\,.\,3+3\,.\,2^4=5-6+48=47.$

$2z^y-3y^x=2\,.\,3^2-3\,.\,2=18-6=12.$

15. $3x^2-2yz-ax+5ax^2y=3\,.\,4^2-2\,.\,3\,.\,2-0+0=48-12=36.$

21. $\sqrt{(3xy)}+\sqrt{(5xz)}+\sqrt{(3yw)}=\sqrt{(3\,.\,1\,.\,3)}+\sqrt{(5\,.\,1\,.\,5)}+0=3+5=8.$

24. $xyp+\frac{2y}{p-y}+2q=3\,.\,4\,.\,8+\frac{2\,.\,4}{8-4}+2\,.\,10=96+2+20=118.$

EXAMPLES. IX. b. Page 66.

31. If p is the greatest, the three numbers are p, $p-1$, and $p-2$.

$\therefore$ the product $=p(p-1)(p-2)$.

But the product also equals y; thus the required equation is

$$p(p-1)(p-2)=y.$$

32. The three consecutive even numbers may be expressed by $2n-2$, $2n$, $2n+2$.

$$\therefore\ (2n-2)+2n+(2n+2)=x.$$

33. The excess of a over $b=a-b$;

$$\therefore\ pq=5(a-b).$$

35. The father's present age is $(a+x)$ years, and 5 years hence his age will be $(a+x+5)$ years.

The son's age 5 years hence will be $(a+5)$ years. But the father's age will then be equal to twice the son's age.

Thus the required equation is $(a+x+5)=2(a+5)$.

If $a=15$, $a+x=2.20-5=35$. $\therefore$ the father's present age is 35.

If $a+x=53$, we have $58=2(a+5)$; whence $a=24$, the son's present age.

36. B's money has been *increased* by the same amount that A's has been *decreased*.

$\therefore$ A has $(p-x)$ pounds, that is, $20(p-x)$ shillings.

B has q shillings $+\,x$ pounds, that is, $(q+20x)$ shillings.

Since A's money is now three times B's, the required equation is

$$20(p-x)=3(q+20x).$$

37. Five years ago the man's age was $p-5$, and the son's age was $q-5$.

$$\therefore\ p-5=7(q-5).$$

EXAMPLES. IX. c. Page 68_A.

1. (iii) Here $h=\frac{2A}{b}$. Since 10 sq. chains = 1 acre we have

$$h=\frac{2\times 10\times 5\cdot 985}{17\cdot 5}=6\cdot 84;$$

$\therefore$ the required height = 6 ch. 84 lks.

2. (iii) Here $h=\frac{3V}{a^2}=\frac{20\times 3}{12}=5$ ft.

3. (iii) Here $s=(5500\times 3)$ ft.; $t=(5\times 60)$ seconds.

$$\therefore\ v=\frac{5500\times 3}{5\times 60}=55 \text{ ft. per sec.}$$

$$55 \text{ ft. per sec.}=\frac{55}{1760\times 3}\times 60\times 60=37\tfrac{1}{2} \text{ miles per hr.}$$

Or thus: $s=\frac{5500}{1760}=\frac{25}{8}$ miles; $t=\frac{1}{12}$ of an hour.

$$\therefore\ v=\frac{25}{8}\div\frac{1}{12}=\frac{25}{8}\times 12=37\tfrac{1}{2} \text{ mi. per hr.}$$

4. (ii) Here $s=402{\cdot}5$, $g=32{\cdot}2$;

$$\therefore\ t^2=\frac{805}{32{\cdot}2}=25;\ \text{whence } t=5.$$

7.
$$\text{Perimeter}=2(\text{length}+\text{breadth})=2(x+y)\text{ ft.}$$
$$\text{Area}=\text{length}\times\text{breadth}=xy\text{ sq. ft.}$$
$$\text{Area of walls}=\text{perimeter}\times\text{height}=2z(x+y)\text{ sq. ft.}$$

12.
$$\text{Area}=\frac{1}{2}(a+b)\times d.$$

If $a=6$ ft. 4 in., $b=7$ ft. 2 in., and $d=4$ ft.

$$A=\frac{1}{2}.\ 13{\cdot}5\times 4=27\text{ sq. ft.}$$

13. Here we have to find N when $D=19$, $Q=17$, and $R=5$.

14. Substitute $N=566$, $Q=37$, $R=11$ in the formula $\frac{N}{D}=Q+\frac{R}{D}$.

15. Substitute $x=5$, $m=3$, $y=15$ in the formula $m(y+x)-x$.

17. If the undivided line contains k units of length, and the several parts of the divided line contain a, b, c units respectively, we have

$$k(a+b+c)=ka+kb+kc.$$

18. (i) Let AO, OB contain a and b units of length respectively; then AB contains $a+b$ units. Then we have to shew that

$$(a+b)^2=(a+b)a+(a+b)b,$$

which is obviously an identity.

(ii) With the same notation as before

$$(a+b)a=a^2+ab.$$

19. (i) This amounts to proving that

$$(a+b)^2=a^2+2ab+b^2.$$

(ii) If the two parts are $a-b$ and b, the whole line is a. We have then to prove that

$$a^2+b^2=2ab+(a-b)^2,$$

or

$$(a-b)^2=a^2-2ab+b^2.$$

21. (i) $\pi\sqrt{\frac{l}{g}}=3{\cdot}1416\times\sqrt{\frac{2{\cdot}0125}{32{\cdot}2}}=3{\cdot}1416\sqrt{\frac{161}{80}\times\frac{5}{161}}$

$$=\frac{3{\cdot}1416}{4}={\cdot}7854.$$

(ii) $\sqrt{2gs}=\sqrt{2\times 32\cdot 2\times 144\cdot 9}=\sqrt{4\times 16\cdot 1\times 9\times 16\cdot 1}$
$=2\times 3\times 16\cdot 1=96\cdot 6.$

(iii) $\frac{1}{2}mv^2=\frac{1}{2}.\ 18\cdot 75\times(5\cdot 6)^2=\frac{1}{2}\times\frac{75}{4}\times\frac{28}{5}\times\frac{28}{5}$
$=294.$

22. $$v^2=\frac{Fgr}{m}=\frac{200\times 32\cdot 2\times 3}{12\cdot 075}=\frac{400\times 16\cdot 1}{4\cdot 025}=1600;$$
$$\therefore\ v=40.$$

23. $$a=\frac{(v+u)(v-u)}{2s}=\frac{60\times 40}{200}=12.$$

24. (i) $$s=\frac{20}{2}(14+964)=10\times 978=9780.$$

(ii) $$25\cdot 2=\frac{12}{2}(a+3\cdot 2)=6a+19\cdot 2;$$
$$\therefore\ 6a=6,\ \text{or}\ a=1.$$

(iii) $$46\cdot 8=\frac{n}{2}(\cdot 6+7\cdot 2)=n(\cdot 3+3\cdot 6);$$
$$\therefore\ n=\frac{46\cdot 8}{3\cdot 9}=12.$$

(iv) $-175\cdot 5=\frac{13}{2}(13\cdot 5+l)$;
$-351=175\cdot 5+13l$;
$\therefore\ 13l=-526\cdot 5$;
that is, $l=-40\cdot 5.$

25. Using the method and arrangement of Ex. 3, page 7, it will be found that the values of y are 4, $5\frac{1}{5}$, $6\frac{2}{5}$, $7\frac{3}{5}$, $8\frac{4}{5}$, 10.

Therefore these are the heights of the wall in feet at the points specified.

MISCELLANEOUS EXAMPLES. II. Page 80.

3. $\sqrt{2xy+4xz}+\sqrt{9y}+\frac{2xyz}{3}=\sqrt{2.3.4+4.3.1}+\sqrt{9.4}+\frac{2.3.4.1}{3}$
$=\sqrt{36}+\sqrt{36}+8=20.$

5. Using Detached Coefficients, as in Art. 45,

$$\begin{array}{l} 1+1+3+5 \\ \underline{1-1-2} \\ 1+1+3+5 \\ \quad -1-1-3-\ 5 \\ \qquad\quad -2-2-\ 6-10 \\ \hline 1+0+0+0-11-10 \end{array}$$

Thus the product is $x^5 - 11x - 10$.

14. $\dfrac{2a+b(2c-a)}{3b-\sqrt{2c^3}} = \dfrac{2.4+3(4-4)}{3.3-\sqrt{2.8}} = \dfrac{8}{9-4} = \dfrac{8}{5}$.

22. $(x+3)(2x+3)-(x+1)(2x+1)=14$;

$\therefore\ 2x^2+9x+9-(2x^2+3x+1)=14$;

$6x+8=14$; whence $x=\mathbf{1}$.

23. The expression $=5.2^3.1-3^2.2^3+2^6.2^5-1.3^4=40-9.8+2048-81$
$=1935$.

24. Removing the brackets,

$$x - \frac{x-13}{9} = \frac{6x+1}{5} + 4 - x.$$

Multiply by 45; thus

$$45x-5(x-13)=9(6x+1)+45(4-x);$$
$$45x-5x+65=54x+9+180-45x;$$
$$40x-9x=189-65;\ \text{whence } x=4.$$

On substituting $x=3$ in the last line, the first side becomes 31×3 or 93, and the second side is 124; thus $x=3$ does not satisfy the equation.

27. By Ex. 2, Art. 53, $a^3+b^3+c^3-3abc$ is the product of $a+b+c$ and $a^2+b^2+c^2-bc-ca-ab$.

Now $a+b+c=1+4-5=0$.

Therefore $a^3+b^3+c^3-3abc$ contains a zero factor, and therefore is itself zero. [Art. 11.]

28. (2) $\dfrac{2x-2}{5}+\dfrac{15}{2}-\dfrac{5x}{2}+\dfrac{19}{10}=\dfrac{3x}{10}-\dfrac{3}{5}$;

Multiply by 10; thus $4x-4+75-25x+19=3x-6$;

$-24x=-96$;

$\therefore\ x=4$.

29. The divisor $= y(y^2+3y-4)-2 = y^3+3y^2-4y-2$.

Using Detached Coefficients, we have

$$
\begin{array}{r}
1+3-4-2\,)\,3+0-37+35+\ 7+0+2\,(\,3-9+2-1 \\
\underline{3+9-12-\ 6\qquad\qquad\qquad} \\
-9-25+41+\ 7\qquad\quad \\
\underline{-9-27+36+18\qquad\quad} \\
2+\ 5-11+0\qquad \\
\underline{2+\ 6-\ 8-4\qquad} \\
-\ 1-\ 3+4+2 \\
\underline{-\ 1-\ 3+4+2}
\end{array}
$$

$\therefore$ the quotient is $3y^3-9y^2+2y-1$.

30. Let x be the number of coins each receives; then the total number of shillings is $\frac{5}{2}x+x$.

$$\therefore\ \frac{5}{2}x+x = 1120\times 20;$$

$$7x = 1120\times 40; \quad \text{whence } x = 6400.$$

That is, A receives 6400 half-crowns, or £800;
B receives 6400 shillings, or £320.

31. The expression $=(-1-2)^2+(-2+3)^2+(3-1)^2$
$=(-3)^2+1^2+2^2 = 9+1+4 = 14$.

32. The product $=(2m^2+8)(m+2)\times 3(m-2)$
$=2\times 3(m^2+4)(m^2-4)$
$=6(m^4-16) = 6m^4-96$.

33. The divisor $=3x^2-6x-6+5x-x^2-15$
$=2x^2-x-21$.

The dividend $=(x^2+x-6)\times(2x-7)$.

Using Detached Coefficients, we have

$$
\begin{array}{r}
1+1-\ 6\qquad \\
\underline{2-7\qquad\qquad} \\
2+2-12\qquad \\
\underline{-7-\ 7+42} \\
2-1-21\,)\,2-5-19+42\,(\,1-2 \\
\underline{2-1-21\qquad} \\
-4+\ 2+42 \\
\underline{-4+\ 2+42}
\end{array}
$$

Thus the quotient is $x-2$.

35. (1) $\frac{3x}{2}-\frac{5}{7}=21x-\frac{2x}{3}-\frac{143}{42}$.

Multiplying by 42, to clear of fractions,

$$63x-30=882x-28x-143;$$
$$91x-882x=30-143;$$
$$-791x=-113;$$
$$\therefore\ x=\frac{1}{7}.$$

(2) $$3x-4-\frac{28x-36}{15}=\frac{24}{5}+\frac{4x-4}{15};$$
$$45x-60-28x+36=72+4x-4;$$
$$45x-32x=68+24;$$
$$13x=92;$$
$$\therefore\ x=7\tfrac{1}{13}.$$

36. Suppose he bought $20x$ at the first price, and $100x$ at the second then the number of shillings he spent was $x\times\frac{4}{3}+x\times\frac{25}{4}$.

Also, he sold $120x$ eggs for $10x\times\frac{5}{6}$ shillings, and gained 27 shillings.

$$\therefore\ \frac{25x}{3}-\left(\frac{4x}{3}+\frac{25}{4}x\right)=27;$$
$$\frac{25x}{3}-\frac{16x+75x}{12}=27;$$
$$100x-91x=12\times27;$$
$$\therefore\ x=36.$$

Hence the number of eggs was 36×120, or 4320.

MISCELLANEOUS EXAMPLES. III. PAGE 126.

3. The expressions may be written

$$(a^3+2ac^2)-(2a^2c+c^3) \text{ and } (a^3+2ac^2)+(2a^2c+c^3).$$

by Art 133, the product is

$$(a^3+2ac^2)^2-(2a^2c+c^3)^2 \text{ or } a^6+4a^4c^2+4a^2c^4-(4a^4c^2+4a^2c^4+c^6),$$

which reduces to a^6-c^6.

7. The value $=\frac{16-9}{5}+\frac{9-4}{6}+\frac{4-16}{7}$

$$=\frac{7}{5}+\frac{5}{6}-\frac{12}{7}=\frac{294+175-360}{210}=\frac{109}{210}.$$

8. Using Detached Coefficients, we have

$$
\begin{array}{l}
1+\frac{3}{2}+\frac{1}{4}\Big)\,1+\frac{9}{4}+\frac{21}{8}+\frac{33}{16}+\frac{5}{16}\Big(\,1+\frac{3}{4}+\frac{5}{4} \\
\qquad\quad 1+\frac{3}{2}+\frac{1}{4} \\
\qquad\quad\quad \frac{3}{4}+\frac{19}{8}+\frac{33}{16} \\
\qquad\quad\quad \frac{3}{4}+\frac{9}{8}+\frac{3}{16} \\
\qquad\qquad\quad \frac{5}{4}+\frac{15}{8}+\frac{5}{16} \\
\qquad\qquad\quad \frac{5}{4}+\frac{15}{8}+\frac{5}{16}
\end{array}
$$

Thus the quotient is $x^2+\frac{3}{4}x+\frac{5}{4}$.

11. (1) By addition, $24x+24y=48$,

$$\therefore\ x+y=2.$$

By subtraction, $2x-2y=-12$,

$$\therefore\ x-y=-6.$$

Whence $x=-2,\ y=4$.

(2) Put $p=19x$, $q=13y$, then the equations become

$$3p+4q=181,\quad 4p-3q=458.$$

Whence $p=95$, $q=-26$. And $x=5$, $y=-2$.

12. Let x miles be the required distance; then the coach goes $\frac{x}{m}$ miles per hour.

And the train goes $\frac{a}{b}$ miles per hour;

$$\therefore\ \frac{a}{b}=\frac{px}{m};\quad \text{whence } x=\frac{am}{pb}.$$

14. (1) $\frac{5x}{7}-\frac{4x}{5}+\frac{3}{5}-\frac{2x}{21}-\frac{1}{3}+1=0.$

Clear of fractions by multiplying by $7\times5\times3$;

thus $75x-84x+63-10x-35+105=0$;

$$\therefore\ -19x=-133;$$

$$\therefore\ x=7.$$

(2) $\frac{10x}{3}-2+\frac{11}{5}+\frac{14x}{15}=\frac{2x}{5}+\frac{7}{5}-7.$

$$\frac{10x}{3}+\frac{14x}{15}-\frac{2x}{5}=-\frac{4}{5}-5;$$

$$50x+14x-6x=-12-75;$$

$$58x=-87;$$

$$x=-1\tfrac{1}{2}.$$

18. Let £x be the sum B wins; then

$$6+x=\frac{5}{9}(50-x); \quad \text{whence } x=14.$$

20. $a(a-1)(a-2)(a-3)=[a(a-3)][(a-1)(a-2)]$

$$=(a^2-3a)(a^2-3a+2)$$

$$=(a^2-3a)^2+2(a^2-3a).$$

Also $\qquad (a^2-3a+1)^2=(a^2-3a)^2+2(a^2-3a)+1.$

22. $\frac{3x}{4}-2-\frac{49}{8}+\frac{21x}{32}=5-\frac{15x}{64}.$

Multiply every term by 64 and transpose; thus

$$48x+42x+15x=7\times 64+49\times 8;$$

$$105x=56\times 15;$$

$$\therefore\ x=8.$$

23. The quotient $=\frac{(3x^2-2xy-y^2)(2x-y)}{x-y}=\frac{(x-y)(3x+y)(2x-y)}{x-y}$

$$=(3x+y)(2x-y)=6x^2-xy-y^2.$$

24. Let x pence represent the price of apples per dozen, and y pence the price of eggs per score.

Then each apple costs $\frac{x}{12}$ pence and each egg costs $\frac{y}{20}$ pence.

Hence, from the given conditions, we have

$$\frac{60x}{12}+\frac{100y}{20}=100, \quad \text{or} \quad x+y=20,$$

and
$$\frac{72x}{12}=\frac{30y}{20}, \quad \text{or} \quad 6x=\frac{3}{2}y.$$

From these equations we obtain $x=4$, $y=16$.

Thus the apples are 4d. per dozen, and the eggs 1s. 4d. per score.

25 $2x^2-13x+15=(2x-3)(x-5),$

$x^2-4x-5=(x+1)(x-5),$

$2x^2-x-3=(2x-3)(x+1);$

$\therefore$ the product $=(2x-3)^2(x+1)^2(x-5)^2;$

$\therefore$ the square root $=(2x-3)(x+1)(x-5).$

26. The expression $=x-y+z-2[x+z+6y-6]+2x-12+18y$

$=x-y+z-2x-2z-12y+12+2x-12+18y$

$=x+5y-z=6+35-8=33.$

27. Using Detached Coefficients, we have

$$\begin{array}{r|l}
3+15+7-9 & 6+57+128-60-130+63\ (2+9-7 \\
 & \underline{6+30+\ 14-\ 18} \\
 & \quad 27+114-\ 42-130 \\
 & \quad \underline{27+135+\ 63-\ 81} \\
 & \qquad -\ 21-105-\ 49+63 \\
 & \qquad \underline{-\ 21-105-\ 49+63}
\end{array}$$

Thus the quotient is $2x^2+9xy-7y^2$.

28. By adding the first and third equations we obtain $5x+9y=37$. Again by adding twice the third equation to the second we obtain $5x+13y=49$. From these two equations we get, by subtraction, $4y=12$; whence $y=3$, $x=2$, $z=0$.

29. The whole work can be done by 1 man in bc days; $\therefore$ 1 man will do $\frac{1}{m}$th of the work in $\frac{bc}{m}$ days, and a men will do $\frac{1}{m}$th of the work in $\frac{bc}{am}$ days.

EXAMPLES. XVIII. b. Page 133.

24. $6-8a-32a^2-18a^3=2(3-4a-16a^2-9a^3),$

$20-35a-95a^2-40a^3=5(4-7a-19a^2-8a^3).$

Using Detached Coefficients, we have

$$\begin{array}{r|r|r|r}
 & 4-\ 7-19-\ 8 & 3-\ 4-\ 16-\ 9 & \\
 & 5 & 4 & \\
4 & 20-35-95-40 & 12-16-\ 64-\ 36 & 3 \\
 & \underline{20-28-48\quad\ } & \underline{12-21-\ 57-\ 24} & \\
-7 & -\ 7-47-40 & 5\quad 7-\ 12 & \\
 & \underline{-\ 7-\ 7\quad\ } & 7\qquad\qquad & \\
-40 & -40-40 & 35-\ 49-\ 84 & -5 \\
 & \underline{-40-40} & \underline{35+235+200} & \\
 & & -284\ |\ -284-284 & \\
 & & 1+\ 1 &
\end{array}$$

Thus the H.C.F. is $1+a$.

25. $9x^2-15x^3-45x^4-12x^5=3x^2(3-5x-15x^2-4x^3)$,

$42x-49x^2-203x^3-84x^4=7x(6-7x-29x^2-12x^3)$.

Using Detached Coefficients, we have

1	$3-5-15-4$	$6-7-29-12$	2
	$3+1-4$	$6-10-30-8$	
-2	$-6-11-4$	$3+1-4$	1
	$-6-2+8$	$3+4$	
	$-3 \mid -9-12$	$-3-4$	-1
	$3+4$	$-3-4$	

Thus the H.C.F. is $x(3+4x)$.

26. Using Detached Coefficients, we have

	$3+0-5+0+0+2$	$2+0+0-5+0+3$	
	2	3	
3	$6+0-10+0+0+4$	$6+0+0-15+0+9$	2
	$6-9+0+3$	$6+0-10+0+0+4$	
	$9-10-3+0+4$	$5 \mid 10-15+0+5$	
	2	$2-3+0+1$	
9	$18-20-6+0+8$	7	
	$18-27+0+9$	$14-21+0+7$	2
7	$7-6-9+8$	$14-12-18+16$	
	$7-14+7$	$-9 \mid -9+18-9$	
8	$8-16+8$	$1-2+1$	
	$8-16+8$		

Thus the H.C.F. is x^2-2x+1.

27. $4x^5-6x^3-28x=2x(2x^4-3x^2-14)$.

	$2+0-3+0-14$	$6+10-17-35-14$	3
	5	$6+0-9+0-42$	
1	$10+0-15+0-70$	$10-8-35+28$	
	$10-8-35+28$	2	
	$2 \mid 8+20-28-70$	$20-16-70+56$	5
2	$4+10-14-35$	$20+50-70-175$	
	$4+0-14$	$-33 \mid -66+0+231$	
5	$10+0-35$	$2+0-7$	
	$10+0-35$		

Thus the H.C.F. is $2x^2-7$

EXAMPLES. XIX. c. PAGE 142.

26. The expression $=\dfrac{a^3(a+x)^3}{(a+x)(a-x)}\times\dfrac{(a-x)^2}{a^2(a+x)(a^2-ax+x^2)}\times\dfrac{a^2-ax+x^2}{a(a+x)^2}$

$$=\frac{a-x}{a+x}.$$

27. The expression $=\dfrac{m(m^2+4mn+4n^2)}{n(3m^2-5mn-2n^2)}\times\dfrac{(m+2n)(m-2n)}{9m^2-3mn+n^2}\times\dfrac{27m^3+n^3}{(m+2n)^3}$

$$=\frac{m(m-2n)}{n(3m+n)(m-2n)}\times\frac{(3m+n)(9m^2-3mn+n^2)}{9m^2-3mn+n^2}$$

$$=\frac{m}{n}.$$

28. The expression

$$=\frac{(1+2x)(1-2x+4x^2)}{(2-x)^2}\times\frac{x(2-x)(2+x)}{(1-2x)(1+2x)}\times\frac{(2-x)(1-2x)}{1-2x+4x^2}$$

$$=x(2+x).$$

29. The expression $=\dfrac{x^2(x-4)^2}{x^2+4x+16}\times\dfrac{(4-x)(16+4x+x^2)}{(4+x)(4-x)}\times\dfrac{(x+4)^2}{x^3(x-4)^3}$

$$=\frac{x+4}{x(x-4)}.$$

30. The expression

$$=\frac{(p+q-r)(p+q+r)}{(p+q+r)^2}\times\frac{p(p+q+r)}{(p-r+q)(p-r-q)}\times\frac{(p-q+r)(p-q-r)}{p(p-q+r)}$$

$$=1.$$

31. The expression

$$=\frac{(a^2+x^2)(a+x)(a-x)}{(a-x)^2}\div\left[\frac{x(a^2+x^2)}{(a-x)(a^2+ax+x^2)}\times\frac{(a^2+x^2)^2-a^2x^2}{x(a^2-ax+x^2)}\right]$$

$$=\frac{(a^2+x^2)(a+x)}{(a-x)}\times\frac{(a-x)(a^2+ax+x^2)}{(a^2+x^2)}\times\frac{a^2-ax+x^2}{(a^2+x^2+ax)(a^2+x^2-ax)}$$

$$=a+x.$$

32. The expression

$$=\frac{a(a+3b)(a+5b)}{(4a-b)(16a^2+4ab+b^2)(a^3+b^3)}\times\frac{(16a^2-b^2)(a^2-b^2)}{(4a+b)(a+5b)}\times\frac{a(a^2-ab+b^2)}{(a+3b)(a-b)}$$

$$=\frac{a(a-b)(a+b)}{(16a^2+4ab+b^2)(a+b)(a^2-ab+b^2)}\times\frac{a(a^2-ab+b^2)}{a-b}$$

$$=\frac{a^2}{16a^2+4ab+b^2}.$$

EXAMPLES. XX. a. PAGE 144.

28. $$(3x-5x^2)^2 = x^2(3-5x)^2;$$
$$6-7x-5x^2 = (3-5x)(2+x);$$
$$4x+4x^2+x^3 = x(4+4x+x^2) = x(2+x)^2.$$
$$\therefore \text{ L.C.M.} = x^2(3-5x)^2(2+x)^2.$$

29. $$14a^4(a^3-b^3) = 14a^4(a-b)(a^2+ab+b^2);$$
$$6a^3b(a-b)(a^2-b^2) = 6a^3b(a-b)^2(a+b);$$
$$21a^2b^2(a-b)^3.$$
$$\therefore \text{ L.C.M.} = 42a^4b^2(a-b)^3(a+b)(a^2+ab+b^2).$$

30. $$m^4+m^2n^2+n^4 = (m^2+mn+n^2)(m^2-mn+n^2);$$
$$m^3n+n^4 = n(m^3+n^3) = n(m+n)(m^2-mn+n^2);$$
$$(m^2-mn)^3 = m^3(m-n)^3.$$
$$\therefore \text{ L.C.M.} = m^3n(m-n)^3(m+n)(m^4+m^2n^2+n^4)$$
$$= m^3n(m-n)^2(m^6-n^6).$$

31. $$(2c^2-3cd)^2 = c^2(2c-3d)^2;$$
$$(4c-6d)^3 = 8(2c-3d)^3;$$
$$8c^3-27d^3 = (2c-3d)(4c^2+6cd+9d^2).$$
$$\therefore \text{ L.C.M.} = 8c^2(2c-3d)^3(4c^2+6cd+9d^2)$$
$$= 8c^2(2c-3d)^2(8c^3-27d^3).$$

EXAMPLES. XXI. b. PAGE 151.

25. The expression $= \dfrac{3(x-2)+x+2}{(x+2)(x-2)^2} = \dfrac{4x-4}{(x+2)(x-2)^2}.$

26. The expression $= \dfrac{x(x+a)-a(x-a)}{ax(x-a)(x+a)^2} = \dfrac{x^2+a^2}{ax(x-a)(x+a)^2}.$

EXAMPLES. XXI. c. PAGE 153.

38. The expression $= \dfrac{x+1}{2x^2(x-2)} + \dfrac{x-1}{2x^2(x+2)} - \dfrac{1}{(x+2)(x-2)}$

$$= \frac{(x+1)(x+2)+(x-1)(x-2)-2x^2}{2x^2(x+2)(x-2)}$$

$$= \frac{x^2+3x+2+x^2-3x+2-2x^2}{2x^2(x+2)(x-2)}$$

$$= \frac{4}{2x^2(x^2-4)} = \frac{2}{x^2(x^2-4)}.$$

39. The expression $=\dfrac{1}{(3x-y)(x-y)}+\dfrac{1}{(x-3y)(x-y)}-\dfrac{3}{(3x-y)(x-3y)}$

$$=\frac{x-3y+3x-y-3(x-y)}{(3x-y)(x-y)(x-3y)}$$

$$=\frac{x-y}{(3x-y)(x-y)(x-3y)}$$

$$=\frac{1}{(3x-y)(x-3y)}.$$

40. The expression $=\dfrac{1}{x-1}+\dfrac{2}{x+1}-\dfrac{3x-2}{(x+1)(x-1)}-\dfrac{1}{(x+1)^2}$

$$=\frac{(x+1)^2+2(x^2-1)-(3x-2)(x+1)-(x-1)}{(x+1)(x-1)(x+1)}$$

$$=\frac{x^2+2x+1+2x^2-2-(3x^2+x-2)-x+1}{(x+1)(x-1)(x+1)}$$

$$=\frac{2}{(x-1)(x+1)^2}.$$

41. The expression $=\dfrac{108-52x-4x(3-x)}{x(3-x)^2}+\dfrac{x(1+x)^2-12(3-x)^2}{x(3-x)^2}$

$$=\frac{108-64x+4x^2+x(1+2x+x^2)-12(9-6x+x^2)}{x(3-x)^2}.$$

$$=\frac{108-64x+4x^2+x+2x^2+x^3-108+72x-12x^2}{x(3-x)^2}$$

$$=\frac{9x-6x^2+x^3}{x(3-x)^2}=\frac{x(9-6x+x^2)}{x(3-x)^2}$$

$$=\frac{(3-x)^2}{(3-x)^2}=1.$$

42. The expression

$$=\frac{(a+b)^2}{(x-a)(x+a+b)}-\frac{a+2b+x}{2(x-a)}+\frac{(a+b)x}{(x+a)(x-a)+b(x-a)}+\frac{1}{2}$$

$$=\frac{(a+b)(a+b+x)}{(x-a)(x+a+b)}-\frac{a+2b+x}{2(x-a)}+\frac{x-a}{2(x-a)}$$

$$=\frac{2(a+b)-a-2b-x+x-a}{2(x-a)}$$

$$=0.$$

43. The expression$=\dfrac{3(x^2+x-2)^2-3(x^2-x-2)^2}{(x^2-4)(x^2-1)}-\dfrac{8x}{x^2-4}$

$$=\frac{3(2x^2-4)2x-8x(x^2-1)}{(x^2-4)(x^2-1)}$$

$$=\frac{4x(3x^2-6-2x^2+2)}{(x^2-4)(x^2-1)}$$

$$=\frac{4x(x^2-4)}{(x^2-4)(x^2-1)}=\frac{4x}{x^2-1}.$$

EXAMPLES. XXI. d. PAGE 156.

27. The expression$=\dfrac{(a-x)^2}{2(a-x)(a+x)}-\dfrac{2ax(a+x)}{(a-x)(a+x)^2}-\dfrac{(x-a)(x+a)}{2(x-a)^2}$

$$=\frac{a-x}{2(a+x)}-\frac{2ax}{a^2-x^2}+\frac{a+x}{2(a-x)}$$

$$=\frac{(a-x)^2-4ax+(a+x)^2}{2(a^2-x^2)}$$

$$=\frac{a^2-2ax+x^2-4ax+a^2+2ax+x^2}{2(a^2-x^2)}$$

$$=\frac{2a^2-4ax+2x^2}{2(a^2-x^2)}=\frac{2(a-x)^2}{2(a^2-x^2)}$$

$$=\frac{a-x}{a+x}.$$

28. The expression$=\dfrac{2}{a+b}-\dfrac{1}{a-b}+\dfrac{3b}{a^2-b^2}+\dfrac{ab}{(a+b)(a^2-ab+b^2)}$

$$=\frac{(2a-2b-a-b+3b)(a^2-ab+b^2)+ab(a-b)}{(a^2-b^2)(a^2-ab+b^2)}$$

$$=\frac{a^3-a^2b+ab^2+a^2b-ab^2}{(a^2-b^2)(a^2-ab+b^2)}$$

$$=\frac{a^3}{(a-b)(a^3+b^3)}.$$

29. The expression$=\dfrac{3+3x+1-x+6}{8(1-x^2)}-\dfrac{1-x}{4(1+x^2)}$

$$=\frac{5+x}{4(1-x^2)}-\frac{1-x}{4(1+x^2)}=\frac{(5+x)(1+x^2)-(1-x)(1-x^2)}{4(1-x^4)}$$

$$=\frac{2+x+3x^2}{2(1-x^4)}.$$

30. The expression $=\frac{1}{x}+\frac{1}{x-1}+\frac{1}{x+1}-\frac{x}{x^2-1}+\frac{3}{x(x^2-1)}$

$$=\frac{x^2-1+x^2+x+x^2-x-x^2+3}{x(x^2-1)}$$

$$=\frac{2x^2+2}{x(x^2-1)}=\frac{2(x^2+1)}{x(x^2-1)}.$$

31. The expression $=\frac{a(a+c)}{c(a^2-c^2)}-\frac{a^2-c^2}{c(a+c)^2}+\ldots-\ldots$

$$=\frac{a}{c(a-c)}-\frac{a-c}{c(a+c)}+\frac{2c}{c^2-a^2}-\frac{3}{a+c}$$

$$=\frac{a(a+c)-(a-c)^2-2c^2-3c(a-c)}{c(a^2-c^2)}$$

$$=\frac{a^2+ac-a^2+2ac-c^2-2c^2-3ac+3c^2}{c(a^2-c^2)}$$

$$=0.$$

32. The expression

$$=\frac{(4a+6b)(a-b)+(6a-4b)(a+b)-(4a^2+6b^2)}{a^2-b^2}+\ldots-\ldots$$

$$=\frac{4a^2+2ab-6b^2+6a^2+2ab-4b^2-4a^2-6b^2}{a^2-b^2}+\ldots-\ldots$$

$$=\frac{6a^2+4ab-16b^2}{a^2-b^2}+\frac{4b^2-6a^2}{a^2+b^2}+\frac{20b^4}{a^4-b^4}$$

$$=\frac{(a^2+b^2)(6a^2+4ab-16b^2)+(a^2-b^2)(4b^2-6a^2)+20b^4}{a^4-b^4}$$

$$=\frac{4ab(a^2+b^2)}{a^4-b^4}=\frac{4ab}{a^2-b^2}.$$

EXAMPLES. XXII. a. PAGE 163.

31. The numerator $=\frac{(a-b)(1+bc)+(b-c)(1+ab)}{(1+ab)(1+bc)}$

$$=\frac{a-b+b-c+bc(a-b)+ab(b-c)}{(1+ab)(1+bc)}$$

$$=\frac{a-c+b(ac-bc+ab-ac)}{(1+ab)(1+bc)}$$

$$=\frac{a-c+b^2(a-c)}{(1+ab)(1+bc)}=\frac{(a-c)(1+b^2)}{(1+ab)(1+bc)}.$$

The denominator $=\dfrac{(1+ab)(1+bc)-(a-b)(b-c)}{(1+ab)(1+bc)}$

$$=\frac{1+ab+bc+ab^2c-ab+b^2+ac-bc}{(1+ab)(1+bc)}$$

$$=\frac{1+ac+b^2(1+ac)}{(1+ab)(1+bc)}=\frac{(1+b^2)(1+ac)}{(1-ab)(1+bc)}.$$

$\therefore$ the fraction $=\dfrac{a-c}{1+ac}.$

32. The first fraction $=\dfrac{a+b+a-b}{a+b-(a-b)}=\dfrac{a}{b}.$

The second fraction $=\dfrac{a^2}{b^2}$, similarly;

$\therefore$ the whole expression $=\dfrac{a}{b}\times\dfrac{b^2}{a^2}=\dfrac{b}{a}.$

33. $$\frac{a-x}{a^2-ax-\dfrac{(a-x)^2}{1-\dfrac{a}{x}}}=\frac{a-x}{a(a-x)-\dfrac{(a-x)^2}{1-\dfrac{a}{x}}}=\frac{1}{a-\dfrac{a-x}{1-\dfrac{a}{x}}}=\frac{1}{a-\dfrac{x(a-x)}{x-a}}$$

$$=\frac{1}{a+x}.$$

34. The first fraction $=\dfrac{2x(x+1)}{\dfrac{x(x-2)}{x-(x-2)}-\dfrac{3}{2}}=\dfrac{2x(x+1)}{\dfrac{x(x-2)}{2}-\dfrac{3}{2}}$

$$=\frac{4x(x+1)}{x^2-2x-3}=\frac{4x(x+1)}{(x+1)(x-3)}$$

$$=\frac{4x}{x-3}.$$

The second fraction $=\dfrac{3(x-1)}{\dfrac{x(x-4)}{x-(x-4)}+\dfrac{3}{4}}=\dfrac{3(x-1)}{\dfrac{x(x-4)}{4}+\dfrac{3}{4}}$

$$=\frac{3(x-1)\times 4}{x^2-4x+3}=\frac{12(x-1)}{(x-1)(x-3)}$$

$$=\frac{12}{x-3}.$$

$\therefore$ the whole expression $=\dfrac{4x-12}{x-3}=4.$

35. $$1+\frac{\dfrac{1}{2x-1}}{1+\dfrac{1}{4x-1}}=1+\frac{4x-1}{4x(2x-1)}=1+\frac{4x-1}{8x^2-4x}$$

$$=\frac{8x^2-1}{8x^2-4x}.$$

the whole fraction $$=\frac{2-4x}{4x-2-\dfrac{4x}{\dfrac{8x^2-1}{8x^2-4x}}}=\frac{2-4x}{4x-2+\dfrac{16x^2(1-2x)}{8x^2-1}}=\frac{1}{-1+\dfrac{8x^2}{8x^2-1}}$$

$$=8x^2-1.$$

36. The first fraction $$=\frac{x^2}{1-\dfrac{1}{x^2+\dfrac{1}{x^2+1}}}=\frac{x^2}{1-\dfrac{x^2+1}{x^4+x^2+1}}$$

$$=\frac{x^2(x^4+x^2+1)}{x^4}=\frac{x^4+x^2+1}{x^2}.$$

The second fraction $$=\frac{x^2-2}{1-\dfrac{1}{x^2-\dfrac{1}{x^2-1}}}=\frac{x^2-2}{1-\dfrac{x^2-1}{x^4-x^2-1}}$$

$$=\frac{(x^2-2)(x^4-x^2-1)}{x^4-2x^2}=\frac{x^4-x^2-1}{x^2}$$

$\therefore$ the whole expression $$=\frac{x^4+x^2+1+x^4-x^2-1}{x^2}=2x^2.$$

EXAMPLES. XXII. c. PAGE 171.

55. The expression $$=\frac{(x+3)(x-2)+6}{(x+3)(x-4)+12}\times\frac{(x+3)(x-3)}{4(x+1)}\times\frac{7(x-4)}{(x-3)(x+3)}$$

$$=\frac{x^2+x}{x^2-x}\times\frac{7(x-4)}{4(x+1)}$$

$$=\frac{7(x-4)}{4(x-1)}.$$

56. The expression $$=(x+2)\left\{\frac{x^2+5x+6}{(x+2)(x-3)}\right\}\left\{\frac{x^2-2x-3}{(x+2)(x+1)}\right\}$$

$$=\frac{(x+2)(x+3)}{(x-3)}\times\frac{(x-3)(x+1)}{(x+2)(x+1)}$$

$$=x+3.$$

57. The expression $=(1+a)^2 \div \left\{1+\dfrac{a}{\dfrac{1-a^3+a}{1+a+a^2}}\right\}$

$$=(1+a)^2 \div \left\{\frac{1+2a+a^2}{1+a-a^3}\right\}$$

$$=1+a-a^3.$$

58. The expression

$$=\frac{\{ax^2+(b-c)x-f+ax^2+(b+c)x-f\}\{ax^2+(b-c)x-f-ax^2-(b+c)x+f\}}{\{ax^2+(b+e)x-f+ax^2+(b-e)x-f\}\{ax^2+(b+e)x-f-ax^2-(b-e)x+f\}}$$

$$=\frac{(2ax^2+2bx-2f)(-2cx)}{(2ax^2+2bx-2f)(2ex)}=-\frac{c}{e}.$$

MISCELLANEOUS EXAMPLES. IV. PAGE 172.

1. The expression $=\dfrac{-4+\sqrt{16+9}}{-64+6(16-9)}=\dfrac{-4+5}{-22}=-\dfrac{1}{22}.$

2. The expression $=\sqrt{3.1(-3)+3(-1)1+3.8.2}=\sqrt{36}=6.$

3. The expression $=a(b-c)[b^2-2bc+c^2-(2b^2-bc+2c^2)+b^2+2bc+c^2]$

$$=abc(b-c)=1.2.3(2-3)=-6.$$

4. The expression $=\sqrt[3]{5(25-9)-16}+\sqrt[4]{3\{4(16-9)-1\}}$

$$=\sqrt[3]{64}+\sqrt[4]{81}=4+3=7.$$

5. The expression $=\sqrt{(1-27+1)(-1+3-3)}\div\sqrt[3]{(-1)(-27)}$

$$=\sqrt{25}\div\sqrt[3]{27}=\frac{5}{3}.$$

6. (1) The expression $=4^3-3.0.4^2+6(1+27)=64+168=232.$

(2) The expression $=1-4(1+3+5)+\sqrt{2+9+25}=1-36+\sqrt{36}$

$$=-29.$$

7. (1) The expression $=x\left\{\dfrac{x}{7}-\dfrac{y}{6}-\dfrac{2}{3}\right\}-\dfrac{2}{7}(xy+y)-\dfrac{1}{7}(x^2-2y)+\dfrac{2}{3}x$

$$=\frac{x^2}{7}-\frac{xy}{6}-\frac{2}{3}x-\frac{2}{7}xy-\frac{2}{7}y-\frac{1}{7}x^2+\frac{2}{7}y+\frac{2}{3}x$$

$$=-\frac{xy}{6}-\frac{2}{7}xy=-7-12=-19.$$

(2) The expression $=x\left\{yz-\frac{y}{z}\right\}-\frac{y}{z}\left\{y-x+3xz^2\right\}-2xz\left\{-y-\frac{y^2}{2xz^2}\right\}$

$$=xyz-\frac{xy}{z}-\frac{y^2}{z}+\frac{xy}{z}-3xyz+2xyz+\frac{y^2}{z}$$

$$=3xyz-3xyz=0.$$

8. The expression $=\frac{4+\{1-2(-2)\}}{4.0+1.3}\times\sqrt{\frac{2-1}{1+4.2}}=\frac{9}{3}\times\frac{1}{3}=1.$

9. The expression $=\frac{2a-2b+c+\sqrt{2a^2+b^2+c^2}}{\sqrt{c^4+a^2b^2+2a^3c}}$

$$=\frac{-2-10+3+\sqrt{2+25+9}}{\sqrt{81+25-6}}=\frac{-9+6}{10}=-\frac{3}{10}.$$

10. (1) The expression $=64+8-6^3-11(-6+3)\left(\frac{9}{2}-\frac{1}{2}\right)$

$$=72-216+33.4=204-216=-12.$$

(2) The expression $=\sqrt[3]{9-4(4+4+1)}-\sqrt[3]{16.\frac{3}{2}+11(-8)1}$

$$=\sqrt[3]{-27}-\sqrt[3]{24-88}=-3+4=1.$$

11. The expression $=l-m-3[\{l-m+m+l\}^2-4l^2]$

$$=l-m-3(4l^2-4l^2)=l-m=1.$$

12. The expression $=\frac{a-b+2a-2b+\frac{b}{2}}{a+ab}$, since $c=0$,

$$=\frac{3a-\frac{5}{2}b}{a(1+b)}=\frac{3+\frac{5}{4}}{\frac{1}{2}}=\frac{17}{2}=8\tfrac{1}{2}.$$

13. The expression $=7(4x-3y)-18\left(x-\frac{4}{3}y\right)-8(3x-2y)+21\left(\frac{16}{3}x-y\right)$

$$=28x-18x-24x+112x-21y+24y+16y-21y$$

$$=98x-2y=\frac{49}{3}+3=19\tfrac{1}{3}.$$

14. (1) The expression $=\frac{z}{x}\left\{y-2z+\frac{x}{z}\right\}-\frac{y}{z}\left[\frac{z^2}{x}-3-\frac{2z^3}{xy}+\frac{x}{z}\right]$

$$=\frac{yz}{x}-\frac{2z^2}{x}+1-\frac{yz}{x}+\frac{3y}{z}+\frac{2z^2}{x}-\frac{xy}{2z}$$

$$=1+\frac{3\times7}{8}-\frac{6\times7}{2\times8}=1.$$

(2) The expression $=2y^2z-\dfrac{3y^2}{5}+3y+\dfrac{y^2}{5}\left[3-10z+\dfrac{xyz}{2}-\dfrac{7xz}{2}\right]$

$$=2y^2z-\frac{3y^2}{5}+3y+\frac{3y^2}{5}-2y^2z+\frac{xy^3z}{10}-\frac{7xy^2z}{10}$$

$$=3y+\frac{xy^2z}{10}(y-7)=3y=21.$$

22. $d^4-4d^2c^2-45c^4=(d^2+5c^2)(d^2-9c^2)=(d^2+5c^2)(d+3c)(d-3c).$

39. $2+x^3-3x^6=(2+3x^3)(1-x^3)=(2+3x^3)(1-x)(1+x+x^2).$

51. $12(a^2b^2-1)+7ab=12a^2b^2+7ab-12=(3ab+4)(4ab-3).$

52. $2(a^4b^2+5)-9a^2b=2a^4b^2-9a^2b+10=(2a^2b-5)(a^2b-2).$

53. $21x^2+2y(5x-8y)=21x^2+10xy-16y^2=(7x+8y)(3x-2y).$

54. $3(6m^2-5n^2)+17mn=18m^2+17mn-15n^2=(9m-5n)(2m+3n).$

55. $c^2-a^2-b^2+2ab=c^2-(a-b)^2=(c+a-b)(c-a+b).$

56. $a^2+2bc-b^2-c^2=a^2-(b-c)^2=(a+b-c)(a-b+c).$

62. $8c^4-2c^2(d+c)^2=2c^2\{4c^2-(d+c)^2\}$

$$=2c^2(2c+d+c)(2c-d-c)=2c^2(3c+d)(c-d).$$

63. $(a^2b^2-1)^2-x^2+2xy-y^2=(a^2b^2-1)^2-(x-y)^2$

$$=(a^2b^2-1+x-y)(a^2b^2-1-x+y).$$

67. $x^4-2x^2-y^2-z^2+2yz+1=(x^2-1)^2-(y-z)^2$

$$=(x^2-1+y-z)(x^2-1-y+z).$$

68. $a^2-16(b-c)^2=\{a+4(b-c)\}\{a-4(b-c)\}$

$$=(a+4b-4c)(a-4b+4c).$$

69. $c-d-4(c-d)^3=(c-d)\{1-4(c-d)^2\}$

$$=(c-d)(1+2c-2d)(1-2c+2d).$$

70. $p^2-16q^2+p-4q=(p+4q)(p-4q)+p-4q$

$$=(p-4q)(p+4q+1).$$

71. $2+128(a+b)^3=2\{1+64(a+b)^3\}$

$$=2\{1+4(a+b)\}\{1-4(a+b)+16(a+b)^2\}.$$

72. $x+3y+x^3+27y^3=x+3y+(x+3y)(x^2-3xy+9y^2)$

$$=(x+3y)(1+x^2-3xy+9y^2).$$

78. $1-(m^2+p^2)-2mp=1-(m^2+2mp+p^2)$

$$=1-(m+p)^2=(1+m+p)(1-m-p)$$

79. $ab(x^2+1)-x(a^2+b^2)=abx^2+ab-a^2x-b^2x$

$$=ax(bx-a)-b(bx-a)$$
$$=(bx-a)(ax-b).$$

80. $9b^2-6bc-16+c^2=(3b-c)^2-16=(3b-c+4)(3b-c-4).$

81. $x^2c^3-c^3+x^2-1=x^2(c^3+1)-(c^3+1)=(c^3+1)(x^2-1)$

$$=(c+1)(c^2-c+1)(x+1)(x-1).$$

82. $3x^2-2ab-x(b-6a)=3x^2-2ab-bx+6ax$

$$=3x(x+2a)-b(x+2a)$$
$$=(3x-b)(x+2a).$$

83. $m^3-n^3-(x^2-mn)(m-n)=(m-n)(m^2+mn+n^2)-(x^2-mn)(m-n)$

$$=(m-n)\{m^2+mn+n^2-x^2+mn\}$$
$$=(m-n)\{(m+n)^2-x^2\}$$
$$=(m-n)(m+n+x)(m+n-x).$$

84. $a(b^2+c^2-a^2)+b(a^2+c^2-b^2)=ab^2+ac^2+a^2b+c^2b-(a^3+b^3)$

$$=ab(a+b)+c^2(a+b)-(a^3+b^3)$$
$$=(a+b)(ab+c^2-a^2+ab-b^2)$$
$$=(a+b)\{c^2-(a-b)^2\}$$
$$=(a+b)(c+a-b)(c-a+b).$$

85. $x^7-x^3+8x^4-8=x^3(x^4-1)+8(x^4-1)$

$$=(x^3+8)(x^2+1)(x^2-1)$$
$$=(x+2)(x^2-2x+4)(x^2+1)(x+1)(x-1).$$

86. The expression $=(x+7)(x+1)(2x-3)(x+1)(2x-3)(x+7)$;

$\therefore$ the square root $=(x+1)(x+7)(2x-3)$.

87. The expression $=(2x+5y)(x-3y)(2x+5y)(2x-5y)(2x-5y)(x-3y)$;

$\therefore$ the square root $=(2x+5y)(x-3y)(2x-5y)$.

88. $13ab^2(x^3-3a^2x+2a^3)=13ab^2(x-a)(x^2+ax-2a^2)$;

$65a^3b(x^2+ax-2a^2)=65a^3b(x+2a)(x-a)$;

$25b^3(x^2-a^2)^2=25b^3(x-a)^2(x+a)^2$;

$\therefore$ L.C.M. $=325a^3b^3(x^2-a^2)^2(x+2a)$.

89. $2(x^4+9)-5x^2(x+1)=2x^4-5x^3-5x^2+18$;

$2x^3(2x-9)+81(x-1)=4x^4-18x^3+81x-81.$

Using Detached Coefficients, we have

	2 − 5 − 5 + 0 + 18	4 − 18 + 0 + 81 − 81	2
	4	4 − 10 − 10 + 0 + 36	
−1	8 − 20 − 20 + 0 + 72	− 8 + 10 + 81 − 117	
	8 − 10 − 81 + 117	5	
−5	− 10 + 61 − 117 + 72	− 40 + 50 + 405 − 585	4
	− 10 + 45 − 45	− 40 + 244 − 468 + 288	
	8 | 16 − 72 + 72	− 97 | − 194 + 873 − 873	
	2 − 9 + 9	2 − 9 + 9	

$\therefore$ H.C.F. $=2x^2-9x+9$.

90. $(2x^4+4x^3)(x^2+2x-8)=2x^3(x+2)(x+4)(x-2)$;

$(2x^3-4x^2)(x^2-2x-8)=2x^2(x-2)(x-4)(x+2)$;

$(x^2-4x)(x^2+2x-8)=x(x-4)(x+4)(x-2)$;

$\therefore$ the required expression is $2x^3(x+2)(x+4)(x-2)(x-4)$.

91. $a(a+c)-b(b+c)=a^2-b^2+ac-bc=(a-b)(a+b+c)$;

$b(b+a)-c(c+a)=b^2-c^2+ab-ac=(b-c)(b+c+a)$;

$c(c+b)-a(a+b)=c^2-a^2+bc-ab=(c-a)(c+a+b)$;

$\therefore$ H.C.F. $=a+b+c$,

L.C.M. $=(a-b)(b-c)(c-a)(a+b+c)$.

92. $(a+b)(a-b)+c(c-2a)=a^2-b^2+c^2-2ac$

$=(a-c)^2-b^2=(a-c+b)(a-c-b)$;

$(a+c)(a-c)+b(b+2a)=a^2-c^2+b^2+2ab$

$=(a+b)^2-c^2=(a+b+c)(a+b-c)$;

$\therefore$ H.C.F. $=a+b-c$.

93. $2a^2-3ab+b^2=(2a-b)(a-b)$;

also the given L.C.M. $=2a^4-3a^3b+a^2b^2-(2a^2b^2-3ab^3-b^4)$

$=a^2(2a^2-3ab+b^2)-b^2(2a^2-3ab+b^2)$

$=(2a^2-3ab+b^2)(a^2-b^2)$

$=(2a-b)(a-b)(a+b)(a-b)$;

$\therefore$ the required expression must contain the factor $a-b$ *twice*, besides the factor $a+b$; hence it must be $(a-b)^2(a+b)$.

94. $x^4-3x^2y^2-4y^4=(x^2-4y^2)(x^2+y^2)$;

$$x^6-64y^6=(x^2-4y^2)(x^4+4x^2y^2+16y^4);$$

$$\begin{aligned}x^5+2x^4y-8x^3y^2-16x^2y^3+16xy^4+32y^5&=x^4(x+2y)-8x^2y^2(x+2y)+16y^4(x+2y)\\&=(x+2y)(x^4-8x^2y^2+16y^4)=(x+2y)(x^2-4y^2)^2;\end{aligned}$$

$\therefore$ H.C.F. $=x^2-4y^2$.

95.
$$(a-c)^2-(b-c)^2=(a-c+b-c)(a-c-b+c)=(a-b)(a+b-2c),$$
$$a^2+b^2-2ac-2bc+2ab=(a+b)^2-2c(a+b)=(a+b)(a+b-2c),$$
$$a^4-b^4=(a^2+b^2)(a+b)(a-b);$$

$\therefore$ L.C.M. $=(a^2+b^2)(a^2-b^2)(a+b-2c)$.

96.
$$a(a-b)^2-ac^2=a\{(a-b)^2-c^2\}=a(a-b+c)(a-b-c),$$
$$a^2b-b(b-c)^2=b\{a^2-(b-c)^2\}=b(a+b-c)(a-b+c),$$
$$(a+c)^2c-b^2c=c\{(a+c)^2-b^2\}=c(a+c+b)(a+c-b);$$

$$\begin{aligned}\therefore\ \text{L.C.M.}&=abc(a-b+c)(a-b-c)(a+b-c)(a+c+b)\\&=abc\{(a-b)^2-c^2\}\{(a+b)^2-c^2\}\\&=abc(a^2+b^2-c^2-2ab)(a^2+b^2-c^2+2ab)\\&=abc\{(a^2+b^2-c^2)^2-4a^2b^2\}\\&=abc(a^4+b^4+c^4-2b^2c^2-2c^2a^2-2a^2b^2).\end{aligned}$$

97. Taking the first two expressions, and using Detached Coefficients to find their H.C.F., we have

	1 − 15 + 75 − 145 + 84	1 − 17 + 101 − 247 + 210	
	1 − 13 + 51 − 63	1 − 15 + 75 − 145 + 84	
−2	− 2 + 24 − 82 + 84	−2 | − 2 + 26 − 102 + 126	
	− 2 + 26 − 102 + 126	1 − 13 + 51 − 63	
	−2 | − 2 + 20 − 42	1 − 10 + 21	
	1 − 10 + 21	− 3 + 30 − 63	−3
		− 3 + 30 − 63	

$\therefore$ $x^2-10x+21$ is the H.C.F.

The four expressions may now be arranged as follows:

$$\begin{aligned}\text{First expression}&=x^2(x^2-10x+21)-5x(x^2-10x+21)+4(x^2-10x+21)\\&=(x^2-5x+4)(x^2-10x+21)\\&=(x-4)(x-1)(x-3)(x-7).\end{aligned}$$

$$\begin{aligned}\text{Second expression}&=x^2(x^2-10x+21)-7x(x^2-10x+21)+10(x^2-10x+21)\\&=(x^2-7x+10)(x^2-10x+21)\\&=(x-5)(x-2)(x-3)(x-7).\end{aligned}$$

Third expression $=x^2(x^2-10x+21)-3x(x^2-10x+21)+2(x^2-10x+21)$

$=(x^2-3x+2)(x^2-10x+21)$

$=(x-2)(x-1)(x-3)(x-7).$

Fourth expression $=x^2(x^2-10x+21)-9x(x^2-10x+21)+20(x^2-10x+21)$

$=(x^2-9x+20)(x^2-10x+21)$

$=(x-5)(x-4)(x-3)(x-7).$

$\therefore$ H.C.F. $=(x-3)(x-7)$; L.C.M. $=(x-1)(x-2)(x-3)(x-4)(x-5)(x-7)$.

98. The expression $=\dfrac{1}{1-x}+\dfrac{x(1-x)}{(1-x)^3}=\dfrac{1}{1-x}+\dfrac{x}{(1-x)^2}$

$$=\frac{1-x+x}{(1-x)^2}=\frac{1}{(1-x)^2}.$$

99. The expression $=\dfrac{2x-7}{(x-3)^2}-\dfrac{2(x+2)}{(x-3)(x+3)}$

$$=\frac{(2x-7)(x+3)-2(x+2)(x-3)}{(x-3)^2(x+3)}$$

$$=\frac{2x^2-x-21-2x^2+2x+12}{(x-3)^2(x+3)}$$

$$=\frac{x-9}{(x-3)^2(x+3)}.$$

100. The expression $=\dfrac{2}{4x^2-1}+\dfrac{1}{(2x+1)^2}=\dfrac{2(2x+1)+2x-1}{(2x+1)^2(2x-1)}$

$$=\frac{6x+1}{(2x+1)^2(2x-1)}.$$

101. The expression $=\dfrac{\{x+1-(x-1)\}\{(x+1)^2+(x+1)(x-1)+(x-1)^2\}}{x(3x^2+1)}$

$$=\frac{2(x^2+2x+1+x^2-1+x^2-2x+1)}{x(3x^2+1)}$$

$$=\frac{2(3x^2+1)}{x(3x^2+1)}=\frac{2}{x}.$$

102. The expression $=\dfrac{(1+x)^2+2(1-x^2)+(1-x)^2}{(1+x)^2(1-x)^2}$

$$=\frac{1+2x+x^2+2-2x^2+1-2x+x^2}{(1-x^2)^2}=\frac{4}{(1-x^2)^2}.$$

103. The expression $=\dfrac{x^2(x^2-2)^2-(x^2-2)^2}{(x-1)(x+1)(x^2-2)^2}=\dfrac{(x^2-1)(x^2-2)^2}{(x^2-1)(x^2-2)^2}=1.$

104. The expression $=\dfrac{1}{6x-2}-\dfrac{3}{6x-2}-\dfrac{1}{1-3x}$

$$=\frac{-2}{6x-2}-\frac{1}{1-3x}=\frac{2}{2(1-3x)}-\frac{1}{1-3x}=0.$$

105. The expression $=\dfrac{x}{9}+\dfrac{2}{3}+\dfrac{4}{x-6}-\dfrac{2x}{3(x-6)}$

$$=\frac{x(x-6)+6(x-6)+36-6x}{9(x-6)}$$

$$=\frac{x^2-6x+6x-36+36-6x}{9(x-6)}$$

$$=\frac{x^2-6x}{9(x-6)}=\frac{x}{9}.$$

106. The expression $=\dfrac{x-(x+y)+x-y}{x^2-y^2}+\dfrac{(y-x)(x-y)+(x^2-xy+y^2)}{xy(x-y)}$

$$=\frac{x-2y}{x^2-y^2}+\frac{x^2-xy+y^2-(x^2-2xy+y^2)}{xy(x-y)}$$

$$=\frac{x-2y}{x^2-y^2}+\frac{xy}{xy(x-y)}=\frac{x-2y+x+y}{x^2-y^2}$$

$$=\frac{2x-y}{x^2-y^2}.$$

107. The expression $=\dfrac{(x-y)^2-4y^2}{x-y}\times\dfrac{(x+y)^2-4x^2}{x+y}\div\left\{\dfrac{3(x^2-y^2)-8xy}{x-y}\right\}$

$$=\frac{x^2-2xy-3y^2}{x-y}\times\frac{y^2+2xy-3x^2}{x+y}\times\frac{x-y}{3x^2-8xy-3y^2}$$

$$=\frac{(x-3y)(x+y)(y+3x)(y-x)(x-y)}{(x-y)(x+y)(3x+y)(x-3y)}$$

$$=y-x.$$

108. The expression

$$=\frac{\{a^4+b^4+ab(a^2+b^2)\}(a-b)^2-\{a^4+b^4-ab(a^2+b^2)\}(a+b)^2}{(a^2-b^2)^2}+\frac{12a^2b^2}{4ab}$$

$$=\frac{(a^4+b^4)\{(a-b)^2-(a+b)^2\}+ab(a^2+b^2)\{(a-b)^2+(a+b)^2\}}{(a^2-b^2)^2}+3ab$$

$$=\frac{2ab(a^2+b^2)^2-4ab(a^4+b^4)}{(a^2-b^2)^2}+3ab$$

$$=\frac{2ab(a^4+b^4)+4a^3b^3-4ab(a^4+b^4)}{(a^2-b^2)^2}+3ab$$

$$=\frac{4a^3b^3-2ab(a^4+b^4)}{a^4+b^4-2a^2b^2}+3ab$$

$$=-2ab+3ab=ab.$$

109. The first fraction

$$=\frac{(ac+bd-ad-bc)[(ac+bd)^2+(ac+bd)(ad+bc)+(ad+bc)^2]}{(a-b)(c-d)}$$

$$=(ac+bd)^2+(ac+bd)(ad+bc)+(ad+bc)^2.$$

Similarly, the second fraction

$$=(ac+bd)^2-(ac+bd)(ad+bc)+(ad+bc)^2.$$

$\therefore$ the difference

$$=2(ac+bd)(ad+bc).$$

110. Here the divisor $=\dfrac{z^2}{a^2+b^2}+\dfrac{a^2+b^2}{a^2b^2}\left\{x^2-\dfrac{2a^2xz}{a^2+b^2}+\dfrac{a^4z^2}{(a^2+b^2)^2}\right\}$

$$=\frac{z^2}{a^2+b^2}+\frac{(a^2+b^2)x^2}{a^2b^2}-\frac{2xz}{b^2}+\frac{a^2z^2}{b^2(a^2+b^2)}$$

$$=\frac{z^2}{a^2+b^2}\left(1+\frac{a^2}{b^2}\right)+\frac{x^2}{b^2}+\frac{x^2}{a^2}-\frac{2xz}{b^2}$$

$$=\frac{z^2}{b^2}-\frac{2xz}{b^2}+\frac{x^2}{b^2}+\frac{x^2}{a^2}$$

$$=\left(\frac{z-x}{b}\right)^2+\frac{x^2}{a^2}$$

$=$ the dividend.

$\therefore$ the whole expression $=1$.

111. The expression

$$=\frac{(x^2+x-1)(x^2-x+1)}{(x^2+1+x)(x^2+1-x)}+\frac{(x+x^2-1)(x-x^2+1)}{(x^2+x+1)(x^2+x-1)}+\frac{(x^2-x+1)(x^2-x-1)}{(x^2+x+1)(x^2-x-1)}$$

$$=\frac{x^2+x-1+x-x^2+1+x^2-x+1}{x^2+x+1}$$

$=1.$

112. The numerator $=\dfrac{(1+x)^2-(1-x)^2+8x}{1-x^2}+\dfrac{4x}{1+x^2}$

$$=\frac{12x}{1-x^2}+\frac{4x}{1+x^2}=\frac{12x(1+x^2)+4x(1-x^2)}{1-x^4}$$

$$=\frac{16x+8x^3}{1-x^4}=\frac{8x(x^2+2)}{1-x^4}.$$

The denominator $=\dfrac{(1+x^2)^2-(1-x^2)^2}{1-x^4}+\dfrac{4x^2}{1+x^4}$

$$=\frac{4x^2}{1-x^4}+\frac{4x^2}{1+x^4}=\frac{8x^2}{(1-x^4)(1+x^4)}.$$

$\therefore$ the whole expression $=\dfrac{8x(x^2+2)}{1-x^4}\times\dfrac{(1-x^4)(1+x^4)}{8x^2}$

$$=\frac{(x^2+2)(x^4+1)}{x}.$$

113. The expression $=\left\{\frac{f^2}{fg-g^2}+\frac{g^2}{fg-f^2}\right\}\times\frac{fg}{f^3-g^3}$

$$=\left\{\frac{f^2}{g(f-g)}+\frac{g^2}{f(g-f)}\right\}\times\frac{fg}{f^3-g^3}$$

$$=\frac{f^3-g^3}{fg(f-g)}\times\frac{fg}{f^3-g^3}=\frac{1}{f-g}.$$

114. The expression $=\dfrac{1}{x-\dfrac{4}{2x+1}}\times\dfrac{x}{2x+1}\div\dfrac{x(x+1)}{2x(x+1)-x-4}$

$$=\frac{2x+1}{2x^2+x-4}\times\frac{x}{2x+1}\times\frac{2x^2+x-4}{x(x+1)}$$

$$=\frac{1}{x+1}.$$

115. The expression $=\dfrac{x}{1+\dfrac{x(1+x)}{1-x^2+x}}\div\dfrac{1+x+x^2}{1+3x+3x^2+2x^3}$

$$=\frac{x(1+x-x^2)}{1+2x}\times\frac{1+3x+3x^2+2x^3}{1+x+x^2}$$

$$=\frac{x(1+x-x^2)(1+2x)(1+x+x^2)}{(1+2x)(1+x+x^2)}$$

$$=x(1+x-x^2).$$

116. The expression $=\dfrac{ab}{a+b}\times\dfrac{3ac+b}{a}-\dfrac{b^2(a^2+b^2)}{(a+b)^3}-\dfrac{2ab^3}{(a+b)^3}$

$$=\frac{b(3ac+b)(a+b)^2-b^2(a^2+b^2)-2ab^3}{(a+b)^3}$$

$$=\frac{3abc(a+b)^2}{(a+b)^3}=\frac{3abc}{a+b}.$$

117. The expression $=\dfrac{\dfrac{2(a^2+b^2)-(b^2-a^2)}{b-a}}{\dfrac{a+b-(b-a)}{(b-a)(a+b)}}\div\dfrac{2b^3+6a^2b}{4ab}$

$$=\frac{(3a^2+b^2)(a+b)}{2a}\times\frac{2a}{b^2+3a^2}=a+b.$$

118. The dividend $=\dfrac{b-c}{(a-b)(c-a)}+\dfrac{c-a}{(a-b)(b-c)}+\dfrac{a-b}{(b-c)(c-a)}$

$$=\frac{(b-c)^2+(c-a)^2+(a-b)^2}{(a-b)(b-c)(c-a)}$$

$$=\frac{2(a^2+b^2+c^2-bc-ca-ab)}{(a-b)(b-c)(c-a)}.$$

$\therefore$ the whole expression $=1$.

EXAMPLES. XXIII. b. PAGE 182.

23. $$\frac{x-a+b}{x-a}+\frac{x-b}{x-2b}=\frac{x}{x-b}+\frac{x-a}{x-a-b}.$$

The equation may be written

$$\frac{(x-a)+b}{x-a}+\frac{(x-2b)+b}{x-2b}=\frac{(x-b)+b}{x-b}+\frac{(x-a-b)+b}{x-a-b},$$

$$1+\frac{b}{x-a}+1+\frac{b}{x-2b}=1+\frac{b}{x-b}+1+\frac{b}{x-a-b};$$

$$\therefore \quad \frac{1}{x-a}+\frac{1}{x-2b}=\frac{1}{x-b}+\frac{1}{x-a-b}.$$

Transposing, $$\frac{1}{x-a}-\frac{1}{x-a-b}=\frac{1}{x-b}-\frac{1}{x-2b},$$

$$\frac{-b}{(x-a)(x-a-b)}=\frac{-b}{(x-b)(x-2b)};$$

$$\therefore \quad (x-a)(x-a-b)=(x-b)(x-2b),$$

$$x^2-2ax+a^2-bx+ab=x^2-3bx+2b^2,$$

$$2bx-2ax=2b^2-ab-a^2,$$

$$2x(b-a)=(b-a)(2b+a);$$

$$\therefore \quad x=\frac{a+2b}{2}.$$

24. $$\left(\frac{x-3a}{a}\right)\left(\frac{3x-a}{a}\right)-\frac{(x-2a)(2x-a)}{a^2}=\frac{(x-a)^2}{a^2}-1,$$

$$3x^2-10ax+3a^2-(2x^2-5ax+2a^2)=x^2-2ax+a^2-a^2,$$

$$-10ax+5ax+2ax=-a^2,$$

$$-3ax=-a^2;$$

$$\therefore \quad x=\frac{a}{3}.$$

25. The equation may be written

$$\frac{b(x+a)}{x^2-b^2}+2+\frac{b-a}{x+b}=2+\frac{2bx}{x^2-b^2}.$$

Clearing of fractions, we have

$$b(x+a)+(b-a)(x-b)=2bx,$$

$$bx+ab+bx-ax-b^2+ab=2bx,$$

$$ax=2ab-b^2;$$

$$\therefore \quad x=\frac{b(2a-b)}{a}.$$

EXAMPLES. XXV. b. PAGE 197.

36. The equation may be written

$$1+\frac{1}{x-3}+3+\frac{1}{x-4}=4+\frac{9}{x+1};$$

$$\therefore \quad \frac{1}{x-3}+\frac{1}{x-4}=\frac{9}{x+1},$$

$$(x-4)(x+1)+(x-3)(x+1)=9(x-3)(x-4),$$

$$x^2-3x-4+x^2-2x-3=9(x^2-7x+12),$$

$$2x^2-5x-7=9x^2-63x+108,$$

$$7x^2-58x+115=0,$$

$$x^2-\frac{58}{7}x+\left(\frac{29}{7}\right)^2=\left(\frac{29}{7}\right)^2-\frac{115}{7};$$

that is,
$$\left(x-\frac{29}{7}\right)^2=\frac{841-805}{49}=\frac{36}{49};$$

whence
$$x-\frac{29}{7}=\pm\frac{6}{7};$$

$$\therefore \quad x=5, \text{ or } \frac{23}{7}.$$

37. We have $(2x-a)a+5a(2x-5a)=2(2x-a)(2x-5a),$

$$2ax-a^2+10ax-25a^2=2(4x^2-12ax+5a^2),$$

$$12ax-26a^2=8x^2-24ax+10a^2,$$

$$8x^2-36ax-36a^2=0,$$

$$2x^2-9ax-9a^2=0,$$

$$x^2-\frac{9ax}{2}+\left(\frac{9a}{4}\right)^2=\left(\frac{9a}{4}\right)^2+\frac{9a^2}{2};$$

that is,
$$\left(x-\frac{9a}{4}\right)^2=\frac{81a^2-72a^2}{16}=\frac{9a^2}{16};$$

whence
$$x-\frac{9a}{4}=\pm\frac{3a}{4};$$

$$\therefore \quad x=3a, \text{ or } \frac{3a}{2}.$$

38. We have $4c(2x-3c)+6c(3x-2c)=7(2x-3c)(3x-2c)$,

$$8cx-12c^2+18cx-12c^2=7(6x^2-13cx+6c^2),$$

$$26cx-24c^2=42x^2-91cx+42c^2,$$

$$42x^2-117cx+66c^2=0,$$

$$14x^2-39cx+22c^2=0,$$

$$x^2-\frac{39cx}{14}+\left(\frac{39c}{28}\right)^2=\left(\frac{39c}{28}\right)^2-\frac{22c^2}{14};$$

that is,
$$\left(x-\frac{39c}{28}\right)^2=\frac{1521c^2-1232c^2}{28^2}=\frac{289c^2}{28^2};$$

whence
$$x-\frac{39c}{28}=\pm\frac{17c}{28};$$

$$\therefore\ x=2c,\text{ or }\frac{11c}{14}.$$

39. We have
$$\frac{a^2b}{x^2}+\left(\frac{x+b}{x}\right)a=\frac{2bx+a^2}{x},$$

$$a^2b+ax(x+b)=(2bx+a^2)x,$$

$$a^2b+ax^2+abx=2bx^2+a^2x,$$

$$x^2(a-2b)+x(ab-a^2)+a^2b=0,$$

$$x^2+\frac{ab-a^2}{a-2b}x+\left\{\frac{ab-a^2}{2(a-2b)}\right\}^2=\left\{\frac{ab-a^2}{2(a-2b)}\right\}^2-\frac{a^2b}{a-2b};$$

that is,
$$\left\{x+\frac{ab-a^2}{2(a-2b)}\right\}^2=\frac{a^2b^2-2a^3b+a^4-4a^2b(a-2b)}{4(a-2b)^2},$$

$$\left\{x+\frac{ab-a^2}{2(a-2b)}\right\}^2=\frac{a^4-6a^3b+9a^2b^2}{4(a-2b)^2};$$

whence
$$x+\frac{ab-a^2}{2(a-2b)}=\pm\frac{a^2-3ab}{2(a-2b)};$$

$$\therefore\ x=a,\text{ or }\frac{ab}{a-2b}.$$

EXAMPLES. XXV. d. PAGE 201B

1.
$$x^3+x^2-x-1=0,$$

$$x^2(x+1)-(x+1)=0;$$

$$\therefore\ (x^2-1)(x+1)=0,\text{ or }(x-1)(x+1)(x+1)=0;$$

$$x=1,\ -1,\ -1.$$

2.
$$x^2(x-2)-(x-2)=0,$$
$$(x^2-1)(x-2)=0;$$
$$\therefore\ x=1,\ -1,\ 2.$$

3.
$$x(x^2-4)=x^2-4;\quad \therefore\ (x^2-4)(x-1)=0;$$
$$\therefore\ x=2,\ -2,\ 1.$$

4. The left-hand side vanishes when $x=1$; hence $x=1$ is one root. The equation may be written

$$x^2(x-1)+8x(x-1)+15(x-1)=0,$$

or
$$(x-1)(x^2+8x+15)=0.$$

Thus the roots are 1, 3, 5.

5. One root is -1, and the equation may be written

$$(x+1)(x^2-x-2)=0.$$

Thus the roots are -1, 2, -1.

6. The equation may be written $x(x^3-3x+2)=0$.

Also $x-1$ is a factor of x^3-3x+2. Thus we obtain

$$x(x-1)(x^2+x-2)=0;$$

whence the roots are 0, 1, 1, -2.

7. By trial $x-2$ is a factor of $x^3-19x+30$. Hence the equation may be written

$$(x-2)(x^2+2x-15)=0.$$

Thus the roots are 2, 3, -5.

8.
$$x^3-39x+70=x^2(x-5)+5x(x-5)-14(x-5)$$
$$=(x-5)(x+7)(x-2).$$

Thus the roots are 5, -7, 2.

9.
$$(x-7)(x+3)(x+4)=0.$$

10.
$$x^3-12a^2x-16a^3=x^2(x-4a)+4ax(x-4a)+4a^2(x-a)$$
$$=(x-4a)(x+2a)^2.$$

Thus the roots are $4a$, $-2a$, $-2a$.

11. $$x^4 + 432a^3x - 108a^2x^2 = x(x^3 - 108a^2x + 432a^3);$$

and $$x^3 - 108a^2x + 432a^3 = x^2(x - 6a) + 6ax(x - 6a) - 72a^2(x - 6a)$$
$$= (x - 6a)(x + 12a)(x - 6a).$$

Thus the equation is $x(x - 6a)(x - 6a)(x + 12a) = 0$,

and the roots are 0, $6a$, $6a$, $-12a$.

12. $$x^2 + 2x - 3\cdot2 = 0;$$
$$\therefore\ x = \frac{2 \pm \sqrt{4 + 12\cdot8}}{2} = -1 \pm \sqrt{4\cdot2}$$
$$= -1 \pm 2\cdot05.$$

13. $$x = \frac{3 \pm \sqrt{9 + 14\cdot04}}{2} = \frac{1}{2}(3 \pm \sqrt{23\cdot04})$$
$$= \frac{1}{2}(3 \pm 4\cdot79) = \frac{7\cdot79}{2}, \text{ or } -\frac{1\cdot79}{2}$$
$$= 3\cdot89, \text{ or } -\cdot89.$$

14. $$x^2 + x - 1\cdot0956 = 0;$$
$$\therefore\ x = \frac{-1 \pm \sqrt{1 + 4\cdot3824}}{2} = \frac{-1 \pm 2\cdot32}{2}$$
$$= \cdot66, \text{ or } -1\cdot66.$$

15. $$x = \frac{36 \pm \sqrt{1296 - 1294\cdot8}}{2} = \frac{36 \pm 2\sqrt{\cdot3}}{2}$$
$$= 18 \pm \cdot55 = 18\cdot55, \text{ or } 17\cdot45.$$

16. $$x = \frac{7 \pm \sqrt{49 - 24\cdot14}}{2} = \frac{7 \pm \sqrt{24\cdot86}}{2} = \frac{1}{2}(7 \pm 4\cdot986)$$
$$= 5\cdot993, \text{ or } 1\cdot007.$$

Thus the roots required are $5\cdot99$, or $1\cdot01$.

17. $$x = \frac{5\cdot5 \pm \sqrt{30\cdot25 - 29\cdot5104}}{2} = \frac{5\cdot5 \pm \sqrt{\cdot7396}}{2}$$
$$= \frac{1}{2}(5\cdot5 \pm \cdot86) = \frac{6\cdot36}{2}, \text{ or } \frac{4\cdot64}{2}$$
$$= 3\cdot18, \text{ or } 2\cdot32.$$

18. $3x^2 - x - \cdot 362 = 0$;

$$\therefore\ x = \frac{1 \pm \sqrt{1 + 4 \times 3 \times \cdot 362}}{6}$$

$$= \frac{1 \pm \sqrt{5 \cdot 384}}{6} = \frac{1}{6}(1 \pm 2 \cdot 32)$$

$$= \cdot 55, \text{ or } -22.$$

19. $2x^2 - 4x + 1 \cdot 73 = 0$;

$$\therefore\ x = \frac{4 \pm \sqrt{16 - 4 \times 2 \times 1 \cdot 73}}{4}$$

$$= \frac{2 \pm \sqrt{4 - 3 \cdot 46}}{2} = \frac{1}{2}(2 \pm \sqrt{\cdot 54})$$

$$= \frac{2 \cdot 73}{2}, \text{ or } \frac{1 \cdot 27}{2}$$

$$= 1 \cdot 4, \text{ or } \cdot 6.$$

20.
$$x = \frac{-a \pm \sqrt{a^2 + 4a^2}}{2} = \frac{a}{2}(-1 \pm \sqrt{5}).$$

Now $\sqrt{5} = 2 \cdot 23607$; $\therefore$ when $a = 12$,

$$x = 6(-1 \pm 2 \cdot 23607) = 6 \times 1 \cdot 23607, \text{ or } 6 \times -3 \cdot 23607$$

$$= 7 \cdot 416, \text{ or } -19 \cdot 416.$$

21. $x^2 - ax + c^2 = 0$;

$$\therefore\ x = \frac{a \pm \sqrt{a^2 - 4c^2}}{2} = \frac{16 \pm \sqrt{16^2 - 4 \times 36}}{2}$$

$$= \frac{1}{2}(16 \pm 4\sqrt{7}) = 8 \pm 2 \times 2 \cdot 6458$$

$$= 13 \cdot 292, \text{ or } 2 \cdot 708.$$

EXAMPLES. XXVII. Page 211.

31. Let $AP = x$, then $a(a - x) = x^2$;

whence
$$x = \frac{1}{2}(-a \pm a\sqrt{5}).$$

When x is negative, P divides AB *externally*; that is BA is produced to P so that $AB \,.\, BP = AP^2$.

[See Hall and Stevens' Geometry Part IV. Prob. 33, Note.]

32. Put $a=6$, and take the positive value of x in Ex. 31.

33. Let $BP=x$; then $8(8+x)=x^2$;

whence $$x=\frac{8\pm8\sqrt{5}}{2}=4(\sqrt{5}+1), \text{ taking the pos. value.}$$

$$\therefore\ BP=3\cdot236\times4=12\cdot9, \text{ to nearest millimetre,}$$

and $$AP=8+BP=20\cdot9.$$

34. Let $AB=a$, $BP=x$; then $a(a+x)=x^2$.

(i) Now $AB^2+AP^2=a^2+(a+x)^2=x^2+2a(a+x)=3x^2$
$=3BP^2$.

(ii) $(AB+AP)^2=(a+a+x)^2=a^2+(a+x)^2+2a(a+x)$
$=3BP^2+2BP^2$, by (i)
$=5BP^2$.

35. $BP=AB\sqrt{2}$; $\therefore AP=AB+BP=3\cdot5(1+\sqrt{2})$.

36. Let $AB=a$, $AP=x$; then $x(x-\overline{a-x})=(a-x)^2$.
Whence $(a-x)a=x^2$, the positive root of which is

$$\frac{a}{2}(\sqrt{5}-1)=2\cdot1\times1\cdot236=2\cdot6. \quad \therefore\ BP=1\cdot6.$$

37. Here $x+y=13$, $xy=36$.

38. Let $AX=x$, $BX=y$, then $AX+BX=x+y=13$.
Also if O is the centre of the circle,

$$xy=AX.XB=(OA-OX)(OA+OX)$$
$$=OA^2-OX^2=OQ^2-OX^2=QX^2$$
$$=36.$$

Thus AX, BX are the roots of the equations in Ex. 37.

EXAMPLES. XXX. b. PAGE 247.

37. $$\frac{3.2^n-4.2^{n-2}}{2^n-2^{n-1}}=\frac{3.2^2.2^{n-2}-4.2^{n-2}}{2^{n-1}(2-1)}=\frac{2^{n-2}(12-4)}{2^{n-1}}$$
$$=\frac{8}{2}=4.$$

38. $$\frac{3^{n+4}-6.3^{n+1}}{3^{n+2}\times7}=\frac{3^{n+1}(3^3-6)}{3^{n+1}\times3\times7}=\frac{21}{3\times7}=1.$$

EXAMPLES. XXXI. g. Page 268.

26. $$\sqrt{a-x}+\sqrt{b+x}=\sqrt{a}+\sqrt{b}.$$

Squaring both sides, we have

$$a-x+b+x+2\sqrt{(a-x)(b+x)}=a+b+2\sqrt{ab},$$

$$2\sqrt{(a-x)(b+x)}=2\sqrt{ab}.$$

Squaring, $$(a-x)(b+x)=ab,$$

$$ab-bx+ax-x^2=ab;$$

that is, $$x^2-(a-b)x=0;$$

$$\therefore\ x=0, \text{ or } a-b$$

27. $$5\sqrt[3]{70x+29}=9\sqrt[3]{14x-15}.$$

Cubing both sides, we have

$$125(70x+29)=729(14x-15),$$

$$7x(1250-1458)=-5(725+2187),$$

$$-7\times 208x=-5\times 2912;$$

$$\therefore\ x=\frac{5\times 2912}{7\times 208}=10.$$

28. $$\sqrt[3]{x^3-3x^2+7x-11}=x-1.$$

Cubing both sides, we have

$$x^3-3x^2+7x-11=x^3-3x^2+3x-1,$$

$$4x=10;$$

$$\therefore\ x=\frac{5}{2}.$$

29. $$\sqrt{8x^3+12x^2+12x-11}=2x+1.$$

Cubing both sides, we have

$$8x^3+12x^2+12x-11=8x^3+12x^2+6x+1,$$

$$12x-11=6x+1;$$

that is, $$6x=12;$$

$$\therefore\ x=2.$$

30. $$\sqrt[3]{1+x}+\sqrt[3]{1-x}=\sqrt[3]{2}.$$

Cubing both sides, we have

$$1+x+1-x+3(1+x)^{\frac{2}{3}}(1-x)^{\frac{1}{3}}+3(1+x)^{\frac{1}{3}}(1-x)^{\frac{2}{3}}=2,$$

$$(1+x)^{\frac{2}{3}}(1-x)^{\frac{1}{3}}+(1+x)^{\frac{1}{3}}(1-x)^{\frac{2}{3}}=0.$$

Divide throughout by $(1-x)^{\frac{1}{3}}(1+x)^{\frac{1}{3}}$; this gives two roots $x=1$, or -1.

Also $$(1+x)^{\frac{1}{3}}+(1-x)^{\frac{1}{3}}=0,$$

or $$(1+x)^{\frac{1}{3}}=-(1-x)^{\frac{1}{3}}.$$

Cubing both sides, we have

$$1+x=-(1-x);$$

that is, $$2x=2,$$

which gives one of the roots already obtained.

EXAMPLES. XXXII. c. PAGE 285.

25. Let the area be represented by A square feet, and the height and base by h and b feet respectively.

Then $h=m\dfrac{A}{b}$, where m is a constant.

When $A=18,\ b=\dfrac{27}{2},\ h=\dfrac{8}{3}$;

$$\therefore\ \frac{8}{3}=m\times 18\times\frac{2}{27}.$$

Whence $m=2$;

$$\therefore\ h=2\frac{A}{b}.$$

Hence when $A=2\frac{2}{3}$, and $b=1\frac{1}{3}$,

$$h=2\times 2=4.$$

26. Let £A represent the constant expenses,

and n .... number of boys,

and £E .... the total expenses.

Then $E=A+mn$, where m is a constant.

Then $$1000=A+m\,.\,150,$$

$$840=A+m\,.\,120.$$

Hence $A=200$ and $m=\dfrac{16}{3}$;

when there are 330 boys

$$E=200+\frac{16}{3}\times 330=1960.$$

MISCELLANEOUS EXAMPLES. V. Page 307.

1. The expression $=\dfrac{a^{\frac{1}{3}}b^{\frac{1}{2}}b^{-2}c^{\frac{1}{3}}}{a^{\frac{2}{3}}b^{-\frac{5}{4}}c^{\frac{1}{4}}}=\dfrac{b^{-\frac{3}{2}+\frac{5}{4}}\,.\,c^{\frac{1}{3}-\frac{1}{4}}}{a^{\frac{1}{3}}}=\dfrac{c^{\frac{1}{12}}}{a^{\frac{1}{3}}b^{\frac{1}{4}}}$

$$=\frac{432^{\frac{1}{12}}}{2^{\frac{1}{3}}3^{\frac{1}{4}}}=\frac{(2^4\,.\,3^3)^{\frac{1}{12}}}{2^{\frac{1}{3}}3^{\frac{1}{4}}}=1.$$

2. $$\frac{x}{x-2y}>\frac{x+y}{x-y} \text{ if } x(x-y)>(x+y)(x-2y),$$

that is, if $$x^2-xy>x^2-xy-2y^2,$$

that is, if $2y^2>0$, which is always the case, since $2y^2$ is positive.

3. Compare Examples XXXII. b., 14–18.

4. Let $$\frac{x}{b-c}=\frac{y}{c-a}=\frac{z}{a-b}=k,$$

then $$x+y+z=k(b-c+c-a+a-b)=0.$$

Also $$(b+c)x+(c+a)y+(a+b)z=k(b^2-c^2+c^2-a^2+a^2-b^2)=0.$$

5. See Art. 294.

6. We have $y=mx+\dfrac{n}{x}$, where m and n are constants.

Then $$7=2m+\frac{n}{2}, \text{ and } -1=m+n, \text{ whence } m=5,\ n=-6.$$

7. The expression $=3\sqrt{5}+2\sqrt{2}-4\sqrt{5}+3\sqrt{2}+\sqrt{5}-\sqrt{2}=4\sqrt{2}$.

8. See Art. 294, Ex. 3.

9. Compare solution of XXXII. a., Ex. 21.

10. We have $\dfrac{a}{1}=\dfrac{b}{1\frac{1}{2}}=\dfrac{c}{1\frac{3}{4}}=\dfrac{a+b+c}{1+1\frac{1}{2}+1\frac{3}{4}}=\dfrac{221}{4\frac{1}{4}}=52$;

$$a=52,\ b=78,\ c=91.$$

11. See Art. 302, Ex. 2.

12. (1) The expression $=x^{a+b-(a-b)}+x^{(a-b)-(a+b)}=x^{2b}+x^{-2b}$

(2) The expression

$$=\frac{(a+b)^{\frac{3}{2}}\times(a+b)^{\frac{1}{2}}\times(a-b)^{\frac{1}{2}}}{(a-b)^{\frac{1}{2}}}=(a+b)^{\frac{3}{2}+\frac{1}{2}}=(a+b)^2.$$

13. The expression $\dfrac{1}{x-a}-\dfrac{1}{x+a}=\dfrac{2a}{x^2-a^2}=\dfrac{2a}{a^2\left(\dfrac{9}{16}-1\right)}=-\dfrac{32}{7a}.$

14. (1) The expression $=\dfrac{a^{-\frac{5}{4}}b^{\frac{15}{4}}}{a^{\frac{5}{12}}}\times\left(\dfrac{b^2}{a^2b^{\frac{1}{2}}}\right)^{\frac{1}{6}}=\dfrac{b^{\frac{15}{4}}b^{\frac{1}{4}}}{a^{\frac{5}{3}}a^{\frac{1}{3}}}=\dfrac{b^4}{a^2}.$

(2) The expression $=\dfrac{2^n\,.\,2^4-2^n\,.\,2}{2^{n+2}\,.\,2^2}=\dfrac{2^n(16-2)}{2^n\,.\,2^4}=\dfrac{14}{16}=\dfrac{7}{8}.$

15. See Art. 290.

16. (1) $\dfrac{a(a+b)}{b(b-a)}=\dfrac{b}{c}\times\dfrac{\dfrac{b^2}{c}+b}{b-\dfrac{b^2}{c}}=\dfrac{b(b+c)}{c(c-b)}.$

(2) The first side $=ab^2+b^3-abc+ac^2+c^3$

$=a\,.\,ac+b\,.\,ac-abc+b^2\,.\,c+c^3=c(a^2+b^2+c^2).$

17. We have $\dfrac{a}{b}=\dfrac{y}{x}$, and $\dfrac{a}{c}=\dfrac{x}{z}.$

$\therefore\ \dfrac{a^2}{ab}=\dfrac{y}{x}$, and $\dfrac{ab}{bc}=\dfrac{x}{z}.$

Also $\dfrac{a}{b}\times\dfrac{a}{c}=\dfrac{y}{z}.$

These three relations are all contained in $x:y:z=ab:a^2:bc$.

18. See Art. 294, Ex. 3.

19. (1) $a(a+b+c+d)=a^2+ab+ac+ad=a^2+ab+ac+bc$

$=a(a+b)+c(a+b)=(a+c)(a+b)$;

$\therefore\ a+c:a+b+c+d=a:a+b.$

(2) $a-b-(c-d)=(a-b)\left[1-\dfrac{c-d}{a-b}\right]$;

but since $\dfrac{a}{b}=\dfrac{c}{d}$, therefore $\dfrac{a-b}{b}=\dfrac{c-d}{d}$, or $\dfrac{c-d}{a-b}=\dfrac{d}{b}.$

$\therefore\ a-b-(c-d)=(a-b)\left(1-\dfrac{d}{b}\right)=\dfrac{(a-b)(b-d)}{b}.$

20. This follows at once from Art. 293, for if $\dfrac{a}{b}$ is a proper fraction $\dfrac{a+x}{b+x}>\dfrac{a}{b}$; and if $\dfrac{a}{b}$ is an improper fraction $\dfrac{a+x}{b+x}<\dfrac{a}{b}.$

23. Here $d=3$. and $28=\frac{7}{2}(2a+6d)$; whence $a=-5$.

24. Here $$145=5(2a+9d),$$
and $$(a+3d)+(a+8d)=5(a+2d);$$
whence $$a=1,\ d=3.$$

25. The expression $=(\sqrt{12}+\sqrt{7})+\sqrt{7}-\sqrt{12}-(\sqrt{28}-1)$
$$=\sqrt{12}+\sqrt{7}+\sqrt{7}-\sqrt{12}-2\sqrt{7}+1$$
$$=1.$$

26. (1) The series is an A.P. whose common difference is 5;
$$\therefore\ s=5(10+9\,.\,5)=275.$$
(2) The series is a G.P. whose common ratio is -2;
$$\therefore\ s=\frac{5[1-(-2)^{10}]}{1+2}=\frac{5}{3}(1-1024)=-1705.$$

27. See solution of Example 4, above.

28. See solution of XXXIII. b., Ex. 24.

29. The series $3+\frac{5}{2}+\frac{25}{12}+\dots$ is a G.P. whose common ratio is $\frac{5}{6}$.

30. If d be the common difference of the corresponding A.P., we have
$$\frac{1}{2}=\frac{1}{3}+2d;\quad \text{whence}\quad d=\frac{1}{12}.$$
$$\therefore\ \text{the 9}^{\text{th}}\text{ term of the A.P.}=\frac{1}{3}+\frac{8}{12}=1.$$

31. The expression $=\dfrac{(a-b)^{\frac{1}{3}}\times(a+b)^{\frac{2}{3}}}{(a^2-b^2)^{\frac{1}{3}}\times(a+b)^{-\frac{2}{3}}}$
$$=\frac{(a-b)^{\frac{1}{3}}\times(a+b)^{\frac{2}{3}}}{(a-b)^{\frac{1}{3}}\times(a+b)^{\frac{1}{3}}\times(a+b)^{-\frac{2}{3}}}=\frac{(a+b)^{\frac{2}{3}}}{(a+b)^{-\frac{1}{3}}}=a+b.$$

32. Here $$s=\frac{n}{2}\{3+(n-1)3\}=\frac{3n^2}{2}.$$

If a be the first of the required terms,
$$a+(a+3)+(a+6)+(a+9)+(a+12)=187\tfrac{1}{2};$$
$$\therefore\ 5a+30=187\tfrac{1}{2};\quad \text{whence}\quad a=\frac{63}{2}.$$

33. Here $$a+7d=2(a+12d)=2a+24d;$$
$$\therefore \quad a+d=2a+18d=2(a+9d);$$
that is, the 2nd term is double the 10th term.

34. (1) Here $$d=a+4x;$$
$$\therefore \quad s=9\{2(a-2x)+17(a+4x)\}=9(19a+64x).$$

(2) The series $\frac{320}{81}-\frac{160}{27}+\frac{80}{9}-\ldots$ is a G.P. whose common ratio is $-\frac{3}{2}$.

$$\therefore \quad s=\frac{320}{81}\frac{\left\{1-\left(-\frac{3}{2}\right)^7\right\}}{1+\frac{3}{2}}=\frac{320}{81}\times\frac{2}{5}\frac{2^7+3^7}{2^7}$$
$$=\frac{2^7+3^7}{81}=\frac{128+2187}{81}=\frac{2315}{81}.$$

35. The series may be written
$$1-\frac{1}{9}+\frac{1}{81}-\ldots \text{ to } n \text{ terms},$$
$$-\left\{\frac{1}{3}-\frac{1}{27}+\frac{1}{243}-\ldots \text{ to } n \text{ terms}\right\};$$
$$\therefore \quad s=\frac{1-\left(-\frac{1}{9}\right)^n}{1+\frac{1}{9}}-\frac{\frac{1}{3}\left[1-\left(-\frac{1}{9}\right)^n\right]}{1+\frac{1}{9}}$$
$$=\frac{9}{10}\{1-(-3)^{-2n}\}\left(1-\frac{1}{3}\right)=\frac{3}{5}\{1-(-3)^{-2n}\}.$$

36. We have $$\frac{2}{2b}=\frac{1}{b-a}+\frac{1}{b-c};$$
$$\therefore \quad (b-a)(b-c)=b(2b-a-c),$$
or $$b^2-ab-bc+ac=2b^2-ab-bc;$$
whence $$b^2=ac.$$

37. We have $$a+(n-1)d=10a,$$
and $$a+(n-2)d=(a+3d)+(a+4d).$$
By subtraction, $d=8a-7d; \quad \therefore \; d=a;$
and from the first equation, $n=10.$

38. (1) This is a G.P. in which $r=-3$.

$$\therefore\ s=\frac{1-(-3)^{2n}}{1+3}=\frac{1}{4}(1-3^{2n}).$$

Also $$l=(-3)^{2n-1}.$$

(2) This series may be written

$$1+5+9+\ldots \text{ to } n \text{ terms},$$
$$-(3+7+11+\ldots \text{ to } n \text{ terms});$$

$$\therefore\ s=\frac{n}{2}\{2+(n-1)4\}-\frac{n}{2}\{6+\overline{n-1}\,4\}$$

$$=\frac{n}{2}(-4)=-2n.$$

Also $$l=-\{3+(n-1)4\}=1-4n.$$

39. Let x and y represent the numbers.

Then $$\frac{x+y}{2}=\sqrt{xy}+2;\quad \frac{2xy}{x+y}=\frac{1}{5}x.$$

From the second equation $$16y=x+y.$$

Substituting $x=9y$ in the first equation we have $5y=\sqrt{9y^2}+2$; whence $$y=1,\ x=9.$$

40. Let the series be $1+r+r^2+r^3+\ldots$;

then $$1=2(r+r^2+r^3+\ldots)$$
$$=\frac{2r}{1-r};$$

$$\therefore\ r=\frac{1}{3}, \text{ and the series is } 1+\frac{1}{3}+\frac{1}{9}+\ldots.$$

41. Here $$\frac{x+y}{2}=\frac{5}{4}\sqrt{xy};\quad \sqrt{xy}-\frac{2xy}{x+y}=\frac{4}{5}.$$

Substitute $x+y=\dfrac{5\sqrt{xy}}{2}$ in the second equation;

then $\sqrt{xy}-\dfrac{4}{5}\sqrt{xy}=\dfrac{4}{5}$, whence $xy=16$, $x+y=10$.

42. The first side $$=\frac{y^2z^2}{x}+\frac{z^2x^2}{y}+\frac{x^2y^2}{z}$$

$$=xz\,.\,\frac{z^2}{x}+\frac{y^4}{y}+\frac{x}{y^2}\,.\,x^2y^2, \text{ since } y^2=xz,$$
$$=z^3+y^3+x^3.$$

EXAMPLES. XXXVII. b. PAGE 328.

13. If there are only 2 bowlers, the remaining 9 men, who do not bowl, must all be included: therefore the number $= {}^4C_2 = \frac{4 \cdot 3}{1 \cdot 2} = 6.$

If there are 3 bowlers, 8 men must be chosen from the 9 who do not bowl: therefore the number $= {}^4C_3 \times {}^9C_8 = 4 \times 9 = 36.$

If there are 4 bowlers, 7 men must be chosen from the 9 who do not bowl; therefore the number $= 1 \times {}^9C_7 = \frac{9 \cdot 8}{1 \cdot 2} = 36.$

Hence the total number of elevens $= 36 + 36 + 6 = 78.$

14. Since neither of the specified men can stand in either of the first or last places, the first place can be filled in $n-2$ ways, and when it has been filled in any one of these ways, the last place can be filled in $n-3$ ways, and therefore the number of ways of filling the first and last places is $(n-2)(n-3)$. The remaining places can then be filled in $\lfloor n-2$ ways; thus the total number of ways of filling the n places is

$$(n-2)(n-3)\lfloor n-2.$$

EXAMPLES XXXIX. b. PAGE 348c.

For Examples 1–26 see Art. 387E.

26. Let $$x = \sqrt[3]{\left(\frac{294 \times 125}{42 \times 32}\right)^2} = \left(\frac{7 \times 125}{32}\right)^{\frac{2}{3}},$$

then $$\log x = \frac{2}{3}(\log 875 - \log 32)$$

$$= \frac{2}{3}(2{\cdot}9420 - 1{\cdot}5051) = {\cdot}9579.$$

$$\begin{aligned} &\text{antilog } {\cdot}957 = 9{\cdot}057 \\ &\text{diff. for } \quad 9 \qquad \underline{\quad 19} \\ &\text{antilog } {\cdot}9579 = 9{\cdot}076 \end{aligned}$$

30. We have $$x^2 = \sqrt[3]{347{\cdot}3} \times \sqrt[5]{256{\cdot}4};$$

$$\therefore 2 \log x = \frac{1}{3}\log 347{\cdot}3 + \frac{1}{5}\log 256{\cdot}4$$

$$= \frac{1}{3}(2{\cdot}5407) + \frac{1}{5}(2{\cdot}4089)$$

$$= {\cdot}8469 + {\cdot}4818 = 1{\cdot}3287;$$

$$\therefore \log x = {\cdot}6643.$$

$$\therefore x = 4{\cdot}616, \text{ from the tables.}$$

31. We have
$$x=\frac{\sqrt[5]{357\cdot8}\times\sqrt[4]{7836}}{\sqrt[3]{32\cdot78}};$$
$$\therefore \log x=\frac{1}{5}\log 357\cdot8+\frac{1}{4}\log 7836-\frac{1}{3}\log 32\cdot78$$
$$=\frac{1}{5}(2\cdot5537)+\frac{1}{4}(3\cdot8941)-\frac{1}{3}(1\cdot5156)$$
$$=\cdot5107+\cdot9735-\cdot5052$$
$$=\cdot9790;$$
$$\therefore x=9\cdot528.$$

32. With the notation of Art. 404,
$$M=35(1\cdot03)^{25};$$
$$\therefore \log M=\log 35+25\log 1\cdot03$$
$$=1\cdot5441+25\times\cdot0128$$
$$=1\cdot5441+\cdot32=1\cdot8641$$
$$\therefore M=73\cdot13;\text{ thus the amount}=£73.$$

33. By Art. 405, we have $V=1000(1\cdot04)^{-17}$;
$$\therefore \log V=3-17\log 1\cdot04=3-\cdot017\times17$$
$$=3-\cdot289=2\cdot711;$$
$$\therefore V=514\cdot0.$$
Thus the present value $=£514$.

34. Here
$$3000=1130(1\cdot05)^n;$$
$$\therefore \log 300=\log 113+n\log 1\cdot05.$$
$$\therefore n=\frac{2\cdot4771-2\cdot0531}{\cdot0212};$$
$$\therefore n=\frac{\cdot424}{\cdot0212}=20.$$

35. Here
$$M=\frac{1}{960}(1\cdot05)^{1000};$$
$$\therefore \log M=1000\log 1\cdot05-\log 960$$
$$=21\cdot2-2\cdot9823=18\cdot2177;$$
$\therefore$, by Art. 381, M has 19 digits in its integral part.

36. Let v = velocity required in ft. per sec.
Then
$$v=\cdot001(1\cdot333)^{25};$$
$$\therefore \log v=\bar{3}+25\log 1\cdot333=\bar{3}+\cdot1249\times25$$
$$=\bar{3}+3\cdot12=\cdot12.$$
$$\therefore v=1\cdot318.$$
Now
$$1\cdot318\text{ ft. per sec.}=\frac{1\cdot318}{1760\times3}\times60\times60\text{ mi. per hour}$$
$$=\cdot89\text{ mi per hr.}$$

CHAPTER XLIV.

EXAMPLES. XLIV. a. PAGE 374.

10. The points (0, 0), (0, 6), (4, 3) are the points O, A, P in Fig. 1. The area of $\triangle OAP = \frac{1}{2} . PK . OA = \frac{1}{2} . 4 . 6$ units of area.

The point (4, 8) is Q. The area of

$\triangle AOQ = \frac{1}{2} . QL . OA = \frac{1}{2} . 4 . 6$ units of area.

This might also have been deduced from the fact that the triangles are on the same base and of equal altitude.

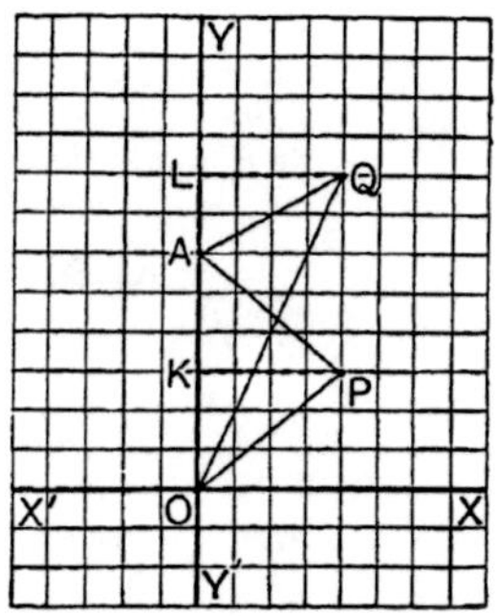

FIG. 1.

13. The points, in the order given, are represented by A, B, C, D, E, F, G, H in Fig. 2. *By measurement* the distance of any one of them from the origin is found to be 5 units.

Also if $P(x, y)$ stands for any one of the points, $OP^2 = x^2 + y^2$. The value of $x^2 + y^2$ is found by calculation to be 25 in each case. Thus $OP = 5$.

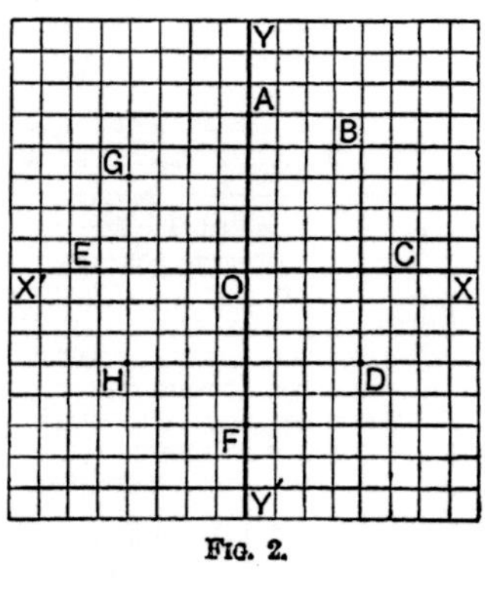

FIG. 2.

14. The points are shewn in Fig. 3; the first series, in the given order, is A, B, C, D, E, and the second series P, Q, R, S, T. The coordinates of the intersection of these lines are seen to be 5 and 8; hence the point is (5, 8).

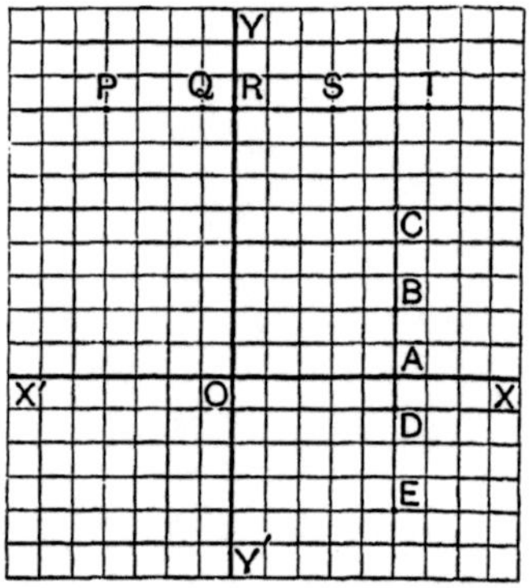

FIG. 3.

15. Proceeding as in Ex. 13, it will be found *by measurement* that each point is at a distance 13 from the origin. Again $OP^2 = x^2 + y^2 = 169$, where P stands for any one of the points. Hence $OP = 13$, and the locus is a circle with centre O and radius 13.

16. It is seen from a figure that these points lie on a line which passes through O; and by measurement it can be shewn that the abscissa and ordinate of any point on the line are equal.

EXAMPLES. XLIV. b. PAGE 377.

For most of these examples a reference to Examples 1, 2, 3 *in* Art. 418 *is sufficient.*

1, 2, and 3. The three lines are parallel, one passing through the origin, and the others making intercepts -4 and 6 on the y-axis.

4, 5, and 6. Three parallel lines.

7, 8, and 9. The first two lines are parallel; the third cuts them at right angles.

10, 11, and 12. The first two lines are parallel; the third cuts them at right angles.

13, 14, and 15. See Fig. 4. **16, 17, and 18.** See Fig. 5.

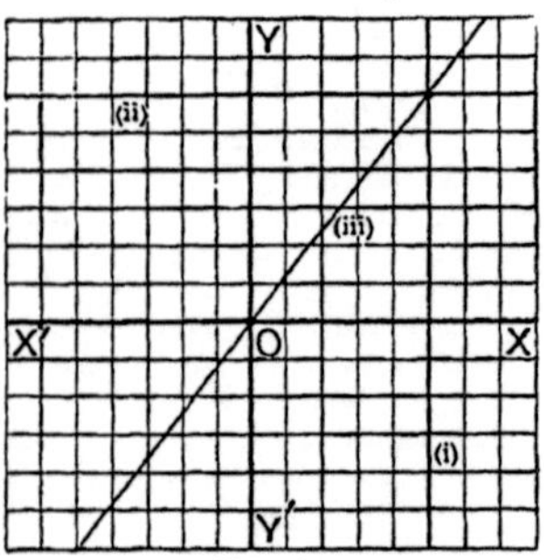

FIG. 4.

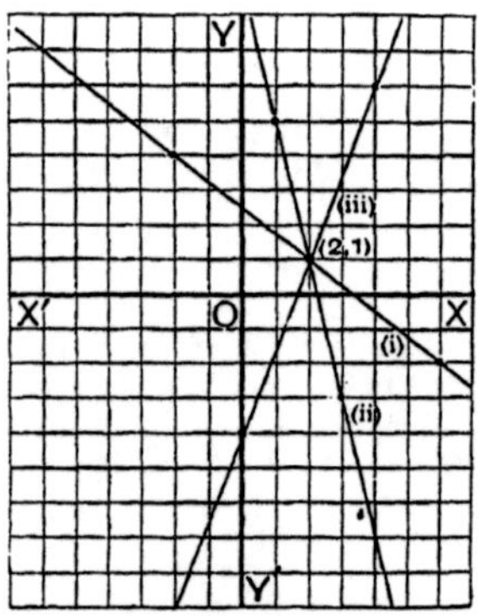

FIG. 5.

23. The lines will be found to be as in the adjoining diagram (Fig. 6) where each of the intercepts on the axes is 6 units. Thus the figure is a square whose area $=4\triangle AOB=72$ units of area.

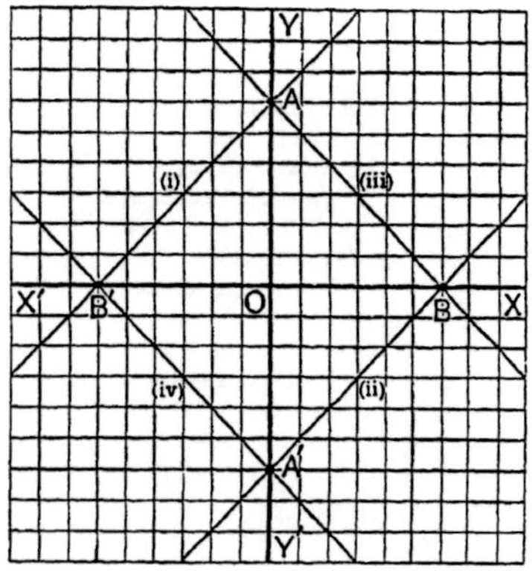

FIG. 6.

24. The figure is given in Fig. 7, the intercepts OA, OA' being each 8 units, and OB, OB' each 4 units. The figure being a parallelogram its area $=4\triangle AOB=4\times 16$ units of area $=0{\cdot}64$ square cms., as we have taken each unit to represent 1 mm.

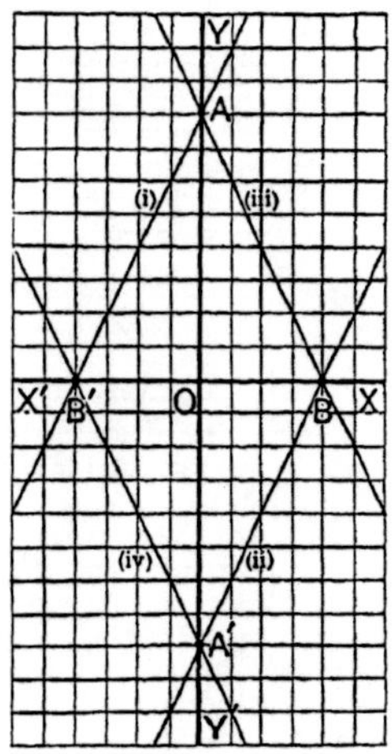

FIG. 7.

EXAMPLES. XLIV. c. PAGE 380.

In Figs. 11–19, pp. 382, 383, corresponding to Examples 1–9, PP' represents the first, and QQ' the second of the given equations. The solution is given by the coordinates of R. The lines in all cases should be drawn by ascertaining two points on each. (See Art. 423.)

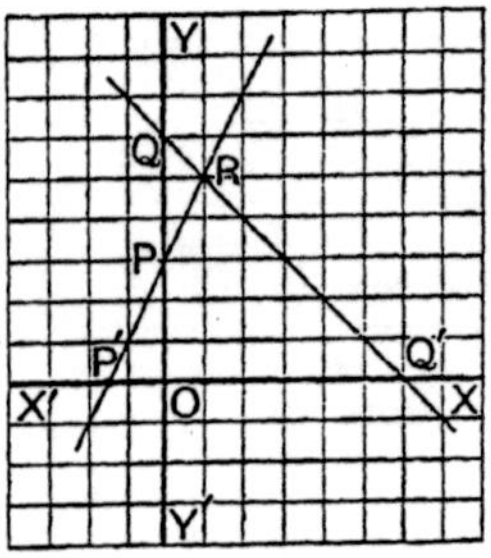

FIG. 8. (Ex. 1.)

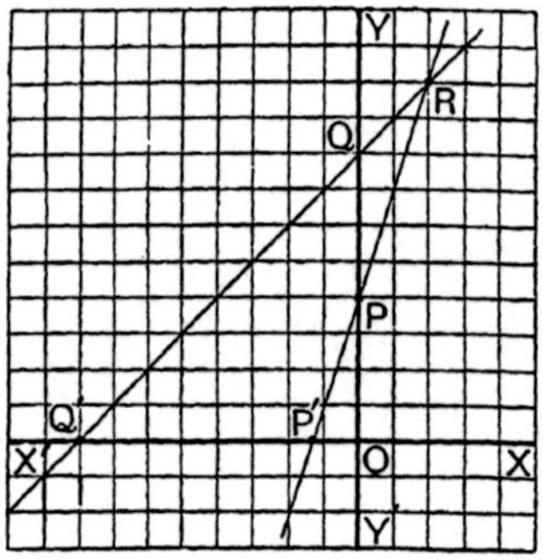

FIG. 9. (Ex. 2.)

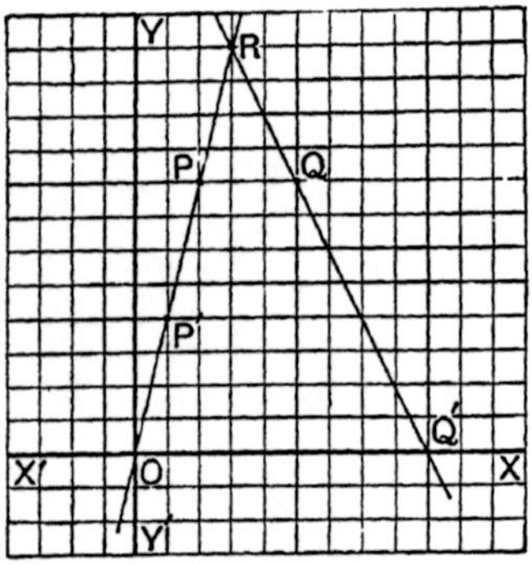

FIG. 10. (Ex. 3.)

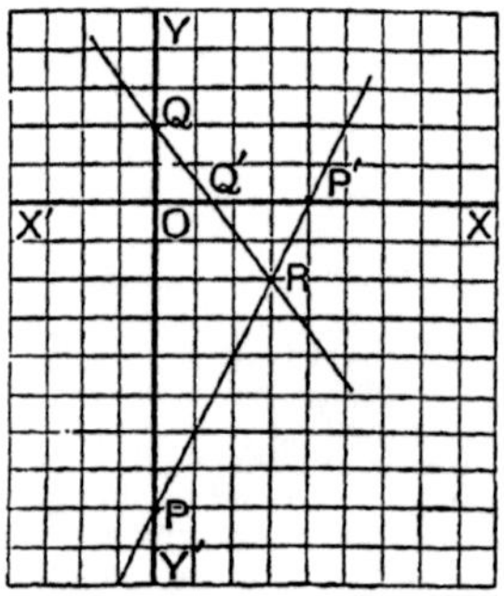

FIG. 11. (Ex. 4.)

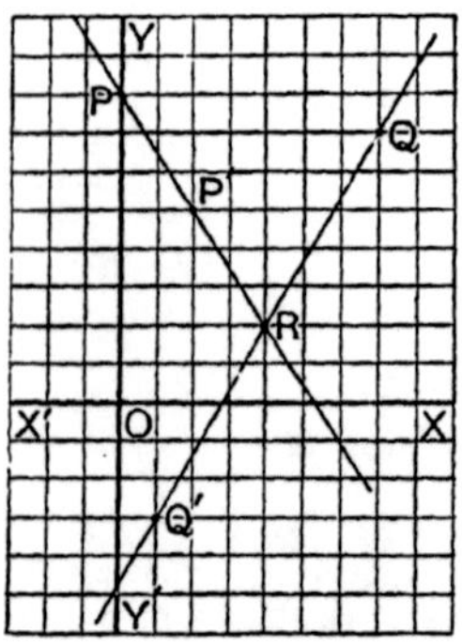

FIG 12. (Ex. 5.)

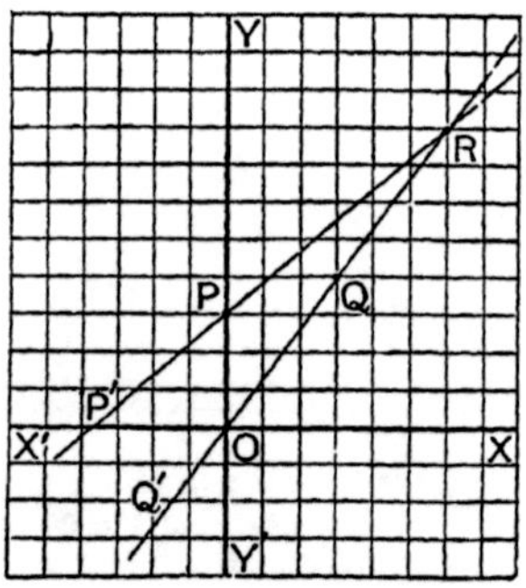

FIG. 13. (Ex. 6.)

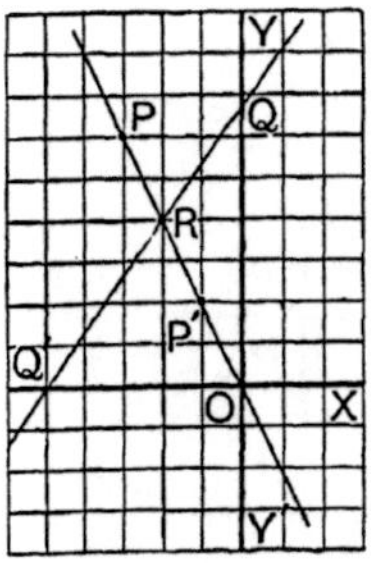

FIG. 14. (Ex. 7.)

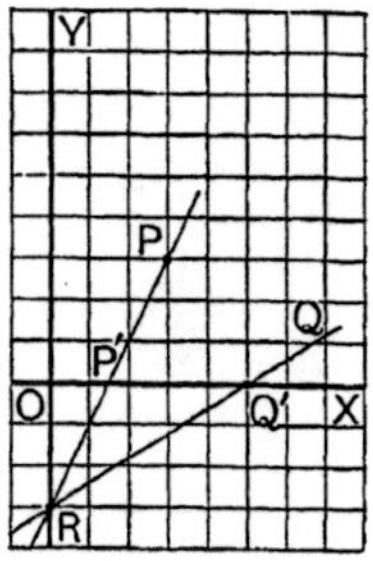

FIG. 15. (Ex. 8.)

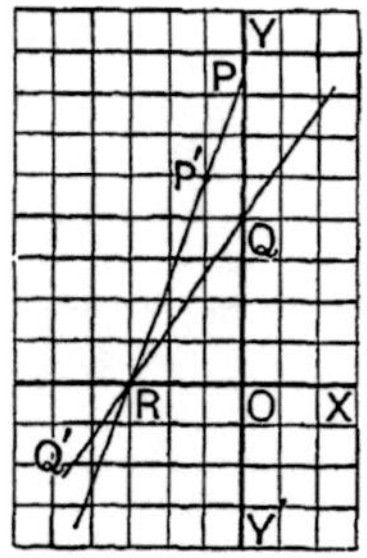

FIG. 16. (Ex. 9.)

10. See solution of Example in Art. 420.

11. (Fig. 17.) Assume $y=ax+b$ as the equation of the line which joins the two points A (1, 1) and B (−3, 4).

Then substituting the coordinates of these points we have

$$1=a+b,\ \ 4=-3a+b.$$

These equations give

$$a=-\frac{3}{4},\quad b=\frac{7}{4}.$$

Hence the equation to the line is

$$y=-\frac{3}{4}x+\frac{7}{4},\ \text{or}\ 3x+4y=7.$$

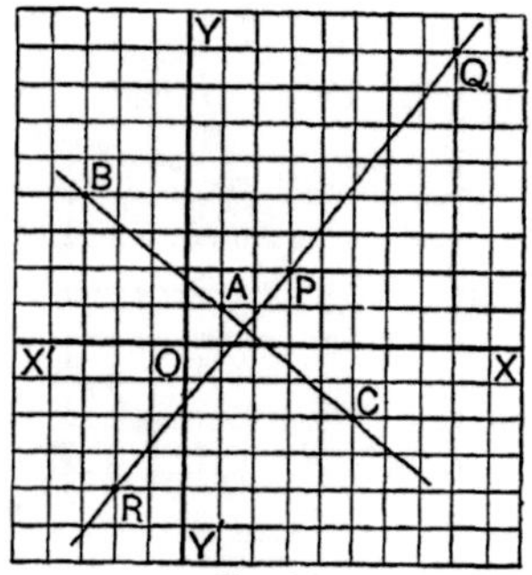

FIG. 17.

It is seen that $x=5$, $y=-2$ satisfies this equation; hence all three points lie on the line $3x+4y=7$.

12. (Fig. 17.) Assume $y=ax+b$ as the equation of the line joining P (3, 2) and Q (8, 8).

Then we have $2=3a+b$; $8=8a+b$; whence $a=\frac{6}{5}$, $b=-\frac{8}{5}$.

Thus the equation becomes $y=\frac{6x}{5}-\frac{8}{5}$, or $6x-5y=8$.

By trial, $x=-2$, $y=-4$ is seen to satisfy this equation; ∴ the point R (−2, −4) lies on the line joining (3, 2) and (8, 8). That is, the three points lie on a line.

To find the intercept on the axis of x, put $y=0$ in the equation of the line; then $6x=8$, or $x=1\frac{1}{3}$. Thus the line cuts the axis of x at the point $(1\frac{1}{3}, 0)$.

EXAMPLES. XLIV. d. PAGE 385.

1. The curves $y=x^2$ and $x=y^2$ are exhibited in Fig. 18. For $y=x^2$ we have the following tabulated values:

x	$-1{\cdot}5$	-1	$-{\cdot}5$	0	·5	1	1·5	2
y	2·25	1	·25	0	·25	1	2·25	4

and for $x=y^2$,

y	$-1{\cdot}5$	-1	$-{\cdot}5$	0	·5	1	1·5	2
x	2·25	1	·25	0	·25	1	2·25	4

It is evident that they cut only at O and P (1, 1). The equation of OP is $x=y$.

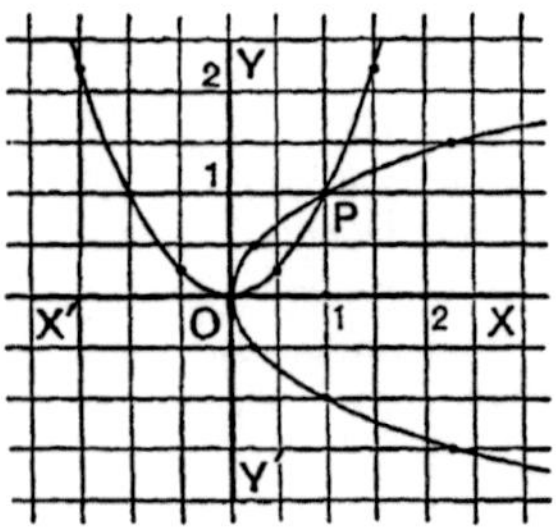

FIG. 18.

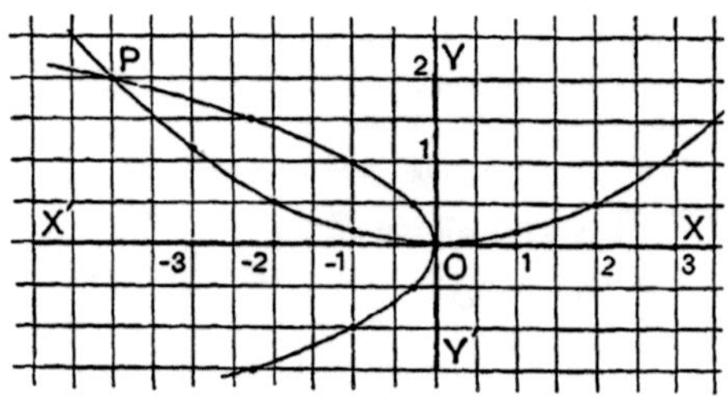

FIG. 19.

2. For $x=-y^2$ we have

y	-2	-1	$-{\cdot}5$	0	·5	1	1·5	2	3
x	-4	-1	$-{\cdot}25$	0	$-{\cdot}25$	1	$-2{\cdot}25$	-4	-9

and for $y=\frac{x^2}{8}$, following the arrangement of Ex. 2, p. 382:

x	-4	-3	-2	-1	0	1	2	3	4
x^2	16	9	4	1	0	1	4	9	16
y	2	1·125	·5	·125	0	·125	·5	1·125	2

The graphs are shewn in Fig. 19. It is seen there are only two points of intersection, O and P, whose coordinates are (0, 0), $(-4, 2)$ respectively.

3. (i) $y^2 = -4x$. (Fig. 20.)

Since y^2 must be positive, the only admissible values of x will be negative. That is, the curve lies entirely in the second and third quadrants. Also it is symmetrical about the axis of x, so that it will be sufficient to plot points in one of these quadrants. For the purpose of calculation write the equation in the form $x = -\frac{y^2}{4}$. Then for positive values of y, we have

y	0	1	2	4	6	8
x	0	− ·25	− 1	− 4	− 9	− 16

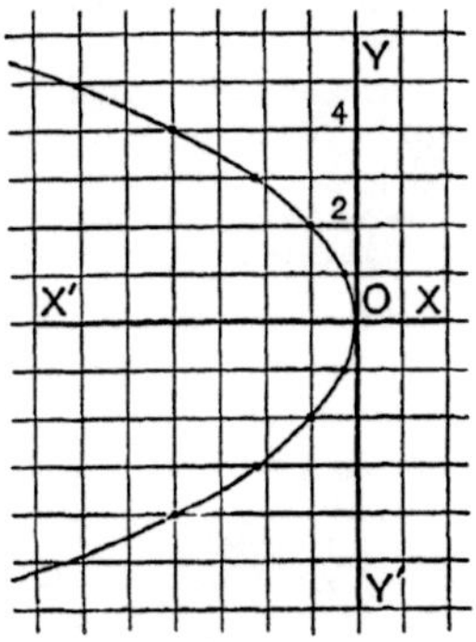

Fig. 20.

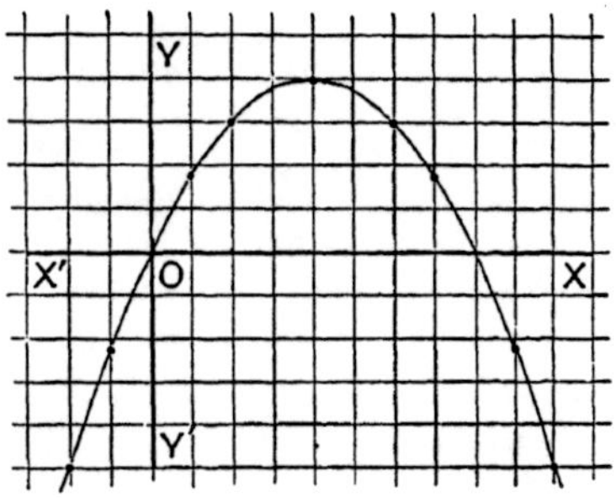

Fig. 21.

3. (ii) $y = 2x - \frac{x^2}{4}$. (Fig. 21.)

The following tabulated values may be used :

x	− 2	0	1	2	4	6	8	10
$2x$	− 4	0	2	4	8	12	16	20
$\frac{x^2}{4}$	1	0	·25	1	4	9	16	25
y	− 5	0	1·75	3	4	3	0	− 5

3. (iii) $y=\frac{x^2}{4}+x-2$. See Fig. 22 on the opposite page.

The following tabulated values may be used:

x	-8	-6	-5	-4	-3	-2	-1	0	1	2	3	4
$\frac{x^2}{4}$	16	9	6·25	4	2·25	1	·25	0	·25	1	2·25	4
y	6	1	-·75	-2	-2·75	-3	-2·75	-2	-·75	1	3·25	6

4. On the same axes as $y=x^2$ (see Ex. 1) draw the graph of the straight line $y=x$. Then the ordinate of the point on $y=x+x^2$ which corresponds to any given abscissa is obtained by adding the ordinates of the points on $y=x^2$ and $y=x$ which have that abscissa. In cases where the ordinate of $y=x$ is negative and numerically greater than that of $y=x^2$, the ordinate of the new curve must be drawn below the line OX and equal to the difference of the two ordinates. The curve will be found to be a parabola passing through the origin and the point $(-1, 0)$.

5. The curve $y=\frac{x^2}{4}+x-2$ is shewn in Fig. 22 on the opposite page.

The line $y=2x-3$ is the line PQ in that figure, and is seen to meet the curve only at the point R (2, 1).

To find the points of intersection algebraically, we must find the values of x for which $2x-3=\frac{x^2}{4}+x-2$. This reduces to $x^2-4x+4=0$, or $(x-2)^2=0$. Hence the only value of x is 2, and substituting this in either of the equations we get $y=1$. Thus (2, 1) is the only common point.

6. (i) To solve the equation $\frac{x^2}{4}+x-2=0$, we must find the abscissæ of all points on the curve $y=\frac{x^2}{4}+x-2$ at which the ordinate is zero. The curve is shewn in Fig. 22, and the abscissæ of the points where the curve cuts the axis of x are found by measurement to be 1·46 and −5·46.

6. (ii) We must first draw the graph of $y=x^2-2x-4$. (Fig. 23.) The tabulated values are

x	-3	-2	-1	0	1	2	3	4	5
x^2	9	4	1	0	1	4	9	16	25
$2x$	-6	-4	-2	0	2	4	6	8	10
y	11	4	-1	-4	-5	-4	-1	4	11

The points whose ordinates vanish are M, M', and their abscissæ are found by measurement to be 3·24 and −1·24 respectively. Thus the roots of the equation $x^2-2x-4=0$ are 3·24 and −1·24.

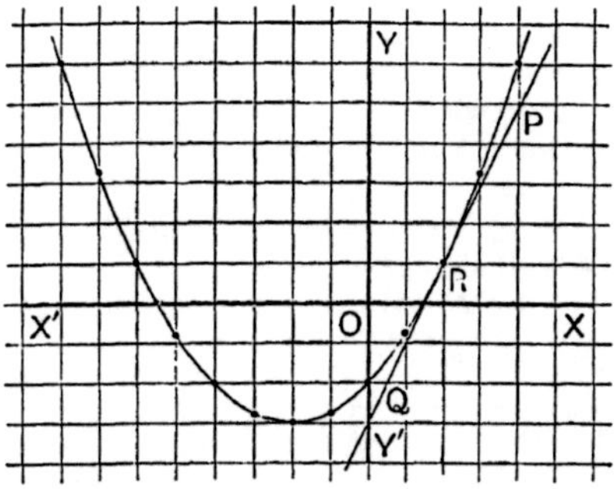

FIG. 22.

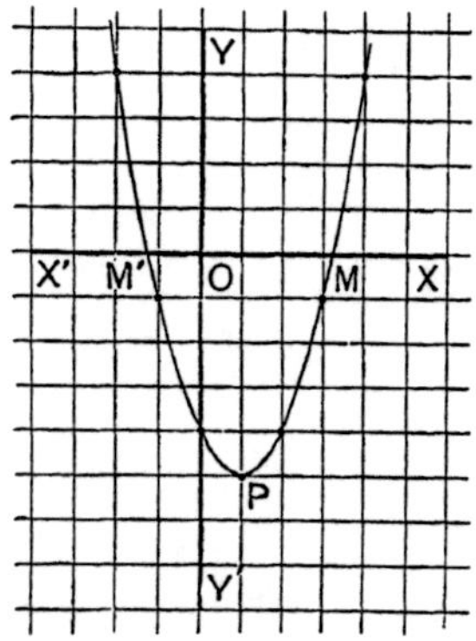

FIG. 23.

6. (iii) Plotting the curve $y = 4x^2 - 16x + 9$, we have

x	·5	1	1·5	2	2·5	3	3·5	4
$4x^2$	1	4	9	16	25	36	49	64
$16x$	8	16	24	32	40	48	56	64
y	2	-3	-6	-7	-6	-3	2	9

Hence the curve is as drawn in Fig. 24.

The ordinates vanish, that is, $4x^2 - 16x + 9 = 0$ at the points M (0·68, 0) and M' (3·32, 0).

Hence the required roots are 0·68 and 3·32.

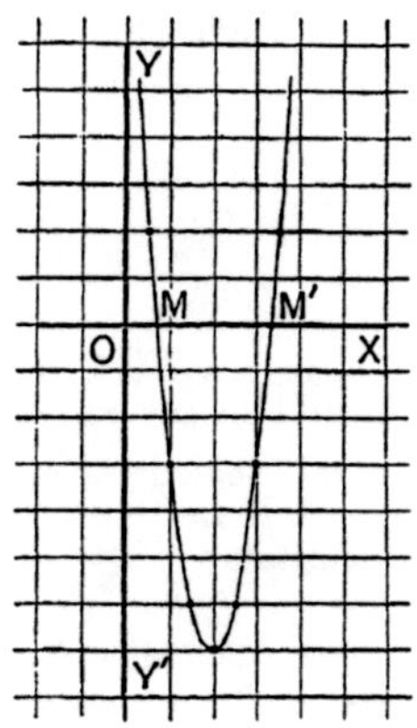

FIG. 24.

7. (i) The curve $y=x^2-2x-4$ has been drawn in Ex. 6 (Fig. 23).

It is seen that the algebraically least ordinate is that of the point P whose value is -5.

Hence the minimum value of y, and therefore of x^2-2x-4, is -5.

(ii) This can be done as above by drawing the graph of $y=5+4x-2x^2$ and proceeding as before, but it can also be immediately deduced from the preceding example.

Since $5+4x-2x^2=-3-2(x^2-2x-4)$, its value must be greatest when x^2-2x-4 is least. Thus the maximum value of

$$5+4x-2x^2=-3-2(-5)=7.$$

8. $y=(x-1)(x-2)$. (Fig. 25.)

x	-2	-1	0	·5	1	1·5	2	3	4	5
y	12	6	2	·75	0	$-$·25	0	2	6	12

Looking for turning points on the curve, we see that the ordinate at the point M is algebraically less than that of any other point in its immediate vicinity. Hence the minimum value of the expression is given by the ordinate of M, which is $-0·25$.

To find the values of x for which $(x-1)(x-2)=5$, we must note the abscissæ of all points on the curve $y=(x-1)(x-2)$ whose ordinates are equal to 5. These are seen to be P, P', and the values of x at these points are 3·79 and $-0·79$.

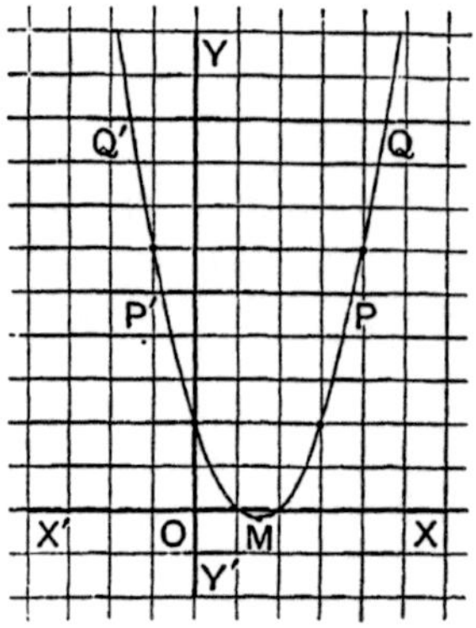

FIG. 25.

Similarly $(x-1)(x-2)=9$ at the points Q, Q' and their abscissæ are found by measurement to be 4·62 and $-1·62$.

9. See Example in Art. 428. The graph of $x^2+y^2=100$ is a circle with centre at the origin and with radius 10. It is cut by the line $x+y=14$ at the points (6, 8), (8, 6). Hence the solution of the equations is $x=6$, $y=8$; or $x=8$, $y=6$.

10. Drawing the graphs of the two equations, it will be found that $x^2+y^2=25$ is a circle with centre at the origin and with radius 5; and that the line $3x+4y=25$ meets it at the point (3, 4) and no where else. That is, the line is the tangent to the circle at that point.

EXAMPLES. XLIV. e. PAGE 394.

Some of the graphs in this Exercise cannot be drawn satisfactorily except on a very large scale. The difficulty may be obviated to some extent by adopting different units of measurement for ordinates and abscissæ as explained in Arts. 437, 438.

1. When x is positive, y is positive; and when x is negative, y is negative. Thus the curve lies entirely in the first and third quadrants.

From the form of the equation it is evident that for any point (x, y) on the curve there is a corresponding point $(-x, -y)$ which satisfies the equation. Thus the following tabulated values may be used, choosing the unit for x five times as great as that for y.

x	0	$\pm 0{\cdot}5$	± 1	$\pm 1{\cdot}5$	± 2	$\pm 2{\cdot}5$	± 3
y	0	$\pm 0{\cdot}125$	± 1	$\pm 3{\cdot}375$	± 8	$\pm 15{\cdot}625$	± 27

The graph is shewn on page 398.

(i) is deduced by changing the sign of every ordinate; thus the curve is similar to that on p. 398 but lies in the second and fourth quadrants.

(ii) is deduced by halving each ordinate.

2. Plot the graphs of $y=x$ and $y=x^3$ on the same axes, and take the difference of the ordinates. These will be the ordinates of $y=x-x^3$.

3. $y=\frac{1}{x^2}$. (Fig. 26.)

Since y is positive whether x is positive or negative, the curve must lie entirely in the first and second quadrants.

Also values of x differing only in sign give the same value of y, hence the curve is symmetrical about the axis of y.

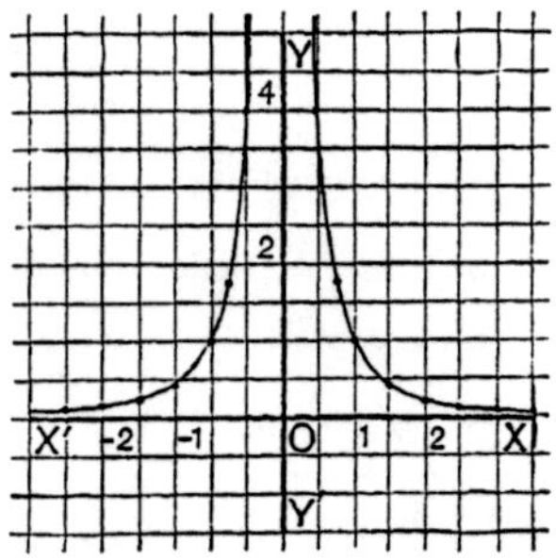

FIG. 26.

When $x=0$, $y=\infty$, and as x increases y diminishes, becoming zero when $x=\infty$. Thus the curve continually approaches nearer and nearer the axes of x and y, which are asymptotes.

The tabulated values for the branch in the first quadrant are

x	0	...	·25	·5	·75	1	1·5	2	2·5	3	...	∞
y	∞	...	16	4	1·77	1	·44	·25	·16	·11	...	0

The second branch can be drawn by symmetry, each negative value of x giving the same value of y as the corresponding positive value of x.

4. The curve $y=\frac{a}{x^2}$ will have the same shape as $y=\frac{1}{x^2}$, lying entirely in the first and second quadrants when a is positive, and in the third and fourth when a is negative.

Any ordinate can be found at once by multiplying the corresponding ordinate of $y=\frac{1}{x^2}$ by a. Thus (since a is integral) the branch which is asymptotic to the axis of y will approach it less rapidly than in the case of $y=\frac{1}{x^2}$.

5. (i) $y=1+\frac{1}{x}$. (Fig. 27.)

Increase each of the ordinates of $y=\frac{1}{x}$ by 1.

Compare Fig. on p. 387.

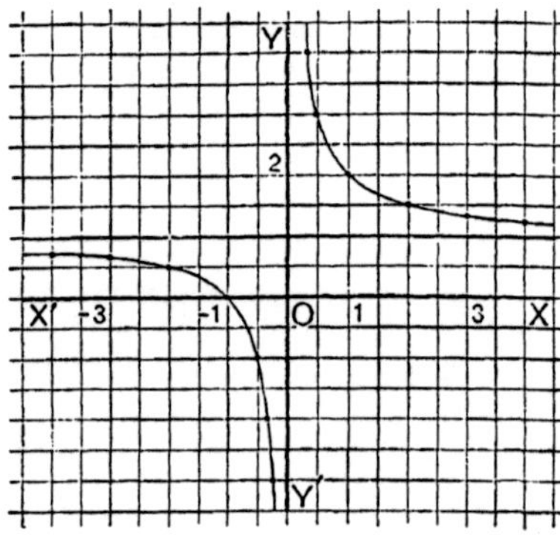

FIG. 27.

(ii) $y=2+\frac{10}{x^2}$. (Fig. 28.)

This may be deduced by increasing the ordinates of $y=\frac{10}{x^2}$ by 2. Compare Ex. 4 above.

6. $y=x^3-3x$. (Fig. 29.) We have for positive values of x

x	0	·2	4	·6	·8	1	1·2	1·4	1·6	1·8	2
y	0	- ·59	- 1·14	- 1·58	- 1·89	- 2	- 1·87	- 1·46	- ·70	·43	2

For negative values of x the values are the same as the above with their signs changed.

The point $(-1, 2)$ is seen to be a turning point of the curve. In the neighbourhood of the point all the ordinates are algebraically less than 2; hence there is a maximum value of y at this point (Art. 427). Similarly at $(1, -2)$ there is a minimum value.

The values which make $x^3-3x=0$ are those which correspond to $y=0$; this occurs at the points O, M, M' and the values of x are, by measurement, 0, $+1·732$, $-1·732$. The figure has been drawn on paper ruled to tenths of an inch and then reduced to half the original size.

In cases where the root is wanted with great accuracy, it is sometimes advisable to enlarge a small portion of the curve in the neighbourhood of the root. Thus in the present case, seeing from the graph that the root lies between 1·7 and 1·8, we might draw the portion of the curve between $x=1\cdot6$ and $x=1\cdot9$ on a separate figure, taking the unit of x ten times as large as above.

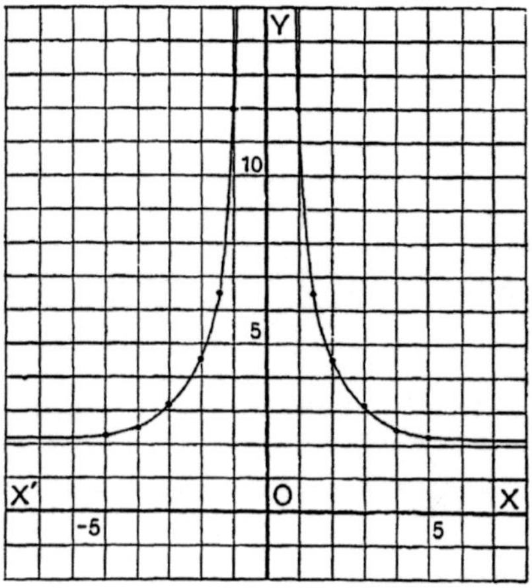

FIG. 28.

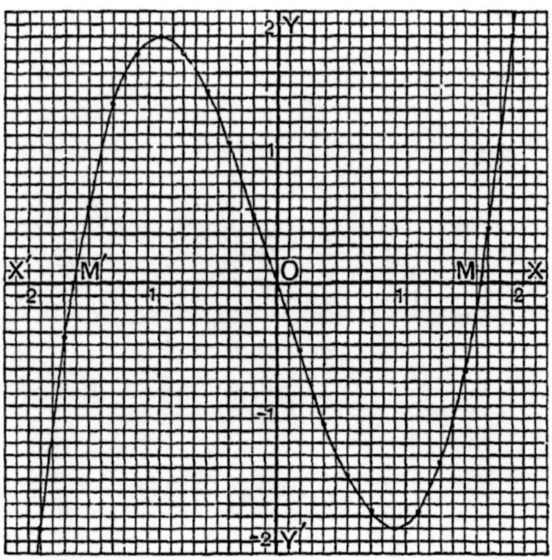

FIG. 29.

7. Solve the equations $3x+2y=16$, $xy=10$. (Fig. 30.)

$xy=10$ gives a rectangular hyperbola, (see Art. 433) of which only one branch is shewn in the figure. It will be found that the line $3x+2y=16$ cuts the curve in the points $(2, 5)$, $\left(\frac{10}{3}, 3\right)$.

8. (i) $y=\frac{15-x^2}{x}$. (Fig. 31.)

The tabulated values are

x	-5	-4	-3	-2	-1	0	1	2	3	4	5	6
y	2	$\cdot 25$	-2	$-5\cdot 5$	-14	∞	14	$5\cdot 5$	2	$-\cdot 25$	-2	$-3\cdot 5$

(ii) $x=\frac{10-y^2}{y}$.

The tabulated values are

y	-5	-4	-3	-2	-1	0	1	2	3	4	5
x	3	$1\cdot 5$	$-\frac{1}{3}$	-3	-9	∞	9	3	$\frac{1}{3}$	$-1\cdot 5$	-3

The curves are two hyperbolas shewn in Fig. 31. They intersect at the points $(3, 2)$, $(-3, -2)$. Thus the required solution is $x=3$, $y=2$, or $x=-3$, $y=-2$.

9. We have $y=\frac{x}{2-x}=\frac{1}{\frac{2}{x}-1}$. Whence by proceeding as in the example of Art. 434 we see that $x=2$, $y=-1$ are asymptotes, and that the curve has two branches, one lying to the left of $x=2$, and above $y=-1$, and the other to the right of $x=2$, and below $y=-1$.

For the first of these branches, choosing values of x between $-\infty$ and 2, we have

x	$-\infty$	...	-2	-1	0	$\cdot 5$	1	$1\cdot 5$	...	2
y	-1	...	$-\cdot 5$	$-\frac{1}{3}$	0	$\frac{1}{3}$	1	3	...	∞

For the other branch, values of x may be taken between 2 and ∞. The curve is shewn in Fig. 32.

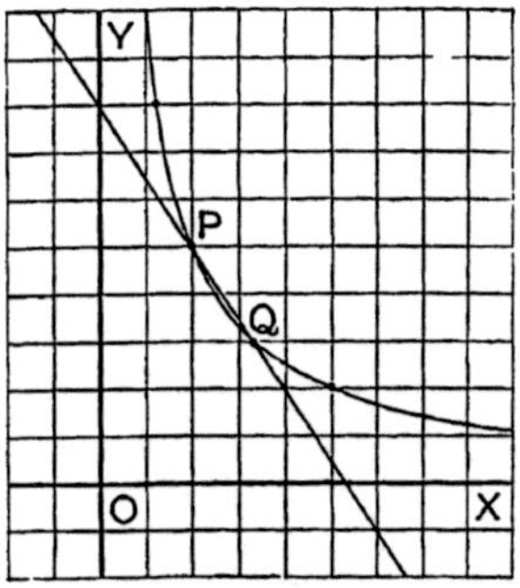

FIG. 30.

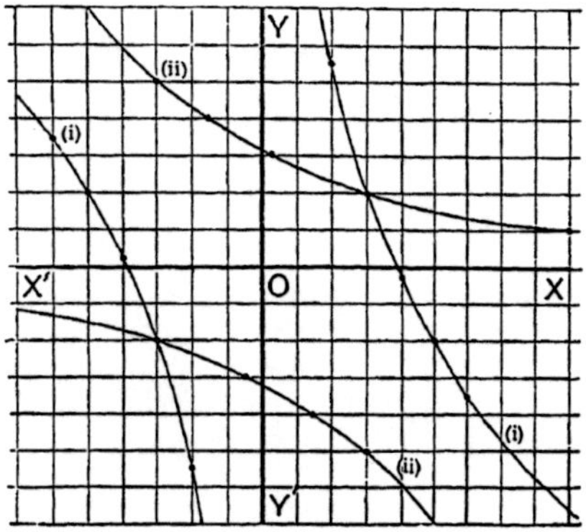

FIG. 31.

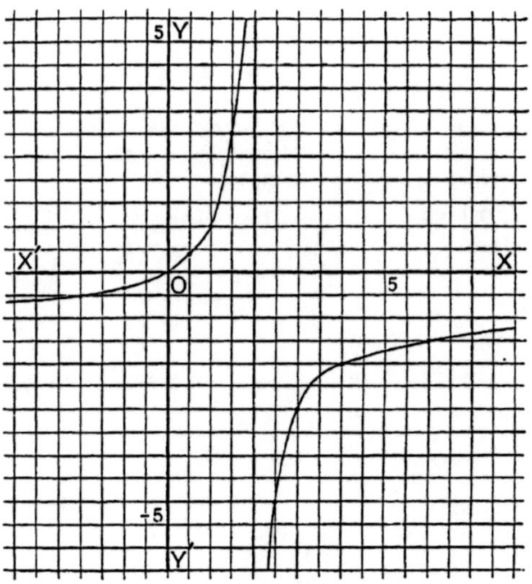

FIG. 32.

[In the following examples much labour in the tabulation of values and in the plotting of points may be avoided by observing the form of the curve in the neighbourhood of the origin, the axes, and infinity, and by the principle of symmetry as explained in Art. 434.]

10. $y=\dfrac{1+x}{1-x}$.

Since this can be written $y=\dfrac{(x+1)}{2-(x+1)}$, the graph could be deduced from that of Ex. 9 by diminishing each abscissa by unity. Otherwise, we have the following table of values:

x	$-\infty$		-3	$-1·5$	-1	0	$·5$	$·75$	1	$1·5$	2	3	5	...	∞
y	-1	.	$-·5$	$-·2$	0	1	3	7	∞	-5	-3	-2	$-1·5$	...	-1

The asymptotes are $x=1$, and $y=-1$. The form of the curve resembles that of Ex. 9.

11. $y=\dfrac{1+x^2}{1-x}=-x-1+\dfrac{2}{1-x}$. (Fig. 33.)

When $x=1$, $y=\infty$; also when x is very little less than 1, y is positive, while when x is very little greater than 1, y is negative. Hence the infinite points on the graph near to the line $x=1$ have positive ordinates to the left, and negative ordinates to the right, of this line. Thus $x=1$ is an asymptote lying to the right of the curve in the first quadrant and to the left in the fourth quadrant.

When $y=0$, we have $x^2+1=0$, which has no real roots; hence the curve never crosses the axis of x. Also when $x=\infty$, the curve approximates to $y+x=-1$, which is another asymptote.

For the upper branch the following values may be used:

x	-9	-5	-3	-2	-1	$-·5$	0	$·25$	$·5$
y	$8·2$	$4·3$	$2·5$	$1·6$	1	$·83$	1	$1·4$	$2·5$

For the lower branch values of x may be taken between 1 and ∞.

12. $y=\dfrac{x^2-15}{x-4}=x+4+\dfrac{1}{x-4}$. (Fig. 34.)

From the second form of the equation we see that when $x=\infty$, $y=x+4$; also when x is positive and very large, the ordinate of the curve is greater than that of the line $y=x+4$; and when x is negative and very large, the ordinate of the curve is less than that of the line. Hence the line is an asymptote, its infinite points lying below the curve when x is positive, and above the curve when x is negative.

When $x=4$, $y=\infty$; thus $x=4$ is an asymptote.

Also when $y=0$, $x^2=15$; or $x=\pm 3·87$.

The following values give the lower branch of the curve, which lies to the left of $x=4$.

x	-4	-2	0	1	2	3	$3{\cdot}5$	...	4
y	$-{\cdot}125$	$1{\cdot}83$	$3{\cdot}75$	$4{\cdot}66$	$5{\cdot}5$	6	$5{\cdot}5$	...	∞

For the upper branch, values of x may be taken from 4 to ∞; it will be found to lie to the right of $x=4$. The graph is as in Fig. 34.

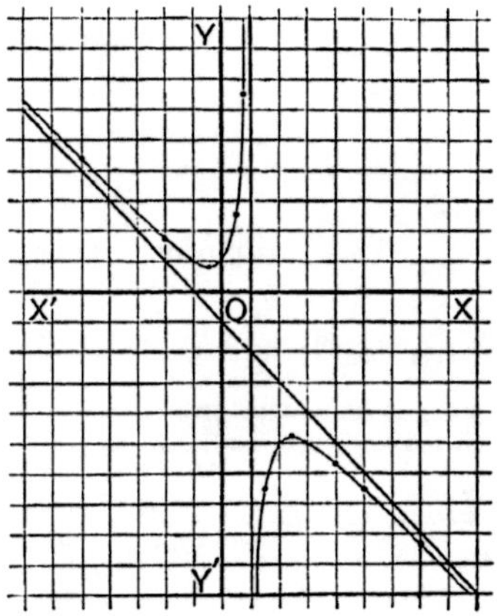

Fig. 33.

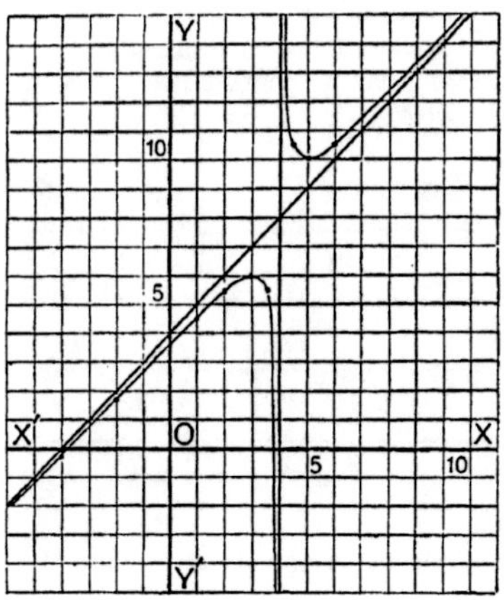

Fig. 34.

13. $y=\dfrac{(x-1)(x-2)}{x-3}=x+\dfrac{2}{x-3}$. (Fig. 35.)

In the figure the unit of x is taken twice as large as the unit of y.

If $x=0$, $y=-\dfrac{2}{3}$; if $y=0$, $x=1$, or 2.

When $x=3$, $y=\infty$; and when $x=\infty$, $y=x$.

Thus the lines $x=3$, and $x=y$ are asymptotes.

x	−2	−1	0	1	1·5	2	2·5	3	3·5	4	4·5	5	7
y	−2·4	−1·5	−·67	0	·17	0	−1·5	∞	7·5	6	5·83	6	7·5

14. $y=\dfrac{x^2+x+1}{x^2-x+1}=1+\dfrac{2x}{x^2-x+1}$. (Fig. 36.)

Putting $x=0$, we get $y=1$;

,, $y=0$, we get $x^2+x+1=0$, which has no real roots.

Writing the equation in the form $y=1+\dfrac{\frac{2}{x}}{1-\frac{1}{x}+\frac{1}{x^2}}$, we see that when $x=\infty$, $y=1$. Thus $y=1$ is an asymptote.

x	−4	−3	−2	−1	−·5	0	·5	1	1·5	2	3	4
y	·62	·54	·43	·33	·43	1	2·33	3	2·71	2·33	1·86	1·61

15. $y=\dfrac{x^2+5x+6}{x^2+1}=1+\dfrac{5(x+1)}{x^2+1}$. (Fig. 37.)

When $y=0$, we have $x^2+5x+6=0$; $\therefore x=-2$, or -3.

Since x^2+1 is always positive, y must also be positive, except when $(x+2)(x+3)$ is negative, that is, when x lies between -2 and -3.

For negative values of x

x	−6	−5	−4	−3	−2·5	−2	−1	−·5
y	·32	·23	·12	0	−·03	0	1	3

For positive values

x	0	·5	1	2	3	4	7	.	∞
y	6	7	6	4	3	2·47	1·8	...	1

The maximum ordinate corresponds to a value of x between $\frac{1}{3}$ and $\frac{1}{2}$.

The line $y=1$ is an asymptote.

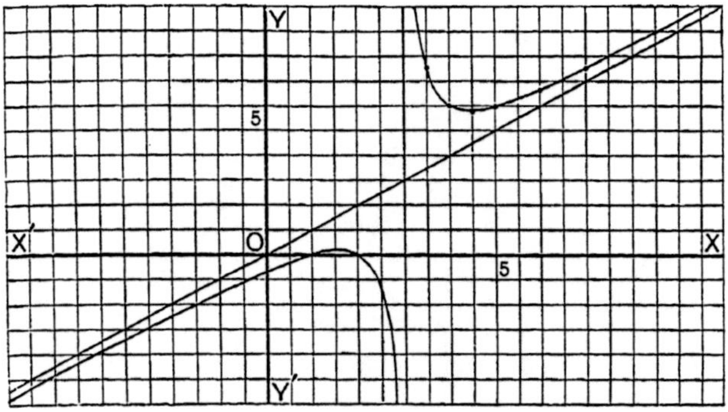

FIG. 35.

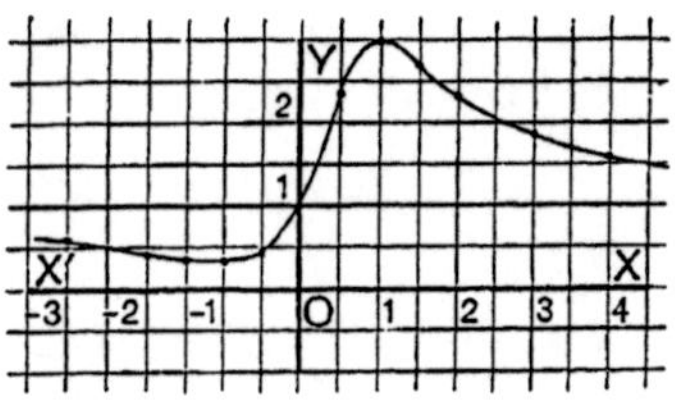

FIG. 36.

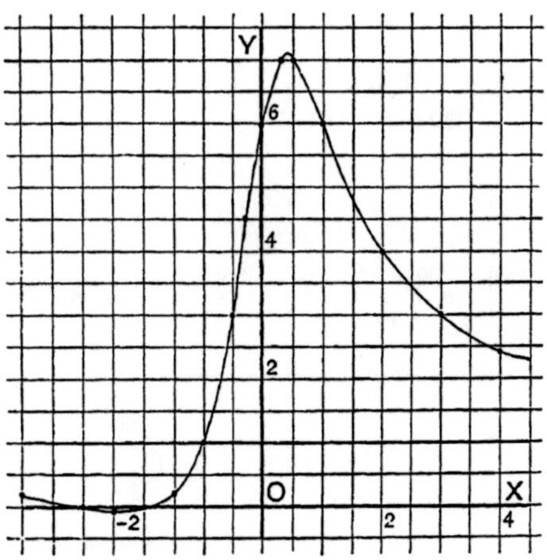

FIG. 37.

16. $y = x^3 - 6x^2 + 11x - 6 = (x-1)(x-2)(x-3)$. (Fig. 38.)

The curve crosses the axis of x where $x = 1$, 2, or 3.

If x is less than 1, y is negative,
if x is between 1 and 2, y is positive,
if x is between 2 and 3, y is negative,
if x is greater than 3, y is positive.

x	-1	0	1	1·5	2	2·5	3	4	5
y	-24	-6	0	·375	0	-·375	0	6	24

Two-tenths of an inch have been taken as unit.

17. $10y = x^3 - 5x^2 + x - 5 = (x-5)(x^2+1)$. (Fig. 39.)

From the equation we see that the curve crosses the axis of x at $x = 5$, that when x is less than 5, y must be negative, and when x is greater than 5, y must be positive.

x	-3	-2	-1	0	1	2	3	4	5	6	7
y	-8	-3·5	-1·2	-·5	-·8	-1·5	-2	-1·7	0	3·7	10

Plotting these points we get the curve as drawn, taking two-tenths of an inch as unit.

18. $y = \dfrac{20}{x^2+2}$. (Fig. 40.)

When $x = \infty$, $y = 0$; thus the axis of x is an asymptote.

The curve is symmetrical about the axis of y (Art. 434); hence it will be sufficient to consider positive values of x.

It will be convenient to take the x-unit three times as great as the y-unit. For the portion of the curve in the first quadrant the following values may be used:

x	0	1	2	2·83	4	4·24	5	∞
y	10	6·66	3·33	2	1·11	1	·74	0

Values of x differing only in sign give the same series of values for y. Hence we obtain the curve in Fig. 40.

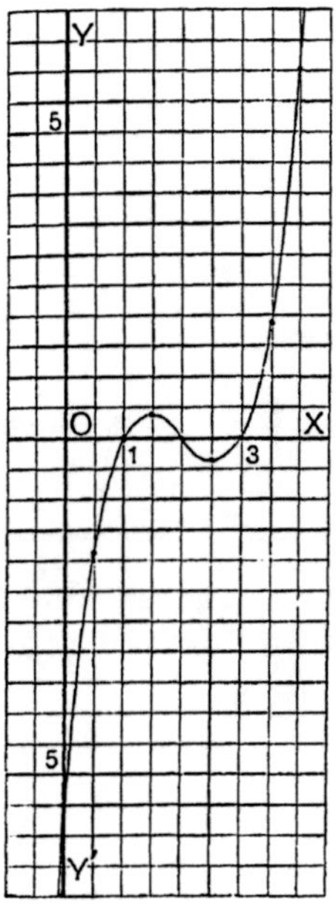

Fig. 38.

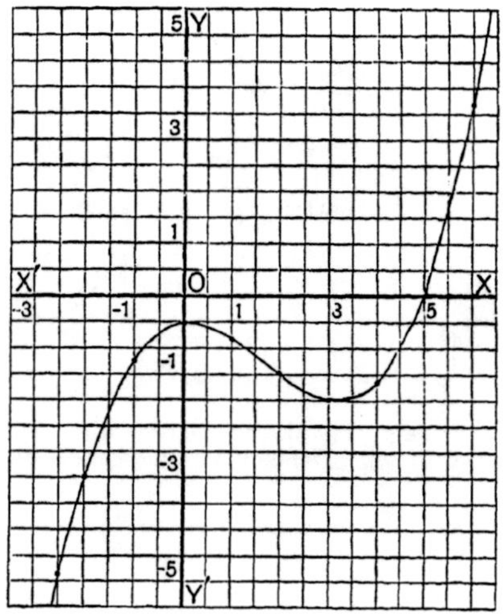

Fig. 39.

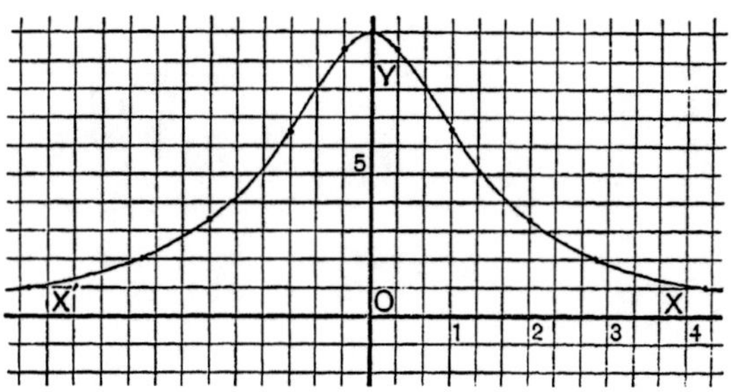

Fig. 40.

19. $y=\frac{40x}{x^2+10}$. (Fig. 41.)

When $x=0$, $y=0$; hence the curve passes through the origin. Also when x is positive, y is positive, and when x is negative, y is negative; hence the curve lies entirely in the first and third quadrants.

x	0	1	2	3	4	5	6	10	15	20
y	0	3·64	5·71	6·32	6·15	5·71	5·22	3·64	2·55	1·95

For negative values of x, the values of y are the same as above with the sign changed. Hence the curve is as drawn.

20. $y=\frac{x(8-x)}{x+5}=13-x-\frac{65}{x+5}$. (Fig. 42.)

When $x=-5$, $y=\infty$; thus $x=-5$ is an asymptote.

When $y=0$, $x=0$, or 8; thus the curve passes through the origin, and cuts the axis of x at (8, 0).

Again when $x=\infty$, $y=13-x$; thus $x+y=13$ is an asymptote.

The following values may be used:

x	−2·5	−2	−1	0	1	3	5	8	10	15
y	−10·5	−6·6	−2·25	0	1·16	1·875	1·5	0	−1·33	−5·25

The curve is a hyperbola, only one branch being shewn in the diagram. The second branch lies entirely in the second quadrant, to the left of $x=-5$, and above $x+y=13$.

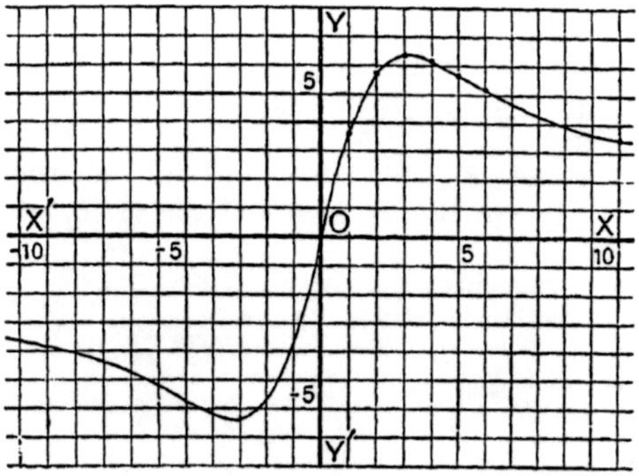

FIG. 41.

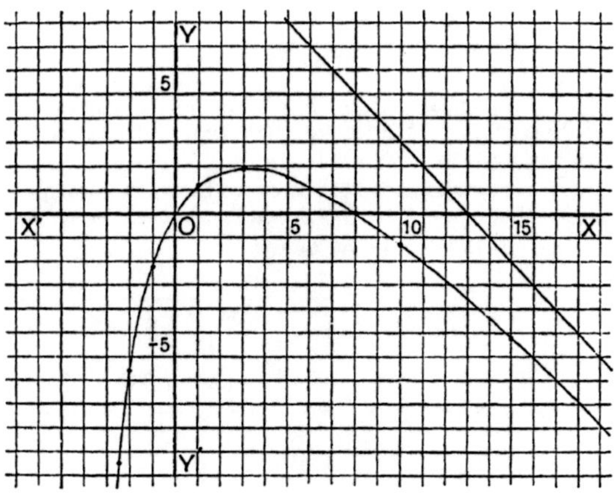

FIG. 42.

21. $y=\frac{(x-2)(x-3)}{x-5}=x+\frac{6}{x-5}$. (Fig. 43.)

In details this example closely resembles Ex. 12, above.

The asymptotes are $x=5$, and $y=x$, and the curve cuts the axis of x when $x=2$, or 3.

For the upper branch of the curve we may take $x=5$, 5·5, 6, 7, 8, 9, 11, ... ; and for the lower branch $x=-7$, -5, -1, 0, 1, 2, 2·5, 3, 4, ...

22. $y=\frac{(x-1)(x-2)(x+1)}{4}$. (Fig. 44.)

The curve will cross the axis of x at the points where $x=-1$, $x=1$, $x=2$. The only portion of the curve that needs discussion will be found to lie between these points.

In the figure, to shew the shape of the curve between these points, we have taken 5 divisions of the paper for the x-unit, and 10 divisions for the y-unit.

The following sets of values may be used :

x	0	·2	·6	1	1·2	1·4	1·6	1·8	2	2·2	2·4
y	·5	·43	·22	0	−·09	−·14	−1·56	−1·08	0	·19	·48

and

x	−·2	−·4	−·6	−·8	−1	−1·4
y	·53	·50	·42	·25	0	−·82

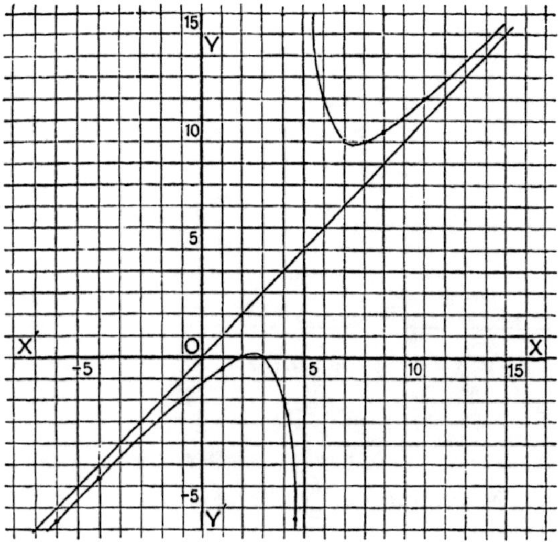

FIG. 43.

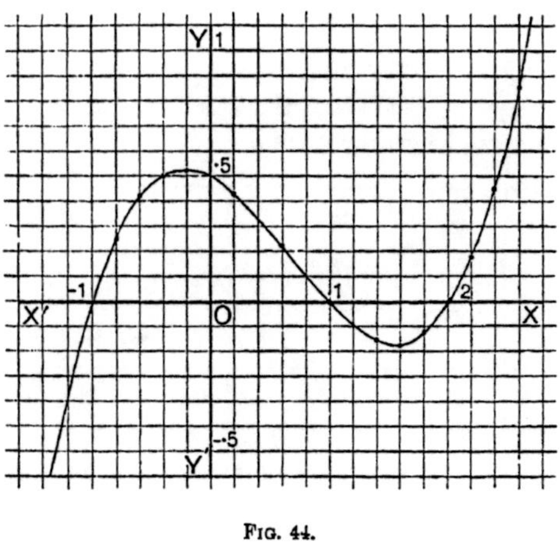

FIG. 44.

23. $y^2=x^2-5x+4=(x-1)(x-4)$. (Fig. 45.)

In order that y may have real values $(x-1)(x-4)$ must be *positive*, hence x must be either greater than 4, or less than 1; that is, there can be no portion of the curve between $x=1$ and $x=4$.

Since no odd powers of y occur the curve is symmetrical about the x-axis.

x	-2	-1	0	·5	1	4	4·5	5	6	7
y^2	18	10	4	1·75	0	0	1·75	4	10	18
y	$\pm4·24$	$\pm3·16$	±2	$\pm1·32$	0	0	$\pm1·32$	±2	$\pm3·16$	$\pm4·24$

When $x=+\infty$, $y=+\infty$; and when $x=-\infty$, $y=-\infty$.

Also $y^2=\left(x-\frac{5}{2}\right)^2-\frac{9}{4}$, or $\left(-x+\frac{5}{2}\right)^2-\frac{9}{4}$.

Thus no positive value of y can be greater than $x-\frac{5}{2}$, and no negative value of y can be less than $-x+\frac{5}{2}$. Hence the curve is as in Fig. 45, the lines $y=x-\frac{5}{2}$, and $y=-x+\frac{5}{2}$ being asymptotes.

24. $4y^2=x^2(5-x)$. (Fig. 46.)

The curve is symmetrical about the axis of x.

No value of x can be greater than 5.

x	5	4	3	2	1	0	-1	-2	-3
$4y^2$	0	16	18	12	4	0	6	28	72
$2y$	0	±4	$\pm4·24$	$\pm3·46$	±2	0	$\pm2·45$	$\pm5·29$	$\pm8·48$
y	0	±2	$\pm2·12$	$\pm1·73$	±1	0	$\pm1·22$	$\pm2·64$	$\pm4·24$

Hence the curve is as drawn, two-tenths of an inch being taken as unit.

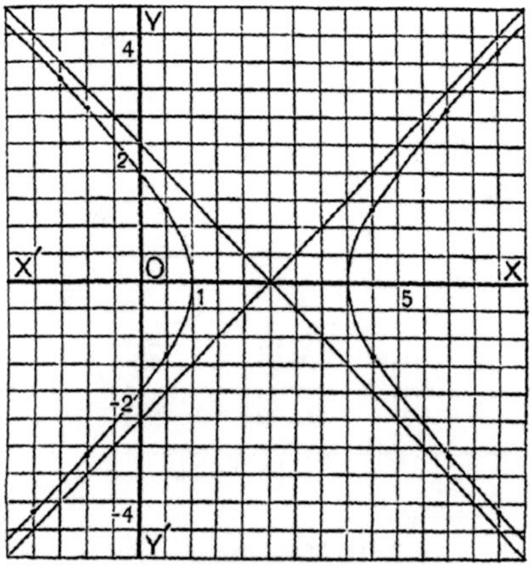

FIG. 45.

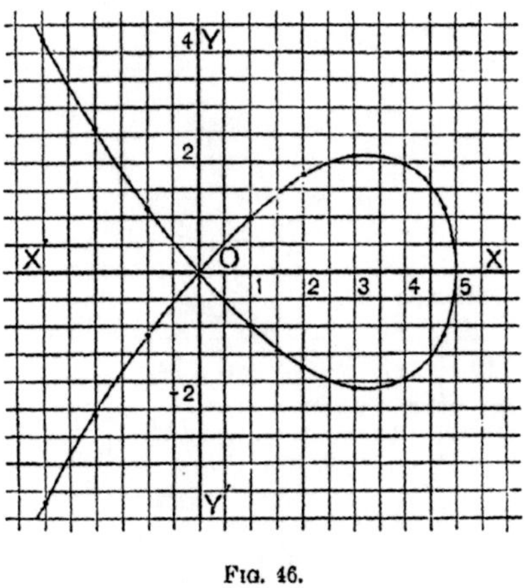

FIG. 46.

25. $y^2 = \dfrac{x(3-x)(x-8)}{x^2+5} = -\dfrac{x(x-3)(x-8)}{x^2+5}$. (Fig. 47.)

Since the denominator of the fraction is always positive there can only be real values of y when $x(x-3)(x-8)$ is negative, that is, when x is negative, or between 3 and 8. Thus there are two separate branches to the curve, one entirely to the left of the y-axis, and the other entirely between $x=3$, and $x=8$.

The curve is symmetrical about the axis of x.

x	8	7	6	5	4	3	1	-2	-4	-6	-10
y^2	0	$\frac{14}{27}$	$\frac{36}{41}$	1	$\frac{16}{21}$	0	6	$\frac{100}{9}$	16	$\frac{756}{41}$	$\frac{156}{7}$
$\log y^2$		$\bar{1}{\cdot}7147$	$\bar{1}{\cdot}9435$		$\bar{1}{\cdot}8819$		·7782			1·2657	1·3480
$\log y$		$\bar{1}{\cdot}8573$	$\bar{1}{\cdot}9717$		$\bar{1}{\cdot}9409$		·3891			·6328	·6740
y	0	·719	·937	1	·873	0	2·45	3·33	4	4·29	4·72

where a few of the values have been obtained by direct evolution.

26. $y^2 = \dfrac{(x+7)(x-4)(x-10)}{x^2+5}$. (Fig. 48.)

For real values of y, x must be either greater than 10, or between -7 and $+4$.

Hence there are two distinct branches of the curve, one to the right of $x=10$, and the other between $x=4$ and $x=-7$.

For one branch we have

x	-7	-5	-4	-3	-2	-1	0	1	2	3	4
y^2	0	9	16	26	40	55	56	36	16	5	0
y	0	± 3	± 4	$\pm 5{\cdot}09$	$\pm 6{\cdot}32$	$\pm 7{\cdot}42$	$\pm 7{\cdot}48$	± 6	± 4	$\pm 2{\cdot}23$	0

and for the second branch

x	10	11	12	14	16	20
y^2	0	1	$\frac{304}{149}$	$\frac{840}{201}$	$\frac{1656}{261}$	$\frac{32}{3}$
$\log y^2$			·3097	·6211	·8025	
$\log y$			·1548	·3105	·4012	
y	0	± 1	$\pm 1{\cdot}43$	$\pm 2{\cdot}04$	$\pm 2{\cdot}52$	$\pm 3{\cdot}26$

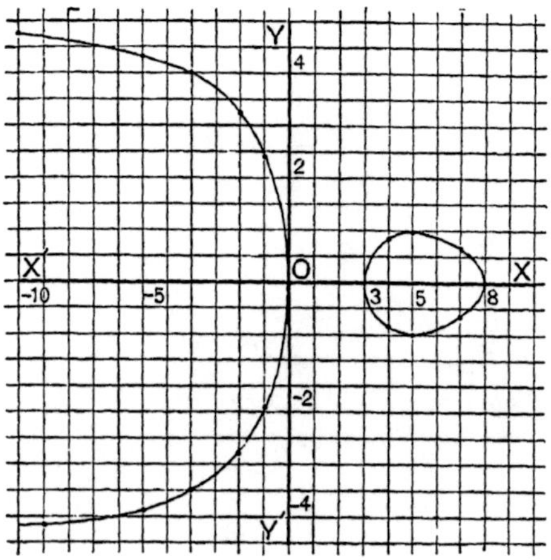

FIG. 47.

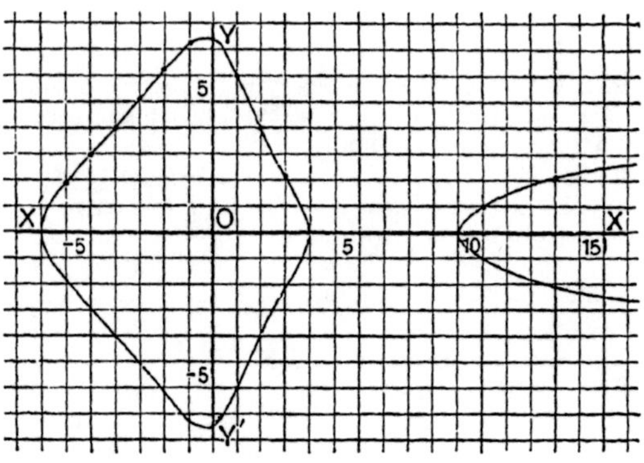

FIG. 48.

27. $y^2 = \dfrac{x^2(49 - x^2)}{50}$. (Fig. 49.)

The curve is symmetrical about both axes.

For real values of y, x must lie between $+7$ and -7.

x	0	± 1	± 2	± 3	± 4	± 5	± 6	± 7
y^2	0	·96	3·6	7·2	10·56	12	9·36	0
y	0	$\pm$·98	$\pm$1·90	$\pm$2·68	$\pm$3·25	$\pm$3·46	$\pm$3·06	0

Hence the curve is as drawn.

28. $y^2 = \dfrac{(81 - x^2)(x^2 - 4)}{100}$. (Fig. 50.)

The curve is symmetrical about both axes.

For real values of y, x^2 must be greater than 4 and less than 81.

Hence one branch of the curve must lie entirely between $+2$ and $+9$, and another branch between -2 and -9.

x	± 2	± 3	± 4	± 5	± 6	± 7	± 8	± 9
y^2	0	3·6	7·8	11·76	14·4	14·4	10·2	0
y	0	$\pm$1·9	$\pm$2·80	$\pm$3·43	$\pm$3·8	$\pm$3·8	$\pm$3·2	0

Hence the curve is as drawn.

29. $y^3 = \dfrac{x(x^2 - 64)}{5}$. (Fig. 51.)

We see by substitution that for every point (a, b) on the curve there is a corresponding point $(-a, -b)$ also on the curve; and consequently there is symmetry in opposite quadrants.

We shall therefore only tabulate the values of y given by positive values of x. With the arrangement of the Example on page 391 we have

x	0	1	2	3	4	5	6	7	8	10
y^3	0	-12·6	-24	-33	-38·4	-39	-33·6	-21	0	72
$\log y^3$		1·1004	1·3802	1·5185	1·5843	1·5911	1·5263	1·3222		1·8573
$\log y$		·3668	·4601	·5062	·5281	·5304	·5088	·4407		·6191
y	0	-2·33	-2·88	-3·21	-3·37	-3·39	-3·23	-2·76	0	4·16

The curve is given on a small scale in Fig. 51.

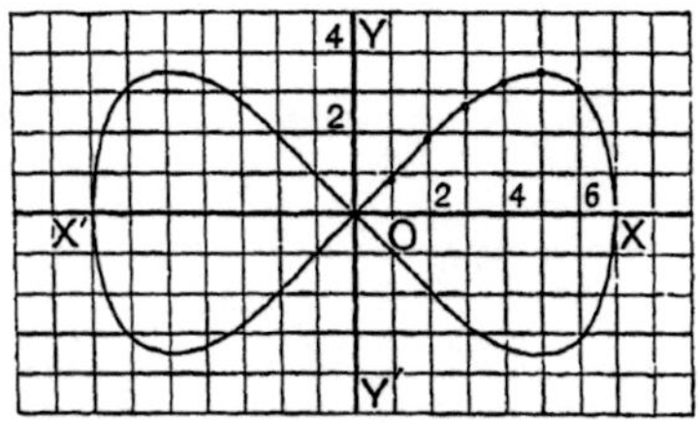

FIG. 49.

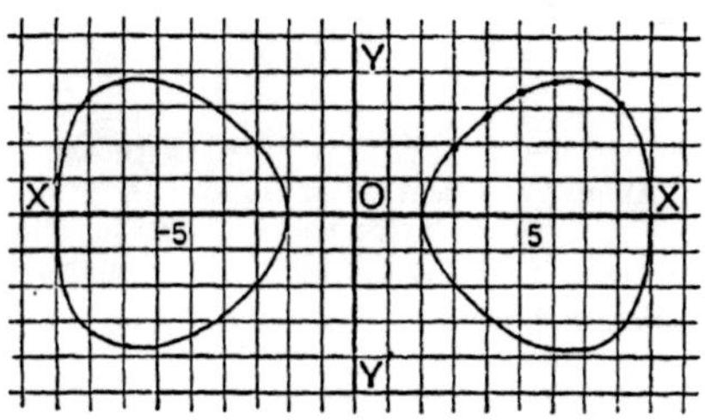

FIG. 50.

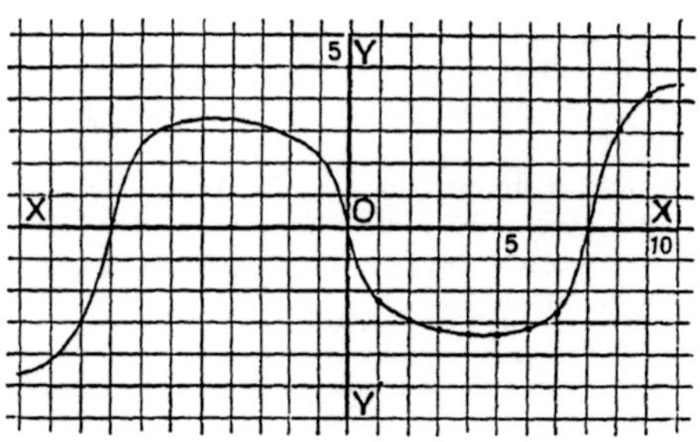

FIG. 51.

(For Examples 29 and 30 paper ruled to centimetres would be found more suitable.)

30. $y^3 = \dfrac{x^2(36 - x^2)}{5}$. (Fig. 52.)

Since no odd powers of x occur, the curve is symmetrical about the axis of y.

x	0	1	2	3	4	5	6	8
y^3	0	7	25·6	48·6	64	55	0	- 358·4
$\log y^3$		·8451	1·4082	1·6866		1·7404		2·5544
$\log y$		·2817	·4694	·5622		·5801		·8515
y	0	1·91	2·95	3·65	4	3·80	0	- 7·10

Hence the curve is as drawn.

31. The graph of $y=x^3$ may be plotted as in the Example on p. 398, the x-unit being taken five times as great as the y-unit. If the graph of $y=2x^2+x-2$ is plotted on the same scale, it will be found to be a parabola which cuts the graph of $y=x^3$ at the points for which $x=-1, 1, 2$.

Now at the common points of the two curves, since the ordinates are the same, we must have

$$x^3=2x^2+x-2, \quad \text{or} \quad x^3-2x^2-x+2=0.$$

Hence the values of x which satisfy this equation are the abscissæ of the common points. That is, the required roots are $-1, 1, 2$.

32. $x^3-4x^2-5x+14=0$. (Fig. 53.)

This may be solved as in Ex. 31, by drawing the graphs of $y=x^3$ and $y=4x^2+5x-14$ on the same axes, and measuring the abscissæ of the common points.

A more convenient figure, however, is obtained by drawing the graphs of $y=x^3-4x^2$ and $y=5x-14$, and proceeding as before.

We notice that $y=x^3-4x^2$ passes through the origin, and cuts the axis of x again at the point (4, 0).

Other values are

x	- 2	- 1·5	- 1	- ·5	0	1	2	3	4	5
y	- 24	- 12·375	- 5	- 1·125	0	- 3	- 8	- 9	0	25

Drawing the line $y=5x-14$, we find that the roots of $x^3-4x^2=5x-14$, or $x^3-4x^2-5x+14=0$, are 1·59, 4·41, -2.

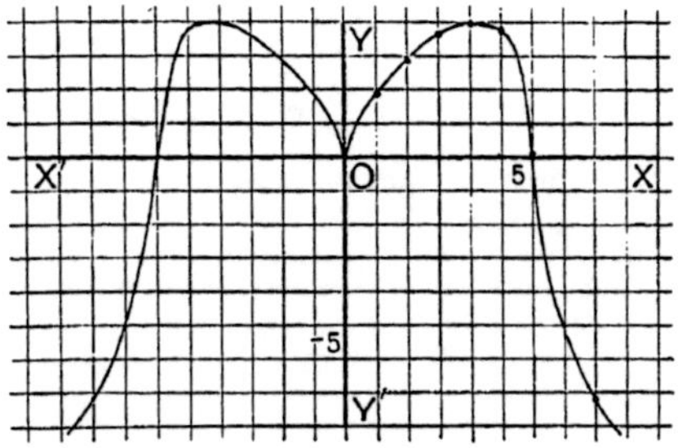

FIG. 52.

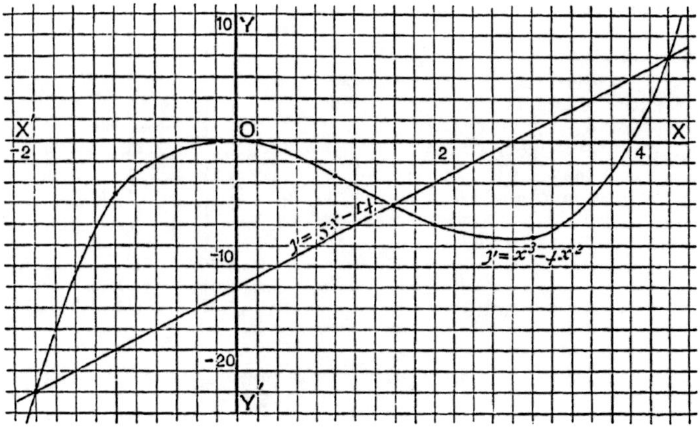

FIG. 53.

EXAMPLES. XLIV. f. PAGE 400.

1. See page 393.

2. For any accuracy this graph should be drawn on a large scale. It will be found that 1″ as the unit of y, and 0·2″ for the unit of x are convenient scales. Suitable values are $x=0$, 1, 4, 9, 16; $y=0$, 1, 2, 3, 4.

3. See pages 396 and 398. **4.** See Example 1, page 396.

5. We must first draw the graph of $y=x^2-2x-8$. (Fig. 54.)
The tabulated values are

x	-3	-2	-1	0	1	2	3	4	5
x^2	9	4	1	0	1	4	9	16	25
$2x$	-6	-4	-2	0	2	4	6	8	10
y	7	0	-5	-8	-9	-8.	-5	0	7

We see that y, that is, x^2-2x-8 vanishes at the points M, M'; and that the corresponding values of x are 4 and -2.

Again, from the figure, at any point on the curve whose abscissa is between the limits 4 and -2, the ordinate is negative, and at all other points positive. Hence for values of x between the limits $x=4$, $x=-2$ the expression x^2-2x-8 is negative, for those values the expression is zero, and for all other values positive.

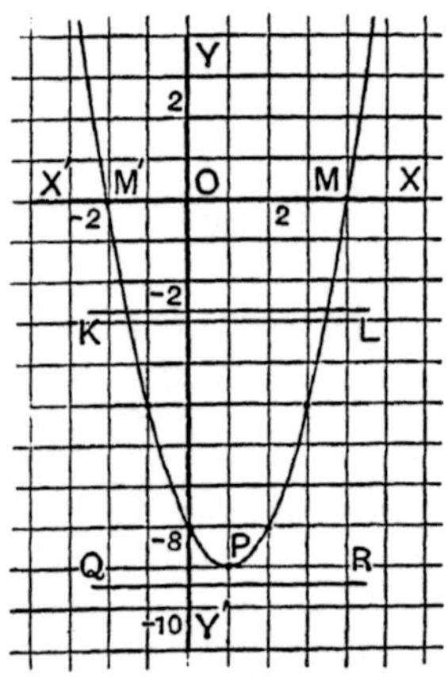

FIG. 54.

The minimum ordinate is that of the point P, and its value is -9. Hence the minimum value of x^2-2x-8 is -9.

6. If $x^2-2x+a=0$ we have $x^2-2x-8=-a-8$; hence the roots of the equation are given by the abscissæ of the points common to the two graphs $y=x^2-2x-8$ and $y=-a-8$.

The first graph has been drawn in Fig. 54, the second is some line, such as KL or QR, drawn parallel to the axis of x and at a distance from it denoted by $-a-8$.

Since the minimum ordinate is -9 it will be seen from the figure that any such line whose distance from the axis is positive, or whose distance is negative and numerically less than 9, will cut the curve in real points; but that a line whose distance is negative and greater than 9 will not meet the curve at all in real points.

Hence the equation cannot have real roots if $-a-8$ is numerically greater than -9. That is, if $a+8$ is greater than 9, or a greater than 1.

7. See solution of Ex. 5.

8. The graphs are shewn in Fig. 55.

To find the roots of $x^2-x-6=0$ by means of the figure we must note the points common to $y=x^2$ and $y=x+6$. These are A, B, and their abscissæ, viz. -2 and 3, are the required roots.

The other equations are solved similarly.

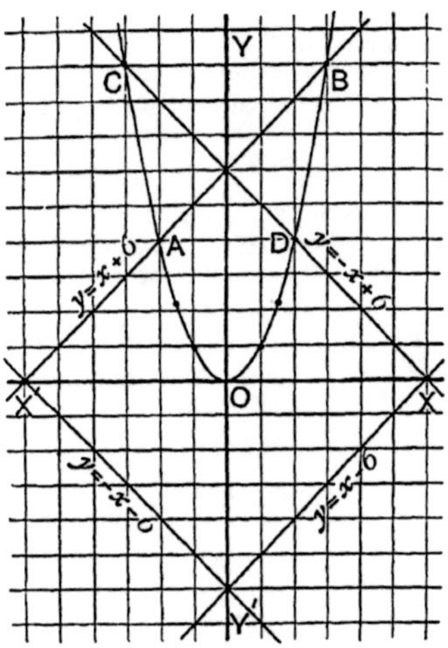

FIG. 55.

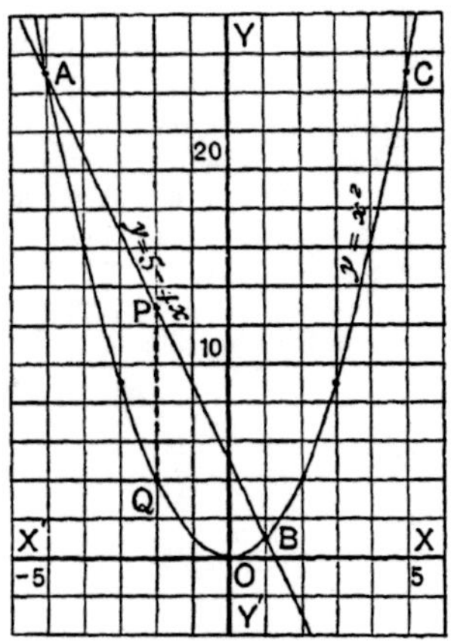

FIG. 56.

9. Using the second method of Ex. 2, Art. 439, we draw the graphs of $y=5-4x$ and $y=x^2$ on the same axes. (Fig. 56.) Then the expression $5-4x-x^2$ measures the difference of the ordinates of points on the two graphs which have the same abscissa. This is positive when the ordinate of the point on the line exceeds that of the point on the curve; from the figure this is seen to be true only of the portion AOB of the curve. The abscissae of A and B are -5 and 1 respectively. Hence $5-4x-x^2$ is positive for all values of x between -5 and 1.

Also the greatest difference between corresponding ordinates is seen to be PQ, which by measurement is equal to 9. Hence the maximum value of $5-4x-x^2$ is 9.

10. We must draw the graphs of $y=x^3$ and $y=3x^2+6x-8$ on the same axes. For the former see page 398; for the latter we have the following set of values

x	-4	-3	-2	-1	0	1	2	3	4
$6x$	-24	-18	-12	-6	0	6	12	18	24
$3x^2$	48	27	12	3	0	3	12	27	48
y	16	1	-8	-11	-8	1	16	37	64

Only a portion of the two curves can be shewn in the diagram. It will be seen that they cut at the points P and Q, the abscissæ of which are 1 and -2. It will be found that there is a third point of section (R) at the point where $x=4$. See Fig. 57.

Thus the required roots are $1, \ -2, \ 4$.

Again between the points Q and P, $y=x^3$ lies above $y=3x^2+6x-8$, i.e. the ordinate of the first curve is algebraically greater than the corresponding ordinate of the second, or $x^3 > 3x^2+6x-8$. Hence between the values $x=-2$ and $x=1$, the expression x^3-3x^2-6x+8 is positive.

Similarly by considering the portions of the curve between P and R we shall find that for values of x between 1 and 4 the expression is negative.

11. We have $x^3+px+q=0$, or $x^3=-px-q$.

Hence the roots of the equation are obtained by observing the abscissæ of the points common to $y=x^3$ and $y=-px-q$.

These are shewn in Fig. 58.

To draw the line $y=-px-q$ it is convenient first to draw a line parallel to it through the origin, viz. $y=-px$. Since p is known to be positive this line must fall in the 2nd and 4th quadrants as OA. Then it can be seen that any line parallel to OA must cut the curve in one point and cannot cut it again. Thus $y=-px-q$ meets $y=x^3$ in one, and only one, real point, i.e. the equation $x^3=-px-q$ has one, and only one, real root.

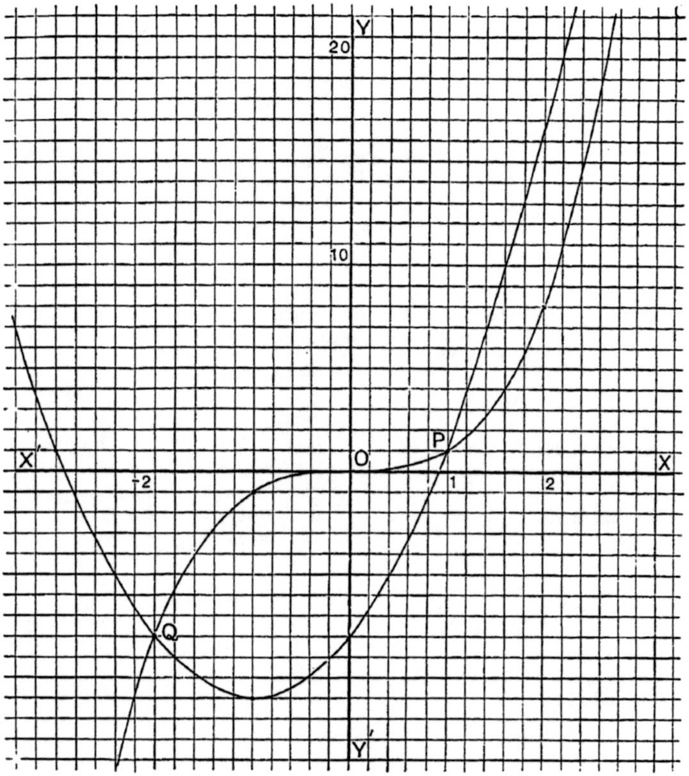

FIG. 57.

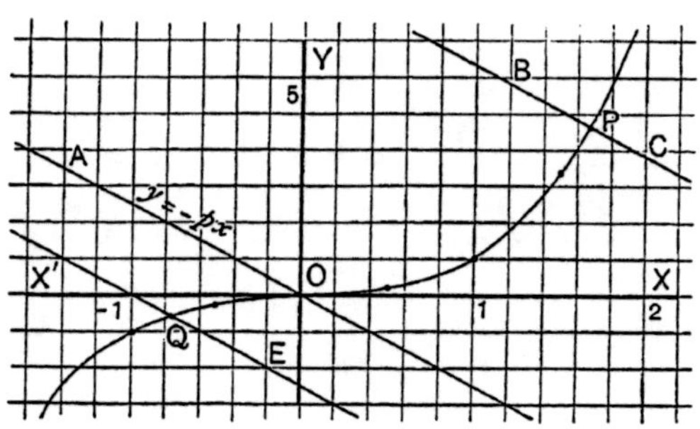

FIG. 58.

12. Take the origin at the point (3, 8), and for units take 1 inch for x and one-tenth of an inch for y. The graph presents no difficulty; it will be found to be a curve resembling a parabola in shape.

When $x=4\cdot75$, $y=26\cdot9$,

,, $x=5\cdot25$, $y=38$.

When $y=12$, $x=3\cdot58$; $\therefore$ $12=2^{3\cdot58}$.

Also since $x=\log_2 y$, we have from above

$$\log_2 26\cdot9+\log_2 38=4\cdot75+5\cdot25$$
$$=10.$$

13. Expressing the given fractional indices in decimal form, it will be found that the corresponding values of x and y are as in the annexed table. Since $x=\log y$, we have to find x when y has the values 1·68, 2·24, 3, 34·3. The first three values of x (at the points P, Q, R) may be read off at once from the graph [Fig. 59]. They will be found to be ·225, ·35, ·477. To find $\log 34\cdot3$, the mantissa is the same as that of $\log 3\cdot43$.

When $y=3\cdot43$ (at the pt. S), $x=\cdot535$. Thus $\log 34\cdot3=1\cdot535$.

x	y
0·0000	1·000
0·0625	1·154
0·1250	1·334
0·1875	1·540
0 2500	1·778
0·3125	2·053
0·3750	2·371
0·4375	2·738
0·5000	3·162
0·5625	3·652

14. From the given values of x we have the following table:

x	0	1	2	3	4	5	6	7	8	9
$(9-x)^2$	81	64	49	36	25	16	9	4	1	0
y	0	64	98	108	100	80	54	28	8	0

With 0·5″ and 0·1″ as units for x and y respectively, the graph presents no difficulty. It is easily seen that the maximum value of y is when $x=3$.

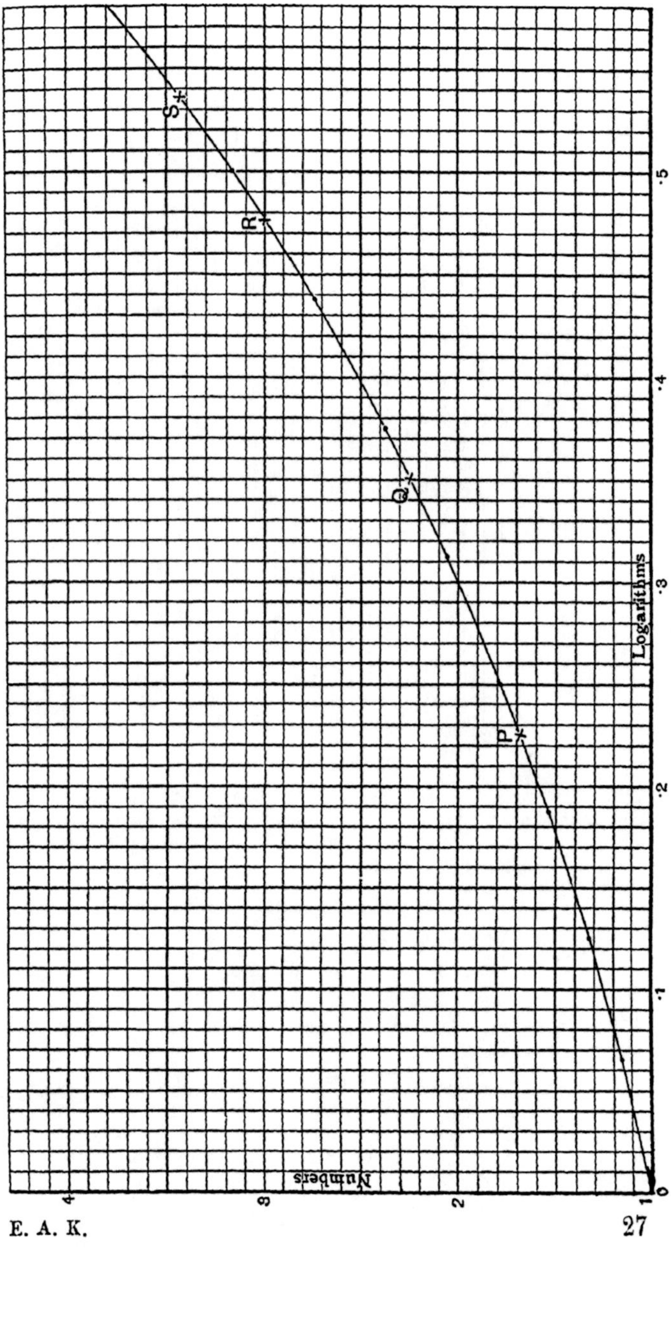

Fig. 59.

15. The corresponding values are found to be

Price in pence	12	16	18	20	22	24	28	33	36
Weight in pounds	22	16·5	14·75	13·25	12	11	9·5	8	7·25

On plotting these points, we have the curve drawn in Fig. 60.

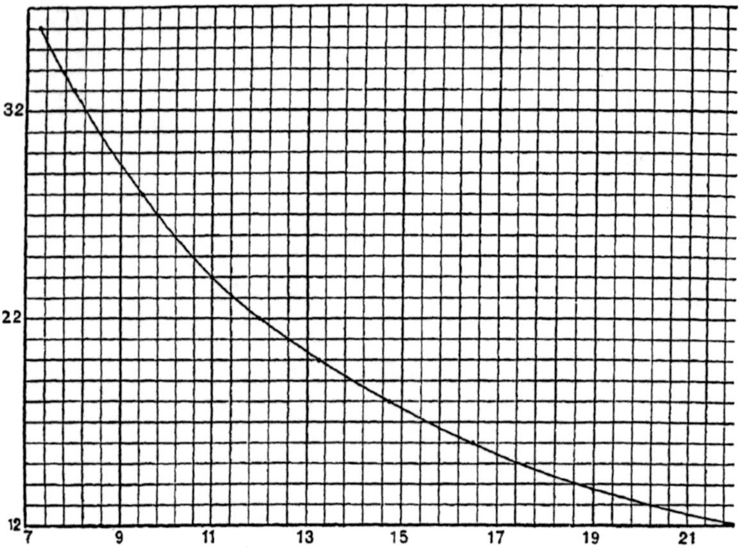

FIG. 60.

Since the product of the price and weight is always the same, viz. 264, we see that the law connecting the values is $xy=264$, where x denotes the weight in lbs. and y the price in pence. Thus the curve is the graph of $xy=264$, a rectangular hyperbola.

16. We have to find the minimum value of y when $y=x+\frac{2\cdot 25}{x}$.

Corresponding values of x and y are

$$x=\ldots \cdot 5,\quad 1,\quad 1\cdot 5,\quad 2,\quad 2\cdot 5,\quad 3,\quad \ldots,$$
$$y=\ldots\ 5,\quad 3\cdot 25,\quad 3,\quad 3\cdot 125,\quad 3\cdot 4,\quad 3\cdot 75,\ \ldots.$$

When these values are plotted, the minimum value of y is 3, corresponding to $x=1\cdot 5$.

17. Use the first method of Art. 439, Ex. 2. Take 1 inch as unit for x, and two-tenths of an inch as unit for y. It will be found that the graph cuts the axis of x at the points whose abscissæ are $-1\cdot 75$ and $1\cdot 25$, and that y is negative between these values. The minimum ordinate is -9, corresponding to $x=-\cdot 25$.

18. Use the method and figure of Art. 440, and find the intersections of

$$\text{(i) } y=x^3, \quad y=-x+2;$$

$$\text{(ii) } y=x^3, \quad y=7x-6.$$

It will be found that in (i) the only point of intersection is at Q, where $x=1$. And in (ii) the intersections are at P, Q, and at the point whose abscissa is -3.

19. The line $x+y=9\frac{1}{2}$ cuts the axes of x and y at a distance $9\frac{1}{2}$ from the origin.

$xy=12$ is a rectangular hyperbola [Art. 431].

To plot $x^2-y^2=32$; when $x=0$, y^2 is negative; thus the curve does not meet the axis of y.

When $y=0$, $x=\pm\sqrt{32}=\pm5{\cdot}7$, and from a table of square roots we obtain the following corresponding values

$$x=5{\cdot}7,\ 6,\ 7,\ 8,\ 9,\ 10,\ 11,\ 12,$$

$$y=0,\ 2,\ 4{\cdot}1,\ 5{\cdot}7,\ 7,\ 8{\cdot}2,\ 9{\cdot}4,\ 10{\cdot}5.$$

From these values a portion of the curve can be plotted, and since it is symmetrical with regard to the axes, it can be completed without further calculation. See Fig. 61.

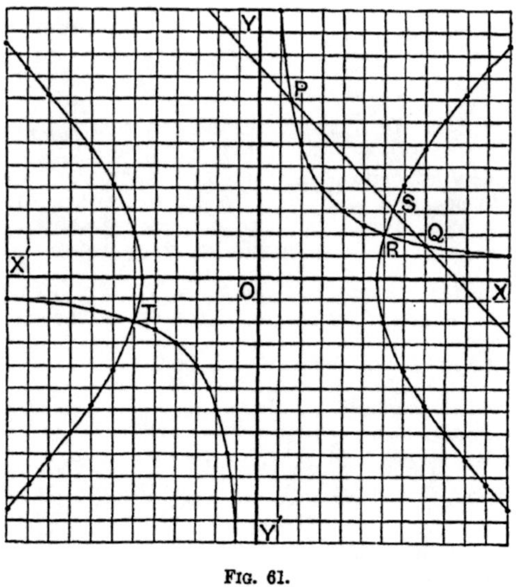

FIG. 61.

The solutions of equations (i) are given by the points P and Q. Equations (ii) have a single solution $x=6{\cdot}43$, $y=3{\cdot}07$ at the point S. Equations (iii) have a pair of solutions shewn by the points R and T.

20. Use the figure and scale of the Example on p. 398.

For distinction, put $y_1=x^3$, and $y_2=3x^2-4$; then we have to find the points on the two graphs at which $y_1=y_2$. As the former curve lies entirely in the first and third quadrants, in tracing the latter the following values will be found sufficient.

$$x=-2,\ -1,\ \ 0,\ \ 1,\ 2,\ \ 3.$$
$$y_1=-8,\ -1,\ \ 0,\ \ 1,\ 8,\ 27.$$
$$y_2=\ 8,\ -1,\ -4,\ -1,\ 8,\ 23.$$

Thus the curves meet when $x=-1$, and when $x=2$, and they will be found to *touch* at the latter point. Thus the roots of the equation are 2, 2, -1.

Let the *numerical* values of y_1 and y_2 be represented by a and b.

For values of $x<-1$, y_1 and y_2 are first both negative and y_2 is the greater; $\therefore\ y_1-y_2$ is negative. When y_2 is positive, y_1-y_2 is still negative.

When x lies between -1 and 0, $y_1=-a$, $y_2=-b$;

$$\therefore\ y_1-y_2=-a+b, \text{ which is positive, since } b>a.$$

When x lies between 0 and 2, at first $y_1=a$, $y_2=-b$;

$$\therefore\ y_1-y_2=a+b, \text{ which is positive,}$$
$$\text{then}\quad y_1=a,\ y_2=b, \text{ and } a>b;$$
$$\therefore\ y_1-y_2 \text{ is positive.}$$

For values of $x>2$, $y_1>y_2$, and both are positive;

$$\therefore\ y_1-y_2 \text{ is positive.}$$

21. See Art. 441. In (i) the circle and rectangular hyberbola will, for different values of a and b, either meet in 4 points or none. In (ii) the straight line $x+y=a$ makes equal intercepts on the axes, either both positive or both negative, and therefore lies entirely in the first or third quadrants. Hence it cannot cut both branches of the hyperbola. Thus for different values of a and b there will be either two solutions or none.

NOTE. When the line $x+y=a$ *touches* either of the curves, there will be two coincident points of section, and a corresponding pair of *equal* roots of the equations.

22. Take 1 inch as x-unit and ·2 of an inch as y-unit; thus the diagram will be on a scale twice as large as that on p. 398.

Put $y_1=x^3$, and $y_2=x^2+3x-3$, and proceed exactly as in Ex. 20.

The curves cut at points P, Q, R, the abscissæ of which are 1, 1·732, and -1·732.

Then it will be found that

for values of x less than -1·732, y_1-y_2 is negative,
.... x between -1·732 and 1, y_1-y_2 is positive,
.... x .... 1 and 1·732, y_1-y_2 is negative,
.... x greater than 1·732, y_1-y_2 is positive.

EXAMPLES. XLIV. g. PAGE 410.

For Examples 1, 2, 3 it will be sufficient to refer to Art. 442, Ex. 1. The graphs may conveniently be drawn on a larger scale.

4. Take 0·1″ and 0·4″ to represent the units of circumference and radius respectively. The points corresponding to the given values of C and r will be found to lie on a straight line. The required values may be read off from the graph.

5. The required values are given by the abscissæ of P and Q (Fig. 62), or from the relation between F and C found below.

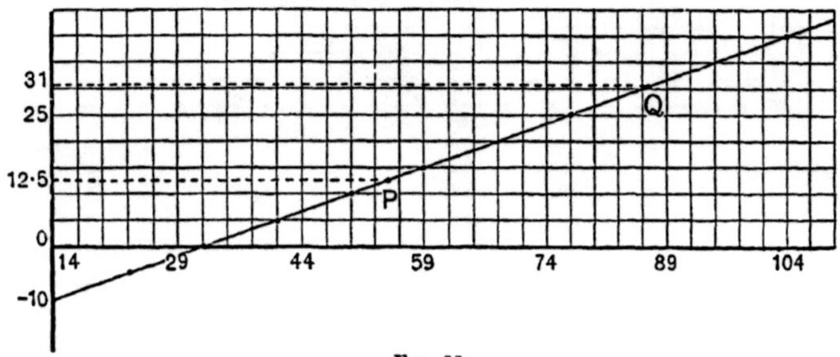

FIG. 62.

Since the graph is a line, its equation must be of the form $F=aC+b$. Substituting the values $C=-10$, $F=14$, and $C=40$, $F=104$, in succession, we have $14=-10a+b$, and $104=40a+b$; whence $a=\frac{9}{5}$, $b=32$. Thus the relation is $F=32+\frac{9}{5}C$.

6. The cost of *printing* x copies $=£\frac{x}{10}$; thus the total cost in pounds of x copies $=100+\frac{x}{10}$, and representing this by y, we have $y=100+\frac{x}{10}$. Since this is a linear equation, its graph will be a straight line determined by any two points on it. Convenient points are (0, 100) and (3000, 400). The required values can then be read off.

7. On plotting the points it will be found that between the ages 10 and 22 the graph is approximately represented by a straight line. Read off the abscissæ of the points on this line whose ordinates are 12 and 20. Compare Ex. 2, p. 443.

8. Representing ages by abscissæ and premiums by ordinates, we find that (on the scale of £2 to an inch) the premiums corresponding to ages 34, 43 are 2·6 and 3·4. Thus the premiums required are £2. 12*s*. and £3. 8*s*. (Fig. 63.)

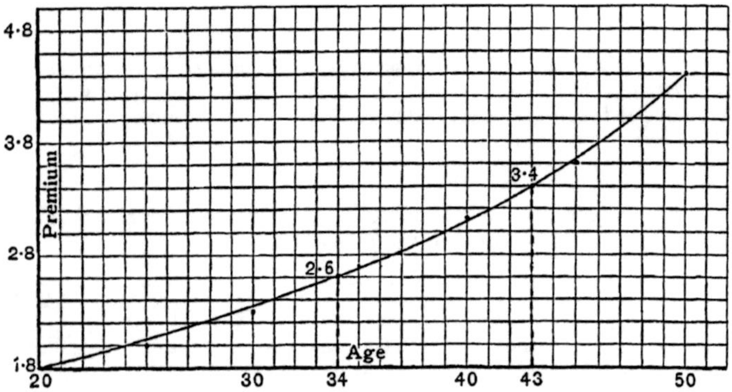

FIG. 63.

9. Representing the lengths from 8 in. to 12 in. by ordinates, and the weights from zero to 25 oz. by abscissæ, we find that the graph is a straight line cutting the axis of length at 8·1. Hence when no weight is attached the length is 8·1 in. Similarly the weight corresponding to a length of 1 foot is 24·375 oz.

10. This example requires a graph on a large scale. Take 1 inch to represent £100 on both axes; then measuring values of A horizontally and values of P vertically, beginning in each case at 2·0, we find that the points lie on a straight line. The ordinates corresponding to abscissæ 3·296 and 5·974 are 3·2 and 5·8. Thus the required principals are £320 and £580.

11. Since the range of unreduced marks is 165, and of reduced marks 150, the actual marks gained must be reduced in the ratio of 150 to 165, or 10 to 11, and then adjusted by the addition or subtraction of some constant number. Thus with the given notation we have $y=a+\frac{10}{11}x$, where a is constant. Substituting 200 for y, and 297 for x, we obtain $a=-70$.

Thus the required equation is $y=\frac{10}{11}x-70$.

The graph being linear presents no difficulty; it may be drawn by joining the points (132, 50) and (297, 200).

12. Plotting the values of t horizontally and values of v vertically, it is easily seen that the points lie on a straight line. We may therefore assume $v=a+bt$. Substituting $v=9$, $t=1$, and $v=45$, $t=10$, we obtain

$$9=a+b, \qquad 45=a+10b.$$

Whence $a=5$, $b=4$. Thus the equation is $v=5+4t$.

From the graph, $t=5{\cdot}75$ when $v=28$; thus the required time is $5\frac{3}{4}$ secs. To find the initial velocity, read off the ordinate of the point where $t=0$.

13. Measure areas vertically, taking two tenths of an inch as unit, and bases horizontally, taking 1 inch as unit, beginning at zero in each case. The points lie on a portion of a parabola passing through the origin, and from the graph we find $A=2{\cdot}49$ when $b=2{\cdot}4$.

14. It will be convenient here to measure values of s vertically *downwards* on a scale of 40 ft. to 1 inch, and values of t horizontally, taking 1 inch to represent 1 second, beginning with zero in each case. The curve is a portion of a parabola passing through the origin. It will be found that when $t=1{\cdot}8$, $s=52$. Again, when $t=3{\cdot}16$, $s=160$.

15. The values are

t	0	$\frac{1}{4}$	$\frac{1}{2}$	$\frac{3}{4}$	1	$1\frac{1}{4}$	$1\frac{1}{2}$	$1\frac{3}{4}$	2	$2\frac{1}{4}$	$2\frac{1}{2}$	...	4
h	0	15	28	39	48	55	60	63	64	63	60	...	0

Hence the graph given in Fig. 64. The maximum height is the ordinate of P, viz. 64 ft., and the "time of flight," represented by AB, is 4 secs.

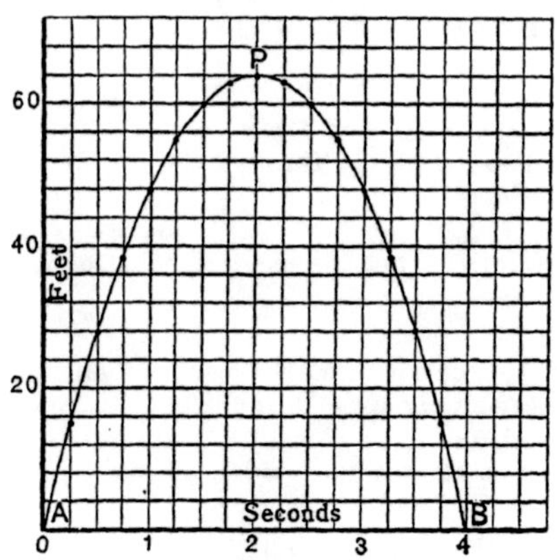

FIG. 64.

16. Plot the values of G horizontally (beginning at 21), taking two tenths of an inch as unit, and the values of P vertically (beginning at 0) with 1 inch as unit.

It will be found that the points lie approximately on a straight line.

If the equation to this line is $P=aG+b$, by substituting values of P and G for *points through which the line passes accurately*, we find, by the method of Art. 443, Ex. 2., that $a=0{\cdot}6$, $b=14{\cdot}4$.

$$\therefore \text{ the required relation is } P=0{\cdot}6\,G-14{\cdot}4.$$

Also the line crosses the horizontal axis when $G=24$.

Here $P=0$; that is, there are no profits for 24 guests.

17. A convenient scale is 3 divisions to a foot and one division to a shilling. The curve is then as drawn in Fig. 65.

18. The dates are 11, 28, 46, 65, 91, 117, 135, 155 days respectively from Mar. 23rd. As the table gives only the nearest degree we cannot expect an exact curve. The approximate shape is shewn in Fig. 66. Reading off the ordinate corresponding to June 8th (77 days) the number of degrees is found to be 93·5.

19. Take two tenths of an inch to represent the unit of v, and one tenth for the unit of p. The curve is given in Fig. 67.

It is a portion of the rectangular hyperbola $pv=180$.

[In Examples 20–22, the figures are inconveniently large. The method, however, is fully explained in Art. 443.]

20. Following the method of Ex. 2 in Art. 443, it will be found that a straight line passes accurately through the 1st, 8th and 9th points, and lies evenly among the rest.

Assume $y=ax+b$, and put $x=3$, $y=2$, and $x=28$, $y=7{\cdot}25$. We thus obtain $a=0{\cdot}21$, $b=1{\cdot}37$. Thus the required equation is $y=0{\cdot}21x+1{\cdot}37$.

21. Plotting the points we find the most probable graph is the straight line which passes accurately through the first, fifth, and last points.

Assuming the relation to be $y=ax+b$, and substituting the coordinates of the first and last points, we find $a=0{\cdot}4$, $b=1{\cdot}6$.

Thus the relation is $y=0{\cdot}4x+1{\cdot}6$.

Putting $x=19$ in this, we have $y=9{\cdot}2$;

.... $y=2{\cdot}8$ .... $x=3$.

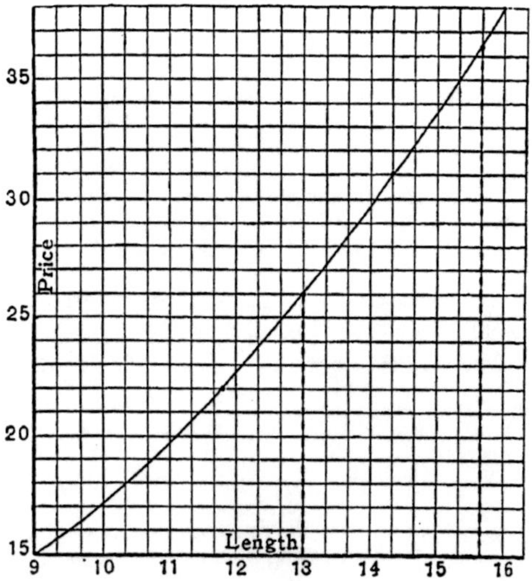

FIG. 65.

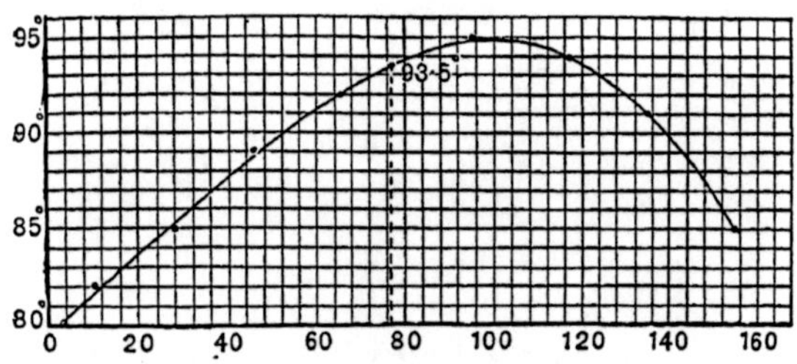

FIG. 66.

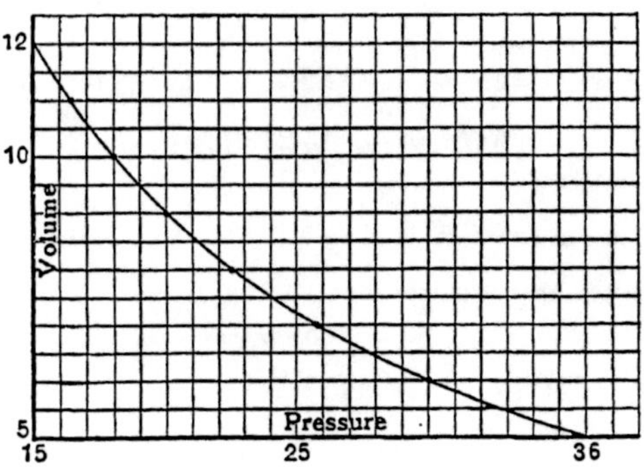

FIG. 67.

22. Beginning values of x at 0, and values of y at 100, let the x-unit be half an inch, and on the y-axis let 1 inch represent 100 units. It will be found that the line which lies most evenly among the plotted points does not pass exactly through any of them. To determine the constants a and b we must choose convenient coordinates of points through which the line is seen to pass. When $x=0$, we find by measurement that the intercept on the y-axis is 118. Hence $b=118$. Again the line passes through the point (4·2, 310);

$$\therefore\ 310=a\times4{\cdot}2+b;\ \text{whence } a=45{\cdot}7.$$

Thus the required equation is $y=45{\cdot}7x+118$.

Putting $x=9{\cdot}9$, we get $y=570{\cdot}43$. Hence the tabulated value 562 is 8·43 in defect.

23. Plotting the points as in Fig. 68, we find that the points lie evenly about a straight line passing through the second, fifth, and last of the given points.

Assuming the relation to be $P=aW+b$ and substituting the coordinates of the second and last points we have

$$3{\cdot}7=25a+b,\quad 11{\cdot}75=82{\cdot}5a+b;$$

whence $a=0{\cdot}14$ and $b=0{\cdot}2$.

Thus the equation is $P=0{\cdot}14W+0{\cdot}2$.

When $P=31{\cdot}7$, we have $0{\cdot}14W=31{\cdot}5$; whence $W=225$.

24. Here for greater convenience values of y are plotted horizontally, and values of x vertically.

The curve is as drawn in Fig. 69, passing accurately through the first, sixth, and last points and lying evenly amongst the others.

Substituting the coordinates of the first and last points in $y=ax^2+b$, we have

$$3{\cdot}25=a+b,\quad 15{\cdot}25=49a+b;$$

whence $a=\frac{1}{4}$ and $b=3$.

Thus the required equation is $y=\dfrac{x^2}{4}+3$.

Putting $y=4$ in this equation, we get $x=2$;
.... $x=6$ .... $y=12$.

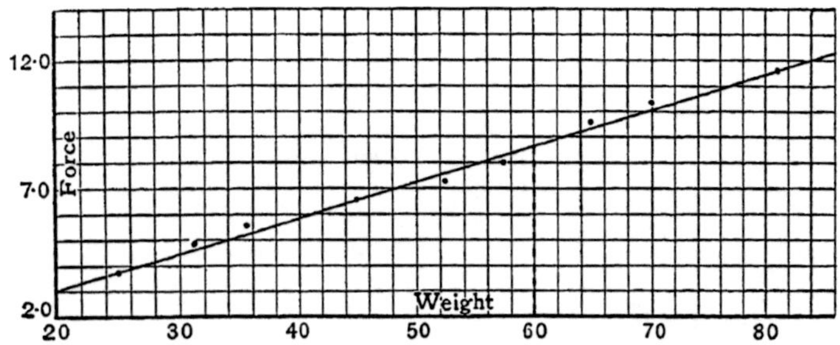

Fig. 68.

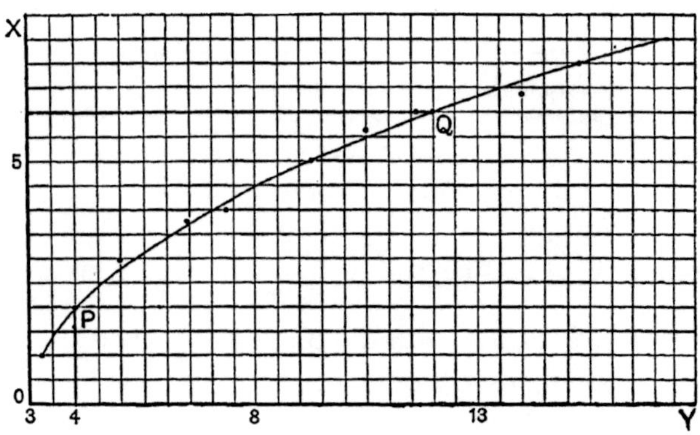

Fig. 69.

25. There are three methods, any one of which may be employed.

(i) Plot the points and draw the hyperbola that most nearly passes through them. Choose two points lying *on* the curve and find a, b by substituting their coordinates.

(ii) Since the equation may be written $\frac{a}{y}+\frac{b}{x}=1$, we can plot the graph for corresponding values of $\frac{1}{x}$ and $\frac{1}{y}$. This will be a straight line; for if we write X, Y for $\frac{1}{x}$ and $\frac{1}{y}$, we get $aY+bX=1$, which is a linear equation in X, Y.

We have the following values:

x	2·75	3	3·2	3·5	4·3	4·5	5·3	6	7	8	10
y	11	9·8	8	6·5	6·1	5·4	5	4·3	4·1	4	3·9
$\frac{1}{x}$	·364*	·333	·312*	·285	·233	·222*	·189	·167	·143	·125*	·1
$\frac{1}{y}$	·091	·102	·125	·154	·164	·185	·200	·233	·244	·250	·256

Plotting these points we find that the nearest line is that which passes through the points marked with an asterisk. (See Fig. 70.)

Substituting the coordinates of the first and last of these we have

$$·364b+·091a=1, \quad ·125b+·25a=1;$$

whence $a=3$, $b=2$.

Substituting $x=3·5$ and $x=7$ in $\frac{2}{x}+\frac{3}{y}=1$, we get the values of y required.

(iii) The equation may also be written $y=a+b\frac{y}{x}$, and we could proceed as in (ii), tabulating corresponding values of y and $\frac{y}{x}$, the latter being found by means of logarithms.

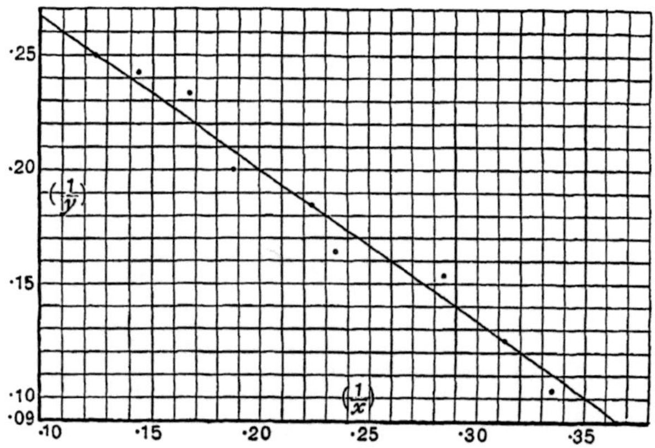

FIG. 70.

26. Proceed as in the Example worked out on page 408.

We have $\log x + n \log y = \log c$, and from the tables in CHAP. XXXIX we have the annexed values of $\log x$ and $\log y$. Plot these on the same scale as that on p. 408, beginning at 1·6 for $\log x$ and 1·4 for $\log y$. The line which lies most evenly among the points will be found to pass through those for which $\log x = 1{\cdot}8$, $\log y = 1{\cdot}544$, and $\log x = 2{\cdot}0$, $\log y = 1{\cdot}477$.

$\log x$	$\log y$
2·000	1·477
1·954	1·492
1·845	1·525
1·778	1·551
1·699	1·576
1·602	1·610

Substituting these values in the equation

$$\log x + n \log y = \log c,$$

we obtain $n = 3$, $\log c = 6{\cdot}432$; whence $c = 27 \times 10^5$.

27. Using the same method and the same scale as in the last example, and beginning values of $\log x$ at 1·4 and values of $\log y$ at 2·0, we obtain a line which passes through the points for which $\log x = 1{\cdot}9$, $\log y = 2{\cdot}05$, and $\log x = 1{\cdot}45$, $\log y = 2{\cdot}71$.

Substituting these values in the equation

$$n \log x + \log y = \log c,$$

we obtain $n = 1{\cdot}5$ approx., and $\log c = 4{\cdot}9$.

Hence $c = 79500$.

GLASGOW: PRINTED AT THE UNIVERSITY PRESS BY ROBERT MACLEHOSE AND CO. LTD.

WORKS BY H. S. HALL, M.A., and S. R. KNIGHT, B.A.

EIGHTH EDITION. Revised and Enlarged.

ELEMENTARY ALGEBRA FOR SCHOOLS (containing a chapter on Graphs). Globe 8vo. (bound in maroon-coloured cloth), 3*s*. 6*d*. With Answers (bound in green-coloured cloth, 4*s*. 6*d*.

ANSWERS TO EXAMPLES IN ELEMENTARY ALGEBRA. Fcap. 8vo. Sewed 1*s*.

SOLUTIONS OF THE EXAMPLES IN ELEMENTARY ALGEBRA FOR SCHOOLS. Crown 8vo. 8*s*. 6*d*.

HIGHER ALGEBRA. A Sequel to Elementary Algebra for Schools. Fifth Edition, revised and enlarged. Crown 8vo. 7*s*. 6*d*.

SOLUTIONS OF THE EXAMPLES IN HIGHER ALGEBRA. Crown 8vo. 10*s*. 6*d*.

ALGEBRA FOR BEGINNERS. Globe 8vo, 2*s*. With Answers. 2*s*. 6*d*.

ALGEBRAICAL EXERCISES AND EXAMINATION PAPERS. With or without Answers. Third Edition, revised and enlarged. Globe 8vo. 2*s*. 6*d*.

ARITHMETICAL EXERCISES AND EXAMINATION PAPERS. With an Appendix containing Questions in LOGARITHMS AND MENSURATION. With or without Answers. Third Edition, revised and enlarged. Globe 8vo. 2*s*. 6*d*.

ELEMENTARY TRIGONOMETRY. Third Edition. Globe 8vo. 4*s*. 6*d*.

SOLUTIONS OF THE EXAMPLES IN ELEMENTARY TRIGONOMETRY. Crown 8vo. 8*s*. 6*d*.

ALGEBRAICAL EXAMPLES Supplementary to Hall and Knight's ALGEBRA FOR BEGINNERS and ELEMENTARY ALGEBRA (Chaps. I.-XXVII.). By H. S. HALL, M.A. With or without Answers. Globe 8vo. 2*s*.

A SHORT INTRODUCTION TO GRAPHICAL ALGEBRA. By H. S. HALL, M.A. Third Edition revised and enlarged. Globe 8vo. 1*s*.

SOLUTIONS OF THE EXAMPLES IN HALL'S GRAPHICAL ALGEBRA. Crown 8vo. 3*s*. 6*d*.

EASY GRAPHS By H. S. HALL, M.A. Crown 8vo. 1*s*.

WORK BY H. S. HALL, M.A., and R. J. WOOD, B.A.

ALGEBRA FOR ELEMENTARY SCHOOLS. Globe 8vo. Parts I., II., and III., 6*d*. each. Cloth, 8*d*. each. Answers. 4*d*. each.

MACMILLAN AND CO., LTD., LONDON.

WORKS BY H. S. HALL, M.A., and F. H. STEVENS, M.A.

A SCHOOL GEOMETRY, based on the recommendations of the Mathematical Association, and on the recent report of the Cambridge Syndicate on Geometry.

Parts I. and II.—*Part I.* Lines and Angles. Rectilineal Figures. *Part II.* Areas of Rectilineal Figures. Containing the substance of Euclid Book I. 1*s.* 6*d.* **Key**, 3*s.* 6*d.*
Part I.—Separately. 1*s.*
Part III.—Circles. Containing the substance of Euclid Book III. 1-34, and part of Book IV. 1*s.*
Parts I., II., III. in one volume. 2*s.* 6*d.*
Part IV.—Squares and Rectangles. Geometrical equivalents of Certain Algebraical Formulæ. Containing the substance of Euclid Book II., and Book III. 35-37. 6*d.*
Parts III. and IV. in one volume. 1*s.* 6*d.*
Parts I.-IV. in one volume. 3*s.* **Key**, 6*s.*
Part V.—Containing the substance of Euclid Book VI. 1*s.* 6*d.*
Parts IV. and V. in one volume. 2*s.*
Parts I.-V. in one volume. 4*s.*
Part VI.—Containing the substance of Euclid Book XI. 1–21, together with Theorems relating to the Surfaces and Volumes of the simpler Solid Figures. 1*s.* 6*d.*
Parts IV., V., VI. in one volume. 2*s.* 6*d.*
Parts I.-VI. in one volume. 4*s.* 6*d.* **Key**, 8*s.* 6*d.*

LESSONS IN EXPERIMENTAL AND PRACTICAL GEOMETRY. Crown 8vo. 1*s.* 6*d.*

A TEXT-BOOK OF EUCLID'S ELEMENTS, including Alternative Proofs, together with Additional Theorems and Exercises, classified and arranged. Third Edition enlarged. Books I.—VI. XI. and XII. Props. 1 and 3. Globe 8vo. 4*s.* 6*d.*

Also in parts, separately, as follows :—

Book I.	1*s.*	Books III. and IV.	2*s.*
Books I. and II.	1*s.* 6*d.*	Books III.—VI.	3*s.*
Books II. and III.	2*s.*	Books IV.—VI.	2*s.* 6*d.*
Books I.—III.	2*s.* 6*d.*	Books V., VI., XI. and XII. 1 and 3.	2*s.* 6*d.*
Books I.—IV.	3*s.*	Book XI.	1*s.*
Sewed, 2*s.* 6*d.*			

A KEY TO THE EXERCISES AND EXAMPLES CONTAINED IN A TEXT-BOOK OF EUCLID'S ELEMENTS. Books I.—VI. and XI. Crown 8vo. 8*s.* 6*d.* Books I.—IV. 6*s.* 6*d.* Books VI. and XI. 3*s.* 6*d.*

AN ELEMENTARY COURSE OF MATHEMATICS. Comprising Arithmetic, Algebra, and Euclid. Globe 8vo. 2*s.* 6*d.*

AN ELEMENTARY COURSE OF MATHEMATICS. Comprising Arithmetic, Algebra, and Geometry. Globe 8vo. 2*s.* 6*d.*

MACMILLAN AND CO., Ltd., LONDON.

2

900806

Made in the USA